ADVANCES IN CHEMICAL SENSORS

Edited by **Wen Wang**

Advances in Chemical Sensors
Edited by Wen Wang

Published by InTech

Janeza Trdine 9, 51000 Rijeka, Croatia

Publishing Process Manager Ivana Zec

Technical Editor Teodora Smiljanic

Cover Designer InTech Design Team

Additional hard copies can be obtained from orders@intechopen.com

Advances in Chemical Sensors, Edited by Wen Wang
p. cm.
ISBN 978-953-307-792-5

For EU product safety concerns:
IN TECH d.o.o., Prolaz Marije Krucifikse Kozulić 3, 51000 Rijeka, Croatia,
info@intechopen.com or visit our website at intechopen.com.

Meet the editor

Wen Wang is a Professor at the Institute of Acoustics, Chinese Academy of Sciences (IOA-CAS), who received his MS degree from the Central South University of China in 2002, and his PhD degree from IOA-CAS in 2005. From 2005 to 2009, he worked as a postdoctoral researcher and research professor at the microsystem laboratory of the Ajou University in Korea. Prof. Wang worked at the Freiburg University supported by the Humboldt Foundation as a guest professor in 2010, and received the "Experienced researcher" award from Humboldt Foundation of Germany for his outstanding research work in SAW technology. He joined the IOA-CAS as an associate professor of acoustical microsystems Lab. and as a professor in 2007 and 2011 respectively. His current research involves acoustic wave devices for sensing applications and wireless sensor, etc. Prof. Wang is the author or co-author of more than 50 papers in a refereed international journal, more than 40 communications in international conferences and over 10 patents.

Contents

Preface

With decades of vigorous research and development, various chemical sensors with excellent performance have been used successfully in areas such as clinical, environmental conservation and monitoring, disaster and disease prevention, and industrial analysis. A chemical sensor is an analyzer that responds to a particular analyte in a selective and reversible way, and transforms chemical information, ranging from the concentration of a specific sample component to total composition analysis, into an analytically useful signal. The chemical information mentioned above may originate from a chemical reaction by a biomaterial, chemical compound or a combination of both attached onto the surface of a physical transducer toward the analyte. Numerous literatures deal with the sophisticated research on chemical sensors by considering the sensor structure, techniques and response mechanism, and leading to quick response, low cost, small size, superior sensitivity, good reversibility and selectivity, and excellent detection limit. Hence, such kind of chemical sensor meets the requirement of practical application.

This book is an attempt to highlight current research advances in chemical sensors on the topics of health, environment, and industry analysis. It is composed of 15 chapters and divided into 5 sections according to the classification following the principles of signal transducer. The current trends, materials design, and principle of detection and monitoring in chemical sensor are introduced in details. Section 1 provides an introduction to optical chemical sensors, and descriptions on the analytical aspects of sensors. Some of the latest research progress regarding the sensor structure and response mechanism is discussed in this section. Section 2 reviews some research achievements of chemical sensor based nanostructure. Section 3 describes the electrical chemical sensor by utilizing conductive and oxide semiconductor materials, and Section 4 performs the chemical sensor trend relating to the artificial chemical sensor. Sensor technologies related to sensor performance improvement by utilizing physical vapor deposition techniques, statistical analysis of the chemical sensor data, and drift correction methods are discussed in Section 5.

It is my pleasure that this collection of up-to-date information and latest research progress on chemical sensor in this book will be of great interest to all those working on chemical sensors.

I would like to acknowledge the hard work and dedication of all the contributing authors. In particular, I would like to thank Ms. Ivana Zec, the publishing process manager of this book, for her great help in proposal collection, evaluation and manuscript editing.

Wang Wen, Ph.D.
Professor
Institute of Acoustics, Chinese Academy of Sciences
Beijing, China, P. R.

Part 1

Optical Chemical Sensor

1

Optical Chemical Sensors: Design and Applications

Aleksandra Lobnik[1], Matejka Turel[1] and Špela Korent Urek[2]
[1]*Universitiy of Maribor, Faculty of Mechanical Engineering,*
[2]*Institute for Environmental Protection and Sensors,*
Slovenia

1. Introduction

Optical sensors, or opt(r)odes, represent a group of chemical sensors in which electromagnetic (EM) radiation is used to generate the analytical signal in a transduction element. The interaction of this radiation with the sample is evaluated from the change of a particular optical parameter and is related to the concentration of the analyte (Blum, 1997). Typically, an optical chemical sensor consists of a chemical recognition phase (sensing element or receptor) coupled with a transduction element (Fig. 1). The receptor identifies a parameter, e.g., the concentration of a given compound, pH, etc., and provides an optical signal proportional to the magnitude of this parameter. The function of the receptor is fulfilled in many cases by a thin layer that is able to interact with the analyte molecules, catalyse a reaction selectively, or participate in a chemical equilibrium together with the analyte. The transducer translates the optical signal produced by the receptor into a measurable signal that is suitable for processing by amplification, filtering, recording, display, etc. (Gründler, 2007; Nagl & Wolfbeis, 2008).

Sensors that have a receptor part based on a biochemical principle are usually called biosensors. The selectivity and sensitivity provided by Nature have been utilized in such sensors, frequently by immobilizing the biologically active compounds, such as enzymes and immunoglobulins, within a receptor part of the sensor (Patel et al., 2010). An effective way of obtaining the biological selectivity is a combination of cell cultures, tissue slices, organs and sometimes of whole living organisms with the transducer.

Optical sensors can be based on various optical principles (absorbance, reflectance, luminescence, fluorescence), covering different regions of the spectra (UV, Visible, IR, NIR) and allowing the measurement not only of the intensity of light, but also of other related properties, such as lifetime, refractive index, scattering, diffraction and polarization (Jerónimo et al., 2007). As an example, a luminescent sensor can be constructed by associating a sensing element, which emits light when in contact with a specific analyte, with a photodiode, which converts the energy of the incident light into a measurable signal.

Optical chemical sensors have numerous advantages over conventional electricity-based sensors, such as selectivity, immunity to electromagnetic interference, and safety while working with flammable and explosive compounds. They are also sensitive, inexpensive, non-destructive, and have many capabilities. Optrodes do not require a reference cell, as is the case in potentiometry. Furthermore, they can easily be miniaturized and allow multiple analyses with a single control instrument at a central site (Lukowiak & Strek, 2009).

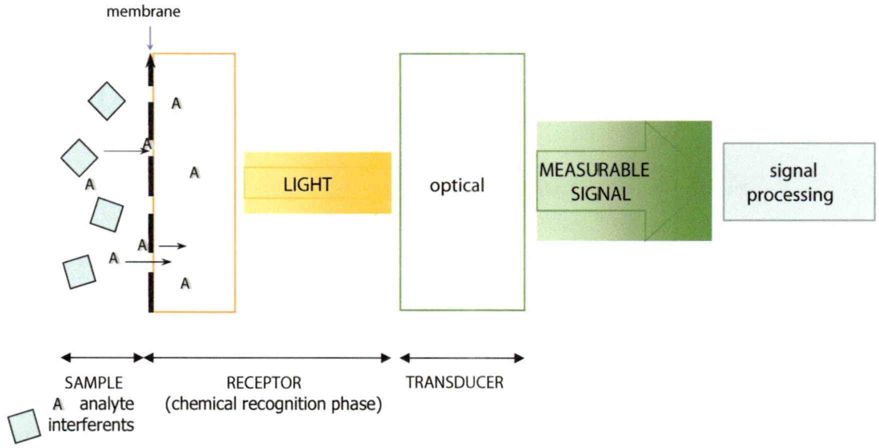

Fig. 1. Schematic representation of the composition and function of an optical chemical sensor

However, besides a number of advantages, optical sensors also exhibit disadvantages: ambient light can interfere with their operation, the long-term stability is limited due to indicator leaching or photobleaching, there may be a limited dynamic range, selectivity may be poor, and a mass transfer of the analyte from the sample into the indicator phase is necessary in order to obtain an analytical signal (Seitz, 1988).

Fiber-optic chemical sensors (FOCSs) represent a subclass of chemical sensors in which an optical fiber is commonly employed to transmit the electromagnetic radiation to and from a sensing region that is in direct contact with the sample. The spectroscopically detectable optical property can be measured through the fiber optic arrangement, which enables remote sensing. In addition to advantages in terms of cheapness, ease of miniaturization, obtaining safe, small, lightweight, compact and inexpensive sensing systems, a wide variety of sensor designs are possible (Jerónimo et al., 2007; Lukowiak & Strek, 2009; Seitz, 1988).

The most common classification of FOCs distinguishes between the intrinsic and extrinsic types of sensors (Seitz, 1988; Wolfbeis, 2008).

- In the intrinsic type of FOCs, the sensing principle is based on the change in light-transmission characteristics due to the change occurring in a fiber property (e.g., refractive index or length) upon the interaction with the analyte or the system being studied. The optical fibre itself has sensory characteristics. This type of sensor is mainly applied to measure physical or physicochemical parameters, such as the pressure, temperature, or enthalpy of reactions.
- In the extrinsic type of FOCs, the optical fiber acts as a transporting media by means of guiding the radiation from the source to the sample or from the sample to the detection system. Extrinsic sensors can be subdivided into a) distal and b) lateral types. The most common are distal-type sensors, in which the indicator is immobilized at the distal end (tip) of the optical fibre. Alternatively, in a lateral sensor, the sensing chemistry can be immobilized along a section of the core of the optical fibre to make an evanescent field sensor.

1.1 Optical detection principles

For sensor applications only part of spectroscopic wavelength range is useful. From the practical point of view the following ways (Fig. 2) in which radiation can interact with an analytical sample are the most useful (Gründler, 2007):

- absorption
- emission (fluorescence or phosphorescence)
- reflexion and refraction

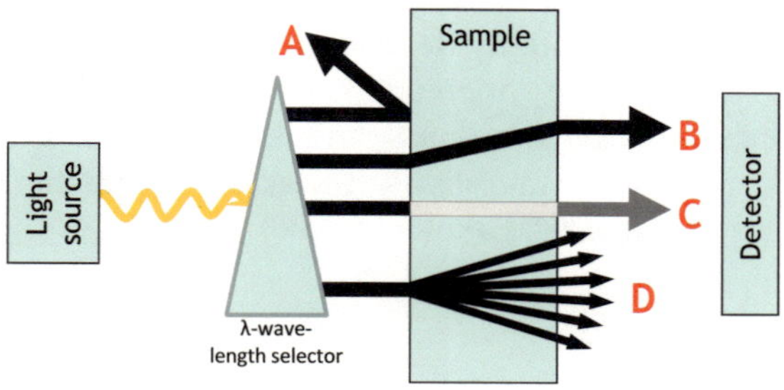

Fig. 2. General arrangement of spectroscopic measurements: A – light reflection, B – light refraction, C – light absorption, D – light emission.

However, the most commonly applied methods in optical sensing are those based on light absorption or light emission. Compared to absorption-based methods, molecular emission (fluorescence, phosphorescence, and generally speaking, luminescence) is particularly important because of its extreme sensitivity and good specificity. The sensitivity of luminescence methods is about 1000 times greater than that of most spectrophotometric methods. In addition, lower limits of detection for the desired analytes can be achieved (Guilbault, 1990; Schulman, 1988; Wolfbeis, 2005). Measuring the emission intensity is also the most popular because the instrumentation needed is very simple and cheap. Nevertheless, measuring the light emission intensity has some disadvantages compared to emission lifetime measurements, in which the sample is excited only by a pulse of EM rather than via continuous illumination, which is the case with intensity-based methods. The precision and accuracy of luminescence intensity-based schemes are greatly affected by fluctuations in the light-source's intensity, detector sensitivity, inner filter effects, indicator concentration (bleaching and leaching), sample turbidity, and sensing layer thickness. However, some of these problems can be minimized or even overcome by measuring luminescence lifetimes instead of intensities. But again, lifetime measurements also have some drawbacks, which are the instrumentation complexity and high costs, along with a limited number of indicator dyes available that show significant analyte-dependent changes in the lifetime (Lippittsch & Draxler, 1993).

Another way to reduce the problems associated with intensity as well as with lifetime detection principles is the use of ratiometric measurements. This technique employs dual-emission or dual-excitation indicators or mixtures of two luminophores, exhibiting separated spectral areas with different behaviour. For example, the ratio of two fluorescent peaks is used instead of the absolute intensity of one peak. The sensors therefore typically contain a reference dye; the advantage of this approach is that factors such as excitation source fluctuations and sensor concentration will not affect the ratio between the fluorescence intensities of the indicator and the reference dye (Arduini et al., 2007; Buck et al., 2004; Cywinski et al., 2009; Doussineau et al., 2009; Frigoli et al., 2009; Sun et al., 2006).

Another important process that occurs in the excited state is the Förster or fluorescence resonance energy transfer (FRET). This process occurs whenever the emission spectrum of one fluorophore, which is the donor, overlaps with the absorption spectrum of another molecule, which is the acceptor. The acceptor must absorb the energy at the emission wavelength(s) of the donor, but does not necessarily have to remit the energy fluorescently itself. The transfer of energy leads to a reduction in the donor's fluorescence intensity and excited state lifetime, and an increase in the acceptor's emission intensity. The rate of energy transfer from donor to acceptor is highly dependent on many factors, such as the extent of the spectral overlap, the relative orientation on the transition dipoles, and, most importantly, the distance between the donor and the acceptor. Due to its sensitivity to distance, FRET has been used to investigate molecular interactions (Demchenko, 2009; Frigoli et al., 2009; Kikuchi et al., 2004; Lakowicz, 2006).

2. Design of an optical sensor

The overall sensor quality is dependent on the total sensor system components, which are defined by the transduction, the sensitive layer, light source, data-acquisition electronics and evaluation software. In the next paragraphs the emphasis is devoted to the components when designing the sensitive sensor layer, namely, to the choice of suitable indicators, polymers, and immobilization techniques. Some selected, recently published applications using new nanomaterials are further presented.

An optical detection system may be based either on a) direct sensing or b) indicator-mediated sensing. In a direct optical sensor, the analyte is detected directly via some intrinsic optical property such as, for example, absorption or luminescence. In an indicator-mediated system, a change in the optical response of an intermediate agent, usually an analyte-sensitive dye molecule (indicator), is used to monitor the analyte concentration (McDonagh et al., 2008; Nagl & Wolfbeis, 2008). The principle of immobilized indicators relies on a large group of optical chemical sensors, because the measuring analytes mostly have no intrinsic optical property or this property is not convenient for their detection.

The reagent immobilised materials (sensitive layers) can be fabricated into several configurations, such as thin films, gels, to be interfaced with optical fibres (Gründler, 2007), in nanoparticles, etc. The most common are thin polymer films or membranes. This "smart" material responds to the species of interest by altering its optical properties (Seitz, 1988; Wolfbeis, 1991). For example, pH is measured optically by immobilizing a pH indicator on a solid support and observing the changes in the absorption or fluorescence of the indicator as the pH of the sample varies with time (Jerónimo et al., 2007; Lobnik, 1998,2006; Turel, 2008; Wolfbeis, 1991).

2.1 Indicators

The basic principle of the indicator chemistry (immobilized in or on the polymer matrix) in an optical chemical sensor is in transforming the measuring concentration of the analyte into a measurable analytical signal. The analyte concentration is measured indirectly, through the alteration of the indicator's optical properties. Various types of indicators are used in optical chemical sensing, such as colorimetric - based on light absorption, and luminescent - based on light emission (Demchenko, 2009; Guilbault, 1990; Lobnik, 2006; Wolfbeis, 1991). However, the latter are of primary importance due to their high interdisciplinarity, great sensitivity, and applicability to different detection principles. A large number of fluorescent synthetic organic products are available nowadays so that a researcher can easily select the proper dye (indicator) corresponding to a particular sensing application in terms of spectroscopic properties and chemical reactivity. On the other hand, the basic organic chemistry also offers great potential as it enables synthesizing tailor-made indicators for specific applications.

2.1.1 Colorimetric indicators

Colorimetric sensing is accomplished using an indicator that changes its colour upon binding the analyte; this change is usually spectroscopically determined, but it is also visibly observed. Among the great variety of organic chromophores, such as azo dyes, nitrophenols, phtaleins, sulfophtaleins, aniline-sulfophtaleins, triphenylmethane dyes, the most popular applications using these materials is to measure the pH parameter. The pK_a of these indicators indicates the center of the measurable pH range, for example, cresol red, bromophenol blue, and bromocresol purple respond to acidic pH ($pH < 7$), while cresol red, naphtolbenzene, and phenolphthalein respond at basic pH ($pH > 7$) (Lobnik, 2006; Wolfbeis, 1991).

Redox indicators are the next example of colorimetric reagents. These are all organic dyestuffs, exhibiting reversible redox reactions. Examples include materials such as anilinic acid, diphenylamine, eriogreen, m-cresol-indophenol, methylene blue, nile blue, etc. (Wolfbeis, 1991; Lobnik, 2006). For example, a redox indicator Meldola blue can be incorporated into a sol-gel layer for an optical sensor measuring hydrogen peroxide in the concentration range of 10^{-8} to 10^{-1} mol/L (Lobnik & Čajlaković, 2001).

Ion sensing is possible using metal indicators that form coloured complexes with metal ions (Kaur & Kumar, 2011; Kim et al., 2009). The so-called ionophores are ligands that selectively bind ions. Chromogenic ionophores are designed to bring about a specific colour change in the interaction with metal cations (Murković Steinberg, 2003). Typical representatives are macrocyclic molecules with an ion-binding cavity, crown ether dyes, etc.

Among the new nano-based materials, nanoparticles (NPs), the colorimetric ones are less common in comparison with the luminescent NPs. However, some examples have recently been reported where gold NPs for sensing Cd^{2+} (Ying et al., 2011), Fe^{3+} (Shu-Pao et al., 2011), Pb^{2+} (Nan et al., 2010), nitrite, nitrate ions were utilized (Weston et al., 2009), and carbon nanotubes were reported for sensing nucleic acids (Ai Cheng, 2007), and nanowires were used for Hg^{2+} (Tsao-Yen, 2011).

2.1.2 Luminescent indicators

The analyte concentration is determined by the change in the emission properties of a luminophore. Luminescence is intrinsically more sensitive than absorption as a sensing technique, so for many applications the literature more often reports on sensing with luminescent probes and sensors. A variety of fluorescent and luminescent materials in the form of molecules, complexes and NPs are available for implementation as the response units into sensing technologies; among them, organic fluorescent dyes are of primary importance. However, there is an increasing application of other materials, such as luminescent metal-ion chelating complexes, fluorescent polymer molecules and especially, from different kinds of NPs (Demchenko, 2009; Basabe-Desmonts et al., 2007;).

As already mentioned, organic dyes are most commonly used in fluorescence sensing. Their advantages are easy availability, a low price, versatility. The best known are fluoresceins, rhodamines, cyanine dyes, Alexa dyes, and BODIPY dyes, which are frequently used for labelling. In addition to these, the environment-sensitive dyes (for example, Nile red), hydrogen-bond responsive dyes (ketocyanine dye), electric-field-sensitive dyes (styryl dyes), supersensitive multicolour ratiometric dyes (3-hydroxychromone dyes), phosphorescent dyes eosin and erythrosine derivatives, optimal FRET pairs (for example Pyrene/Coumarin) are known fluorescent reporters (Demchenko, 2009,2010; Lakowicz, 2006).

Luminescent metal complexes, with europium(III) (Eu^{3+}) and terbium(III) (Tb^{3+}) ions being the most used, represent advantages since they show longer lifetimes, large Stokes' shifts and, therefore, enable eliminating the light-scattering effects and short-lived background luminescence (Turel et al., 2009,2010), thus significantly increasing the sensitivity of the analysis. Transition-metal complexes that exhibit phosphorescence are also formed by ruthenium (Balzani et al., 2000), osmium, and rhenium ions and there are those based on porphyrin complexes (with Pt and Pd ions) (Papkovsky & O'Riordan, 2005).

Interesting alternatives to fluorescent indicators are represented by the dye-doped NPs, NPs made of organic polymer, silica-based NPs, dendrimeres, quantum dots, noble metal NPs, fluorescent conjugated polymers and visible fluorescent proteins (Borisov&Klimant, 2008; Demchenko, 2010; Wolfbeis, 2005).

2.2 Immobilization techniques

A method for indicator immobilization into a suitable polymer matrix also has an important influence on the sensing characteristics. The following possibilities are usually applied:

- Impregnation – the indicator is immobilized in the polymer matrix through physical adsorption, chemisorption or electrostatic bonding. The polymer thin film is dipped into a saturated indicator solution and the solvent is then left to evaporate (Wolfbeis, 1991).
- Covalent bonding – the indicator is covalently bonded to the polymer matrix. This may be achieved by a) choosing the indicator that contains a functional group for covalent bonding to the polymer, which is at the same time insensitive for the target analyte, or b) polymerizing the indicator to certain monomers to form a copolymer (Baldini et al., 2006; Lobnik et al., 1998).
- Doping – the indicator is entrapped in the matrix during the polymerization process, where the indicator is simply added to the starting polymer solution (Lobnik et al., 1998).

Covalent immobilization enables the sensor having good stability (no leaching, crystallization and evaporation of components) and a longer operational lifetime. The disadvantage is that the covalent bonding often lowers the sensitivity for the analyte and prolongs the response time of the sensor (Lobnik et al., 1998). Although the impregnation technique is widely used and lowpriced, it is used first of all for test strips and in gas sensors due to its low stability (indicator leaching). Doping is one of the most used immobilizations as it is not restricted to certain indicators and polymers. The sensor stability (in terms of indicator leaching) is better compared to impregnation and worse compared to covalent bonding. The response time is better than in covalent immobilization.

2.3 Polymers

Polymer chemistry is an extremely important part of optical sensor technology. Both the light guide (including its cladding and coating) and the sensing chemistry of indicator-mediated sensors are made from organic or inorganic polymers (Baldini et al., 2006; McDonagh et al., 2008; Orellana et al., 2005; Wolfbeis, 1991). The choice of polymer is governed by the permeability of the polymer for the analyte, its stability and availability, its suitability for dye immobilization, its compatibility with other materials used in the fabrication of optrodes, and its compatibility with the sample to be investigated. The polymer micro-environment has a strong effect on the spectral properties of the immobilized indicator, pKa value, luminescence lifetime, binding constant, etc. (Lobnik & Wolfbeis, 2001; Wolfbeis, 1991). Consequently, the choice of polymer material has a pronounced influence on the sensor performance and its characteristics, such as selectivity, sensitivity, working range, calibration, response time, (photo)stability (Orellana et al., 2005; Korent et al., 2007). The response time, for example, will be governed by the diffusion coefficients of the gases or liquids, and the quenching efficiency by the solubility of the gas in the polymer.

However, although most authors have compiled a considerable amount of data on various polymers, numerous new materials are available for which no data exist. It is also known that copolymers and polymer mixtures do not necessarily display the properties that may be expected from averaging the data of the pure components.

On the other hand, nano-sized (polymer) materials pose new technological and analytical challenges in many different sensor designs to improve industrial process monitoring (air and water quality), food-quality surveillance, and medical diagnostics, and to provide the reliable, real-time detection of chemical, biological, radiological and nuclear hazards for military and anti-terrorism applications – all this by enabling improved sensor characteristics, such as sensitivity, selectivity and response time, along with dramatically reduced size, weight and power requirements of the resulting monitoring devices compared to the conventional, macroscaled alternatives (Basabe-Desmonts et al., 2007; Borisov & Klimant, 2008; Demchenko, 2010).

2.3.1 Hydrophobic polymers

Silicones have excellent optical and mechanical properties, and unique gas solubility. The main applications of silicone materials is in sensors for oxygen and other uncharged quenchers, such as sulfur dioxide and chlorine, and as gas-permeable covers in sensors for carbon dioxide or ammonia. Silicones cannot be easily plasticized by conventional plasticizers, but form copolymers, which may be used instead. Blackened silicone is a most useful material for optically isolating gas sensors in order to make them insensitive to the optical properties of the sample (Baldini et al., 2006; Wolfbeis, 1991).

Poly(vinil chloride) (PVC), poly(methyl methacrylate) (PMMA), polyethylene, poly(tetrafluoroethylene) (PTFE), polystyrene (PS), and ethylcellulose comprise another group of hydrophobic materials that efficiently reject ionic species (Amao, 2003). Except for polystyrene, they are difficult to chemically modify so that their function is confined to that of a "solvent" for indicators, or as a gas-permeable cover. For example, PMMA and PDMS have been selected as the optimum matrix for oxygen sensing.

2.3.2 Hydrophilic polymers

Hydrophilic polymers provide a matrix that corresponds to an aqueous environment. Hydrophilic supports are characterized by a large number of hydrogen-bridging functions, such as hydroxyl, amino, or carboxamide groups, or by anionic groups (mainly carboxyl and sulfo) linked to the polymer backbone. Typical examples are the polysaccharides (cellulose), polyacrylates, polyacrylamides, polyimines, polyglycols, and variety of so-called hydrogels. Depending on the degree of polymerization and cross-linking, they are water-soluble or water-insoluble. The ions can diffuse quite freely, but the possible water uptake (10-1000%) can cause significant swelling of the polymer. Swelling of the matrix affects the optical properties of the sensors and, consequently, the signal changes. They display poor compatibility with hydrophobic polymers, such as silicone and polystyrene. Most hydrophilic polymer membranes are easily penetrated by both charged and uncharged low-molecular-weight analytes, but not by large proteins, and have found widespread application as support for indicators (Baldini et al. 2006; Wolfbeis, 1991). Hydrophilic matrices have been widely used for pH sensing.

2.3.3 Hydrophobic/hydrophilic polymers

Glass is widely used for manufacturing optical fibres. Its surface may be made either hydrophilic or hydrophobic by treatment with a proper surface-modification reagent. Surface derivatization is usually performed with reagents, such as amino-propyl-triethoxysilane, which introduces a free amino group onto the surface of the glass to which dyes or proteins may be covalently attached. Glass does not measurably swell, but is difficult to handle because of its brittleness.

Sol-gel forms an attractive alternative to conventional glass (Baldini et al. 2006; Lobnik & Wolfbeis, 2001). By changing parameters, such as the sol pH, precursor type and concentration, water content, and curing temperature, materials of the desired porosity and polarity can be produced. The versatility of the process facilitates tailoring of the physicochemical properties of the material in order to optimize sensor performance. The basic process involves the hydrolysis and polycondensation of the appropriate metal alkoxide solution to produce a porous glass matrix. The reagent is entrapped in such a matrix and the analyte can diffuse to it. By altering its polarity, the sol-gel matrix makes it possible to sense either ions or gas molecules (Lobnik & Wolfbeis, 1998; Lobnik & Čajlaković, 2001; Murković, 2003; Pagliaro, 2009; Turel et al., 2008).

2.4 Effect of nanodimensions on sensor characteristics

The sensor characteristics can be tuned not only by the choice of the indicator and polymeric support but also by merely reducing the size (< 100 nm). This is because materials that are smaller than the characteristic lengths associated with the specific phenomena often display new chemistry and new physics that lead to new properties that depend on size. Perhaps one of the most intuitive effects is due to the change in the surface-to-volume ratio. When the size of the structure is decreased, this ratio increases considerably and the surface phenomena predominate over the chemistry and physics in the bulk. Therefore, the sensor characteristics, such as sensitivity (Chu & Lo, 2009) and response time (Waich et al., 2008), can be dramatically improved.

Nanoparticles containing indicator dyes can either be used directly as nanosensors (NSs) or as the components of optical sensor materials. In the sensor matrix, more than one component can be encapsulated, thus allowing a synergistic approach to be employed (Aylott, 2003). Most of the NSs reported so far have used fluorophores as the sensitive and selective indicators. For making quantitative measurements in the intracellular environment, the so-called PEBBLE (Probes encapsulated by biologically localized embedding) NSs have been designed as they are small enough to be inserted into living cells with a minimum of physical perturbation (Buck et al., 2004). PEBBLEs have many advantages over widely used fluorescence-dye-based methods, such as: a) the increased number of analytes that can be measured because NSs are not limited to using a single fluorophore and can utilize cooperative interactions between ionophores, enzymes, reporter dyes, etc. (e.g., pH-sensitive and oxygen-sensitive beads can be incorporated into one polymer), b) the matrix protects the intracellular environment from any potentially toxic effects of the sensing dye, c) the matrix protects the sensing dye from potential interferences in the cellular environment, e.g., non-specific binding proteins and organelles, d) no selective sequestration of the NSs into cellular compartments or leaking from, or being pumped out of, cells, e) enhanced ability to carry out ratiometric measurements, and f) the in-vitro calibration of NSs is valid for in-vivo measurements (Aylott, 2003).

NSs provide advantages, such as an improved sensitivity, response time and ability to perform in-vivo measurements. However, the down side of using "free" NSs for in-vivo measurements needs to be considered. The prime concern is the retention of these particles in the body and the harmful effect in the long run (Sounderya & Zhang, 2008) since NPs can be responsible for a number of material interactions that could lead to toxicological effects (Nel et al., 2006). In any case, the optical properties that can be controlled at the nanoscale are of great interest in the field of optical sensor designing (Borisov & Klimant, 2008). Some optical chemical NSs rely on quantum dots (Asefa et al. 2009), metal beads (Shtykov & Rusanova, 2008) and other materials; however, most of them make use of indicators embedded in polymer beads (Lapresta et al., 2009; Zenkl & Klimant, 2009) and sol–gels (Arduini et al., 2007; Hun & Zhang, 2007; Sun et al., 2006).

3. Selected applications

Optical chemical sensors provide the opportunity to continuously monitor chemical species and have thus found numerous applications in areas such as the chemical industry, biotechnology, medicine, environmental sciences, personal protection, etc. Books and reviews presenting various optical sensing schemes (fiber optics, capillary waveguides, microsystems and microstructures, refractive index-based, surface plasmon resonance-based, biosensing, etc.) and various applications (sensing gases, vapours, humidity, pH, ions, organic chemicals, certain bacteria, DNA, etc. in medical and chemical analyses, molecular biotechnology, marine and environmental analysis, industrial production monitoring, bioprocess control, automotive industry) have been published in recent years. The most comprehensive studies include work by Baldini et al., 2006; Gauglitz, 2005; Gründler, 2007; McDonagh et al., 2008; Nagl & Wolfbeis, 2008; Wolfbeis, 2005; Wolfbeis, 2008.

The broad variety of applications of optical chemical sensors would deserve a special chapter devoted only to the applications. However, due to the limited space, we had to restrict our contribution to selected materials. Since the nano-world continues to rapidly enter our lives in many different ways, the following pages will survey the recently developed optical chemical nanosensors. The applications are selected for sensors or probes based on sol-gel and polymer NPs that have their dimensions ≤ 100 nm. Among the many optical methods employed in nanosensing, fluorescence has attracted particular attention because it is sensitive, offers several techniques to be used to explore various parameters that can serve as an analytical information; therefore, the luminescent approach is selected for the two types of nanoparticles mentioned.

3.1 Polymer-based nanoparticles

Polymer NPs are usually obtained by microemulsion polymerization. Microemulsions are clear, stable, isotropic liquid mixtures of oil, water and surfactant, sometimes in combination with a cosurfactant. The microemulsion polymerization of monomers may be achieved by incorporating a monomer in any of the water and oil phases of the system (Pavel, 2004). The two basic types of microemulsions, direct (oil dispersed in water, o/w) and reversed (water dispersed in oil, w/o) are frequently used. In w/o microemulsion nanodroplets of oil surrounded by the surfactant are dispersed in the continuous bulk water phase. The size of the synthesized particles is determined by the size of those droplets (Košak et al., 2004,2005).

NPs can also be prepared by the precipitation method, which is based on the use of two miscible solvents (Borisov et al., 2009; Higuchi et al., 2006). The nanobeads are formed by diluting the polymer solution with a poor solvent. Gradually, evaporation of the good solvent at room temperature causes precipitation of the polymer solute as fine particles. Using this method, NPs can be prepared from a variety of polymers (e.g., engineering plastics, biodegradable polymers and electro-conductive polymers, etc.). The diameter of the particles can be controlled by changing the concentration of the solution, and the mixing ratio of the good solvent and the poor solvent, respectively. It should be emphasized that this process does not require the addition of surfactants (and their subsequent removal) as in the case of NSs prepared via polymerization (Borisov et al., 2009).

An indicator can be added to the mixture of monomers to be entrapped in the bead during polymerization. Both physical (Borisov et al., 2009) entrapment and covalent (Sun et al., 2006) coupling are used. Physical entrapment of an indicator in NPs is preferred because of its simplicity and reproducibility. In contrast to bulk sensor films (typically several microns thick), in nanosensors many indicator molecules are located close to the surface so that leaching can become a serious problem. To avoid leaching, covalent binding can be used. However, in this case, both the dye and the beads require having a reactive group through which a covalent bond can be formed between the polymer and the indicator. This situation is often undesirable because excess reactive groups on the surface of the beads may compromise their properties and often make them more prone to aggregation (Borisov et al.,

2008b). Swelling is another widespread method for the encapsulation of indicators in NPs (Méallet-Renault, 2004; Frigoli et al., 2009). This method is only useful for hydrophobic materials that are not swellable in water. It is essential to use water-insoluble indicators for this method otherwise leaching can occur over time.

Nanosensors based on hydrophobic materials even allow the monitoring of hydrophilic species with acceptable response times (in contrast to monolithic films based on the identical composition), due to small diffusion distances. Polymers with polar properties (such as polyacrylonitrile) have a large surface-to-volume ratio and are therefore especially useful for ion sensing. Indicators of an amphiphilic nature are often located on the surface of the bead, allowing a response even to hydrophilic analytes.

3.2 Sol-gel based nanoparticles

Sol-gels (inorganic silica beads and organically modified silica - Ormosils) are very popular materials for designing optical nanosensors (Jain et al., 1998; Rossi et al., 2005; Shibata et al. 1997). This is due to the fact that the beads can easily be manufactured, are porous to allow an analyte to diffuse freely inside, are robust, and are biocompatible, making them suitable for intracellular measurements. Compared with polymer NPs, silica NPs possess several advantages. Silica NPs are easy to separate via centrifugation during particle preparation, surface modification, and other solution-treatment processes because of the higher density of silica (e.g., 1.96 g/cm^3 for silica vs. 1.05 g/cm^3 for polystyrene). Silica NPs are more hydrophilic and biocompatible, they are not subject to microbial attack, and no swelling or porosity change occurs with changes in the pH (Jain et al., 1998).

Nanoparticles based on sol-gel materials can be prepared by two general synthetic routes: the Stöber (Rossi et al., 2005; Shibata et al., 1997) and reverse microemulsion processes (Bagwe et al., 2004; Santra et al., 2005; Wang et al., 2005). In a typical Stöber-based protocol, a silica alkoxide precursor (such as tetraethyl orthosilicate, TEOS) is hydrolyzed in an ethanol and ammonium hydroxide mixture. The hydrolysis of TEOS produces silicic acid, which then undergoes a condensation process to form amorphous silica particles. In general, the lower the concentration of water and ammonia, the smaller are the particles. Indicators are typically entrapped inside the pores of the beads. However, that does not always prevent them from leaching into solution. Therefore, similar to the polymer beads, covalent coupling is often preferred.

Dye-doped silica NPs can also be synthesized by hydrolyzing TEOS in a reverse-micelle or water-in-oil (w/o) microemulsion system. In a typical w/o microemulsion system, water droplets are stabilized by surfactant molecules and remain dispersed in the bulk oil. The nucleation and growth kinetics of the silica are highly regulated in the water droplets of the microemulsion system, and the dye molecules are physically encapsulated in the silica network, resulting in the formation of highly monodispersed dye-doped silica NPs (Schmidt et al., 1999). Polar dye molecules are used in the w/o microemulsion system to increase the electrostatic attraction of the dye molecules to the negatively charged silica matrix, so that dye molecules are successfully entrapped inside the silica matrix. Water-soluble inorganic dyes, such as ruthenium complexes, can be readily encapsulated into nano-particles using this method (Brasola et al., 2003; Frasco & Chaniotakis, 2009; Méallet-Renault et al., 2004; Ramazzo et al., 2005). Various trapping methods can be used, such as introducing a hydrophobic silica precursor (Tapec et al., 2002), using water-soluble dextran-molecule-conjugated dyes, and synthesizing in acidic conditions (Frasco & Chaniotakis, 2009). The unique advantage of the w/o microemulsion method is that it produces highly spherical and monodisperse NPs of various sizes. It also permits the trapping of a wide variety of inorganic and organic dyes as well as other materials, such as luminescent quantum dots.

3.3 Core-shell systems for sensing chemistry

The core-shell nanostructures based on polymers and sol-gel materials are well known (Arduini et al., 2007; Burns et al., 2006; Frigoli et al., 2009; Sounderya & Zhang, 2008). The versatility matrix allows the sensing chemistry to be incorporated either into the core or into the shell of a bead, which for some NSs can slightly tune the sensitivity and helps to minimize interferences (Borisov et al., 2008a). Moreover, the encapsulation of two different fluorescent dyes in the core-shell structure, one analyte-sensitive dye and other analyte-insensitive incorporated into the core of the bead used as reference, allows ratiometric detection. This method overcomes the problems (fluctuations in light source intensity of the instrument, photobleaching of the analyte-sensitive dye, drifts in the optoelectronic setup and background fluorescence) associated with intensity-based measurements. Different approaches can be used to prepare ratiometric sensors, as shown in Fig. 3.

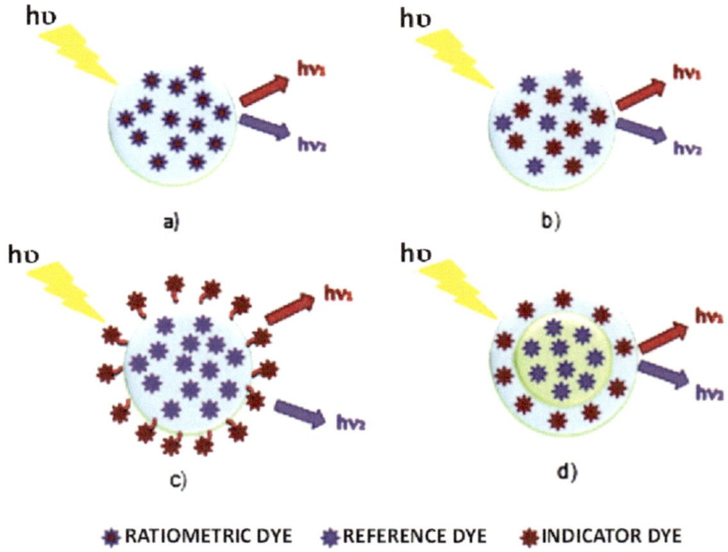

Fig. 3. Types of ratiometric approaches: a) ratiometric indicator dye is immobilized in the NP; b) indicator and reference dye are immobilized inside the NP; c) reference dye is immobilized (covalently attached or doped) in the NP, and the indicator dye is covalently bound to the NP surface; d) in the core-shell system, the reference dye is immobilized in the core and the indicator dye is immobilized in the shell.

3.4 Sensors based on polymer and sol-gel nanoparticles

In the following section the recently developed photoluminescence-based optical chemical NSs will be described, highlighting the most important sensor characteristics, such as the indicator and matrix used, dynamic range, response time, limit of detection (LOD), indicator leaching, operational lifetime, interferences, etc.

3.4.1 Nanosensors for pH

The determination of pH has great importance for a wide range of applications, e.g., in the medical, environmental and biotechnological fields. The most common pH sensors are electrochemical devices (glass electrodes). Although they can be reliable analytical tools in many situations, their use for intracellular measurements is limited. In the field of biology and medicine, sensors able to monitor pH in real time are in great demand, as they can be used for the general understanding of biological processes and for biomedical diagnostics (Doussineau et al., 2009). Recently, a lot of work has been done on optical pH-sensitive micro- (Borisov et al., 2008a; Doussineau et al., 2009; Schulz et al., 2009 ; Hornig et al., 2008) and nano-sensors (Allard & Larpent, 2008; Burns et al., 2006; Gao et al., 2005; Gao et al., 2007; Kim et al., 2006; Peng et al., 2007; Sun et al., 2006;). Here, the latter will be discussed, since they are less than 100 nm in size. Table 1 summarizes pH nanosensors.

Material	Indicator	pK_a or dynamic range	Response time	Leaching	Operational lifetime	Effect of ionic strength	Ref.
Sol-gel (TEOS[1])	RITC[2]	5.0-10.0	5 min	not given	not given	< 5% signal decrease	(Gao et al., 2005)
Sol-gel (TEOS)	Fluorescein	4-8	5 min	not given	5 weeks	< 4.5% signal decrease	(Gao et al., 2007)
Sol-gel (ICTES[3])	NVP[4]	4-8	not given	not given	>6 months	not given	(Kim et al., 2006)
Sol-gel (TEOS, APTES[5])	FITC[6]	4-7 0.05 pH units	1s	30% signal decrease	not given	not given	(Peng et al., 2007)
Sol-gel (TEOS, APTES)	FITC	6.4	not given	no leaching	not given	not given	(Burns et al., 2006)
Polystyren	FITC	5.6	few s	no leaching	>6 months	change in pKa	(Allard & Larpent, 2008)
Polyacrylamide	AF[7]	5.8-7.2	0.4 s	9% of the dye	not given	not given	(Sun et al.,2006)

[1]TEOS – Tetraethoxysilane
[2]RITC - Rhodamine-β-Isothiocyanate
[3]ICTES - (3-isocyanatopropyl)triethoxysilane)
[4]NVP - Naphtalenylvinyl-pyridine
[5]APTES – Aminoproplytriethoxysilane
[6]FITC - Fluorescein isothiocyanate
[7]AF – Aminofluorescein

Table 1. Nanosensors/probes for pH.

Several sol-gel pH NSs were prepared using a microemulsion procedure (Allard & Larpent, 2008; Gao et al., 2005; Gao et al., 2007; Kim et al., 2006; Peng et al., 2007; Sun et al., 2006) and one by Stöber (Burns et al., 2006). Gao et al., 2007, designed pH sensors using the doping technique for the immobilization of indicators in sol-gel NPs. Although leaching can be a serious problem, in this case it was not tested. The pH dynamic range of sensing NPs (60 ± 4 nm) is between 5.0 and 10.0 (Gao et al., 2005) and between 4.0 and 8.0 for NSs (30 ± 4 nm) based on indicator Fluorescein (Gao et al., 2007). The response time (5 minutes) for both sensors is relatively long compared to other pH NSs (Table 1), even compared to macro-sized pH optical sensors (Jerónimo et al., 2007).

Kim et al., 2006, designed ormosil NPs (33±6 nm) in which the ratiometric pH-responsive dye is covalently immobilized. Peng et al., 2007, and Burns et al., 2006, developed two-fluorophore-immobilized NPs sensors. The first author used fluorescein isothiocyanate as an indicator dye that was covalently crosslinked with the network formed by the hydrolysis of tetraethoxysilane and aminoproplytriethoxysilane, while tris(2,2'-bipyridyl) dichlororuthenium(II) hexahydrate (reference) was entrapped simultaneously inside the same NSs (42 nm) by electrostatic interaction. The second author synthesized a nanosensor (70 nm) via a modified Stöber method, incorporating a reference (tetramethylrhodamineisothiocyanate) and an indicator dye (fluorescein isothiocyanate) covalently bound to the matrix in the core-shell architecture. In this case the pH sensor is composed of a shell of covalently bound indicator-dye molecules surrounding a core of sequestered, covalently bound, reference-dye molecules. What is interesting is that although both authors used covalent immobilization, Peng et al., 2007, reported an extensive dye leakage.

Two different approaches have been used for the development of polymeric ratiometric pH nanosensors. In the first case a reference dye (Rhodamine B derivate) and a pH-sensitive dye (Aminofluorescein) were covalently embedded in polyacrylamide NPs (50 nm) (Sun et al., 2006). Allard & Larpent, 2008, presented a pH sensor based on polystyrene NPs, in the 20 nm range, with a hydrophobic reference dye (1,9-diphenylanthracene) embedded within the particle and the indicator dye (Fluorescein isothiocyanate) covalently attached to the surface. In this case the pH sensor may retain many of the drawbacks related to free fluorescent dyes because the indicator molecules are not protected from the cellular environment. It would be interesting to know, if the fotostability of the covalently attached indicator to the NP surface is improved compared to the free indicator.

3.4.2 Nanosensors for oxygen

Oxygen is undoubtedly one of the most important analytes. In particular, a determination of the dissolved oxygen is of utmost importance in many fields of science and technology (Papkovsky & O'Riordan, 2005), including biotechnology (Kostov et al., 2001), biology (Schmaelzlin et al., 2005), marine science (Schroeder et al., 2007), and in medicine (Kimura et al., 2007). Recent interest in methods for measuring dissolved oxygen has been focused mainly on optical sensors, due to their advantages over conventional amperometric electrodes in that they are faster and do not consume oxygen (McDonagh et al., 2002).

The principle behind the operation of these sensors is the reduction in the luminescent intensity as a consequence of the oxygen quenching of the emitting state. The sensor optode, either in typical microparticles type (Borisov et al., 2008a; Waich et al., 2008) or in recent NP type (Borisov et al., 2009; Cywinski et al., 2009), consists of a dye entrapped in a matrix with a high permeability to oxygen. Borisov et al., 2009, showed that different polymer NPs (size 44-55 nm), depending on their gas permeability, embedded with iridium (III) coumarin, could be used as oxygen NSs. However, the characterisation of these sensors should be included. Cywinski et al., 2009, prepared ratiometric polystyrene-based oxygen NSs (20.9 ± 7.1 nm). Platinum(II)mesotetra(pentafluorophenyl)porphine was used as an indicator dye, whereas N,N-bis(1-hexylheptyl)perylene-3,4:9,10-bis-(dicarboximide) was used as a reference dye.

3.4.3 Nanosensors for ions

Table 2 summarizes the recently developed ion NSs. Lead (Pb^{2+}) is one of the most toxic heavy metals, and it is particularly dangerous in terms of its effects on children. Therefore, the determination of lead in low concentrations is important. Arduini et al., 2007, developed a fluorescent nanosensor for the detection of Pb^{2+} ions. Silica core-shell NPs were produced from 50 nm diameter "pure" silica cores and a 5-nm-thick shell in which the indicator dansyl dye was covalently linked to a silica network. Then the surface of the shell was functionalized with (mercaptopropyl) triethoxysilane. Surface thiol groups play a key role in the recognition of the analyte. Selectivity tests show that Cu^{2+} is a serious interferent. On the other hand, no effect is observed after the addition of Zn^{2+}, Ca^{2+}, and Mg^{2+}, while Ni^{2+} and Co^{2+} produce much less quenching than Pb^{2+}.

Copper (Cu^{2+}) is an essential trace element that is required in various biological processes (hemopoiesis, metabolism, growth, and immune system (Uauy et al., 1998)). On the other hand, excess Cu^{2+} concentrations are extremely toxic. The detection of trace amounts of copper ions is of increasing importance in light of its environmental and biomedical implications. NSs for the detection of Cu^{2+} ions typically combine two components: a metal chelating molecule designed to bind the target ion selectively and a fluorophore as a readout system. The chelating molecules and the fluorophore are both covalently linked on to the surface of the NP (Brasola et al., 2003; Ramazzo et al., 2005) or the fluorophore is entrapped inside the NP (Méallet-Renault et al., 2004; Frigoli et al., 2009). The use of two different metal chelating ligands, picolinamide (Brasola et al., 2003) and (2-pyridinmethyl)-glycinamide (Ramazzo et al., 2005), and the same fluorescent dye (danslyamide) for the surface modification of silica NPs were reported. Méallet-Renault et al., 2004, described the concept of associating an efficient fluorescent reporter, a BODIPY derivative, and an excellent metal-chelating receptor, cyclam, in polymer NPs. The hydrophobic dye is entrapped within the particle core and the receptor, covalently linked to the polymer backbone, is mainly located near the surface. The detection is based on the quenching of fluorescence by energy transfer from the dye to the receptor-target complex.

Core-shell-type dual-fluorescent NPs (NPs) in the 16-nm-diameter range with a selective ligand (cyclam) attached to the surface and two fluorophores, 9,10-diphenyl-anthracene (donor, D) and pyrromethene PM567 (acceptor, A), embedded within the polymer core were synthesized (Frigoli et al., 2009). Dual-dye-doped fluorescent NPs permit the design of cascade FRET-mediated sensing devices. The two sequential FRET processes, the energy transfer from D to A, and quenching of the resulting sensitized emission of A by copper complexes that are formed at the NP surface, provide an efficient means to sense copperions by monitoring the quenching of the acceptor dye upon irradiation at the donor excitation wavelength. This sensor is highly selective towards other divalent cations. Zheng et al., 2010, have developed a Cu^{2+} sensor based on a fluorescent ligand (N-(quinoline-8-yl)-2-(3-triethoxysilyl-propylamino)-acetamide) that acts as both a binding and readout system.

Zinc (Zn^{2+}) is the second most abundant trace element in humans after iron. Most of the Zn^{2+} found in the body is bound to proteins such as carbonic anhydrase or zinc finger proteins. Zinc has been implicated in Alzheimer's and Parkinson's disease (Cuajungco & Lees, 1997; Frederickson et al., 2000) and as a neuromodulator (Choi & Koh, 1998). Since the full extent of zinc's purpose in the body remains unclear, the development of nanosensors is needed to further elucidate the role of zinc in vivo. Sumner et al., 2002, designed a polyacrylamide PEBBLE sensor, containing immobilized Newport Green as the indicator dye and Texas Red-dextran as the reference dye. The maximum response to zinc (percentage increase in the "intensity ratio") is 50%, which is less than observed with the Newport Green dye in solution (increase of approx. 250%). It is obvious that the matrix is affecting the indicator dye's sensitivity, but is on the other hand preventing the indicator dye from reacting to non-specific bindings of proteins (ex. bovine serum albumin). This is a prime advantage of the PEBBLEs compared to free dyes, especially for cellular analysis. Teolato et al., 2007, selected the 6-methoxy-8-(p-toluensulfonamido)-quinoline (TSQ) as the active unit for the preparation of Zn^{2+} sensing silica NPs via the Stöber–Van Blaaderen procedure. The Coumarin 5 was used as a reference dye covalently linked to a silica network. The sensor suffers from the Cu^{2+} interferences and, to a minor extent, from Cd^{2+}.

Iron (Fe^{3+}) is an integral metal ion in several physiological processes involving electron transfer and oxidation. Numerous enzymes use Fe^{3+} as a cofactor for hydroxylation, oxygen transport, DNA synthesis, etc. (Frausto da Silva & Williams, 1991). Because of iron's vital role in a number of proteins, deficiency during the developmental stages can lead to an irreversible loss of motor skills and has been linked to behavioral alterations in iron-deficient rats (Burdo & Connor, 2003). The accumulation of iron in the central nervous system has been implicated in a number of diseases. Therefore, the development of sensors that allow in-situ measurements would be beneficial. A PEBBLE sensor based on Alexa Fluor 448 (indicator dye) and Texas Red (reference) immobilized in polyacrylamide NPs for Fe^{3+} detection has been introduced (Sumner & Kopelman, 2005). What is surprising is that the reference dye is quenched by 20% in the presence of 1 μM copper.

Changes in intracellular ion concentrations accompany many processes in living cells, including transport, signalling, and enzyme function (Nuccitelli, 1994). A magnesium-sensitive PEBBLE, which contains immobilized Coumarin 343 as the indicator dye, and Texas Red-dextran as a reference dye, has been reported (Park et al., 2003). The interference from calcium is a major complication of current strategies aimed at the measurement of measuring magnesium concentrations in biological environments (Gotoh et al., 1999). The results show that Coumarin 343 is insensitive to Ca^{2+}. A fluorescence-based calcium nanosensor was described (Hun & Zhang, 2007) that exploits silica NPs doped with calcein as an indicator dye for Ca^{2+} determination in blood serum. It is surprising that the effect of pH on the fluorescent intensity of the nanosensor was not tested in the clinically important pH range (6.6-7.8).

Analyte	Material	Indicator	LOD (M)	Response time	Leaching	Operational lifetime	Interferences	Ref.
Pb^{2+}	Sol-gel (TEOS, MPS[8])	Dansyl-amide	1.1×10^{-6}	not given	not given	>30 days	Cu^{2+}, Co^{2+}, Ni^{2+}	(Arduini et al., 2007)
Cu^{2+}	Sol-gel	Dansyl-amide	not given	not given	not given	not given	Ni^{2+}	(Brasola et al., 2003)
Cu^{2+}	Sol-gel	Dansyl-amide	3.0×10^{-8}	not given	not given	not given	Ni^{2+}	(Ramazzo et al., 2005)
Cu^{2+}	latex	BODIPY-ciklomat	5×10^{-9}	not given	no leaching	not given	no	(Méallet-Renault et al., 2004)
Cu^{2+}	Styren	DAP[9]-PM567[10]-ciklomat	20×10^{-9}	not given	no	>6 months	no	(Frigoli et al., 2009)
Cu^{2+}	Sol-gel	QlOEt[11]	3.8×10^{-7}	not given	not given	not given	Zn^{2+}	(Zheng et al., 2004)
Zn^{2+}	Polyacrylamide	Newport Green	4×10^{-6}	<4 s	intensity signal drop for 3% aft. 2 days	not given	K^{+}	(Sumner et al., 2002)
Zn^{2+}	Sol-gel (TEOS)	TSQ[12]	2.6×10^{-7}	not given	not given	not given	Cu^{2+} Cd^{2+}	(Teolato et al., 2007)
Fe^{3+}	Polyacrylamide	Alexa Fluor 448	1×10^{-6}	10 min	not given	not given	Cu(I)	(Sumner & Kopelman, 2005)
Mg^{2+}	Polyacrylamide	Coumarin 343	340×10^{-6}	<4 s	intensity signal drop for less than 11.8 % after 24-h	not given	pH	(Park et al., 2003)
Ca^{2+}	Sol-gel (TEOS)	calcein	7.48×10^{-8}		intensity signal drop for 17% after 12-h	not given	pH	(Hun & Zhang, 2007)

[8]MPS – (mercaptopropyl)triethoxysilane
[9]DAP - 9,10-diphenyl-anthracene
[10]PM567 – pyrromethene
[11]QlOEt - N-(quinoline-8-yl)-2-(3-triethoxysilyl-propylamino)-acetamide
[12]TSQ - 6-methoxy-8-(p-toluensulfonamido)-quinoline

Table 2. Nanosensors/probes for ions.

3.4.4 Nanosensors for other molecules

The characteristics of NSs for the detection of glucose, naproxen and dipicolinic acid are summarized in Table 3.

Analyte	Material	Indicator	LOD (M)	Response time	Leaching	Operational lifetime	Interferences	Ref.
naproxen	Polyacryl -amide	mDMASP[13]	0.5×10^{-3}	not given	not given	not given	cAMP[14]	(Lapresta et al., 2009)
glucose	Sol-gel (TEOS[1], APTES[5])	Eu complex	4.4×10^{-6}	20 min	not given	not given	no in the 5% range of signal change	(Gao et al., 2009)
dipicoli-nic acid	Sol-gel (TEOS, APTES)	Eu-EDTA[15] complex	0.2×10^{-9}	30 s	not given	not given	no	(Ai et al., 2009)
dipicoli-nic acid	Sol-gel (TEOS, APTES)	Tb-EDTA complex	10.3×10^{-9}	not given	not given	not given	no	(Taylor & Lin, 2009)

[13]mDMASP - (E)-4-[4-(dimethylamino)styryl]-1-[4-(methacryloyloxymethyl)benzyl]pyridinium chloride
[14]cAMP - Adenosine 3:5-cyclic monophosphate
[15]EDTA - Ethylenediamine tetraacetic acid

Table 3. Nanosensors/probes for other molecules

A nonenzymatic sensor for glucose based on amino-functionalized luminescent silica NPs (LSNPs) doped with the europium(III) mixed complex, Eu(TTA)3phen with 2-thenoyltrifluoroacetone (TTA) and 1,10-phenanthroline(phen) was reported (Gao et al., 2009). Lapresta et al., 2009, developed a fluorescent polyacrylamide nanosensor for the determination of non-steroidal, anti-inflammatory drug naproxen. The fluorescent monomer (E)-4-[4-(dimethylamino)styryl]-1-[4-(methacryloyloxymethyl)benzyl]-pyridinium chloride (mDMASP) was used as an indicator. Upon indicator immobilization in polymer NPs the sensitivity and LOD were reduced. The high LOD thus enables those NSs only to be used in wastewater or industrial routine tests. For the detection of dipicolinic acid, a chemical marker for bacterial spores, silica NSs have been developed based on Eu-complex (Ai et al., 2009) and Tb-complex (Taylor & Lin, 2009). In both cases the EDTA ligand is covalently attached to the surface of silica NPs doped with reference dyes, such as fluorescein isothiocyanate (Ai et al., 2009) or Ru-complex (Taylor & Lin, 2009). Interferences resulting from the nonselective binding of aromatic ligands (benzoic acid, m-phthalic acid, o-phthalic acid) to a Tb-complex should be tested. Moreover, it is surprising that a Ru-complex, a common indicator for oxygen sensors (Li et al., 2007; Wu et al., 2010; Zhang et al., 2008) has been used as a reference, but the possible effect of oxygen on its fluorescence properties has not been tested.

4. Summary and trends

The design of an optical sensor is primarily governed by the application for which the sensor is to be developed. And because of the existence of many optical principles as well as of many sensor components that can be altered, optical chemical sensors (OCSs) can be applied to a huge number of applications. OCSs may either be very simple and cost-effective devices or enable advanced multisensor applications. In part by the sensitive layers, it is selectivity, sensitivity, stability, and reversibility that are the requirements for sensor systems that must be provided. Besides, the user normally expects the following sensor characteristics: a relatively high signal-to-noise ratio, short response times, low limits of detection, high sensitivity, low cost, a possibility for on-line, in-situ applications.

However, the most critical limitations of optical sensors result from unavailable or inadequate (unspecific) molecular probes. The detection limit that in certain cases is expected to lie in the 10^{-13} to 10^{-14} mol/L concentration range could also be improved. A preferred solution for improving the selectivity would be the use of sensor arrays (Paolesse et al., 2011), rather than a single sensor. It additionally turns out that most of the improvements that can be expected in optical detection methods are in the area of sensitive layers, which also have to provide reversibility and greater stability.

Another step in achieving improvements in sensor characteristics is the use of nanoscale materials for OCS, which has emerged as one of the most important research areas over the past decade. Nanomaterials exhibit highly tunable size- and shape-dependent chemical and physical properties show a unique surface chemistry, thermal and electrical properties, a high surface area and a large pore-volume per mass unit area. Because of their unique and advantageous features they can help to improve the sensitivity, response time and detection limit of sensors. Nevertheless, although zero-dimensional materials are applied in sensor design, the factors such as matrix, indicator and immobilization technique are still indispensable tools in achieving the desired sensor's characteristics.

However, as more and more types of smart nanomaterials with unique and tunable properties continue to be invented, increasing numbers of efficient and selective NSs are expected to emerge. Sensors capable of detecting very low analyte concentrations for non-invasive and non-degenerative analyses in a wealth of different promising applications will be produced. It is also believed that the cost of individual sensors will be radically reduced and sensors will be more easy to use for the end-users. Other important challenges, which are likely to move future trends in the research and development of OCS, such as the development of fully engineered monitoring systems requiring modules to collect the target molecules and bring them to the sensor, the nanosensor array to carry out the detection, a mechanism to refresh and regenerate the nanosensor as needed, and data-management capabilities to communicate and display the information as well as the sensors' long-term stability, wireless networks, optical materials and components, are expected to be fulfilled.

5. References

Ai Cheng L.; Jian-Shan Y.; Swee N.T.; Daniel P.P.; Fwu-Shan S.; Chew K.H. & Tit M.L. (2007). Carbon nanotube-based labels for highly sensitive colorimetric and aggregation-based visual detection of nucleic acids. *Nanotechnology,* Vol.18, pp. 455102.1- 455102.9

Ai, K.L.; Zhang, B.H. & Lu, L.H. (2009). Europium-Based Fluorescence Nanoparticle Sensor for Rapid and Ultrasensitive Detection of an Anthrax Biomarker. *Angew Chem Int Ed,* Vol.48, pp. 304-308

Allard, E. & Larpent, C. (2008). Core-shell type dually fluorescent polymer nanoparticles for ratiometric pH-sensing. *J Polym Sci, A: Polym Chem,* Vol.46, pp. 6206-6213

Amao, Y. (2003). Probes and Polymers for Optical Sensing of Oxygen. *Microchim Acta,* Vol.143, pp. 1-12

Arduini, M.; Mancin, F.; Tecilla, P. & Tonellato, U. (2007). Self-organized fluorescent nanosensors for ratiometric Pb^{2+} detection. *Langmuir,* Vol.23, pp. 8632-8636

Asefa, T.; Duncan, C.T. & Sharma, K.K. (2009). Recent advances in nanostructured chemosensors and biosensors. *Analyst,* Vol.134, pp. 1980-1990

Aylott, J.W. (2003). Optical nanosensors - an enabling technology for intracellular measurements. *Analyst,* Vol.128, pp. 309-312

Bagwe, R.P.; Yang, C.; Hilliard, L.R. & Tan, W. (2004). Optimization of Dye-Doped Silica Nano-particles Prepared Using a Reverse Microemulsion Method. *Langmuir,* Vol.20, pp. 8336-8342

Baldini, F.; Chester, A.N.; Homola, J. & Martellucci, S. (1st ed.). (2006), *Optical Chemical Sensors,* Springer, ISBN 1402046103, Netherlands

Balzani, V.; Ceroni, P.; Gestermann, S.; Kauffmann, C.; Gorka, M. & Vogtle, F. (2000) Dendrimers as fluorescent sensors with signal amplification. *Chem Commun,* No.10, pp. 853–854

Basabe-Desmonts L., Reinhoudt D.N. & Crego-Calama M. (2007). Design of fluorescent materials for chemical sensing. *Chem Soc Rev,* Vol.36, pp. 993-1017

Blum, L.J. (1997). *Bio- and Chemi-Luminescent Sensors,* World Scientific Publishing Company, Incorporated, ISBN 9810228996, Singapore

Borisov, S.M. & Klimant, I. (2008). Optical nanosensors - smart tools in bioanalytics. *Analyst,* Vol.133, pp. 1302-1307

Borisov, S.M.; Mayr, T. & Klimant, I. (2008a). Poly(styrene-block-vinylpyrrolidone) beads as a versatile material for simple fabrication of optical nanosensors. *Anal Chem,* Vol.80, pp. 573-582

Borisov, S.M.; Mayr, T.; Karasyov, A.A.; Klimant, I.; Chojnacki, P.; Moser, C.; Nagl, S.; Schaeferling, M.; Stich, M.; Kocincova, A.S. & Wolfbeis, O.S. (2008b). New Plastic Microparticles and Nano-particles for Fluorescent Sensing and Encoding. *Springer Ser Fluoresc,* Vol.4, pp. 431-463

Borisov, S.M.; Mayr, T.; Mistlberger, G.; Waich, K.; Koren, K.; Chojnacki, P. & Klimant, I. (2009). Precipitation as a simple and versatile method for preparation of optical nanochemosensors. *Talanta,* Vol.79, pp. 1322-1330

Brasola, E.; Mancin, F.; Ramazzo, E.; Tecilla, P. & Tonellato, U. (2003). A fluorescence nanosensor for Cu^{2+} on silica particles. *Chem Commun,* No.24, pp. 3026-3027

Buck, S.M.; Koo, Y-E.L.; Park, E.; Xu, H.; Philbert, M.A.; Brasuel, M.A. & Kopelman, R. (2004). OptochemicalNanosensor PEBBLEs: Photonic Explorers for Bioanalysis with Biologically Localized Embedding. *Curr Opin Chem Biol,* Vol.8, pp. 540-546

Burdo, J.R. & Connor, J.R. (2003). Brain iron uptake and homeostatic mechanisms: An over-view. *Biometals,* Vol.16, pp. 63-75

Burns, A.; Sengupta, P.; Zedayko, T.; Baird, B. & Wiesner, U. (2006). Core/shell fluorescent silica nanopartictes for chemical sensing: Towards single-particle laboratories. *Small,* Vol.2, pp. 723-726

Choi, D.W. & Koh, J.Y. (1998). Zinc and brain injury. *Annu Rev Neurosci,* Vol.21, pp. 347-375

Chu, C.S. & Lo, Y.L. (2009). Highly sensitive and linear optical fiber carbon dioxide sensor based on sol-gel matrix doped with silica particles and HPTS. *Sens Actuators, B,* Vol.143, pp. 205-210

Cuajungco, M.P. & Lees, G.J. (1997). Zinc metabolism in the brain: Relevance to human neurodegenerative disorders. *Neurobiol Dis,* Vol.4, pp. 137-169

Cywinski, P.J.; Moro, A.J.; Stanca, S.E.; Biskup, C. & Mohr, G.J. (2009). Ratiometric porphyrin-based layers and nanoparticles for measuring oxygen in biosamples. *Sens Actuators, B,* Vol.135, pp. 472-477

Demchenko, A.P. (2009). Design and properties of fluorescence reporters, In: *Introduction to Fluorescence Sensing,* pp. 119-196, Springer, ISBN 9781402090028, Netherlands

Demchenko, A.P. (2010). *Advanced Fluorescence Reporters in Chemistry and Biology II,* Springer, ISBN 9783642046995, Germany

Doussineau, T.; Smaihi, M. & Mohr, G.J. (2009). Two-Dye Core/Shell Zeolite Nanoparticles: A New Tool for Ratiometric pH Measurements. *Adv Funct Mater,* Vol.19, pp 117-122

Frasco, M.F. & Chaniotakis, N. (2009). Semiconductor quantum dots in chemical sensors and biosensors. *Sensors,* Vol.9, pp. 7266-7286

Frausto da Silva, J.J.R. & Williams, R.J.P. (1991). *The biological Chemistry of the Elements: The Inorganic Chemistry of Life,* Clarendon Press, ISBN 0198555989, Oxford

Frederickson, C.J.; Suh, S.W.; Silva, D. & Thompson, R.B. (2000). Importance of zinc in the central nervous system: The zinc-containing neuron. *J Nutr,* Vol.130, pp. 1471S-1483S

Frigoli, M.; Ouadahi, K. & Larpent, C. (2009). A Cascade FRET-Mediated Ratiometric Sensor for Cu(2+)Ions Based on Dual Fluorescent Ligand-Coated Polymer Nanoparticles. *Chem Eur J,* Vol.15, pp. 8319-8330

Gao, F.; Wang, L.; Tang, L.J. & Zhu, C.Q. (2005). A novel nano-sensor based on rhodamine-beta-isothiocyanate - Doped silica nanoparticle for pH measurement. *Microchim Acta,* Vol.152, pp. 131-135

Gao, F.; Tang, L.J.; Dai, L. & Wang, L. (2007). A fluorescence ratiometricnano-pH sensor based on dual-fluorophore-doped silica nanoparticles. *Spectrochim Acta, Part A,* Vol.67, pp. 517-521

Gao, F.; Luo, F.B.; Chen, X.X.; Yao, W.; Yin, J.; Yao, Z. & Wang, L. (2009). A novel nonenzymatic fluorescent sensor for glucose based on silica nanoparticles doped with europium coordination compound. *Talanta,* Vol.80, pp. 202-206

Gauglitz, G. (2005). Direct optical sensors: principles and selected applications. *Anal Bioanal Chem,* Vol.381, pp. 141-155

Gotoh, H.; Kajikawa, M.; Kato, H. & Suto, K. (1999). Intracellular Mg2+ surge follows Ca2+ in-crease during depolarization in cultured neurons. *Brain Res,* Vol.828, pp. 163-168

Gründler, P. (2007). *Chemical Sensors, An Introduction for Scientists and Engineers,* Springer-Verlag, ISBN 9783540457428, Berlin Heidelberg

Guilbault, G. (1990). *Practical Fluorescence.* Marcel Dekker Inc., ISBN 0824783506, New York

Higuchi, T.; Yabu, H. & Shimomura, M. (2006). Simple preparation of hemispherical polysty-rene particles. *Colloids Surf A,* Vol.284/285, pp. 250-253

Hornig, S.; Biskup, C.; Grafe, A.; Wotschadlo, J.; Liebert, T.; Mohr, G.J. & Heinze, T. (2008). Bio-compatible fluorescent nanoparticles for pH-sensoring.*Soft Matter,* Vol.4, pp. 1169-1172

Hun, X. & Zhang, Z.J. (2007). Preparation of a novel fluorescence nanosensor based on calcein-doped silica nanoparticles, and its application to the determination of calcium in blood serum. *Microchim Acta,* Vol.159, pp. 255-261

Jain, T.K.; Roy, I.; De, T.K. & Maitra, A. (1998). Nanometer silica particles encapsulating active compounds: a novel ceramic drug carrier. *J Am Chem Soc,* Vol.120, pp. 11092-11095

Jerónimo, P.C.A; Araújo, A.N. & Montenegro, M.C.B.S.M. (2007). Optical sensors and biosensors based on sol-gel films. *Talanta,* Vol.72, pp. 13-27

Kaur, N. & Kumar, S. (2011). Colorimetric metal ion sensors. *Tetrahedron,* doi: 10.1016/j.tet.2011.09.003

Kikuchi K., Takakusa H. & Nagano T. (2004). Recent advances in the design of small molecule-based FRET sensors for cell biology. *TrAC,* Vol.23, pp.407-415

Kim, D.S.; Chung, Y.M.; Jun, M. & Ahn, K.H. (2009). Selective Colorimetric Sensing of Anions in Aqueous Media through Reversible Covalent Bonding. *J Org Chem,* Vol.74, pp. 4849-4854

Kim, S.; Pudavar, H.E. & Prasad, P.N. (2006). Dye-concentrated organically modified silica nanoparticles as a ratiometric fluorescent pH probe by one- and two-photon excitation. *Chem Commun,* No.19, pp. 2071-2073

Kimura, S.; Matsumoto, K.; Mineura, K. & Itoh, T. (2007). Cerebral oxygen metabolism in idio-pathic-normal pressure hydrocephalus. *J NeurolSci,* Vol.258, pp. 60-68

Korent, Š.M.; Lobnik, A. & Mohr, G.J. (2007). Sol-gel-based optical sensor for the detection of aqueous amines. *Anal Bioanal Chem,* Vol.387, pp. 2863-2870

Kostov, Y.; Harms, P.; Randers-Eichhorn, L. &Rao, G. (2001). Low-cost microbioreactor for high-throughput bioprocessing. *Biotechnol Bioeng,* Vol.72, pp. 346-352

Košak, A.; Makovec, D. & Drofenik, M. (2004). The preparation of MnZn-ferrite nanoparticles in water-CTAB-hexanolmicroemulsions. *Nanotechnology (Bristol),* Vol.15, pp. S160-S166

Košak, A.; Makovec, D.; Drofenik, M. & Žnidaršič, A. (2005). The synthesis of spinel-ferrite nanoparticles using precipitation in microemulsions for ferrofluid applications. *J Magn Magn Mater,* Vol.289, pp. 32-35

Lakowicz J.R. (2006). *Principles of Fluorescence Spectroscopy,* Springer, ISBN 13: 978038731278, USA

Lapresta-Fernandez, A.; Cywinski, P.J.; Moro, A.J. & Mohr, G.J. (2009). Fluorescent polyacrylamide nanoparticles for naproxen recognition. *Anal Bioanal Chem,* Vol.395, pp. 1821-1830

Li, L.; Gao, X.K.; Lv, B.Q.; Zhou, Z.D. & Xiao, D. (2007). An oxygen sensor based on fluorescence quenching of Ru(bpy)(2)(dpp)(2+) immobilized on sol-gel-derived porous silica coatings. *Sensor Lett,* Vol.5, pp. 441-444

Lippittsch, M.E. & Draxler, S. (1993). Luminescence decay-time-based optical sensors: principles and problems. *Sens Actuators B,* Vol.11, pp. 97-101

Lobnik, A. & Wolfbeis, O.S. (1998). Sol-gel based optical sensor for dissolved ammonia. *Sens Actuators B,* Vol.51, pp. 203-207

Lobnik, A.; Oehme, I.; Murkovic, I. & Wolfbeis, O.S. (1998). pH optical sensors based on sol-gels: Chemical doping versus covalent immobilization. *Anal ChimActa,* Vol.367, pp. 159-165

Lobnik, A. & Čajlaković, M. (2001). Sol-gel based optical sensor for continuous determination of dissolved hydrogen peroxide. Sens Actuators B, Vol.74, pp. 194-199

Lobnik, A. & Wolfbeis, O.S. (2001). Probing the Polarity of Sol-Gels and Ormosils via the Absorption of Nile Red. *J Sol-Gel SciTechnol,* Vol.20, pp. 303-311

Lobnik, A. (2006). Absorption-based sensors. In: *Optical chemical sensors,* Baldini, F.; Chester, A.N.; Homola, J. &Martellucci, S., pp.77-98, Springer, ISBN1402046103, Netherlands

Lukowiak, A. & Strek, W. (2009). Sensing abilities of materials prepared by sol-gel technology. *J Sol-gel Sci Technol,* Vol.50, pp. 201-215

McDonagh, C.; Bowe, P.; Mongey, K. & MacCraith, B.D. (2002). Characterisation of porosity and sensor response times of sol–gel-derived thin films for oxygen sensor applications. *J Non-Cryst Solids,* Vol.306, pp. 138-148

McDonagh, C.; Burke, C.S. & MacCraith, B.D. (2008). Optical chemical sensors. *Chem Rev,* Vol.108, pp. 400-422

Méallet-Renault, R.; Pansu, R.; Amigoni-Gerbier, S. & Larpent, C. (2004). Metal-chelating nano-particles as selective fluorescent sensor for Cu^{2+}. *Chem Commun,* No.20, pp. 2344-2345

Murković Steinberg, I.; Lobnik, A. & Wolfbeis, O.S. (2003). Characterisation of an optical sensor membrane based on the metal ion indication Pyrocatechol Violet. *Sens Actuators B,* Vol.90, pp. 230-235

Nagl, S. & Wolfbeis, O.S. (2008). Classifiction of optical Chemical Sensors and Biosensors Based on Fluorescence and Phosphorescence. *Springer Ser Fluoresc,* Vol.5, pp. 325-346

Nan, D.; Qian, C.; Hong, Z.; Yimin, Y.; Lixi, Z.; Yujian, H.; Kaixiang, X. & Guangwei, W. (2010). Colorimetric Assay for Determination of Lead (II) Based on Its Incorporation into Gold Nanoparticles during Their Synthesis. *Sensors,* Vol.10, pp. 11144-11155

Nel, A.; Xia, T.; Mädler, L. & Li, N. (2006). Toxic potential of materials at the nanolevel, *Science,* Vol.311, pp. 622-627

Nuccitelli, R. (1994). *A Practical Guide to the Study of Calcium in Living Cells (Methods in cell biology).* Academic Press, ISBN: 0125228104, Inc., San Diego

Orellana, G.; Moreno-Bondi, M.C.; Garcia-Fresnadillo, D. & Marazuela, M.D. (2005). The Interplay of Indicator, Support and Analyte in Optical Sensor Layers, *Springer Ser Chem Sens Biosens,* Vol.3, pp. 189–225.

Pagliaro, M. (2009). *Silica-Based Materials for Advanced Chemical Applications,* RSC Publishing, ISBN: 9781847558985, UK

Paolesse, R.; Monti, D.; Dini, F. & Di Natale, C. (2011). Fluorescence Based Sensor Arrays. *Top Curr Chem,* Vol.300, pp. 139-174

Papkovsky, D.B. & O'Riordan, T.C. (2005). Emerging applications of phosphorescent metalloporphyrins. *J Fluoresc,* Vol.15, No.4, pp. 569–584, ISSN: 1053-0509

Park, E.J.; Brasuel, M.; Behrend, C.; Philbert, M.A. & Kopelman, R. (2003). Ratiometric optical PEBBLE nanosensors for real-time magnesium ion concentrations inside viable cells. *Anal Chem,* Vol.75, pp. 3784-3791

Patel, P.N.; Mishra, V. & Mandloi, A.S. (2010). Optical Biosensors: Fundamentals & Trends. *JERS,* Vol.I, pp. 15-34

Pavel, F.M. (2004). Microemulsion Polymerization. *J Dispersion Sci Technol,* Vol.25, pp. 1–16

Peng, J.F.; He, X.X.; Wang, K.M.; Tan, W.H.; Wang, Y. & Liu, Y. (2007). Noninvasive monitoring of intracellular pH change induced by drug stimulation using silica nanoparticle sensors. *Anal Bioanal Chem,* Vol.388, pp. 645-654

Ramazzo, E.; Brasola, E.; Marcuz, S.; Mancin, F.; Tecilla, P. & Tonellato, U. (2005). Surface modification of silica nanoparticles: a new strategy for the realization of self-organized fluorescence chemosensors. *J Mater Chem,* Vol.15, pp. 2687-2696

Rossi, L.M.; Shi, L.; Quina, F.H. & Rosenzweig, Z. (2005). Stöber Synthesis of Monodispersed Luminescent Silica Nanoparticles for Bioanalytical Assays. *Langmuir,* Vol.21, pp. 4277-4280

Santra, S.; Bagwe, R.P.; Dutta, D.; Stanley, J.T.; Walter, G.A.; Tan, W.; Moudgul, B.M. & Mericle, R.A. (2005). Synthesis and Characterisation of Fluorescent, Radio-Opaque, and Paramagnetic Silica Nanoparticles for Multimodal Bioimaging Applications. *Adv Mater,* Vol.17, pp. 2165-2169

Schmaelzlin, E.; Van Dongen, J.T.; Klimant, I.; Marmodee, B.; Steup, M.; Fisahn, J.; Geigenberger, P. & Loehmannsroeben, H.-G. (2005). An optical multifrequency phase-modulation method using microbeads for measuring intracellular oxygen concentrations in plants. *Biophys J,* Vol.89, pp. 1339-1345

Schmidt, J.; Guesdon, C. & Schomäcker, R. (1999). Engineering aspects of preparation of nanocrystalline particles in microemulsions. *J Nanoparticle Res,* Vol.1, pp. 267–276

Schroeder, C.R.; Polerecky, L. & Klimant, I. (2007). Time-resolved pH/pO2 mapping with luminescent hybrid sensors. *Anal Chem,* Vol.79, pp. 60-70

Schulman, S.G. (1988). *Molecular Luminescence Spectroscopy, Methods and Applications, Part 2,* John Wiley & Sons Inc., ISBN 0471636843, USA

Schulz, A.; Hornig, S.; Liebert, T.; Birckner, E.; Heinze, T. & Mohr, G.J. (2009). Evaluation of fluorescent polysaccharide nanoparticles for pH-sensing. *Org Biomol Chem,* Vol.7, pp. 1884-1889

Seitz, W.R. (1988). Chemical Sensors Based on Immobilized Indicators and Fiber Optics. *Crit Rev Anal Chem,* Vol.19, pp. 135-173

Shibata, S.; Taniguchi, T.; Yano, T. & Yamane, M. (1997). Formation of Water-Soluble Dye-Doped Silica Particles. *J Sol-Gel Sci Technol,* Vol.10, pp. 263-268

Shu-Pao, W.; Yi-Pu, C. & Yi-Ming, S. (2011). Colorimetric detection of Fe^{3+} ions using pyrophosphate functionalized gold nanoparticles. *Analyst,* Vol.136, pp. 1887-1891

Shtykov, S.N. & Rusanova, T.Y. (2008). Nanomaterials and nanotechnologies in chemical and biochemical sensors: Capabilities and applications. *Russ J Gen Chem,* Vol.78, pp. 2521-2531

Sounderya, N. & Zhang, Y. (2008). Use of Core/Shell Structured Nanoparticles for Biomedical Applications. *Recent Patents on Biomedical Engineering,* Vol.1, pp. 34-42

Sun, H.; Scharff-Poulsen, A.M.; Gu, H. & Almdal, K. (2006). Synthesis and characterization of ratiometric, pH sensing nanoparticles with covalently attached fluorescent dyes. *Chem Mater,* Vol.18, pp. 3381-3384

Sumner, J.P.; Aylott, J.W.; Monson, E. & Kopelman, R. (2002). A fluorescent PEBBLE nanosensor for intracellular free zinc. *Analyst,* Vol.127, pp. 11-16

Sumner, J.P. & Kopelman, R. (2005). Alexa Fluor 488 as an iron sensing molecule and its application in PEBBLE nanosensors. *Analyst,* Vol.130, pp. 528-533

Tapec, R.; Zhao, X.J.J. & Tan, W.H. (2002). Development of organic dye-doped silica nano-particles for bioanalysis and biosensors. *J Nanosci Nanotechnol,* Vol.2, pp. 405-409

Taylor, K.M.L. & Lin, W.B. (2009). Hybrid silica nanoparticles for luminescent spore detection. *J Mater Chem,* Vol.19, pp. 6418-6422

Teolato, P.; Rampazzo, E.; Arduini, M.; Mancin, F.; Tecilla, P. & Tonellato, U. (2007). Silica nano-particles for fluorescence sensing of Zn-II: Exploring the covalent strategy. *Chem Eur J,* Vol.13, pp. 2238-2245

Tsao-Yen W.; Hsin-Yun C.; Yen-Fei L.; Yu-Lun H. & Chih-Ching H. (2011). Selective Tellurium Nanowire-based Sensors for Mercury(II) in Aqueous Solution. *J Chin Chem Soc,* Vol.58, pp.1-7

Turel, M.; Čajlaković, M.; Austin, E.; Dakin, J.P.; Uray, G. & Lobnik, A. (2008). Direct UV-LED lifetime pH sensor based on a semi-permeable sol-gel membrane immobilized luminescent Eu3+ chelate complex. *Sens Actuators B,* Vol.131, pp. 247-253

Turel, M.; Dürkop, A.; Yegorova, A.; Scripinets, Y.; Lobnik, A. & Samec, N. (2009). Detection of nanomolar concentrations of copper (II) with a Tb- quinoline-2-one probe using luminescence quenching or luminescence decay time. *Anal Chim Acta,* Vol.644, pp. 53-60

Turel, M.; Duerkop A.; Yegorova A.; Karasyov A.; Scripinets Y. & Lobnik A. (2010). Microtiterplate phosphate assay based on luminescence quenching of a terbium complex amenable to decay time detection. *Anal Chim Acta,* Vol. 675, pp. 42–48,

Uauy, R.; Olivares, M. & Gonzalez, M. (1998). Essentiality of copper in humans. *Am J Clin Nutr,* Vol.67, pp. 952S-959S

Waich, K.; Mayr, T. & Klimant, I. (2008). Fluorescence sensors for trace monitoring of dissolved ammonia. *Talanta,* Vol.77, pp. 66-72

Wang, L.; Yang, C. & Tan, W. (2005). Dual-Luminophore-Doped Silica Nanoparticles for Multiplexed Signaling. *Nano Lett,* Vol.5, pp. 37-43

Weston, L.D.; Min Su H.; Jae-Seung L. & Chad A.M. (2009). Colorimetric Nitrite and Nitrate Detection with Gold Nanoparticle Probes and Kinetic End Points. *J Am Chem Soc,* Vol.131, pp. 6362–6363

Wolfbeis, O.S. (1991). *Fiber Optic Chemical Sensors and Biosensors,* CRC Press, ISBN 0849355087, Boca Raton

Wolfbeis, O.S. (2005). Materials for fluorescence-based optical chemical sensors. *J Mater Chem,* Vol.15, pp. 2657-2669

Wolfbeis, O.S. (2008). Fiber-Optic Chemical Sensors and Biosensors. *Anal Chem,* Vol.80, pp. 4269–4283

Wu, X.D.; Song, L.T.; Li, B. & Liu, Y.H. (2010). Synthesis, characterization, and oxygen sensing properties of Ru(II) complex covalently grafted to mesoporous MCM-41. *J Lumin,* Vol.130, pp. 374-379

Ying, X.; Hong, Z.; Zhijiao W.; Xiangjun L.; Yujian H. & Zhuobin Y. (2011). Colorimetric detection of Cd2+ using gold nanoparticles cofunctionalized with 6-mercaptonicotinic acid and L-Cysteine. *Analyst,* Vol.136, pp. 3725-3730

Zenkl, G. & Klimant, I. (2009). Fluorescent acrylamide nanoparticles for boronic acid based sugar sensing - from probes to sensors. *Microchim Acta,* Vol.166, pp. 123-131

Zhang, H.R.; Li, B.; Lei, B.F. & Li, W.L. (2008). Oxygen-sensing materials based on [Ru(bpy)(3)](2+) covalently grafted MSU-3 mesoporous molecular sieves. *J Lumin,* Vol.128, pp. 1331-1338

Zheng, J.N.; Xiao, C.; Fei, Q.; Li, M.; Wang, B.J.; Feng, G.D.; Yu, H.M.; Huan, Y.F. & Song, Z.G. (2010). A highly sensitive and selective fluorescent Cu^{2+} sensor synthesized with silica nanoparticles. *Nanotechnology,* Vol.21, p. 045501

2

Optical Sensors Based on Opal Film and Silica Nanoparticles Modified with a Functional Dye

Ivan Boldov[1], Natalia Orlova[2], Irina Kargapolova[2], Alexandr Kuchyanov[1], Vladimir Shelkovnikov[2] and Alexandr Plekhanov[1]
[1]*Institute of Automation and Electrometry, Siberian Branch, Russian Academy of Sciences,*
[2]*Novosibirsk Institute of Organic Chemistry, Siberian Branch, Russian Academy of Sciences, Russia*

1. Introduction

Chemical sensing using optics is under extensive research all over the world and many optical chemical sensors are finding increasing application in industry, environmental monitoring, medicine, biomedicine and chemical analysis (Baldini et al., 2006). Optical sensors can be used as fiber optics microsensors, as planar coatings in bioreactors, in microtiterplate format, in disposable single-shot device, and as planar membranes that can be imaged using sensing camera. The spectral range extends from the UV to the infrared, and from absorption to emission and to surface plasmon resonance (Narayanaswamy & Wolfbeis, 2004). Hence, a variety of schemes are conceivable. The purpose of this chapter is to introduce the concept of highly sensitive and selective optical sensors for different type reagent based on opal film and silica nanoparticles modified with a functional dye.

In the last two decades photonic crystals have much potential for creation of new optical functional materials (Joannopoulos et al., 2008). Photonic crystals are inhomogeneous media with a periodically varying relative permittivity. The capability of such periodic structures to form photonic band gaps changes the paradigm of controlling the propagation of light. For this reason, the features of the behavior of refracted light in the allowed bands and at the edge of the band gaps currently attract a lot of attention. A number of unusual properties of the propagation of light in photonic crystals such as the refraction of light in negative media (Foteinopoulou et al., 2003) and the effect of a superprism and the self-collimation of light (Kosaka et al., 1998) were recently revealed. It was proposed to use these effects for creating a supersensitive light beam splitter (Baba & Nakamura, 2002; Wu et al., 2003) and for controlling an optical flux (Chen et al., 2004). Moreover, schemes with photonic crystals can underlie supersensitive optical chemical sensors (Yakimansky et al., 2009). For such applications, it is important to determine and analyze regions with a strong angular dispersion in optical systems with photonic crystals and to study the behavior of light at the interface of photonic crystals with other optical media. Extension of sensory capabilities of photonic crystals by the chemical modification of silica nanoparticles with functional dyes offers promise as a high selective fluorescent sensor and many ways to miniaturize and integrate photonic devices on high functionality optical chips.

We show state-of-the-art fabrication of photonic crystals in the visible optical region based on a single-crystal opal platform (Plekhanov et al., 2006). We have investigated a new effect appearing in the displacement of the photonic bang gap on the background of the spectrum of backward diffracted and reflected Bragg waves at the grazing incident of white-light beam on the glass - photonic crystal opal film interface. The physical basis for observable effect is such characteristics of the photonic crystals as a strong angular dispersion and the dependence of the spectral position of photonic bang gap on the amount and type of analyte. It was found that the spectrum of the Bragg backward reflection and refraction manifested the photonic bang gap, which changes its position under insignificant change of the concentration of vapor of a range of substances (isopropyl alcohol, dibutylamin, tributylamine, water, ammonia), filling the photonic crystal. We consider this effect in second part of the chapter.

Quick analysis of the presence of a particular analyte is a fairly common task. In this case, as a rule do not need ultra-high sensitivity and reversibility. The implementation of spectroscopic schemes into a useful sensing scheme has been hampered though, by the lack of appropriate materials including silica nanoparticles with functional organic dyes. This chapter will also address the progress made in the past years. We have developed the method for the chemical modification of silica nanoparticles with a pyrylium dye under the sol–gel synthesis conditions. The films of silica particles modified with a pyrylium dye gave luminescence in visible spectral range upon treatment with a solution of butylamine.

During the last 15 years, fiber optics sensors have been developed and evaluated to assess air, water and soil contamination level. These sensors utilize the optical properties of chemicals or products of chemicals or biological reactions to assess contaminants to monitor process quality, provide compliance monitoring and evaluate real-time insitu contaminant levels in uncontrolled and remediated hazardous waste sites. The coating of the thin-film nanostructures based on silica nanoparticles on the surface of optical fiber allows significantly extend the scope of fiber-optic sensors (Boldov et al., 2011). We have realized for the first time a fiber-optic chemical sensor for butylamine.

2. Features of the manifestation of a stop band in the spectrum of light diffracted at a glass–opal interface

In this section, we report the results of the investigation of a new effect of the displacement of the photonic band gap (stop band) against the unchanged spectrum of diffracted white light at the glass–thin opal film interface and the possibilities of using this effect to create optical chemical sensors and to control spectral light fluxes (Plekhanov et al., 2009).

Artificial opal films consisting of spherical silicon dioxide nanoparticles were used as photonic crystals in the experiments. The experiments were performed with monolayer and multilayer films and heterostructures consisting of several types of layers with various sphere diameters from 220 to 280 nm. Photonic crystal films were grown by the movable meniscus method from the suspension of monodisperse spherical particles of SiO_2 on glass prisms; these samples allow for the investigation of the spectra for all of the angles of incidence of light on the glass–photonic crystal interface. The moving meniscus method makes it possible to grow photonic crystals with the structure ordered over the entire area. In addition, monolayer opal films, which had a domain structure, were deposited by the method of the short-term immersion of a prism into a suspension.

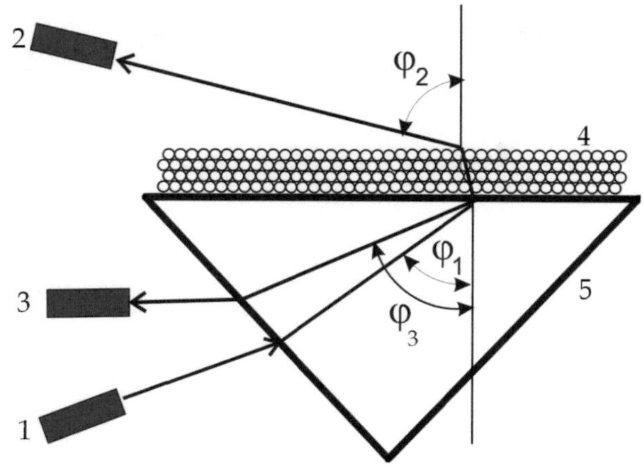

Fig. 1. Scheme of the interaction of light with the opal film photonic crystal on the glass substrate: 1 – halogen lamp; 2, 3 - spectrometer; 4 – photonic crystal film; 5 – glass substrate.

A white-light beam from a halogen lamp was incident on a face of a glass prism and, then, on the glass–photonic crystal interface (see the interaction scheme in Fig. 1.). Here, we consider refracted and reflected Bragg waves for which the directions of tangential projections of wave vectors are opposite to the corresponding projection of the wave vector of the incident wave. Refracted (2) and backward reflected (3) waves were observed after the incidence of the white-light beam (1) from the glass with the reflective index n_g = 1.51 at the interface between this medium and the three-dimensional photonic crystal film deposited on its surface. The angles φ_1 and φ_3 are measured in the glass and the angle φ_2 is measured in the air between the normal vector to the growth plane of the photonic crystal and the wave propagation direction. The angular dependences of the spectra of these waves were detected by an «Avantes» spectrometer whose optical fiber receiver was mounted on the arm of a goniometer. The angular resolution was $\Delta\varphi \approx 1°$.

The dispersion of light was examined for various synthesized photonic crystals at various angles of incidence φ_1 of white light on the (glass–photonic crystal) interface. The effect of the strong angular dispersion $\varphi_{2,3}(\lambda)$ in refracted light and light reflected back to the glass was observed for the angles of incidence φ_1 of white light larger than ≈ 41.5 , i.e., under the condition of the total internal reflection at the (photonic crystal–air) interface. For this reason, the observation of this dispersion is impossible at the incidence of light on the photonic crystal from air. Figure 2 shows the angles φ_2 and φ_3 of Bragg waves refracted and reflected from the (glass–photonic crystal) interface, respectively, as functions of the wavelength λ at φ_1 = 51.

The maximum angular dispersion, i.e., the maximum derivative $\partial\varphi_{2,3}(\lambda)/\partial\lambda$, is observed in the wavelength range λ = 580–590 nm for the opal monolayer. It is apparently caused by a strong change in the curvature of the isoenergetic surface (Kosaka et al., 1998), the normal vector to which determines the direction of the group velocity of the light wave with a given wavelength. The investigation of the photonic crystal samples containing two to hundreds of opal layers shows a number of qualitative differences. Even for several opal layers for the refracted and reflected waves, the monotonic dependence $\varphi_{2,3}(\lambda)$ was observed in a wide wavelength range and for reflection angles up to $\varphi_3 \approx 12°$. As the number of opal ball layers increases, the intensity of refracted and reflected light increases in the directions satisfying the Bragg condition. Moreover, with an increase in the number of the layers, a dip and a peak appear in the spectra of the waves passed through the photonic crystal and reflected from it at the wavelength coinciding with the position of the stop band of the photonic crystal. For the photonic crystal heterostructures consisting of two types of layers with different diameters of the balls, two spectral features corresponding to the stop bands of different photonic crystal layers appear in the spectrum of the waves.

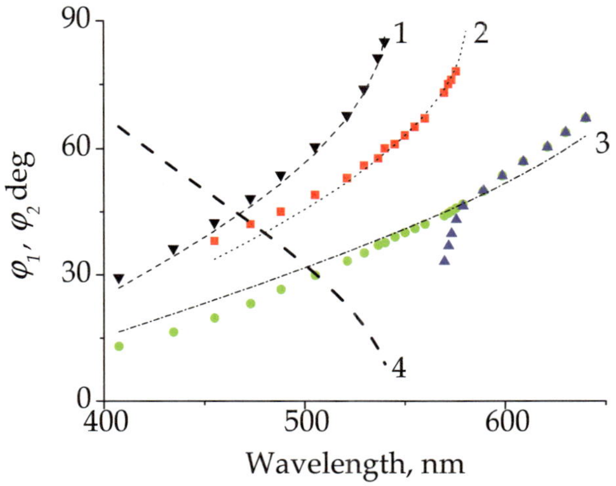

Fig. 2. Angles of refraction, φ_2, and reflection, φ_3, of Bragg waves at the (glass–photonic crystal) interface versus the wavelength λ at the angle of incidence of white light φ_1 = 51° for the (▲) reflection in the case of the opal monolayer, (●) reflection in the case of the photonic crystal, (■) refraction in the case of the opal monolayer, and (▼) refraction in the case of the photonic crystal. The solid curve is the position of the stop band of the photonic crystal calculated by Eq. (2). The dashed (1), dotted (2), and dash-dotted (3) curves are the approximation by Eq. (1) with n_{eff} = 1.36, 1.11, and 1.51, respectively.

It was revealed that the spectra of reflected and refracted Bragg waves have third order azimuthally anisotropy and were observed only for the three directions of the incident beams, $[\bar{2}11]$, $[11\bar{2}]$ and $[1\bar{2}1]$, lying in the (111) plane (the notation refers to a face-centered cubic lattice). A comparison shows that the light scattering spectra from the photonic crystals obtained by different methods mentioned above are similar to each other, but anisotropy was not observed in the case of the photonic crystal with the domain structure formed in the immersion method. This indicates that the spatial anisotropy of the effect is attributed to the hexagonal structure of the photonic crystal in the (111) plane and is responsible for the unchanged angular dispersion of white light.

The scattering of light at the glass-photonic crystal interface is described by the Bragg diffraction equation following from the condition that the tangential component of the wave vector of the incident wave is equal to the sum of the tangential components of the wave vector of the refracted wave and the corresponding vector of the crystal lattice. For the first unit cell of the photonic crystal, this equation has the following form and describes the angular dispersion:

$$\varphi_3(\lambda) = \arcsin\left(\frac{\lambda/a - n_g \sin\varphi_1}{n_{eff}}\right), \tag{1}$$

where φ_1 and φ_3 are the angles of incidence and reflection of the light wave, respectively; λ is the light wavelength in vacuum; n_g and n_{eff} are the refractive indices in the homogeneous and periodic media, respectively; and a is the lattice period in the direction of the incidence of light. The angular dispersion $\varphi_2(\lambda)$ for the refracted light is described by Equation (1) with the change $\varphi_3 \rightarrow \varphi_2$ and Snell's law for the passage of light from the photonic crystal to air.

The curves calculated by Equation (1) well approximate the experimental data. The dash-dotted curve (3) (Fig. 2.) is the approximation by Equation (1) for the Bragg wave reflected from the opal monolayer with $n_{eff} = n_g = 1.51$ (because the incident and reflected waves propagate in glass). The dotted curve (2) (Fig. 2.) is the approximation for the refracted Bragg wave with $n_{eff} = 1.11$ (this n_{eff} value is taken for the boundary layer (Kalinin et al., 2006). For the multilayer opal film with the number of layers beginning with two, the spectral dependence $\varphi_2(\lambda)$ is also described by Equation (1), but with the refractive index of the photonic crystal $n_{eff} = 1.36$ (the dashed curve (1) in Fig. 2.). The calculations were performed taking into account the refraction of reflected light at the glass–photonic crystal interface and the electron microscopy data from which the nanoparticle size was determined as $D = 254$ nm.

The most important property of the optical scheme under consideration is the manifestation of the properties of the photonic crystals and diffraction at the glass–photonic crystal interface. The experiments show that the introduction of a CCl_4 immersion liquid, which has a refractive index close to the value for quartz, to the photonic crystal does not shift the spectrum of Bragg waves; i.e., the angular dependence of the spectrum of the halogen lamp is almost insensitive to the refractive index of the photonic crystal. For this reason, this spectrum is used as a reference against which the spectral position of the stop band of the photonic crystal is measured.

The curve 2 in Fig. 3. shows the spectrum of the refracted Bragg wave. The stop band of the photonic crystal is observed in this spectrum as a dip. The position of the stop band is determined by jointly solving Equation (1) and the equation determining the behavior of the stop band when light is incident from air on the photonic crystal film at the angle $\varphi = -\varphi_2$:

$$\lambda = \sqrt{\frac{8}{3}} D n_{eff} \cos(\varphi). \tag{2}$$

The position of the stop band of the photonic crystal extracted from our experimental data is in good agreement with that calculated by Equation (2). The solid curve (4) in Fig. 2. corresponds the position $\lambda(\varphi)$ of the stop band of the photonic crystal that has the effective refractive index $n_{eff} = 1.36$ and consists of SiO_2 balls with $D = 254$ nm. The point of the intersection of this curve with the experimental dependence $\varphi_2(\lambda)$ for the wave refracted in the photonic crystal corresponds to the angle φ_2 at which the pronounced dip is seen in the spectrum of the refracted wave (see Fig. 3.).

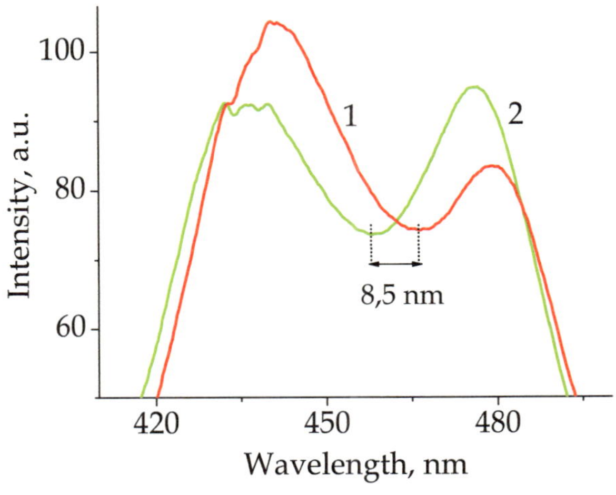

Fig. 3. Spectrum of the refracted Bragg wave with the stop band (curve 2) before and (curve 1) after the action of ammonia vapors with a density of 0.2 mg/m^3.

We found that the position of the stop band in the unchanged spectrum of reflected and refracted Bragg waves for light incident from glass on the interface with the photonic crystal is highly sensitive to a small change in the concentration of vapors of water, ammonia, ethanol, or isopropyl alcohol. The observed effect is reversible and can be used to create optical chemical sensors.

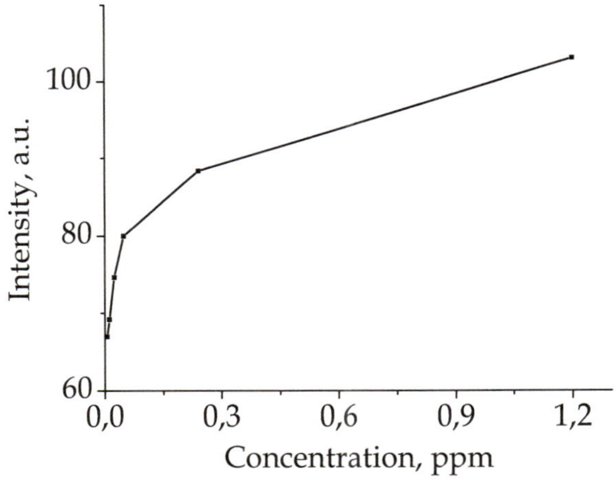

Fig. 4. Effect of ammonia concentration on the intensity of the reflected Bragg wave.

The variation in analyte concentration is measured by the signal from the photodiode (position 3 in Fig. 1.). The photodiode was located so that it recorded the signals in the wing of stop band. In this case the signal from the photodiode was proportional to the change in concentration of the analyte. At certain angles of observation we have observed a maximum in the spectrum of the reflected wave associated with the presence of the stop band. The intensity of reflected light is several times higher than refractive one, so the implementation of an optical chemical sensor in this geometry is preferable. In this case a functional diagram of the sensor contains two photodiodes measuring the intensity of long- and short-wavelength wings of Bragg maximum. The maximum are shifted in wavelength with a change of analyte concentration. It leads to an increase in a signal of a photodiode and a decrease in another. Electric circuit comparing the signals from the photodiodes allows to significantly increase the sensitivity of registration and to exclude fluctuations in the intensity halogen lamp. Figure 4 shows the effect of ammonia concentration on the intensity of the reflected Bragg wave.

It is well known that siloxane (Si–O–Si) and silanol (Si–OH) groups are present on the surface of silica nanoparticles. Their extremal density can reach five OH groups per nanometer squared. The presence of a mobile hydrogen atom in polar hydroxyl groups gives rise to the effective interaction with the molecules of the gas and liquid phases by two ways. The first interaction is electrostatic attraction between hydroxyl groups on the silica surface and the surrounding dipole molecules. In addition, hydrogen bonds can appear between hydroxyl groups and the surrounding molecules if these molecules have an undivided electron pair. For this reason, water and ammonia, being the most polar molecules among those listed above, manifest the strongest electrostatic interaction and form the strongest hydrogen bonds with the silica surface with a high probability of the formation of monolayers or cluster island films on the surface of nanoparticles. Ethanol and isopropyl alcohol also form hydrogen bonds owing to the presence of an oxygen atom and a mobile hydrogen atom, but these bonds are weaker than those formed by water or ammonia, because the polarity of alcohols is smaller. Nonpolar carbon tetrachloride molecules, which do not have atoms capable of forming hydrogen bonds, almost do not interact with the silica surface.

Indeed, the experiments show that the maximum spectral shift ≈8 nm of the stop band of the photonic crystal is observed under the action of water or ammonia vapors. Figure 3 shows the spectra of the refracted Bragg waves for the photonic crystal in air without (curve 2) and with (curve 1) ammonia vapors with an NH_3 density of about 0.2 mg/m^3 . The data were obtained at the angle of incidence of light $\varphi_1 = 51°$. The estimates show that the ammonia molecular monolayer uniformly covering the surface of the opal balls gives rise to a change in n_{eff} in ammonia vapors by $\Delta n_{eff} \approx 0.004$, which corresponds to the spectral shift of the center of the stop band by about 1 nm. However, it should be taken into account that silica balls can consist of globules (Baryshev et al., 2007) and their specific surface can be larger by an order of magnitude, which was not taken into account in the estimates.

A smaller spectral shift of 1–2 nm was observed for ethanol and isopropyl alcohol vapors and was absent for carbon tetrachloride vapors; this observation corresponds to a decreasing sequence of the dipole moments of these molecules. The revealed selectivity can be enhanced by the controlled chemical modification of the surface of silica nanoparticles as a carrier. For example, the amount of absorbed water is determined by the type of the functional group and decreases in the series SO_3H > $N(CH_3)_3Cl$ > NH_2 > OH (Sakai et al., 1989). By fixing more specific complex organic molecules on the surface (Orlova et al., 2009), it is possible to significantly expand the functionality of a sensor based on the described effect, when the interaction of selective functional groups on the surface of an element with an identified element gives rise to its adsorption and results in a change in the number of the physical parameters of the modified photonic crystals.

3. The synthesis and physical-chemical properties of silica nanoparticles modified with pyrylocyanine dye

The luminescence properties of organic dyes are enhanced in the mesoporous nanostructures of silica and the light resistance of these dyes increases (Collinson, 2002; Sokolov et al., 2007). Thus, the binding of luminescent functional dyes with silica nanoparticles is promising for the use of modified silica nanoparticles as luminescence sensors. We have developed a method for the preparation of solid sensors for amine-type reagents based on luminescent silica nanoparticles modified with a functional pyrylocyanine dye.

It is known that pyrylocyanine dyes interact with primary amines to form pyrydocyanines (Balaban et al., 1982). Pyrydocyanines exhibit luminescence, which is different from the luminescence of the starting pyrylocyanines; this can be used for the luminescence detection of organic amines (Höfelschweiger, 2005).

3.1 The synthesis of silica nanoparticles modified with pyrylocyanine dye

The method of modified nanoparticles formation includes three steps.

The first step is a synthesis of an oligomer precursor containing trimethoxysilyl groups and a pyrylium dye. The covalent bond between the dye and the nanoparticle occurred through a precursor, which was obtained by the copolymerization of an oxiranyl derivative of alkoxysilane and a dye containing the oxirane group. A pyrylocyanine dye capable of covalently binding to the skeleton of a silica nanoparticle was synthesized for the modification of silica nanoparticles. The oxirane group was introduced into the aldehyde moiety of the dye molecule (Liang et al., 2002).

To avoid premature epoxide ring opening, we proposed a reaction scheme for the functionalization of nanoparticles, which involved the initial reaction of aldehyde **2** with an epoxy derivative of trialkoxysilane. The epoxide ring opening reaction in aldehyde **2** in the presence of trimethoxy-[2-(7-oxabicyclo-[4.1.0]hept-3-yl)-ethyl]-silane **5** was performed with the addition of a $BF_3 \cdot Et_2O$ (boron trifluoride etherate) catalyst. As a result, precursor **6** was formed; because of the presence of an aldehyde group, this precursor retained its ability to undergo subsequent condensation with 2,4,6-trimethylpyrylium perchlorate **3** (Fig. 5.) to give precursor **7.**

The second step is a silica clusters preparation.

The silicate sol was obtained by hydrolysis and condensation of tetraetoxysilane $Si(OEt)_4$. The process is shown on Fig. 6. The first stage is the hydrolysis catalyzed by the acid. The second stage is the condensation. The stages are alternate to form -Si-O-Si- containing clusters and further nanoparticles.

The third step is preparation of silica particles modified with pyrylocyanine dye.

Oligomer precursor **7** was introduced into a reaction with a hydrolyzed tetraethoxysilane sol. The resulting colloid solution was centrifuged, and the solvent was decanted. The precipitate was washed with acetonitrile to remove free dye; ethanol was added, and the precipitate was repeatedly dispersed to a colloid solution on a «Bandelin» ultrasonic homogenizer. As a result, a colloid solution of silica nanoparticles modified with pyrylocyanine **4** was obtained.

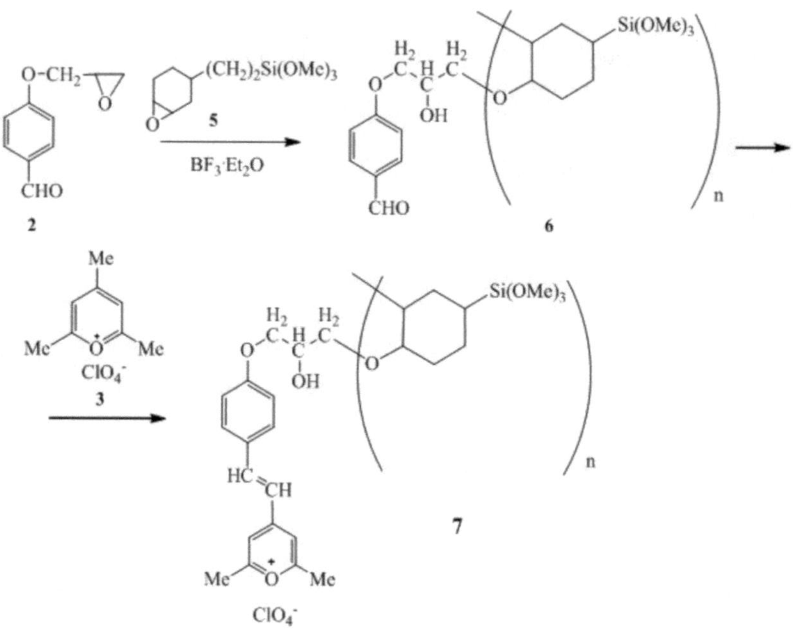

Fig. 5. The synthesis of an oligomer precursor containing trimethoxysilyl groups and a pyrylium dye.

a) $Si(OEt)_4 + H_2O \xrightarrow{HCl} HO{-}Si(OEt)_3 + EtOH$

b) $2\,HO{-}Si(OEt)_3 \longrightarrow (EtO)_2Si(OH){-}O{-}Si(OEt)_3 + EtOH$

$\phantom{b)\ 2\,HO{-}Si(OEt)_3} \searrow (EtO)_3Si{-}O{-}Si(OEt)_3 + H_2O$

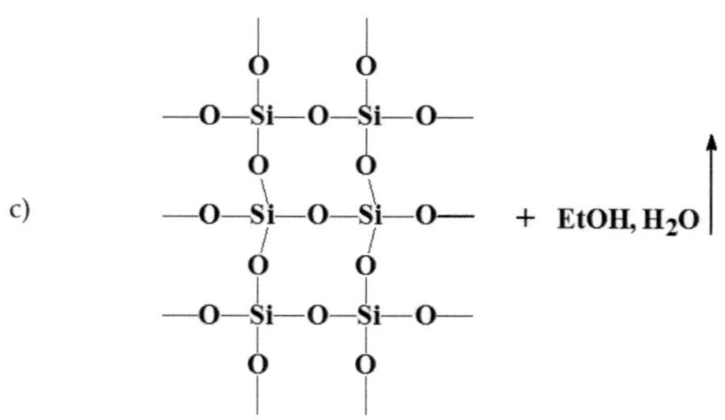

Fig. 6. The silica clusters preparation: (a) the hydrolysis catalyzed by the acid; (b) the condensation; (c) the alternating -Si-O-Si- containing clusters and further nanoparticles.

The resulting compounds were characterized by NMR, IR, and UV spectroscopy and spectrofluorometry. The NMR spectra were measured on an AC-200 instrument (Bruker), and the IR spectra were measured on a Vector 22 instrument (Bruker). The UV spectra were recorded on an 8453 spectrophotometer (Hewlett-Packard), whereas the fluorescence spectra were measured on a Cary Eclipse fluorimeter (Varian). The films of nanoparticles were studied by luminescence confocal microscopy on an LCM 750 microscope (Carl Zeiss) and by electron microscopy on a TM-1000 microscope (Hitachi).

To prepare films, the colloid solution of modified silica nanoparticles was supported onto a glass substrate by casting, washed with water and dried in air. An inhomogeneous opaque light orange film was formed.

3.2 The physical-chemical properties of silica nanoparticles modified with pyrylocyanine dye

Depending on the molar ratio between an aromatic aldehyde and a silicon-containing oxirane introduced into the reaction, different numbers of aldehyde and trimethoxysilyl groups were formed in precursor **6**. Based on the calculation of the relative signal intensities of these groups in the ^1H NMR spectra of copolymer **6**, we found that the number of reacted aldehyde molecules correlated with the initial oxirane : aldehyde ratio (see the Table 1.). Depending on the initial oxirane: aldehyde ratio, the precursor contained from 8 to 50 methoxysiliyl groups covalently bound to an aldehyde molecule.

Aldehyde amount dissolved in 1 ml of oxirane, g	Oxirane : aldehyde molar ratio	$(MeO)_2Si : H_{ald}$ integral intensity ratio
0.05	15 : 1	50 : 1
0.1	7.5 : 1	35 : 1
0.2	3.75 : 1	8 : 1

Table 1. Oxirane: aldehyde rations in copolymer 6, as found from the ^1H NMR spectra.

To determine the degree of modification of the resulting silica nanoparticles with pyrylocyanine **4**, we calculated the number of SiO structural units per dye molecule. The calculation was performed based on the ratio between the average molecular weight of a colored nanoparticle, which was obtained from absorbance data in the absorption spectra of a colloid solution

$$M_w = \frac{l\varepsilon[C]}{D}, \quad (3)$$

where [C] is the substance concentrationin g/l, and the molecular weight of the dye taken the same molar absorption coefficients of the dye in both species. As a result, we found that on average a dye molecule accounted for 48 SiO structural units.

3.2.1 Luminescent properties of the resulting modified silica particles

The luminescence response of a sensor based on silica nanoparticles modified with pyrylocyanine depends on the reaction of the dye with primary amines to result in luminescent pyrydocyanines. Therefore, we performed a reaction with a 10% solution of butylamine in ethanol on the modified colloid particles in solution and in a film. After the addition of 0.2 ml of the butylamine solution to 10 ml of a colloid solution of modified silica particles, blue fluorescence was observed under UV illumination. Figure 7 shows the spectrum of this fluorescence. A portion of a film of silica nanoparticles modified with pyrylocyanine **4** on a glass substrate was treated with a butylamine solution, and the other portion of the film remained untreated.

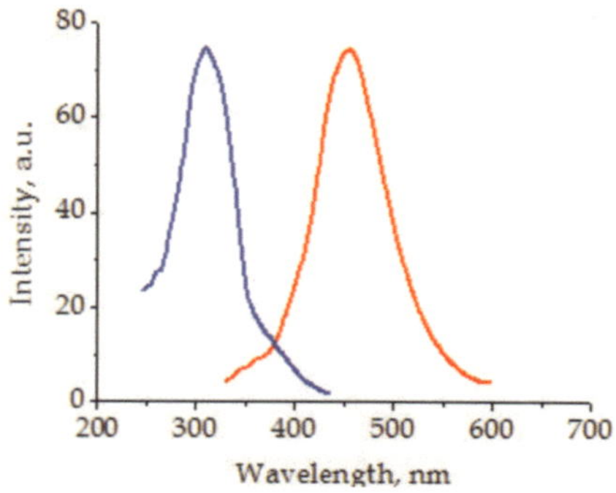

Fig. 7. Excitation (1) and luminescence (2) spectra of the reaction product of colloid particles modified with pyrylocyanines 4 with butylamine in ethanol.

Figure 8 shows the luminescence photographs obtained on a confocal microscope and the luminescence spectra of individual microregions in the untreated and treated films. The luminescence of nanoparticles modified with pyrylocyanine **4** exhibited a maximum at 600 nm; it changed to the luminescence of a pyrydocyanine dye with a maximum at 490 nm after treatment with butylamine.

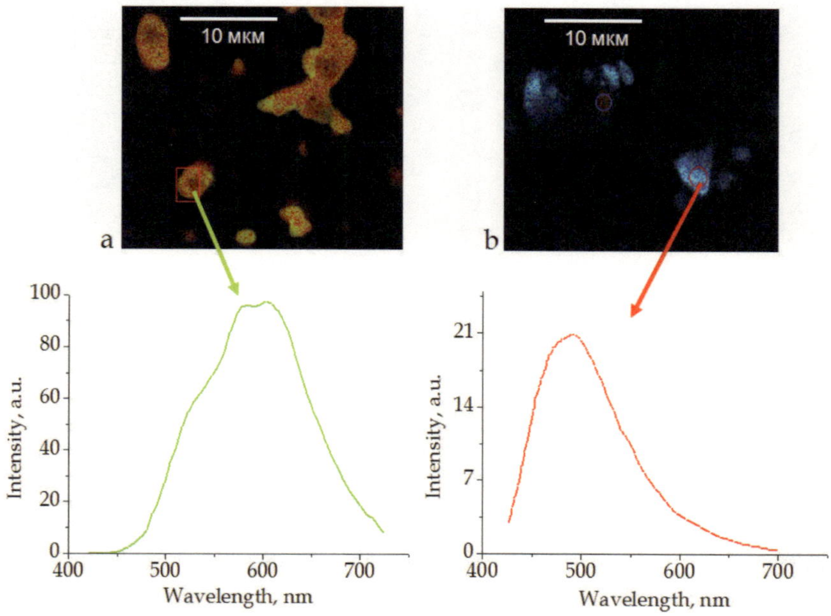

Fig. 8. Luminescence photographs and luminescent spectra of individual microregions of (a) an untreated film and (b) a film of silica nanoparticles modified with pyrylocyanine **4** treated with a 10% solution of butylamine in ethanol obtained on an LCM 750 confocal microscope upon laser excitation at a wavelength of 407 nm.

From a comparison of the luminescence spectra of a pyrydocyanine dye bound to silica nanoparticles in a film and the luminescence spectrum of this dye in a colloid solution, it follows (cf. Figs. 7 and 8) that, in the film, the luminescence spectra of the pyrydocyanine dye were broadened and shifted to the long-wavelength region, as compared with those of the colloid solution. The light resistance of silica particles modified with pyrylocyanine dye **4** in both a colloid solution and a film was studied under irradiation with monochromatic light at an intensity of 0.8 mW/cm^2 at a dye absorption maximum. The light source was a DKSSh-1000 lamp with an MDR-2 monochromator. No spectral changes occurred upon exposing the dye-modified nanoparticles both in a colloid solution and in a film at total exposures of 0.5–1 J/cm^2.

3.2.2 Change in film morphology upon amine treatment

The treatment of a nanostructured film of silica particles modified with a pyrylium dye with a solution of butylamine resulted in a change in the film morphology. Figures 9a and 9b show the micrographs, which were obtained on a Hitachi TM-1000 electron microscope, of a film of silica nanoparticles modified with pyrylocyanine **4** before and after treatment with a solution of butylamine, respectively. As can be seen in these micrographs, the treatment with a solution of butylamine resulted in a partial agglomeration of the initial loose nanostructured film to form spherical structures of micrometer sizes. In this case, it can be seen that some incompletely formed spherical microstructures were also structured at the nanolevel.

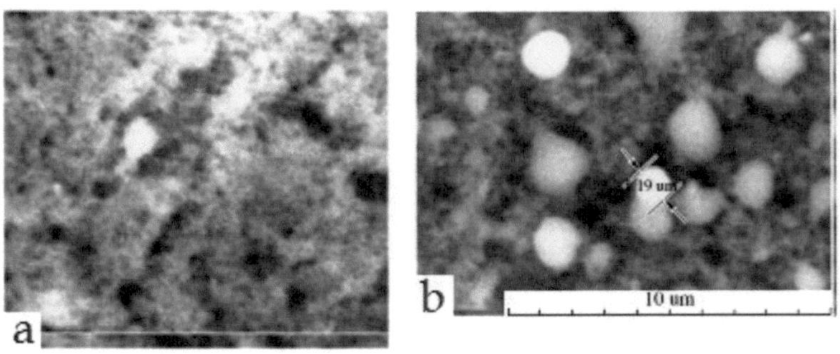

Fig. 9. Electron micrographs of a film obtained from a colloid solution of silica nanoparticles modified with dye 4 on a magnification of 10000X (a) before and (b) after treatment with butylamine.

4. Fiber-optic sensor for butylamine based on modified silica nanoparticles

Fiber optics is one of the most dynamically developing field of physics. The application of fiber optics in a range of sensors has several important advantages. Insensitivity to electromagnetic and radiation interference, chemical and thermal stability, remote sensing, relatively low cost makes these types of sensors indispensable in some cases. The coating of the thin-film nanostructures based on silica nanoparticles on the surface of optical fiber allows significantly extend the scope of fiber-optic sensors (Janotta et al., 2003; Xiao et al., 2005; Yan et al., 2009) It is known that amines are usually toxic and in some cases are strong carcinogens. One of the significant shortcomings of existing chemical sensors for amine-type compounds is the lack of selectivity. This chapter reports the results of experimental studies of selective sensor for butylamine based on nanostructure thin-film coated at the end of an optical fiber. This film consists of silica nanoparticles of 8-10 nm modified with polymethine dyes. Pyrydocyanines exhibit luminescence, which is different from the luminescence of the starting pyrylocyanines. This effect was inventive incentive for the synthesis of pyrylocyanine **2** and pyrylocyanine **3** dyes, which selectively react with butylamine. Therefore, the binding of luminescent functional dyes with silica nanoparticles is promising for the use of optical chemical sensors. The pyrylocyanine dye have a covalent bond with the skeleton of silica nanoparticles. The method of chemical modification of silica nanoparticles in a sol-gel synthesis is described in (Orlova et al., 2009).

«Avantes AvaSpec-2048TEC» spectrometer with a spectral resolution of 1 nm was used in the studies of luminescence properties of the thin films, which are consist of modified silica nanoparticles. «Newport LQC 405 - 85E» semiconductor laser was used as a source of exciting light with a wavelength of 407 nm (Fig. 10.). The thin-film nanostructure is deposited on the end of a freshly cleaved optical fiber with a diameter of 600 μm by dropping into the solution of modified silica nanoparticles or deposition and subsequent drying a drop of the solution at the end of the fiber. We have experimentally found that the film of modified silica nanoparticles is durable to the various kinds of mechanical effects and demonstrates a high radiation resistance.

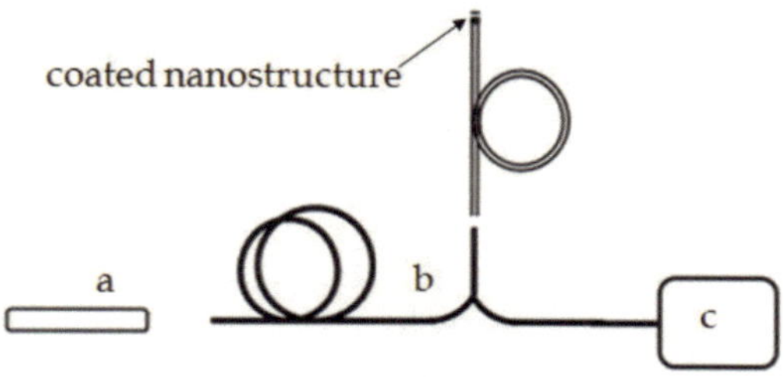

Fig. 10. Optical scheme of measuring; a - semiconductor laser, b - reflection probe, c - spectrometer.

A typical luminescence spectrum of the silica nanoparticles film modified with pyrylocyanine **3** dye has a pronounced peak at 660 nm. The film gives a green-yellow luminescence with a maximum at 560 nm after a treatment with 10% solution of butylamine (Fig. 11.).

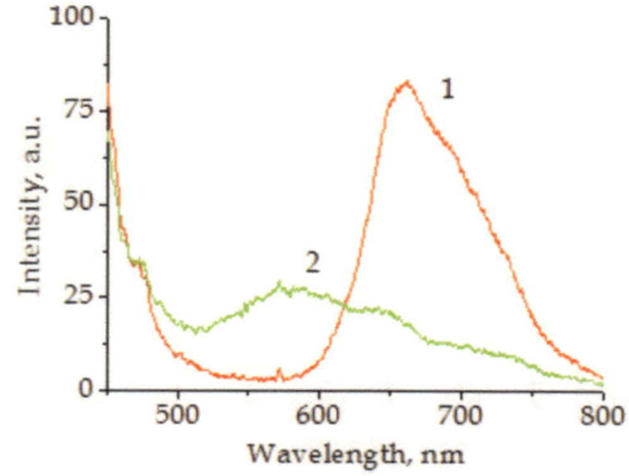

Fig. 11. Luminescence spectrum of the modified silica nanoparticles thin-film with pyrylocyanine **3** before (1) and after (2) treatment with butylamine solution.

It should be noted that the luminescence effect of the modified silica nanoparticles film after the treatment with other amino liquids leads only to the quenching of luminescence in the red spectrum, while the yellow-green luminescence does not appear. The minimum concentration of butylamine in ethanol was 10^{-3} mol/l in which possible to record the luminescent responses from the silica nanoparticles film modified with pyrylocyanine **3** dye.

In contrast to the silica nanoparticles film modified with pyrylocyanine **3** dye, the films modified with pyrylocyanine **2** do not have the luminescence before the treatment. The maximum of luminescence at 580 nm was exhibited after treatment with butylamine solution. On exposure to other amino substances, the luminescence peak does not appear which also shows selectivity for the analyte. To increase the amplitude of the luminescence response from the modified silica nanoparticles film coated on the end of optical fiber as a porous selective mirror was applied. The mirror is an artificial opal photonic-crystal film. The maximum of reflection of this mirror coincides with the peak of pyrylocyanine luminescence in modified silica nanoparticles after the treatment with butylamine solution.

It was achieved by choosing the diameter of monodisperse silica spheres of about 260 nm. The porosity of photonic crystal film allows butylamine and other substances easily penetrate into the modified silica nanoparticles film. The photonic crystal opal film was grown by moving meniscus method on the end of the optical fiber (Kalinin et al., 2006). For this purpose optical fiber is fixed in the suspension of monodisperse silica spheres on the surface under an angle of 15 degrees. The reflection spectrum of «photonic crystal mirror» deposited on the end of the optical fiber is practically coincided with the luminescence band of pyrylocyanine **2** dye in the region of 580 nm.

The experiments have demonstrated that the coating of the «photonic crystal mirrors» on the modified silica nanoparticles film leads to an increase in the amplitude of the reflected signal in 3 - 4 times (Fig. 12.). The photonic crystal film was a highly ordered hexagonal close-packed structure of the monodisperse silica spheres, and the crystallographic direction (111) was normal to the surface. The presence of «photonic crystal mirror» does not affect the rate of change of the optical response of sensory film.

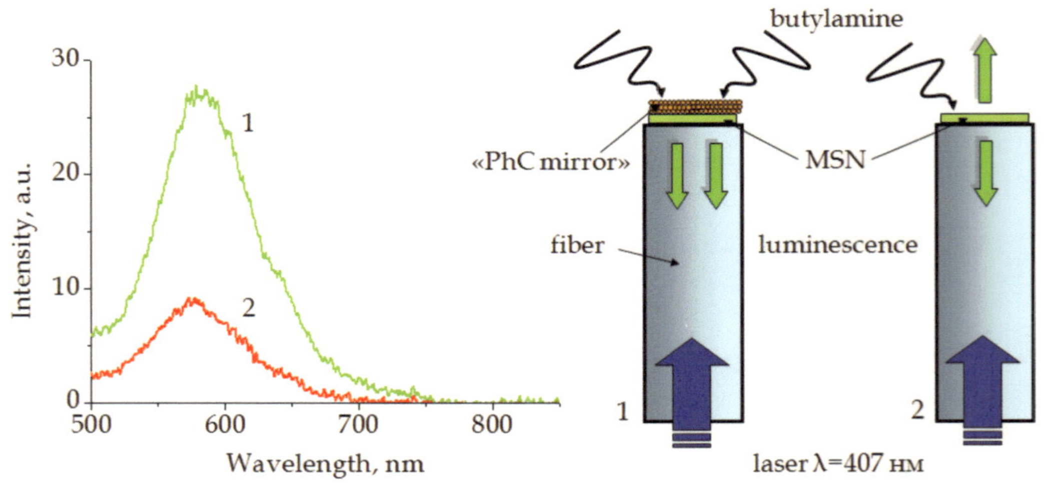

Fig. 12. Luminescence spectrum of modified silica nanoparticles with pyrylocyanine **2** with (1) and without (2) «photonic crystal mirror».

The further increase in the amplitude of a luminescence of modified silica nanoparticles might be enhanced due to the presence of nearby metallic nanoparticles such as Ag or Au (Geddes et al., 2003). We have found that the addition of Ag nanoparticles with a diameter of 5 - 7 nm into the modified silica nanoparticles film leads to an increase the luminescence in 2 - 3 times (see Fig. 13.).

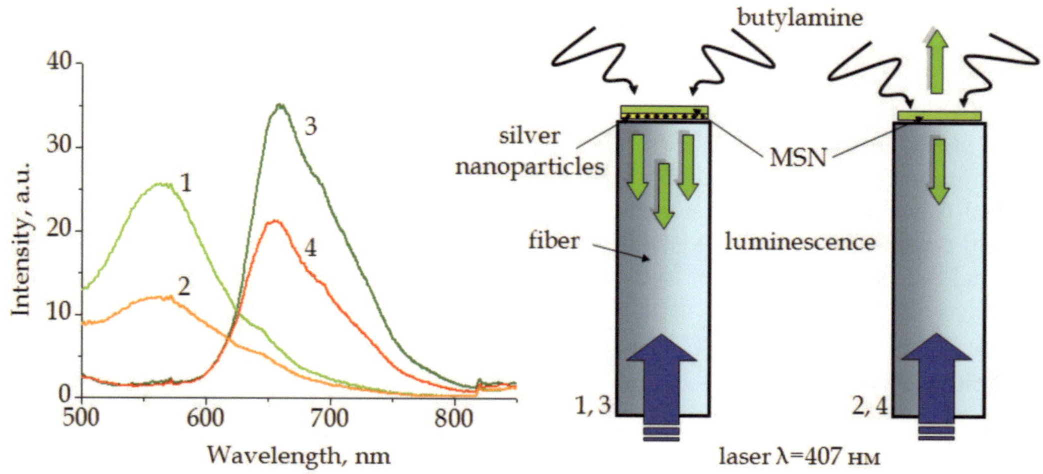

Fig. 13. Luminescence spectrum of modified silica nanoparticles with pyrylocyanine 3 before (3,4) and after (1,2) treatment with butylamine solution; with (1,3) and without (2,4) injection of silver nanoparticles.

5. Conclusion

This chapter proposed concept of highly sensitive and selective optical sensors for different type reagent based on negative light diffraction and refraction at a glass-opal interface and modified silica nanoparticles. The optical fiber butylamine chemical sensor based on selectively altering the luminescent properties of modified silica nanoparticles thin-film. The sensor component is a modified silica nanoparticles covalently bound with pyrylocyanine dye and deposited at the end of the optical fiber. A method for the chemical modification of silica nanoparticles with a pyrylium dye under sol-gel synthesis conditions was developed. The films of silica particles modified with a pyrylium dye were obtained; these films emitted luminescence in visible range of spectra. Upon the treatment of the films with a solution of butylamine, a fluorophore from the pyrydocyanine series was formed on the modified silica nanoparticles; this fluorophore emitted luminescence shifted in shortwave range of spectra. The film samples based on silica nanoparticles modified with pyrylocyanines exhibited good light resistance. We confirmed that our sensor has high sensitivity for actual use, and 10% solution of butylamine concentration less than 10^{-3} mol/l can be detected. It was shown that the addition of silver nanoparticles into the film of modified silica nanoparticles and the coating it the photonic crystal film as a porous selective mirror on the end of the fiber can increase the sensors sensitivity about 10 times. We assumed that the improving the sensing element by doping it in photonic crystal film with metallic nanoparticles allow us to develop a highly sensitive selective optical chemical sensors remote monitoring.

We found that the position of the stop band in the unchanged spectrum of reflected and refracted Bragg waves for light incident from glass on the interface with the photonic crystal is highly sensitive to a small change in the concentration of vapors of water, ammonia, ethanol or isopropyl alcohol. The observed effect is reversible and can be used to create optical chemical sensors. Thus, the revealed effect of the spectral shift of the stop band of photonic crystals against the chromatic refraction background is well described by the Bragg refraction and the reflection of waves at the glass–photonic crystal interface and can be useful for creating compact, highly sensitive, optical chemical or biological sensors. At the same time, the observed effect of the shift of the stop band of photonic crystals with respect to the unchanged angular spectrum under a small change in the refractive index of the medium filling space between the photonic crystal balls can be used in the schemes of optical sensors, as well as for controlling the spectrum of refracted and reflected light in an optical demultiplexer by an external field. The implementation of the analyzed effect in optical schemes, e.g. with the use of optical fiber technologies, in multiple violated total internal reflection elements, or in intracavity devices, will make it possible to increase the sensitivity of the identification of an element by several orders of magnitude.

6. Acknowledgment

We are grateful to Kalinin D. for placing silica nanoparticle suspensions at our disposal and to Chubakov V. for assistance in the work.

7. References

Baba T. & Nakamura M. (2002). Photonic Crystal Light Deflection Devices Using the Superprism Effect. *IEEE Journal of Quantum Electronics*, Vol. 38, No. 7, pp. 909-914, ISSN 018-9197

Balaban A., Fischer G., Dinulescu A., Koblik A., Dorofeenko G., Mezheritski V. & Schroth W. (1982). Advances in Heterocyclic Chemistry, In: *Pyrylium Salts: Synthesis, Reactions and Physical Properties*, Katritzky A., pp. 429-434, N.-Y. et al. Acad. Press, ISBN 0-12-020652-8

Baldini F., Chester A., Homola J. & Martellucci S. (Eds.). (2006). *Optical Chemical Sensors*, Springer, ISBN 140204609X, Netherlands

Baryshev A., Khanikaev A., Inoue M., Lim P., Sel'kin A., Yushin G. & Limonov M. (2007). Resonant Behavior and Selective Switching of Stop Bands in Three-Dimensional Photonic Crystals With Inhomogeneous Components. *Physics Review Letters*, Vol. 99, No. 6, pp. 063906-1-063906-4

Boldov I., Kuchyanov A., Plekhanov A., Orlova N., Kargapolova I. & Shelkovnikov V. (2011). Fiber-Optic Sensor for Butylamine. *Journal of Physics: Conference Series*, Vol. 291, No. 1, pp. 1-4

Chen L., Kuo C. & Ye Z. (2004). Guiding Optical Flows by Photonic Crystal Slabs Made of Dielectric Cylinders. *Physics Review E*, Vol. 69, No. 6, pp. 066612-1-066612-6

Collinson, M. (2002). Recent Trends in Analytical Applications of Organically Modified Silicate Materials. *Trends in Analytical Chemistry*, Vol. 21, No. 1, pp. 30-38

Foteinopoulou S., Economou E. & Soukoulis C. (2003). Refraction in Media With a Negative Refraction Index. *Physics Review Letters*, Vol. 90, No. 10, pp. 107402-1-107402-4

Geddes C., Cao H., Gryczynski I., Gryczynski Z., Fang J. & Lakowicz J. (2003). Metal-Enhanced Fluorescence Due to Silver Colloids on a Planar Surface: Potential Applications of Indocyanine Green to in Vivo Imaging. *Physical Chemistry,* Vol. 107, No. 18, pp. 3443-3449

Höfelschweiger B. (2005). The Pyrylium Dyes: a New Class of Biolabels. Synthesis, Spectroscopy, and Application as Labels and in General Protein. Assay, *PhD Dissertation*

Janotta M., Katzir A. & Mizaikoff B. (2003). Sol-Gel-Coated Mid-Infrared Fiber-Optic Sensors. *Applied Spectroscopy,* Vol. 57, No. 7, pp. 823-828

Joannopoulos J., Johnson S., Winn J. & Meade R. (2008). *Photonic Crystals: Molding the Flow of Light,* Second Edition, Princeton University Press, ISBN 978-0-691-12456-8, Singapore

Kosaka H., Kawashima T., Tomita A., Notomi M., Tamamura T., Sato T. & Kawakam S. (1998). Superprism Phenomena in Photonic Crystals. *Physics Review B,* Vol. 58, No. 16, pp. 10096-10099

Liang J., Yeh J., Wang C., Liou S., Tsai C. & Chen I. (2002). The New Generation Dihydropyridine Type Calcium Blockers, Bearing 4-Phenyl Oxypropanolamine, Display α-/β-Adrenoceptor Antagonist and Long-Acting Antihypertensive Activities. *Bioorganic & Medicinal Chemistry,* Vol. 10, No. 3, pp. 719-730

Narayanaswamy R. & Wolfbeis O. (Eds.). (2004). *Optical Sensors for Industrial, Environmental and Clinical Applications,* Springer, ISBN 3-540-40888-X, Berlin

Orlova N., Kargapolova I., Shelkovnikov V. & Plekhanov A. (2009). Luminescent Silica Nanoparticles Modified With a Functional Pyrylocyanine Dye. *High Energy Chemistry,* Vol. 43, No. 7, pp. 602–606

Plekhanov A., Kalinin D. & Serdobintseva V. (2006). Nanocrystallization of Single Crystal Opal Films and the Spectral Characteristic of Related Photonic Structures. *Nanotchnologies in Russia,* Vol. 1, No. 1-2, pp. 245-251

Plekhanov A., Kuch'yanov A. & Zabolotskii A. (2009). Features of the Manifestation of a Stop Band in the Spectrum of Light Diffracted at a Glass–Opal Interface. *Journal of Experimental and Theoretical Physics Letters,* Vol. 90, No. 8, pp. 565–568

Sakai Y., Sadaoka Y., Matsuguchi M., Morigaa N. & Shimada M. (1989). Humidity Sensors Based on Organopolysiloxanes Having Hydrophilic Groups. *Sensors and Actuators,* Vol. 16, No. 4, pp. 359-367

Sokolov I., Kievsky Y. & Kaszpurenko J. (2007). Self-Assembly of Ultra-Bright Fluorescent Silica Particles. *Small,* Vol. 3, No. 3, pp. 419-423

Wu L., Mazilu M. & Krauss T. (2003). Beam Steering in Planar-Photonic Crystals: From Superprism to Supercollimator. *Journal of Lightwave Technology,* Vol. 21, No. 2, pp. 561-566, ISSN 0733-8724

Xiao H., Zhang J., Dong J., Luo M., Lee R. & Romero V. (2005). Synthesis of MFI Zeolite Films on Optical Fibers for Detection of Chemical Vapors. *Optics Letters,* Vol. 30, No. 11, pp. 1270-1272

Yakimansky A., Menshikova A., Shevchenko N., Shabsels B., Bazhenova A., Sel'kin A., Sazonov S., Vedernikov A., Gromov S., Sazhnikov V. & Alfimov M. (2009). From Polymeric Nanoparticles to Dye-Containing Photonic Crystals: Synthesis, Self-Assembling, Optical Features, and Possible Applications. *Polymers for Advanced Technologies,* Vol. 20, No. 6, pp. 581-588

Yan H., Wang M., Ge Y. & Yu P. (2009). Colloidal Crystals Self-Assembled on The End Face of Fiber: Fabrication and Characterizations. *Optical Fiber Technology,* Vol. 15, No. 3, pp. 324-327

3

Some Methods for Improving the Reliability of Optical Porous Silicon Sensors

Tanya Hutter[1] and Shlomo Ruschin[2]
[1]*Department of Chemistry, University of Cambridge,*
[2]*Department of Physical Electronics, Faculty of Engineering, Tel-Aviv University,*
[1]*United Kingdom*
[2]*Israel*

1. Introduction

The area of chemical sensors is one of the fastest growing both in research and in commercial fields. Most of the research work in this area is concentrated towards reducing the size of sensors and the identification and quantification of multiple species. Quick response, minimum hardware requirement, good reversibility, sensitivity, and selectivity are expected from an excellent sensor, and hence there is a need for further research. The applications of chemical sensors include quality and process control, biomedical analysis, medical diagnostics, environmental pollution control, continuous and long term monitoring of pollutants and hazardous substances. There are however several outstanding problems hindering applications of chemical sensors based on optical readout. It is widely accepted that in many instances the sensitivity is not the limitation of the sensor. Indeed many sensors display over-sensitivity at the expense of specificity and are vulnerable to noise. In real applications the environment in which sensors are located is not sterile: additional substances present will cause spurious readouts, and moreover, the substances may react over the sensor's surface and readouts will be then cross-correlated.

Methods to overcome these hindrances are addressed in this chapter using porous silicon as a testbed. They are based on signal or statistical analyses of the data acquired and take advantage of the possibility of multi-sensing with properly modified sensor arrays. Several new configurations for optical multi-sensing are presented in this chapter, and demonstrated for ammonia detection. The methods are however generic for optical remote sensing of multiple gas components and can be applied to industrial and environmental gas supervision as well as biomedical applications. The detection of ammonia is based on chemical reaction between ammonia-sensitive dye and the detected molecules. These reactions are accompanied by changes in optical absorption spectrum within the VIS-NIR range. pH sensitive dyes have been widely used for ammonia detection (Malins et al., 1999; Malins et al., 1998; Potyrailo et al., 1994). For these sensors, the response to ammonia vapour is highly dependent upon the level of humidity in the environment since the acid–base reaction of the dye materials is always mediated by water, and therefore calibrating for atmospheric humidity is required for determining true values of ammonia concentration.

2. Porous slicon

Porous silicon (PSi) is typically produced by electrochemical etching of bulk crystalline silicon. The porosity of the produced material is directly proportional to the applied current density during the etching process, providing therefore a simple means to control its refractive index (Vial and Derrien, **1995**). Various optical components such as Bragg and rugate reflectors have been successfully fabricated with porous silicon (Lorenzo et al., 2005). PSi can be used as smart transducer material in sensing applications, and in particular in the detection of vapours (De Stefano et al., 2004a). Upon exposure to chemical substances, several physical quantities, such as refractive index, photoluminescence, and electrical conductivity change drastically. A key feature of a physical transducer, being sensitive to chemical molecules, either in vapour and liquid state, is a high surface area to volume ratio. Moreover, PSi is an available, low cost material, compatible with standard microelectronics processes, so that it can be successfully employed in the realization of smart sensors and microsystems.

2.1 Fabrication

The two main characteristics of porous silicon are the thickness of the porous layer and its refractive index distribution, which depend on electrochemical anodization time and the current density (Pavesi and Mulloni, 1998). Anodization of porous silicon is usually performed at constant or controlled current density. The Si wafer acts as the anode, a platinum spiral wire as the cathode and the electrolyte is composed of an HF solution. The current density applied during the formation process determines the porosity of the porous silicon layer. Generally, higher current density will produce higher porosity, resulting in lower refractive index. Porosity is defined as the fraction of void within the volume of the porous silicon layer. Pore diameters that can be obtained are in the range of 30 Å to 1 μm and porosities are between 10 and 90% (Searson and Macaulay, 1992). It is acceptable to divide the porous silicon into three categories based on the size of its pores; macroporous (for pore diameters larger than 50 nm), mesoporous (for pore diameters between 2 and 50 nm) and microporous (for pore diameters less than 2 nm). The morphology of the pores is affected by several factors such as crystallographic orientation of the Si wafer, doping, resistivity and current density.

Aging, i.e., the slow spontaneous oxidation of PSi, poses a big disadvantage for practical application of porous silicon as sensor. Due to the aging effect, the structural and optical properties of PSi show continuous change with the storage time. One of the common methods to passivate the PSi structures is thermal oxidation, which results in an increase of the hydrophilicity (wetting) ability due to the formation of polar silicon dioxide (SiO_2). The refractive index is also affected by oxidation, it decreases after the oxidation. If PSi is fully oxidized then its refractive index value is between that of air and that of silica.

2.2 Optical properties - reflectance

A porous medium will exhibit optical properties different than those of the same material in bulk. If the typical feature sizes (e.g. pore size) are much smaller than the wavelengths of the incident electromagnetic field, the field in the porous medium encounters an effective dielectric function. Therefore the reflection from PSi layer displays basically a Fabry-Pérot interference spectrum. The Fabry–Pérot interference phenomenon for thin film is depicted in Fig. 1a; light travelling from one medium encounters a thin film with refractive index n_{eff} and thickness d. Thin film interference involves interference between light reflected from the top interface (air-PSi) of the film and light reflected from the bottom interface (PSi-Si) of the film. Typical interference fringes measured from PSi thin film are illustrated in Fig. 1b. The Fourier transform of the optical reflectivity spectrum provides a simple means to monitor the optical thickness ($n_{eff}d$) of the porous silicon layer (Anglin et al., 2004; Pacholski et al., 2005b). An example is presented in inset of Fig. 1b, where Fourier transform spectrum of the PSi film produces a peak whose position is dependent the optical thickness of the porous layer.

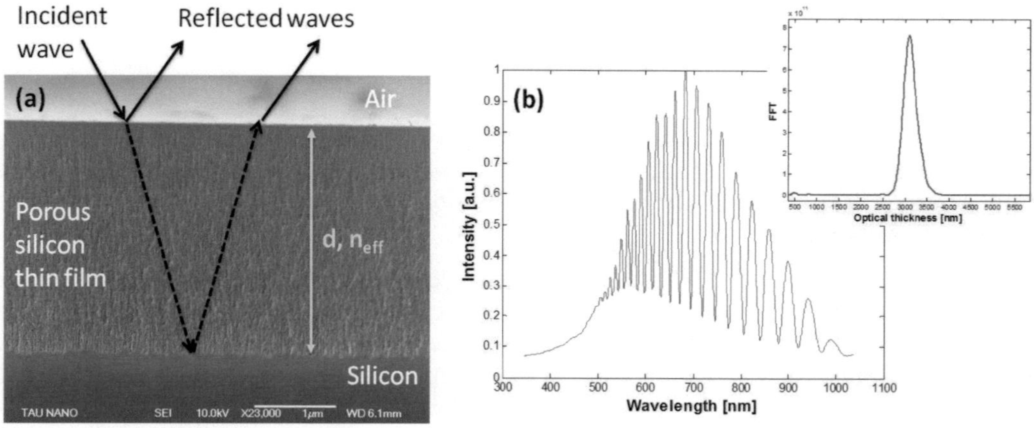

Fig. 1. (a) SEM image of porous silicon cross section. Arrows show light beam reflection from the thin-film with thickness d and refractive index n_{eff}. (b) Reflection spectrum of porous silicon thin-film showing Fabry-Perot interference phenomenon. Inset displays corresponding Fourier transform, where the position of the peak is proportional to the optical thickness.

The refractive index of PSi layer can be determined as a function of the porosity and the wavelength through reflectance measurements. Reflectivity maxima are located at wavelengths determined by the Fabry–Pérot relationship (Eq. 1), where m is the spectral order of the fringe at wavelength λ, θ is the angle, n_{eff} the refractive index of the film and d is the film thickness. The position of the interference maxima for two consecutive maxima, m and m+1 satisfies:

$$2n_{eff}d\cos\theta\cdot\left(\frac{1}{\lambda_m}-\frac{1}{\lambda_{m+1}}\right)=1 \qquad (1)$$

If the thickness of the layer is known independently, then the refractive index can be obtained. It is expected that the refractive index of PSi layer should be lower than of bulk Si and higher than air. PSi is a mixture of the substrate silicon and air, therefore the refractive index decreases with increased porosity. The Bruggemann Effective Medium Approximation (Khardani et al., 2007) (Eq. 2) can be successfully applied in order to find the effective index of porous film.

$$(1-p)\frac{\varepsilon_{Si}-\varepsilon_{eff}}{\varepsilon_{Si}+2\varepsilon_{eff}}+p\frac{\varepsilon_{air}-\varepsilon_{eff}}{\varepsilon_{air}+2\varepsilon_{eff}}=0 \quad (2)$$

where p is the film porosity which describes the volumetric fraction of embedding medium (air), ε_{Si} and ε_{air} are the dielectric functions of silicon and the embedding medium, and ε_{eff} is the effective dielectric function for PSi.

The effective dielectric constant of the PSi film undergoes a significant change when vapours enter the pores and replace the air. This effect of capillary condensation occurs in pores with radii below the Kelvin radius, which is given by:

$$r_K=-\frac{2M\gamma\cos\theta}{\rho N_A kT\ln(\frac{P}{P_0})} \quad (3)$$

where N_A is the Avogadro constant, k is the Boltzmann constant, T is the absolute temperature, M is the molecular weight, ρ is the liquid density, r is the radius of the capillary, θ is the contact angle, γ is the surface tension, P_0 is saturation pressure and P is the equilibrium vapour pressure.

When porous silicon is exposed to analytes in the gas phase, capillary condensation causes an increase of its effective refractive index n_{eff}, because air (n=1) is replaced by the condensed analyte vapour (n>1). The increase in the n_{eff} causes red-shift of the maxima of the fringes. Water vapour readily diffuses to the inner regions of the pore structure and adsorbs on the (oxidized) silicon skeleton. Several humidity sensors based on this principle have been proposed in the literature, some designed to detect humidity through changes of the capacitance (Das et al., 2003), or through changes in optical properties of PSi (Oton et al., 2003). Optical changes of PSi have been also used to detect organic vapours by measuring the optical peak-shifts after exposure to vapours of several liquid mixtures. It was possible to detect the presence of chemical substances (Anderson et al., 2003; De Stefano et al., 2004c; Gao et al., 2000; Snow et al., 1999) and also quantitatively characterize binary mixture composition (De Stefano et al., 2004b; Letant and Sailor, 2001).

3. Increasing reliability in porous silicon optical sensors

Reliable detection of chemical and biological materials depends on the selectivity and sensitivity of the system. Although both properties are desirable in any sensing scheme, they often contradict each other. In general, a trade-off exists between probability of true detection and probability of misdetection (Khodarev et al., 2003). Prioritizing these properties is a matter of the specific implementation conditions. Most attempts to produce sensors able of multi-analyte detection usually result in complicated systems that are not totally immune to interfering substances. As a result, new methods that are capable of multi-sensing are required, while maintaining the simple measurement setup and the sensor's cheap price and easy fabrication.

For a single sensor, selectivity is achieved though specific receptor that selectively binds the analyte of interest, i.e. "lock-and-key" design. In order to be able to detect selectively different analytes, the surface of PSi can be functionalized chemically using surface treatments such as oxidation or carbonization (Salonen et al., 2002), or by attaching chemical or biological receptors (Lugo et al., 2007; Ocampo et al., 2011). In sensor array configurations, the binding elements do not necessarily need to be highly selective to any particular analyte, and the discrimination is achieved using the collective output with signal processing, thus a specific pattern is generated for each analyte. Such patterns may also be obtained from equilibrium or kinetic responses, with the latter often providing additional discrimination and increased sensitivity (Mescheder et al., 2007). The number of sensors in an array has a direct effect on the quality of the discrimination by improving the signal-to-noise ratio, and helps to avoid ambiguity in interpreting the signals if background gases are changing in concentration. Good reviews about chemical cross-reactive sensor arrays can be found in the literature (Albert et al., 2000; Anzenbacher et al., 2010).

After a short overview of different means to increase the reliability PSi sensors, this section is divided to three main sub-sections. Different configurations for multi-sensing using PSi with emphasis on the reliability improvement are presented in each one. Such methods are not limited to porous silicon sensors and can similarly be implemented to other types of sensors.

The work of Islam *et. al.* presents an example for improving selectivity of porous silicon sensor. They demonstrate discrimination of organic vapours using an array of porous silicon sensors having different porosity and pore morphology (Islam et al., 2006). Sensitivity of PSi sensors for different vapours varies significantly with the variations of pore dimensions, and thus facilitates discrimination through the analysis of the array's collective output. The sensitivity of the PSi depends on the physical parameters of the vapours, e.g. surface tension, molecular weight, dimensions, vapour pressure and dipole moment. In this work, the output was analyzed using neural network based pattern recognition for organic vapours, and showed unique fingerprint for each vapour in a mixture. The recognition can be further enhanced either by increasing the number of sensors in the array or by using the data of the response and the recovery time, which are unique for each vapour. Recently, a technique to produce porous silicon samples with gradually varying porosity was reported (Park et al., 2010). It is based on inserting a Si wafer gradually (or by stages) into a HF solution during the anodization process to produce pore-size and layer thickness gradient or various multilayers, on a single substrate. Such lateral pore gradient distribution can also be used as size-exclusion matrix, and is expected to open up application areas involving optical electronic nose systems.

In another example of improving the selectivity, the changes in reflectivity and PL spectra of PSi chips were recorded during the injection of the analyte (Létant et al., 2000). A series of solvent vapours, ethyl esters, and perfumes were investigated and discrimination obtained with PSi sensors was been found to be as good as those obtained with commercial chemiresistive metal oxide sensors. In a similar example, three independent quantities of PSi were measured: the electrical conductance, the photoluminescence intensity, and the wavelength of the optical resonance for different vapours (Baratto et al., 2002). It was shown that it is possible to distinguish between a pollutant like NO_2 and interfering gases like humidity and ethanol by coupling the measurement of optical and electrical quantities. In another example, a gas sensor with two PSi chips that have different hygroscopic properties was used (Jalkanen et al., 2010). Here, the optical and electrical signals were monitored simultaneously, and both sensor types were able to differentiate between the tested analytes.

3.1 Multivariable data based methods

Multivariable data analysis is used to obtain statistical information for a large number of parameters or measurements; it enables identification of the dominant patterns in the data, such as groups and trends. A common method is principal components analysis (PCA), which results in the objects being described by the principal components rather than by the original variables in the data table. PCA is a linear technique that is used to extract the useful information and the relationship between objects and variables, and reveal groupings among sets of cases. It identifies the orthogonal directions of maximum variance in the original data, in decreasing order, and projects the data into lower dimensional space formed by a subset of the highest-variance components. Usually the number of relevant principal components is much less than the number of variables, and a substantial reduction of the number of dimensions can be achieved. In the case of two or three relevant principal components it is possible to represent the data in 2D and 3D plots in which the axes represent the principal components. PCA is a widely used technique in sensor arrays and in 'electronic noses' (Marquis and Vetelino, 2001; Santonico et al., 2008; Uttiya et al., 2008) for evaluation of the array response.

In this section we demonstrate this method for a single porous silicon sensor coated with a sensitive dye for the determination of ammonia concentration in varying humidity. The sensitivity and the response of the sensor to ammonia are strongly affected by the humidity levels, and therefore it is very important to monitor both simultaneously. When the PSi chip is sensitized by pH sensitive dye, humidity induces a red-shift of the fringes (Fig. 2a) and ammonia induces absorption at wavelengths between 500 and 650 nm in the reflected spectrum (Fig. 2b) (also the absorption sensitivity to ammonia is increased by the presence of humidity). Consequently, the reflected spectrum for certain ammonia concentration is different at different relative humidity levels. By monitoring changes in the whole spectrum, we attempt to achieve differentiation using PCA. The results for ammonia gas at 0, 25 and 50% relative humidity (RH) are presented in Fig. 3. This PCA calibration plot shows clear separation of the ammonia measurements at different humidity levels. It might be possible to increase the selectivity if one also monitors the time-dependent curves, as response and recovery, in addition to the equilibrium condition.

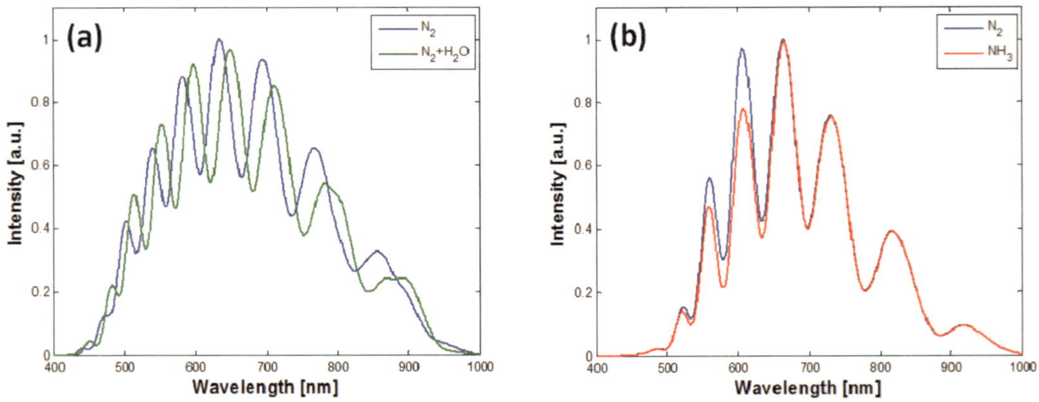

Fig. 2. The effect of (a) humidity and (b) ammonia on the reflection spectrum of PSi sample immobilized with pH dye.

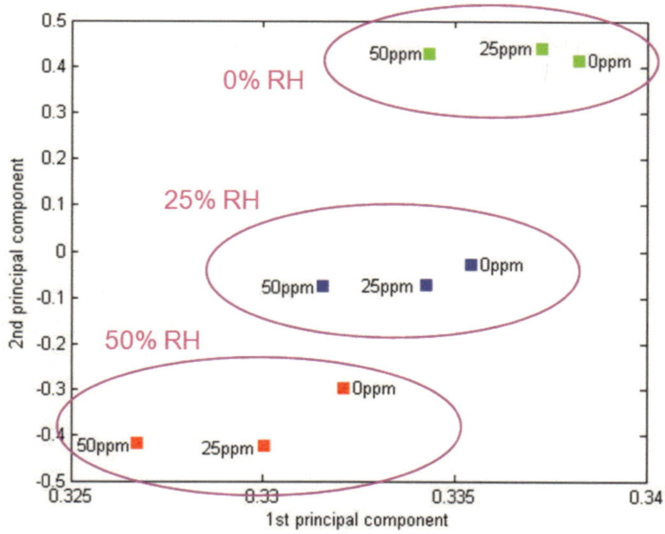

Fig. 3. Calibration plot for ammonia at different humidity conditions, each marked group refers to certain humidity level.

3.2 Muli-sensing via optical multiplexing

In this section we describe an optical non-imaging method for multi-sensing using thin film interference and a single light source and detector (Hutter and Ruschin, 2010). Porous silicon can be successfully employed as a transducer, however any other materials which induce similar spectral interference changes may be used. The method presented here is generic for optical remote sensing of multiple gas components. The device arrangement consists of sectioned porous silicon sensors where each component is sensitive to a different substance. In addition, the different sections can differ in film thickness or refractive index. The combined device is illuminated with a single white-light beam in a non-imaging and distance-independent configuration. The combined reflection is monitored, and the sectioned sensors are designed to provide frequency-encoded patterns which are processed using a fast Fourier transform (FFT) algorithm, allowing real-time monitoring of the effect of several gases by means of single light beam. Fourier transform spectroscopy in PSi samples was reported by M.J. Sailor *et. al.* (Pacholski et al., 2005a; Pacholski et al., 2006) who describe self-compensating interferometric biosensor composed of two layers with different pore diameters for separate sensing of biomolecules. Light reflection from such sensor displays a complex interference spectrum from layers stacked one on top of the other, whose components were resolved by means of Fourier transform. Same setup has been used to detect organic vapours and humidity, thus compensate for fluctuations in humidity (Ruminski et al., 2008).

In an alternative way, the different porous silicon sections are placed one beside the other, on one plane. This arrangement allows the exposure of each section to the same chemical environment and avoids possible partial blocking of stacked films. Each section in the array is made of porous silicon with a different functionality, like a different surface modification or immobilized indicator dye. This arrangement is scalable, with the possibility of increasing the number of porous silicon sections. In each sample, the PSi film differs either in thickness or porosity, therefore the thin-film reflection spectrum of each of the array components varies in periodicity. This periodicity distinction enables to encode the spectrum in such a way that the distinct variations can be straightforwardly analyzed using spectral data analysis methods such as FFT. The schematic representation of such device arrangement is illustrated in Fig. 4. White light is collimated to illuminate the entire sample and then reflected back from all the sensors simultaneously. Light is collected by a lens and focused into a single reading fibre which feeds the light into a spectrometer. Since no imaging is implemented in the light collection, the obtained spectrum consists of many overlapping interference patterns each reflected from a different sensor section. The non-imaging arrangement is especially attractive for remote sensing or cases where several sensing units are monitored with a common optical source and detector. In order to separate the information regarding each of them, an FFT algorithm is applied on the combined spectrum as seen in Fig. 5. It is desirable that peaks in the Fourier-transformed spectrum for each porous silicon sensor will be well separated along the Fourier co-ordinate, to allow good discrimination. The location of the peak depends on the refractive index (porosity) and the thickness of the layers, and these parameters are easily controlled in the electro-chemical processing of porous silicon.

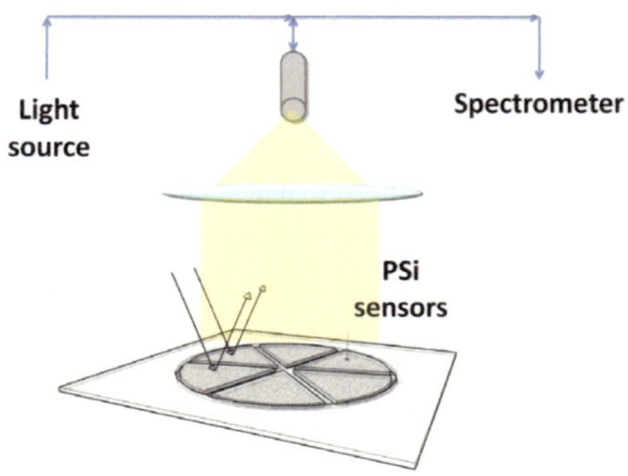

Fig. 4. Schematic presentation of a multi-sensor device based on PSi thin films. Each section is designed to have a specific purpose in the sensing mechanism.

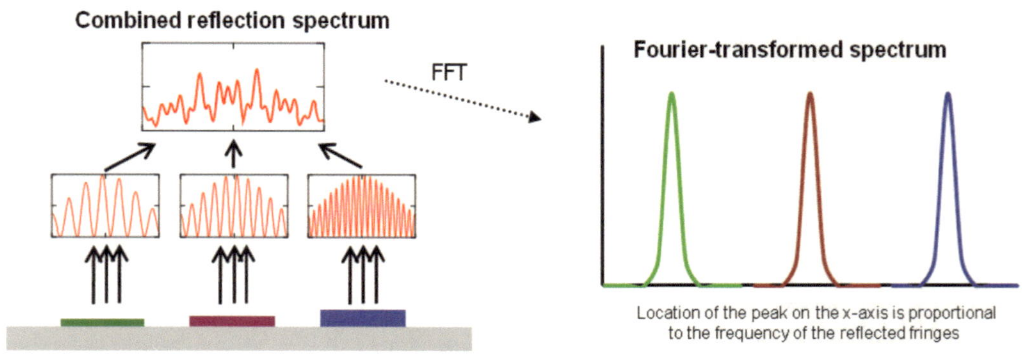

Fig. 5. Illustration of a spectral encoding principle. Each thin film is designed to have a specific reflection spectrum. The combined reflection spectrum is measured and FFT algorithm is then applied. The Fourier-transformed spectrum showing peaks, whose location corresponds to the reflected frequency of each thin film.

It should be noted, that more than one peak is induced in the Fourier domain by a single thin-film (overtones). Avoiding overtone overlap should be taken into account when designing the chip.

Implementation example:

A simple demonstration of such device was fabricated by us, in which the sample was sectioned into two parts: one for water vapour and one for ammonia. Correspondingly, one half was made of oxidized porous silicon and the other one was made of oxidized porous silicon with a chemical pH indicator dye immobilized inside the pores. As discussed earlier, the oxidized half is highly sensitive and reversible towards water vapour, and therefore used as a humidity sensor. The pH indicator dye which was immobilized into the second half, responds to both ammonia concentration and humidity. The independent determination of water vapour concentration is therefore necessary, and by this method this is achieved by a single measurement. The combined use of both sensors enables the separate determination of these two components. The sensor is interrogated with a single white-light beam, and changes in the reflectivity sensor are monitored. The combined interference pattern from both PSi sections is observed and measured under various concentrations of ammonia and water vapour. The structure is designed to frequency encode the reflectivity spectrum and Fourier-transform data-analysis is applied in order to enable simple discrimination between different ammonia and water vapour concentrations. Illustration of the method steps is shown in Fig. 6. The measurements, calibration and analysis of data were performed according to the following sequence: as a first step for each set of measurements, the reflected reference spectrum (measured in dry nitrogen only) was subtracted from all the measured spectra. In the next step, the x-coordinate of reflected spectra was inverted from wavelength (nm) to wavenumbers (nm^{-1}) and a linear interpolation was applied in order to obtain an evenly spaced data set in the inverted new x-axis. Finally a FFT algorithm was applied. Each porous silicon section produces a characteristic peak in the Fourier domain. The position of the peak depends on the porous layer properties, and therefore is unique for each set of fringes. In our case we applied FFT on the difference spectra, i.e. spectrum when exposed to vapour minus the reference spectrum at dry nitrogen.

A schematic illustration of the peaks in Fourier domain due to humidity and dry ammonia is shown in Fig. 6c. When water vapour is introduced into the system, a red-shift of the fringes for both PSi sensors causes peaks to occur in the normalized spectrum and in the Fourier domain at the original periodicity value. The more water vapour infiltrates the pores, the bigger the red shift, the bigger the amplitude of the normalized spectrum and therefore the higher the peak. When dry ammonia is introduced, there is absorbance at 550-650 nm, and the total area of the graph spectrum is reduced. As a consequence, an additional peak at low (DC) frequencies (0-1 nm in the Fourier spectrum) occurs. The magnitude of this peak is therefore sensitive to the ammonia concentration. The oxidized PSi sensor without the pH sensitive dye, shows no response to dry ammonia, and therefore its corresponding peak is absent from the Fourier domain. For a certain ammonia concentration, the reason of the increase of the DC peak with humidity is that the sensitivity of the dye to ammonia is affected by the presence of water vapour.

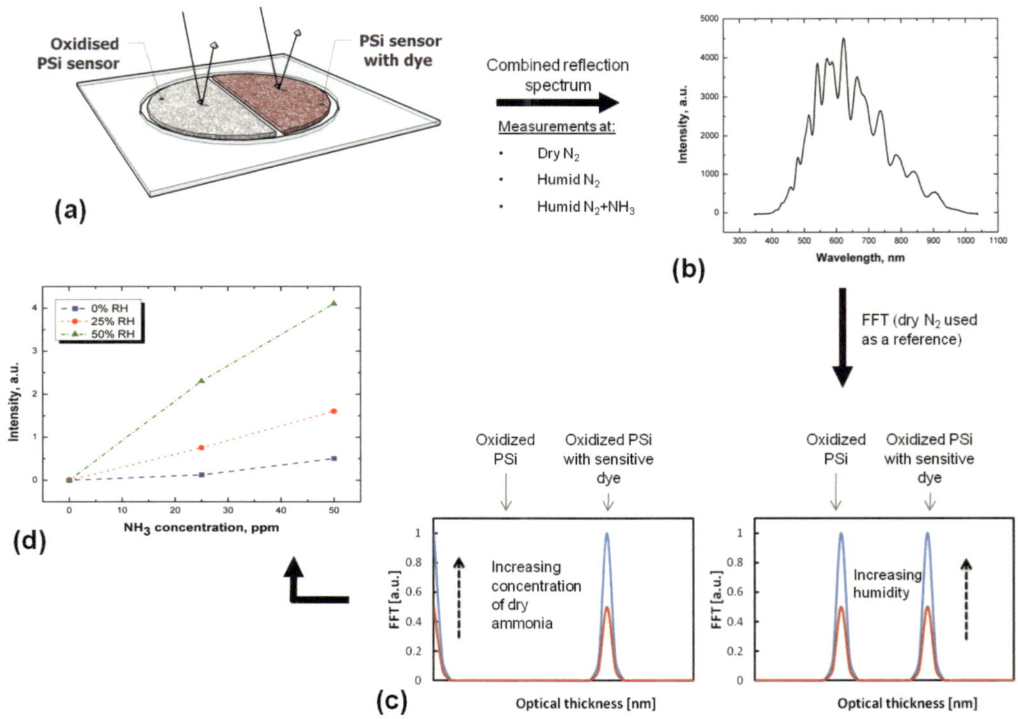

Fig. 6. Illustration of the presented method for ammonia measurement in changing humidity environment. (a) Two PSi sections with different functionality. (b) Measured combined spectrum from two PSi sections, showing interference arising from two thin-films. (c) Schematic illustration of Fourier domain of the two sections once exposed to dry ammonia (left) and once to humidity with no ammonia (right). (d) Experimental result demonstrating the ability of the method to determine ammonia at different relative humidity (RH) values.

In this section a device consisting of sectioned porous silicon, able of simultaneous sensing of different chemical substances in gaseous phase was presented. Each section differs from the other in one or more of its fabrication parameters, namely: thickness, porosity or chemical sensitization. The method was applied to the determination of ammonia in a humid environment, by simultaneous monitoring of two PSi sensors with single light beam. In a broader context, the device provides an *in-situ* method of chemical analysis by gathering information about changes in both the absorption and refractive index in a chemical reaction. Here it is possible to compensate for illumination fluctuations by placing an additional inert PSi thin film. A drawback of this method is that device calibration is required for the determination of the ammonia concentration in changing humidity environment.

3.3 Sensor arrays based on indicator gradient

In this section, we present another method for improving the detection reliability for a given chemical in the presence of other chemical agents or perturbations that generate spurious signals (Hutter et al., 2011). The method provides a thresholding parameter which can be tuned by the user according to his priority considerations between sensitivity and selectivity. We demonstrate the method using a dye immobilised porous silicon optical detector array aimed at identifying ammonia gas at varying humidity conditions and unstable illumination.

In general, a chemical or biological sensor uses a probe (or receptor) to react with the analyte of interest, leading to a measurable signal. The signals' reliability may depend on the number of probes available to react with the analyte, while the precise type of such dependency differs between different sensor types and working conditions. In the method presented here, the available probes are chosen in a manner such that sensors are of the same type but generate a signal varying sensitivity, approximately linear with respect to the probe concentration on their surface. As will be shown in the following, the resulting signal vector accommodates for sensitivity and selectivity towards the analyte, and does not require training or calibration for its detection. We note that the linearity assumption is not strict but will limit the method for analyte concentrations that either saturate completely the sensor or are below its minimum detectable limit. A schematic drawing of such a sensor array is depicted in Fig. 7. The demonstrated array is comprised of *n* optical sensors that are similar in all physical parameters other than the probe concentration. A common light source is applied to all sensors, and the reflected light from each sensor is measured at a single wavelength by a dedicated detector. The detector array output is a vector of signals *s*. The vector of output signals *s* is correlated with the probe concentration gradient *p*. The system's direct internal light source I_{dir} is used for all sensors, where the detector array is exposed to additional undetermined external light I_{ext}. The output of the correlator *y* compared to a threshold θ; when it is exceeded, the analyte presence, is inferred.

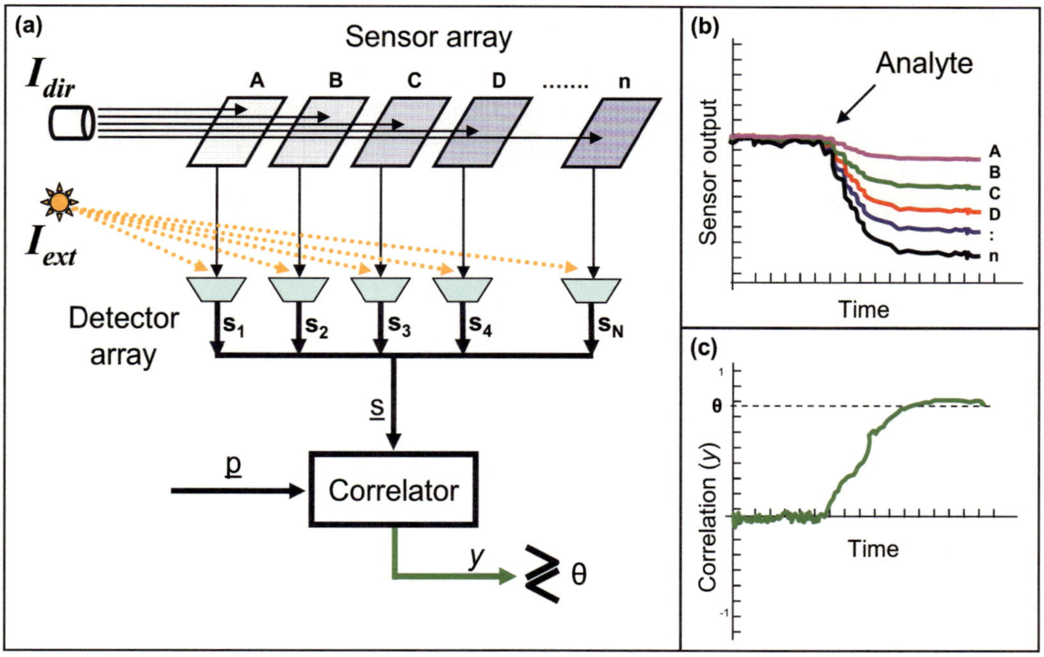

Fig. 7. (a) Sensor array detector illuminated with a light source for the optical readout, and also subjected to some external illumination. (b) Output signal of each sensor. (c) Output of the correlator.

The offset of each sensor is canceled by using the baseline steady state signal as a reference (e.g., averaging s_i over time after initialization, when there is no analyte in the system). We therefore assume for simplicity that the sensitivity vector $\underline{h}$ is given by the product of the probe concentration vector $\underline{p}$ with some unknown transduction factor T:

$$\underline{h} = \mathrm{T} \cdot \underline{p} \tag{5}$$

The array output $\underline{s}$ is:

$$\underline{s} = I_{dir} \cdot \underline{h} \cdot x + \underline{v} \tag{6}$$

where x is the analyte concentration, and $\underline{v}$ vector of the sum of external light source intensity fluctuations at the detector (denoted I_{ext}) and other time dependent noises. We note that T may incorporate many different effects such as aging or the presence of other chemicals that may enhance or degrade the sensor's signal. Since in most cases both T and x are unknown, and coupled in the common case of a single sensor, their efficient estimation cannot be achieved without a thorough calibration process of the factor T which in many cases depends on a series of environmental conditions. We emphasize that the method developed here does not require the a-priori knowledge of T, while preserving the ability to infer the presence or absence of the analyte, which is sufficient in many practical situations.

The detection algorithm should provide a decision, based on the sensor array output, regarding whether the analyte is present or not. As we are unable to estimate a priori the concentration of the analyte, we provide a metric for the array output fit to the expected behavior when the analyte is present. This metric will be compared to a threshold, which can be set per user specification. When the metric exceeds this threshold, a signal indicating the presence of the analyte will be asserted. Lowering this threshold will raise the probability of false alarms, therefore decreasing the specificity of the system while improving its sensitivity, whereas raising it will lower the probability of misdetection, at the cost of lower sensitivity. For the system modeled here, we used the normalized correlation coefficient (NCC) as the metric (Pratt, 2001). In this approach the output vector of the sensor array $\underline{s}$ is correlated with the sensors probe quantity gradient vector $\underline{p}$ and normalized with the norm of both vectors:

$$y = \begin{cases} \dfrac{\sum_{i=1}^{n}\left(s_i - \bar{s}\right)\left(p_i - \bar{p}\right)}{\sqrt{\sum_{i=1}^{n}\left(s_i - \bar{s}\right)^2 \sum_{i=1}^{n}\left(p_i - \bar{p}\right)^2}} & \sum_{i=1}^{n}\left(s_i - \bar{s}\right)^2 \neq 0 \\ 0 & \sum_{i=1}^{n}\left(s_i - \bar{s}\right)^2 = 0 \end{cases} \tag{7}$$

With

$$\bar{s} = \frac{1}{n}\sum\nolimits_{i=1}^{n} s_i \tag{8}$$

And

$$\bar{p} = \frac{1}{n}\sum\nolimits_{i=1}^{n} p_i \tag{9}$$

y assumes values between 1 and -1, and it can be determined in real time. As the analyte is introduced into the system, its value increases.

Regarding the immunity to light source fluctuations and transduction changes, in the calculation of y, we subtract the mean of signal vector $\underline{s}$. When the detectors of the array are exposed to I_{ext}, e.g. to environmental lighting, a uniform additive signal is expected for all elements of $\underline{s}$. The NCC should be invariant with this addition. Fluctuations in I_{dir} or T may also create noisy output which is not compensated by the subtraction of average. However, since all sensor outputs are scaled by I_{dir} and T as in Eq. 6, normalizing by the norm of $\underline{s}$ makes y independent of these factors, therefore independent of their changes over time, as long as the changes do not take place within a single measurement event, taken from all detectors.

Implementation example:

The method was implemented in a PSi ammonia gas detector, using bromocresol purple, which is a pH sensitive dye. The array is composed of four oxidzed porous silicon sensors, each coated with a different concentration of the dye (sensor *A* coated with the lowest amount, and sensor *D* with the highest amount). The resulting signal vector is collected by measuring the absorption at a single wavelength at each sensor. Ammonia gas induces absorption at 500 - 650 nm in the reflected spectrum, in our study we have chosen a wavelength of 557 nm. As stated, each sensor was coated with different amounts of the sensitive dye. Threrefore, the magnitude of the signal should be proportional to the quantity of the coated dye, i.e. the sensitivity of the sensor increases as the quantity of the sensitive dye increases (Fig. 8a). When humidity is present, water vapors condense inside the pores via capillary condensation causing an increase in the refractive index of the porous layer. This increase in optical thickness causes a red-shift in the fringes of the reflected light from the sensor. According to that effect, the signal should not be proportional to the quantity of the coated dye (Fig. 8b). On the other hand, in case of ammonia in humid environment (Fig. 8c), the increase in the output signal is still expected to increase with the qunatity of dye. According to the graph of Fig. 8b, the signal generated from sensor *D* is lower than expected; this performance degradation may be due to defective coating or saturation. We deliberately did not omit this sensor's output from the subsequent analysis, to show the robustness of the method against failure of a single sensor in the array.

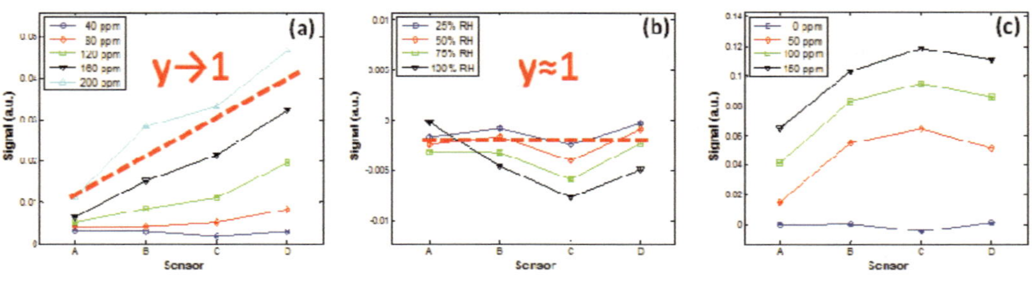

Fig. 8. Signal generated from four sensors at a presence of (a) different ammonia concentrations (b) different humidity conditions, and (c) ammonia at 25% RH.

Fig. 9a shows the value of the correlator parameter y vs. ammonia concentrations in dry air and in the presence of 25% RH (relative humidity). Fig. 8 shows the probability density function of y, showing a good separation. The choice of threshold value of the NCC will vary according to the measurement scenario and practical considerations. Setting the threshold for detection between 0.6 and 0.65 will ensure the positive detection of ammonia with negligible chance of false detection due to pure humidity. The value of y for humid ammonia at high concentrations is lower than that of dry ammonia in spite of the fact that the sensitivity is lower for the dry case. This is probably due to sensor D which generated lower signals than expected (see Fig. 8c). We note that even in this case the system retains its reliability.

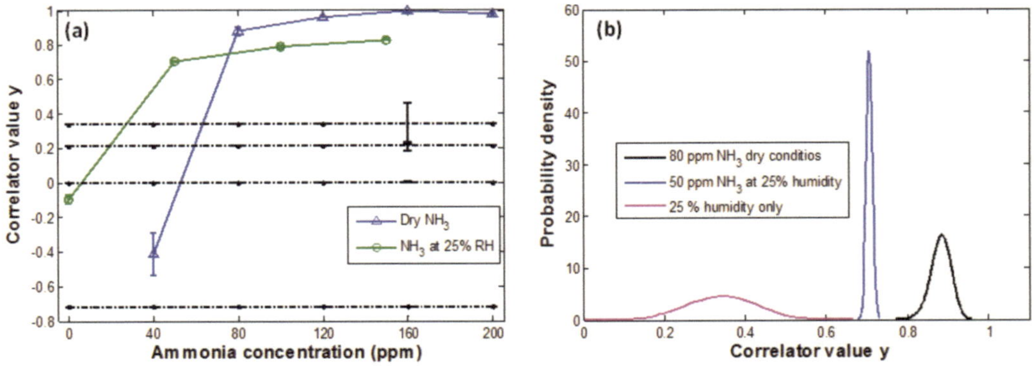

Fig. 9. (a) The NCC for dry ammonia and ammonia at 25 % RH. The horizontal lines correspond to the correlation coefficient at 25, 50, 75 and 100% RH. (b) The probability plot as a function of correlation coefficient.

One of the main features of the method proposed here is improved selectivity. We note that this feature comes at the expense of sensitivity and accurate estimation of the analyte concentration. The sensitivity of a single sensor to changes in concentration levels was found to be below 1 ppm. We trade this superior property with the discrimination ability, as the high sensitivity in any single sensor does not facilitate discriminative ability at a single wavelength. Using the data from each of the sensors, the overall sensitivity is degraded. Using the gradient array, we gain immunity to humidity generated signals; with the normalization, we gain immunity to light source fluctuations. Another advantage of the proposed method is its immunity to tolerable fluctuations in the transduction factor T.

To summarise this section:

- The gradient method is aimed at identifying ammonia gas at a single wavelength at varying humidity, unstable illumination and other disruptive effects. This feature suggests the possibility of using a laser source to monitor the sensor, with clear benefits in signal magnitude and range.
- It is best suited to function as an alarm generator working in a yes/no mode for the presence of a given hazardous substance.
- Simple correlation of the signal vector with an expected gradient-like response enables detection of ammonia (linearity assumption).
- No calibration or training is required.
- It improves immunity to false signals created by changes in illumination intensity and the humidity of the environment.
- Trade off - increased selectivity, decreased sensitivity.

- Optimization may be achieved by increasing the number of sensors, optimizing dye quantities, using a more selective dye to ammonia or changing the working wavelength.

The method of detection is generic and applicable to further types of sensor arrays where the probe concentration can be controlled.

4. Conclusion

In this chapter several optical methods for improving the reliability and specificity of the detection of one or more chemical in the presence of other chemical agents that generate spurious signals were presented. Optical properties of nano-structured porous silicon were employed for this purpose. The methods were demonstrated for detection of ammonia in gaseous phase, which is important for industrial or clinical applications.

5. References

Albert, K. J., Lewis, N. S., Schauer, C. L., Sotzing, G. A., Stitzel, S. E., Vaid, T. P. and Walt, D. R. (2000). *Chemical Reviews* 100, 2595-2626.

Anderson, M. A., Tinsley-Bown, A., Allcock, P., Perkins, E. A., Snow, P., Hollings, M., Smith, R. G., Reeves, C., Squirrell, D. J., Nicklin, S. and Cox, T. I. (2003). *Physica Status Solidi a-Applied Research* 197, 528-533.

Anglin, E. J., Schwartz, M. P., Ng, V. P., Perelman, L. A. and Sailor, M. J. (2004). *Langmuir* 20, 11264-11269.

Anzenbacher, J. P., Lubal, P., Bucek, P., Palacios, M. A. and Kozelkova, M. E. (2010). *Chemical Society Reviews* 39, 3954-3979.

Baratto, C., Faglia, G., Sberveglieri, G., Gaburro, Z., Pancheri, L., Oton, C. and Pavesi, L. (2002). *Sensors* 2, 121-126.

Das, J. O., Dey, S., Hossain, S. M., Rittersma, Z. M. C. and Saha, H. (2003). *Ieee Sensors Journal* 3, 414-420.

De Stefano, L., Moretti, L., Lamberti, A., Longo, O., Rocchia, M., Rossi, A. M., Arcari, P. and Rendina, I. (2004a). *Nanotechnology, IEEE Transactions on* 3, 49-54.

De Stefano, L., Moretti, L., Rendina, I. and Rossi, A. M. (2004b). *Physica Status Solidi a-Applied Research* 201, 1011-1016.

De Stefano, L., Rendina, N., Moretti, L., Tundo, S. and Rossi, A. M. (2004c). *Applied Optics* 43, 167-172.

Gao, J., Gao, T. and Sailor, M. J. (2000). *Applied Physics Letters* 77, 901-903.

Hutter, T., Horesh, M. and Ruschin, S. (2011). *Sensors and Actuators B: Chemical* 152, 29-36.

Hutter, T. and Ruschin, S. (2010). *IEEE Sensors Journal* 10, 97-103.

Islam, T., Das, J. and Saha, H. (2006). In "Sensors, 2006. 5th IEEE Conference on", pp. 1085-1088.

Jalkanen, T., Tuura, J., Mäkilä, E. and Salonen, J. (2010). *Sensors and Actuators B: Chemical* 147, 100-104.

Khardani, M., Bouaicha, M. and Bessais, B. (2007). *Physica Status Solidi C - Current Topics in Solid State Physics,* Vol 4 No 6 4, 1986-1990.

Khodarev, N. N., Park, J., Kataoka, Y., Nodzenski, E., Hellman, S., Roizman, B., Weichselbaum, R. R. and Pelizzari, C. A. (2003). *Genomics* 81, 202-209.

Letant, S. E. and Sailor, M. J. (2001). *Advanced Materials* 13, 335-338.

Lorenzo, E., Oton, C. J., Capuj, N. E., Ghulinyan, M., Navarro-Urrios, D., Gaburro, Z. and Pavesi, L. (2005). *Applied Optics* 44, 5415-5421.

Lugo, J. E., Ocampo, M., Kirk, A. G., Plant, D. V. and Fauchet, P. M. (2007). *Journal of New Materials for Electrochemical Systems* 10, 113-116.

Létant, S. E., Content, S., Tan, T. T., Zenhausern, F. and Sailor, M. J. (2000). *Sensors and Actuators B: Chemical* 69, 193-198.

Malins, C., Doyle, A., D. MacCraith, B., Kvasnik, F., Landl, M., Simon, P., Kalvoda, L., Lukas, R., Pufler, K. and Babusik, I. (1999). *Journal of Environmental Monitoring 1*, 417-422.

Malins, C., Landl, M., Simon, P. and MacCraith, B. D. (1998). *Sensors and Actuators B: Chemical* 51, 359-367.

Marquis, B. T. and Vetelino, J. F. (2001). *Sensors and Actuators B: Chemical* 77, 100-110.

Mescheder, U., Bauersfeld, M. L., Kovacs, A., Kritwattanakhron, J., Muller, B., Peter, A., Ament, C., Rademacher, S. and Wollenstein, J. (2007). In "Solid-State Sensors, Actuators and Microsystems Conference, 2007. TRANSDUCERS 2007. International", pp. 1417-1420.

Ocampo, M., Doti, R., Faubert, J. and Lugo, E. (2011). In "Biosensors - Emerging Materials and Applications". InTech, ISBN: 978-953-307-328-6.

Oton, C. J., Pancheri, L., Gaburro, Z., Pavesi, L., Baratto, C., Faglia, G. and Sberveglieri, G. (2003). *Physica Status Solidi a-Applied Research* 197, 523-527.

Pacholski, C., Sartor, M., Sailor, M. J., Cunin, F. and Miskelly, G. M. (2005a). *Journal of the American Chemical Society* 127, 11636-11645.

Pacholski, C., Sartor, M., Sailor, M. J., Cunin, F. d. r. and Miskelly, G. M. (2005b). *Journal of the American Chemical Society* 127, 11636-11645.

Pacholski, C., Yu, C., Miskelly, G. M., Godin, D. and Sailor, M. J. (2006). *Journal of the American Chemical Society* 128, 4250-4252.

Park, S. H., Lee, K. W. and Kim, Y. Y. (2010). *Thin Solid Films* 518, 2860-2863.

Pavesi, L. and Mulloni, V. (1998). *Journal of Luminescence* 80, 43-52.

Potyrailo, R. A., Golubkov, S. P., Borsuk, P. S., Talanchuk, P. M. and Novosselov, E. F. (1994). *Analyst* 119, 443-448.

Pratt, W. K. (2001). "Digital Image Processing".

Ruminski, A. M., Moore, M. M. and Sailor, M. J. (2008). *Advanced Functional Materials* 18, 3418-3426.

Salonen, J., Laine, E. and Niinisto, L. (2002). *Journal of Applied Physics* 91, 456-461.

Santonico, M., Pittia, P., Pennazza, G., Martinelli, E., Bernabei, M., Paolesse, R., D'Amico, A., Compagnone, D. and Di Natale, C. (2008). *Sensors and Actuators B: Chemical* 133, 345-351.

Searson, P. C. and Macaulay, J. M. (1992). *Nanotechnology* 3, 188.

Snow, P. A., Squire, E. K., Russell, P. S. J. and Canham, L. T. (1999). *Journal of Applied Physics* 86, 1781-1784.

Uttiya, S., Pratontep, S., Bhanthumnavin, W., Buntem, R. and Kerdcharoen, T. (2008). In "Nanoelectronics Conference, 2008. INEC 2008. 2nd IEEE International", pp. 618-623.

Vial, J.-C. and Derrien, J. (1995). "Porous Silicon Science and Technology". Springer Verlag, Berlin.

4

Optochemical Sensor Systems for *In-Vivo* Continuous Monitoring of Blood Gases in Adipose Tissue and in Vital Organs

Merima Čajlaković[1], Alessandro Bizzarri[1], Gijs H. Goossens[2], Igor Knez[3], Michael Suppan[1], Ismar Ovčina[3] and Volker Ribitsch[1]

[1]*Joanneum Research Forschungsgesellschaft mbH, MATERIALS - Institute for Surface Technologies and Photonic, Sensorsystems, Graz,*
[2]*Department of Human Biology, NUTRIM School for Nutrition, Toxicology and Metabolism, Maastricht University Medical Centre, Maastricht,*
[3]*Department of Cardiac Surgery, University Clinic of Surgery, Medical University Graz, Graz,*
[2]*The Netherlands*
[1,3]*Austria*

1. Introduction

Sensors capable of determining both oxygen and carbon dioxide partial pressure, as well as pH in blood, are of particular interest with respect to the medical care of critically ill patients. Those parameters are informative indicators of the conditions of a living system and can be valuable in determining the physiologic status of critically ill patients. Tissue oxygenation represents the balance between local oxygen delivery and consumption. Under resting conditions the level varies between and within organs; it is lower in more metabolically active tissues, such as brain (Hou et al., 2003), liver (Vollmar et al., 1992), renal cortex and medulla Whitehouse et al., 2006) and higher in tissues with low metabolic rates, such as the bladder (Rosser et al., 1995, 1996). Changes in blood carbon dioxide levels are associated with myocardial acidosis and alkalosis, ventricular fibrillation and reduced myocardial contractility in addition to providing an indication of cardiac resuscitability (Maldonado et al., 1993; Tang et al., 1996). The respiratory gases themselves reflect cardiopulmonary homeostasis. This is the ability of the cardiopulmonary system to maintain a constant relationship between respiration in the cells of the body, the supply of oxygen from the lungs to the blood and the elimination of carbon dioxide by the lungs. In this way, partial pressures of O_2 and CO_2 are kept at their normal physiological values. Thus, the body constantly consumes oxygen and produces carbon dioxide, and there is a normal balance between the absolute contents and the partial pressures of these gases in the blood (Hahn, 1998). Although this balance can be upset by many clinical or physiological factors, the cardiopulmonary function has a remarkable ability to compensate for these changes, especially to maintain the gases at their normal values. The magnitude of the arterial blood pCO_2 is the key clinical measure of the effectiveness of the lungs to expel CO_2. Therefore, it is clinically important to measure the inspired and expired partial pressures of both oxygen and carbon dioxide in the critically ill adult and infant patient. It is possible either to take a blood sample and analyse this in the laboratory, what is routinely done in hospitals, or to perform the measurement *in vivo*. Although a number of methods are used in the clinical setting to measure blood gases, including blood gas analysis and electrodes (Astrup, 1986; Harsten, 1988; Suzuki, 1999), some practical issues restrict their use, e.g. the distance of the analyzer from the bedside, the need for special training of clinicians and staff for proper blood sample collection and handling, potential contamination of blood samples by the biological medias and the sensitivity of the method. Thus, there is huge need and interest to develop more sensitive and practical methodologies such as optical techniques, especially in the field of *in vivo* intravascular blood-gas analysis (Suzuki, 2001; Tsukada, 2003; Tusa, 2001; Wolfbeis, 1991). *In vivo* gas sensors should respond rapidly, preferably with a response time of the order of seconds, if the measurements are to have any significant physiological meaning. So far, the only commercially available sensors for the monitoring of adipose tissue oxygen partial pressure pO_2 are needle-type sensors. Those suffer of the typical drawbacks of a point measurement, which is strongly dependent on the position of the sensor tip inside the adipose tissue and show high fragility due to the small dimension of the sensor heads.

Combination of the optical sensor with a microdialysis technique is a powerful tool for online monitoring of metabolism in tissues (Cooney, 1997, 2000; Ungerstedt, 1991). This is mainly due to the fact that the subcutaneous adipose tissue is easily accessible and the risk of infection is low. The microdialysis probe, which is a hollow fibre which functions as an artificial blood vessel, is inserted into the tissue and is perfused with a dialysis solvent. Exchange of substances between the microdialysis probe and the extracellular fluid (sample medium) occurs over a semi-permeable membrane. Low molecular weight substances in the interstitial fluid surrounding the microdialysis catheter are allowed to diffuse across the membrane and to equilibrate with the perfusion solution. The interstitial fluid leaves the probe via a side arm from which it is collected (Goossens et al., 2011). Optical sensing approach combined with microdialysis have been already used for monitoring of dissolved gases and pH in a flow loop (Cooney, 1997, 2000). Pasic et al. (Pasic et al., 2006) have described a microdialysis-based glucose-sensing system with an integrated fibre-optic hybrid sensor. An important advantage of this system is that for example adipose tissue pO_2 is measured over a much larger area compared to commercially available needle-type oxygen sensors.

In our first work we reported on optical capillary sensors for continuous measurement of oxygen and carbon dioxide in subcutaneous tissue using microdialysis for extraction of interstitial fluid in subcutaneous tissue (Bizzarri, 2006; Cajlakovic, 2006). The adopted sensing scheme for both sensors was phase fluorometry, which is based on the dynamic quenching due to oxygen of the photo-excited luminophore Platin(II) meso-tetra(pentafluorophenyl) porphyrin (Pt-TFPP). In the case of carbon dioxide sensor an ion-pairing approach combined with dual luminophore referencing as a detection technique was used for the preparation of carbon dioxide sensitive layers. The analysis of extracted sample by means of the external instrumentation is properly designed for rapid analysis of small volumes (less than microliters) and allowed a less invasive technique than the intravascular approach (Bizzarri, 2006; Cajlakovic, 2006). Recently, we have presented an improved optochemical measurement system for continuous monitoring of adipose tissue oxygen tension in biological fluid using a miniaturized flow-through cell instead of a capillary-based sensor coupled to a microdialysis catheter (Bizzarri, 2007; Cajlakovic, 2009). The effectivness and accuracy of the sensors were investigated *in vitro* using physiological solution (ringer) and *in vivo* tests in order to examine the magnitude of changes in tissue gas tension in response to evolving shock and resuscitation: 1) on test animals, where hemorrhagic schock model has been adopted to simulate critically condition and 2) on healthy volunteers, where infusion of adrenaline caused the changes in tissue oxygenation. Adipose tissue oxygen tension may be involved in the development of insulin resistance (impaired action of the hormone insulin) in humans. Therefore the continuous monitoring of adipose tissue oxygen tension (pO_2) in individuals affected by obesity and/or type 2 diabetes mellitus is highly interesting. Recent cell culture and preclinical experiments suggest that obesity is accompanied by low levels of oxygen in adipose tissue (Ye et al., 2007). However, this concept was recently challenged by Goossens et al. (Goossens et al., 2011), demonstrating increased adipose tissue PO2 in obese insulin resistant compared to lean insulin sensitive men. The optochemical sensor technology combined with microdialysis is particularly suitable for the development of minimally invasive instrumentation for measurement of tissue pO_2 (Goossens et al., 2011). In the present chapter, we present an improved microdialysis-based oxygen-sensing instrumentation with integrated planar sensitive membrane inserted in a miniaturised flow-through-cell (volume app. 5µl) for the continuous monitoring of adipose tissue oxygen tension in humans. This instrumentation is suitable for measurement at a flow rate as low as 2 µl/min with a response time, including the flow time from the catheter to the sensitive membrane, of about 10 min. The sensor system has been validated in *in vivo* experiments, in which adipose tissue oxygen tension in humans has been monitored under resting conditions and after the local administration of a vasoconstrictor and vasodilator (Goossens et al., 2011). As expected, local administration of the vasoconstrictor in adipose tissue decreased both adipose tissue blood flow and adipose tissue pO_2. Subsequently, local infusion of the vasodilator increased both adipose tissue blood flow and pO_2 values (Goossens et al., 2011). Obtained results showed good agreement between the observed adipose tissue oxygen tension and the expected behavior due to the induced alteration of adipose tissue blood flow. Thus, this system is suitable for in vivo monitoring of pO_2 in adipose tissue.

The second part of this contribution includes the description of the optochemical measurement system for simultaneously monitoring of tissue O_2 and CO_2 partial pressure in brain, heart and liver by means of miniaturized implanted fiber optical sensor. pO_2 and pCO_2 measurements are based on the determination of the luminescence lifetime via phase modulation fluorometry. The instrumentation unit consists of two independent channels for measuring pO_2 and pCO_2. The O_2 sensor is based on the luminescent dye Pt-pentafluorophenyl-porphyrin, while for CO_2 sensor the Resonant Energy Transfer (RET) scheme is used to convert the intensity of the analyte dependent dye indicator into long luminescence lifetime information, which can be measured by phase modulation fluorometry (Lakowicz, 1983). Therewith it was for the first time possible to measure continuously and simultaneously O_2 and CO_2 content in brain, heart and liver during open-heart surgery. Fiber optical sensors have been tested in an experimental animal model during cardiac surgery using two different methods as extra corporal circulation: conventional cardiopulmonary bypass (CPB) and minimal extracorporeal circulation systems (mini-EEC). CPB is the most accepted surgical technique to treat cardiac pathologies. It is a safe and established technique with a low mortality rate (Pentila et al.,

2001). However, besides neurological events, which may at least be partially related to the use of CPB, and the inflammatory response, the ideal myocardial protection is still a matter of scientific and clinical debate (Chang, 2002; Immer, 2005; Levy, 2003). Mini-ECC system consists of a centrifugal pump, an oxygenator and a modified suction system. The main advantage of mini-ECC is the reduction of tubing length (reduction of the priming volume). It was designed to reduce blood tubing, the contact between blood and air and priming volume during cardiac surgery. The main issue of developing miniaturised extracorporeal circulation systems is the strong need for a CPB concept for cardiac surgery which will cause fewer traumas to blood components and reduce the extensive volume shifts and bubble formation during cardiopulmonary bypass (Kofidis et al., 2008). A closed circuit, where the patient himself/herself serves as a reservoir with a minimized priming volume, could be the ideal alternative to reduce the adverse effects of CPB (Kamiya et al., 2006). In a complex, technically demanding, prospective, randomized and blinded experimental setting, an extracorporeal circulation has been performed using mini-EEC versus standard CPB techniques with main emphasis on differences in organ specific oxygen and carbon dioxide partial pressures and metabolism. *In-vivo* experiments on animals with implanted optical O_2 and CO_2 sensors were performed in order to be able to establish if there is any difference in O_2/CO_2 metabolism when using two types of extracorporeal circulation, CPB and mECC, respectively.

2. The continuous monitoring of adipose tissue oxygen tension (pO_2) in humans

2.1 Design of an optical oxygen sensor

The luminophore Platin(II) meso-tetra(pentafluorophenyl) porphyrin (Pt-TFPP) was used because of its long lifetime (~ 50 ms), strong absorption in green region of the spectrum, which is compatible with green light-emitting diodes (LEDs), its highly emissive metal-to-ligand charge transfer state and its high photostability. The dye was entrapped in a hydrophobic polymer of polystyrene (4 w. % dye/g solid polystyrene in chloroform). Afterwards the sensor cocktail was spread onto a transparent substrate; either a polyethylene terephthalate foils PET (Mylar foil with thickness 100 µm) or PET foils (50 µm thickness) coated with Al_2O_3. Knife coating and screen printing techniques have been used for production. After coating, the oxygen sensitive membranes have been dried at 60 °C for eight hours. In the case of knife coating a successive mechanical reduction of the sensitive spot to a diameter of 2 mm has been applied. This procedure was not necessary in case of the screen printed sensitive membranes.

2.2 Measurement principle

The luminescence lifetime of a luminophore is an intrinsic property and is independent of for example detector sensitivity and fluctuations in light intensity therefore phase fluorometry was used as sensing scheme. In this case the luminophore is excited by sinusoidally modulated light at a given frequency ω and the emitted sinusoidal luminescence is also modulated at the same frequency. The lifetime of the luminophore causes a time lag between the absorbance and emission, expressed by the phase angle θ and a decreased intensity relative to the incident light, called demodulation. The relationship between the lifetime τ and the phase shift ϕ for a single exponential decay is given by equation (1) and is illustrated in Fig. 1:

$$\tau = \frac{tan\phi}{2\pi f} \tag{1}$$

where f is the modulation frequency. An optimal frequency need to be selected because signal-to-noise (SNR) ratio decreases and phase sensitivity increases with increasing modulation frequency. The used luminophore was Pt-TFPP, whose luminescence is dynamic quenched (reduced) in the presence of oxygen and subsequently the lifetime as well. The optimal frequency for this dye was chosen 4.5 kHz. The instrumentation measures the phase shift of the luminescence with respect to the excitation signal. This is related to the luminescence lifetime of the dye, which in turn depends on the pO_2 of the tissue in contact with the surface of the sensor probe. The pO_2 can be calculated from the measured phase shift of the luminescence using the so called false-light model (Bizzarri et al., 2006) as described by equation (2):

$$\phi = \phi_0 \left[\frac{f_0}{1 + K \cdot pCO_2} + 1 - f_0 \right] \tag{2}$$

where ϕ is the phase shift measured at a given oxygen partial pressure pO_2, ϕ_0 is the phase shift in absence of O_2, K is the quenching parameter and f_0 is the false-light parameter (the fraction of the luminescence emitted by the sensor, which is quenched by oxygen).

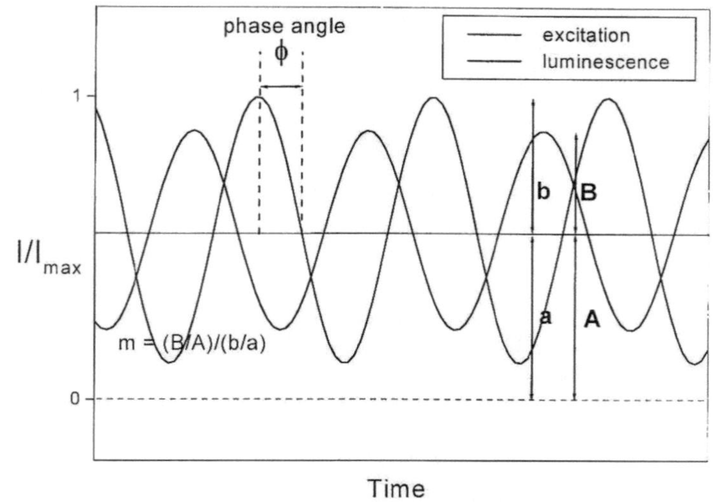

Fig. 1. Principle of phase fluorimetric technique. Luminescence emission is delayed in phase angle Φ and demodulated relative to the excitation (m).

2.3 Miniaturised flow-through cell and optoelectronic measurement unit

The fibre optic measurement system consists of an optoelectronic read-out module for measurements with optical fibres (core diameter 1000 μm) and of a miniaturised flow-through cell (volume app. 5 µl), where a planar oxygen sensitive membrane (diameter of the sensitive spot 2 mm) is allocated. The measurement set-up is shown in Fig. 2. The flow through cell is equipped with inlet tubing of inner diameter of 0.15 mm and outer diameter of 0.30 mm (Rometch GmbH). The outlet tubing and the flow chamber have an inner diameter of 2.0 mm. The outer diameter of the outlet tubing is 3.0 mm as presented in Fig. 3. The inner diameter of the inlet was chosen to be 0.15 mm in order to match the inner diameter of the inlet of the microfluidic catheter. This geometry has been chosen to reduce the risk of overpressure during measurement, which would cause an overestimation of the dissolved pO_2. The length of the inlet is 20 mm whereas the outlet is 32 mm long. The tubings are made of stainless steel in order to prevent or to reduce the permeation of ambient oxygen to non-significant values.

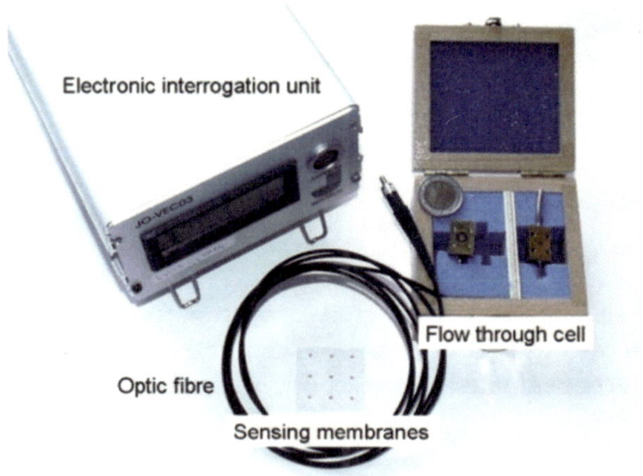

Fig. 2. Measurement system for planar sensing membranes. The flow-through cell has a size of 20 x 13 x 4.8 mm (length x width x heigth). The measuring chamber of the flow-through-cell has a diameter of circa 1.8 mm and a height of about 2.8 mm, thus resulting in a volume of about 5 µl. The sensing membranes have a round sensitive spot of 1.8 mm in diameter.

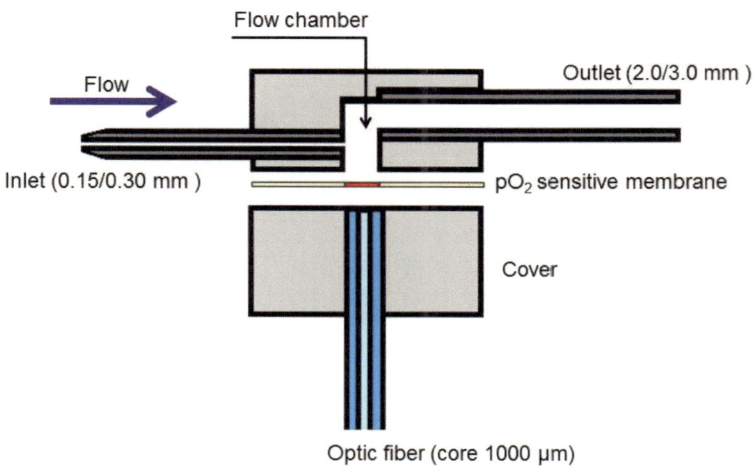

Fig. 3. Schematic draw of the flow through cell

The optic set-up is integrated in the processing electronic unit and is based on a collimating optics, which converts the light of the excitation source (a Nichia NSPE590S LED) into a parallel beam. This is then focused by a GRIN lens into an optic fibre, which guides the excitation light to the oxygen sensitive layer inside the flow-through cell. The luminescence collected by the same optic fibre is deflected by a dichroic filter at 45 ° (DC Red, Qioptiq Photonics GmbH & Co. KG) towards the detector a silicon photodiode (Hamamatsu S5106). A set of optical filters (SCHOTT BG39, thickness 2 mm, for the excitation source and SCHOTT RG630, thickness 2 mm, for the photodiode) is used to separate the luminescence signal from the excitation light and other optical background signals (e.g. luminescence emitted by materials close to the O_2 sensitive membrane). A reference light source (Nichia NSPE590S LED) is used to compensate the reading of the measuring unit for changes of the electronic parameters of the components with fluctuations of the operating conditions (e.g. forward current through the LED and temperature). The main advantage of using fiber optic technology is the possibility to separate the optoelectronic instrumentation from the flow-through cell, thus allowing an extreme miniaturization of the cell itself. The new flow-through-cell has size of only 20 x 13 x 4.8 mm, which makes it more flexible for use in *in vivo* experiments and use in hospitals. In order to prevent disconnection of the fluidic when the patient moves the flow-through cell was mounted on a belt for fixed positioning on the abdomen of the subject during *in vivo* experiment, as it is shown in Fig. 4.

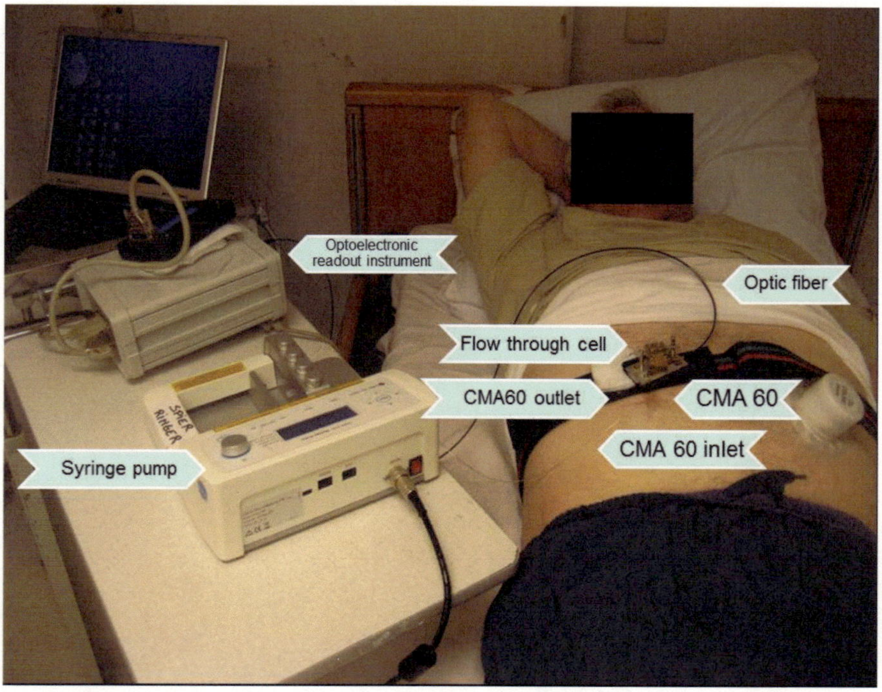

Fig. 4. Use of the optochemical oxygen sensor instrumentation during a typical in vivo experiment. The main components are the flow-through-cell, which is fixed near the umbilicus of the volunteer close to the position of insertion of the CMA 60 microdialysis catheter. The luminescence emitted by the oxygen sensitive membrane is collected by an optic fibre cable and guided to the optoelectronic read-out unit. The microdialysis probe is perfused using a microinfusion syringe pump. These experiments were carried out in the laboratory of Dr. G.H. Goossens (Maastricht University Medical Centre, The Netherlands).

For the extraction of interstitial fluid microdialysis catheter CMA60 was used. The outlet tubing of catheter was first cut as close as possible to the microdialysis membrane (at a distance of about 2 mm) and then connected with the flow through cell by means of steel tubing (diameters 0.15/0.30 mm). The length of the steel tubing was standardised at 90 mm in each experiment. In this way the ambient oxygen diffusion through the permeable tubing materials of the catheter can be neglected. This has been proven in oxygen recovery rate measurements, which have been carried out in laboratory. These measurements have been already exhaustively described in previously reported work (Cajlakovic et al., 2009). The standardization of the length of all the connection tubing allows that the flow time from the catheter to the sensitive membrane is in all experiments the same.

2.4 Applications and practical systems

2.4.1 Sensor performance and laboratory calibration protocol

The membranes have been produced by knife coating technique onto PET substrate and by means of screen printing onto PET substrate coated with Al_2O_3 as oxygen barrier.

All sensor membranes have been firstly laboratory characterized in Ringer solution (physiological solution used as perfusate in many clinical procedures) at temperature held constant at 37 °C. The temperature regulation was realised by using a water bath thermostat (Lauda RC 6 CP). The measurement medium has been equilibrated with a gas mixture of nitrogen and oxygen at defined concentrations (typically 0, 5, 10 and 21 % O_2) using a mass flow controller (Bronkhorst, The Netherlands). A paramagnetic gas purity analyser (Servomex, U.K.) was served as reference instrumentation. The solution was then pumped by means of a peristaltic pump into a miniaturised flow-through cell at the flow rate of 2.0 µl/min (Minipuls 3, Gilson, Germany). One example of a typical sensor response curve, phase angle as a function of different O_2 concentrations, is shown in Fig. 5.

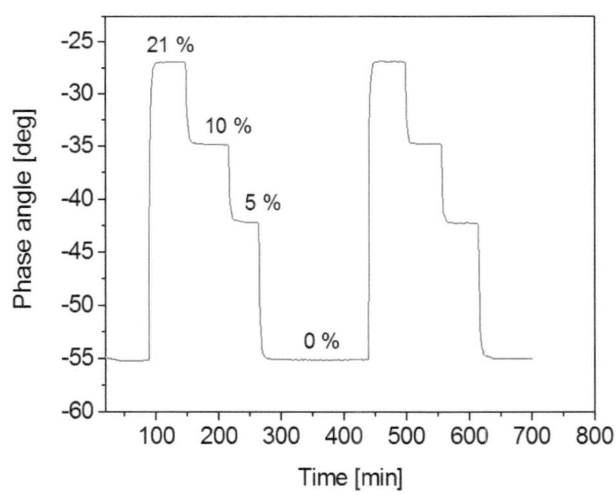

Fig. 5. Example of the experimental protocol used for the characterisation of the miniaturised planar O_2 sensitive membrane. The measurement is carried out in Ringer solution equilibrated at defined O_2 concentrations (0, 5, 10 and 21 % O_2) at the constant temperature of 37 °C.

Two cycles have been performed, whereas the first complete cycle of dissolved O_2 was used for sensor calibration purpose, and the following cycle is used to check the specifications of the oxygen measurement (Table 1).

	Specifications
Working range	0-300 mmHg
Resolution	1 mmHg
Accuracy	< 7% or ±1 mmHg

Table 1. Achieved specifications of the O_2 measurement with planar sensitive membranes compared to the optimum requirements defined in the project

Both types of membranes perform satisfactorily with the same measurement specifications.

2.4.2 Delay time

The time needed for the extracted interstitial fluid to reach the sensitive oxygen membrane in the flow-through cell is indicated as delay time. This delay time is defined as the time required for the sensor output to change from its previous state to a final settled value within a tolerance band of the new value. It is a common practice to measure the delay time as the time needed for the sensor to reach the 90 % of the final output. This is usually indicated as the t_{90} of the sensor. Fig. 6 shows the delay time t90 for an oxygen change from air saturated solution (circa 21 % O_2) to oxygen free solution (about 0 %).

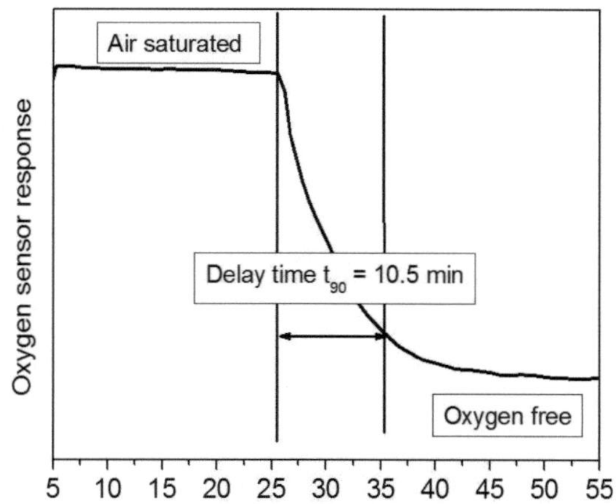

Fig. 6. Measured delay time between 21 and 0 % of O_2 in case of one oxygen sensitive membrane

To perform this measurement, two different glass vessels filled with Ringer solution has been equilibrated with air and nitrogen. The catheter CMA 60 connected by steel tubing to the flow through cell has been first immersed in the vessel containing air saturated Ringer solution and the peristaltic pump was started with flow rate of 2 µl/min. After a stable value of the oxygen sensor instrumentation has been achieved, the pump has been shortly stopped to change the catheter in the oxygen free Ringer solution and the flow started again. The time needed to achieve the 90% of the final value as been recorded. Six different membranes of both types have been investigated and the averaged t_{90} could be estimated as 11.7 ± 4.6 min. Even if, the two different types of sensitive membranes (based on uncoated or Al_2O_3 coated PET substrate) perform satisfactorily with regard to the measurement specifications, it could be observed that, in single membranes based on the uncoated PET substrate, the response time increased above the specified time of 10 min, due to ambient oxygen diffusion through microscopic damages of the coated surface, caused by the mechanical reduction of the sensitive spot.

2.4.3 Clinical validation in humans

2.4.3.1 Validation protocol

To validate the sensor instrumentation for *in vivo* use in humans, a healthy, normal-weight male subject participated in an experiment carried out at Maastricht University Medical Centre in The Netherlands. The subject was asked to refrain from drinking alcohol and to perform no strenuous exercise for a period of 48 hours before the *in vivo* experiment. On arrival, a microdialysis catheter (CMA 60, CMA microdialysis AB, Stockholm, Sweden) was inserted in the abdominal subcutaneous adipose tissue 6 cm right from the umbilicus after the skin was anesthetized by a lidocaine/prilocaine cream (EMLA, AstraZeneca BV, Zoetermeer, The Netherlands). After insertion, the probe was perfused with Ringer solution (Baxter BV, Utrecht, The Netherlands), supplemented with 50 mM ethanol, at a flow rate of 2.0 µl/min (CMA400 micro infusion pump, CMA microdialysis AB, Stockholm, Sweden). Afterwards, the system was allowed to equilibrate. After pO_2 had reached stable values, the probe was consecutively perfused with a vasoconstrictor and vasodilator, each for 60 min. There was a wash-out period of 120 min (Ringer infusion) before the infusion of the vasodilator was started.

2.4.3.2 Results and discussion of *in vivo* measurement of adipose tissue oxygen tension in humans

A stable baseline pO_2 of 52.5 mmHg was reached after 2-3 hours (Fig. 7, time -30 to 0 min). As hypothesized, local administration of a vasoconstrictor in adipose tissue (t0-60 min) decreased both adipose tissue blood flow and adipose tissue pO_2 (37.3 mmHg, Figure 7). During the wash-out period (t60-180min) the probe was perfused with Ringer solution, and pO_2 values returned to near-baseline values again (53.5 mmHg). Subsequently, local infusion of a vasodilator increased both adipose tissue blood flow and pO_2 values to 74.8 mmHg (t180-240min). This experiment clearly showed that pO_2 values can be reliably measured *in vivo* in humans, and can be modified by changes in tissue blood flow. These results have been extensively described elsewhere (Goossens et al., 2011).

The rapid increase of the measured pO_2 at the beginning of each drug administration (spikes in Fig.7) is caused by ambient oxygen, which has been introduced in the fluidic during the change of the infusion solution.

After the *in vivo* experiments several oxygen sensitive membranes were recalibrated in the laboratory, and the deviations of the pO_2 values obtained by the second calibration were calculated and plotted against the actual pO_2 values (Fig. 8). Importantly, the change in the calibration of the sensitive membrane after the *in vivo* experiment causes very small discrepancy (<0.5 mmHg) from the actual pO_2 values. This result is excellent considering that the oxygen sensitive membranes have been recalibrated about one year after the first laboratory calibration and after in vivo experiments of a typical duration of 5 – 8 hours.

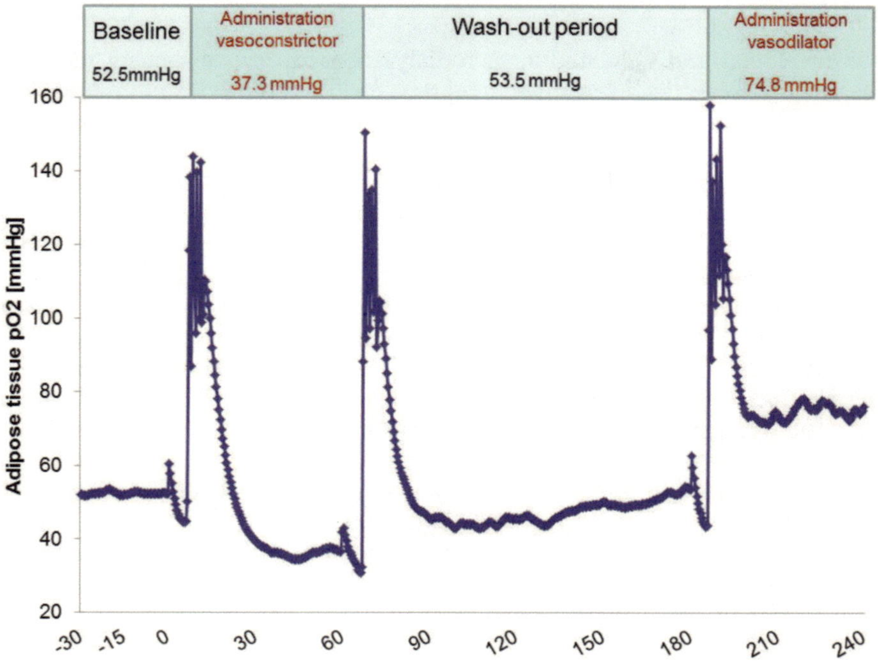

Fig. 7. Measured abdominal adipose tissue pO_2 with optochemical sensitive membrane under stable baseline conditions (local Ringer administration up to time 0), during local administration (2 µl/min) of a vasoconstrictor between 0 and 60 min, followed by wash-out period with Ringer infusion (from 60 to 180 min). At 180 min an infusion of a vasodilator was started up to 240 min. The experiment was carried out in the laboratory of Dr. G.H. Goossens (Maastricht University Medical Centre, The Netherlands) (Goossens et al., 2011).

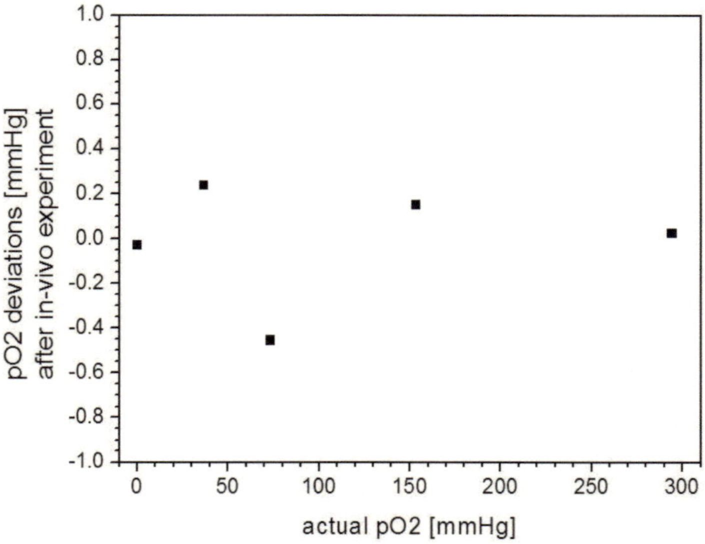

Fig. 8. Deviations of the pO_2 values measured for single oxygen sensitive membrane after in vivo experiments compared to the calibration values.

2.5 Conclusion

We have developed and validated a microdialysis-based oxygen-sensing instrumentation with integrated miniaturized sensitive membrane for the continuous monitoring of adipose tissue oxygen tension in humans. In comparison to previously reported instrumentation (Cajlakovic et al., 2009), this system can be used at a flow rate as low as 2 µl/min with a response time of about 10 min (including the flow time from the catheter to the sensitive membrane). The production technology of the oxygen sensitive membranes has been improved, allowing the pO_2 measurement in the range between 0 and 300 mmHg with accuracy better than ±1 mmHg and resolution better than 1 mmHg. The system has been validated *in vivo*, where adipose tissue oxygen tension has been monitored in humans under resting conditions and during local administration of vasoactive compounds. Local administration of the vasoconstrictor in adipose tissue decreased adipose tissue pO_2 and subsequently, local infusion of the vasodilator increased pO_2 in adipose tissue (Goossens et al., 2011). The agreement between the observed adipose tissue oxygen tension and the expected behaviour after modulation of adipose tissue blood flow allows concluding that the system is suitable for in vivo monitoring of pO_2 in humans. Several sensitive membranes were recalibrated after in vivo experiments in the laboratory and deviations smaller than ± 0.5 mmHg were observed. This indicates that the membranes have an excellent storage and operating stability, and that contact with biological fluid does not affect the read-out of the optochemical measurement instrumentation.

2.6 Acknowledgement

The in vivo experiments in humans were carried out at Maastricht University Medical Centre, Maastricht, The Netherlands, under supervision of Dr. G. H. Goossens.

3. Optochemical measurement system for simultaneously monitoring of O_2 and CO_2 partial pressure in vital organs by means of miniaturized implanted fibre optical sensor

The measurement system is composed of two channels allowing simultaneously measuring of O_2 and CO_2 partial pressures (pO_2 and pCO_2). The optoelectronic measurement unit for O_2 is same as described in chapter 2 (Paragraph 2.3). For this application optical fibres were coated with sensitive layers (O_2 and CO_2). In the case of using an optical fibre as support for the sensitive layer, pulses of light emitted by a green (in the case of O_2 sensor) or blue LED (in the case of CO_2 sensor) are carried via an optical fiber to excite the dye at the sensor tip. On interaction with measured parameter the luminescence is quenched and is transmitted back within the same fibre to the detection unit. The instrumentation measures the phase shift of the luminescence with respect to the excitation signal.

3.1 Optoelectronic measurement unit for CO_2

As in the case of the O_2 measurement channel a measurement system for CO_2 is based on a collimating optics to convert the light of excitation source (Nichia NSPB500) into a parallel beam, which can be focused by a GRIN lens into the optic fibre. The luminescence collected by the optic fibre is also transformed by the same GRIN lens into a parallel beam, which is deflected by a dichroic filter at 45 ° (Unaxis B51) with respect to the optical axis to the silicon photodiode as a photo-detector (Hamamatsu S5106). Dichroic filters transmit selected ranges of the visible spectrum and reflect the complementary ranges, thus they may be used to separate the radiation emitted by the excitation source from the luminescence. Additional glass optical filters (SCHOTT BG12, thickness 2 mm for the excitation source and SCHOTT KV550, thickness 3 mm as emission filter) are used to separate the luminescence from stray light and optical background due to the excitation source. For the compensation of the reading of the measurement unit for changes of the electronic parameters of the components with fluctuations of the operating conditions (e.g. forward current through the LED and temperature) a reference light source is used (HLMA-KL00-I0000 (Avago Technologies US Inc.). The excitation light of the LED was sine wave modulated at a frequency *f* of 20 kHz.

3.2 Optical fiber tip micro fabrication by chemical etching

The most efficient configuration of optical fibre for collecting the luminescence emission and to reduce the optical background was achieved by coating the sensitive layers onto the sharpened tip of optic cables with core diameter of 400 and 200 µm. For this purpose the fiber tips have been micro-machined by chemically etching process (tube etching) of multi-mode optical fibers (Si core and hard polymer cladding, Thorlabs). The schematic illustration of the forming of a sharpened fiber tip is presented in Fig. 9. After the etching process was completed, the tip of the optical fiber was rinsed with distilled water and cleaned with acetone. The height of the obtained cone tip is about 1.5 mm. All the etching experiments were carried out at room temperature. The obtained form of an optical fibre efficiently resulted in improved signal-to-noise ratio of about one order of magnitude for both types of sensors thus resulted in an increased luminescence signal.

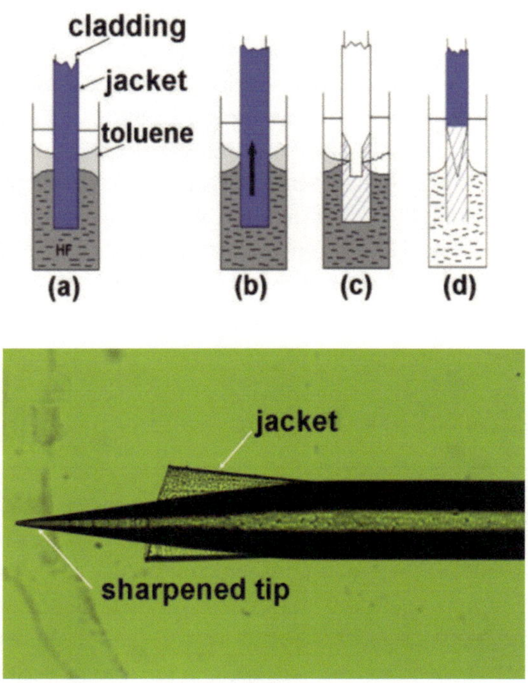

Fig. 9. Top: (a) immersion of a fiber tip into the etchant-organic solution (40 % hydrofluoric acid-toluene); (b) pull out of the fibre until a meniscus is formed at the interface; (c) the meniscus height reduces progressively until the portion of the fiber below the organic solvent is completely etched (d). Below: formed a cone-shaped tip with an angle between 15 - 20 °.

3.3 Preparation of O_2 fiber optical sensor

The sensitive cocktail is composed of Pt-TFPP dissolved in polystyrene (4 w. % dye/g solid polystyrene in chloroform). The optical fibres were coated by dipping only the sharpened tip of the fibre into a sensitive cocktail. After coating, the oxygen sensors have been dried at 60 °C for eight hours.

3.4 Preparation of CO_2 sensing cocktail

The carbon dioxide sensitive cocktails were prepared by mixing donor and acceptor dye, along with a phase transfer agent, tetraocytlammonium hydroxide (TOAOH) in ethylcellulose solution and coating a sensitive thin film onto the tip of etched optical fiber. The polymer solution of ethyl cellulose (EC) was prepared by dissolving 4.4 g of EC in the 40 g of a toluene/ethanol mixture (80/20, v/v), giving a viscous 11 % (w/w) solution. Ru(dpp) nanobeads incorporated in oxygen impermeable polymer, polyacrilonitrile (PAN) have been suspended in polymer (EC) solution. Acceptor was prepared by dissolving 2.03 mg of phenol red (PR) in 1.5 ml of toluene/ethanol mixture to which 20 µl of TOA-OH (20 % solution in methanol) solution was added. The rate of energy transfer and therefore the sensitivity of the sensor are almost independent of the concentration of the donor in the matrix. However, an increase in donor concentration enhances the luminescence intensity and therefore the signal-to-noise ratio. In contrary, the concentration of the deprotonated form of acceptor is a decisive factor upon which the rate of energy transfer strongly depends since the overlap integral is proportional to the concentration of the acceptor in Förster equation. Owing to this, the signal change depends not only on the degree of protonation by carbon dioxide but also on the overall concentration of the acceptor in the sensor membrane. Therefore, the final cocktail composition consisted of 13.8 mmol Ru(dpp) and 3.08 mmol PR per kg of solid polymer. The cocktails were saturated with carbon dioxide before coating of the fiber tip. The resulting fiber optical sensors were dried for 4 hours at 70 °C before an additional ion impermeable layer of Teflon with added black graphite as an optical and protective isolation was applied. The dried sensors were stored in desiccator over sodium carbonate before use. Before the calibration was proceeded the optical fibers have been exposed to solution equilibrated with CO_2 (20 %) and N_2 in order to obtain faster stable response of the sensor.

3.5 Measuring principle of CO_2 optical sensor

For the realisation of an optical lifetime-based CO_2 sensor, whose luminescence and lifetime are modulated by pH dependent efficiency of resonance energy transfer, some criteria need to be considered at selection suitable donor-acceptor pairs: (a) a pH-sensitive indicator is combined with long-lived transition metal complex; (b) an absorbance spectrum of deprotonated form of analyte dependent pH-indicator has to overlap with the luminescence spectrum of the inert luminophore, added as a donor so that efficient energy transfer takes a place; (c) the distance between donor and acceptor does not exceed Förster distance (in the range 1-10 nm), (d) lifetime of the donor in microsecond range (e) high quantum yield of the donor and (f) excitation of the acceptor and donor at a single band of wavelength and detection of donor luminescence at a common wavelength using a single photodetector. In order to convert a change in absorption spectrum of acceptor in presence of CO_2 into lifetime information, a long-lifetime luminescent donor need to be co-immobilised in the same matrix. Depending on the pH, the acceptor changes its colour and deactivates the excited state of the donor. This effect changes both the luminescence intensity and the lifetime of the donor. Ru(II) diphenyl-phenanthroline (Ru(dpp)) was selected as a donor due to the high quantum yield (0.6) and long lifetime in range of microseconds (~ 5.5 μs). Its absorption spectrum perfectly overlaps the emission of bright blue LEDs and its red emission spectrum overlaps the absorption of the deprotonated form of pH-indicators in red region of spectrum. Among the investigated pH indicators the main problem was their poor photostability (for example for the case of cresol purple and bromothymol blue) and due to the pronounced photobleaching these indicators turned out to be not suitable for the fiber optic measurement system. Even the strategy of optimising the ratio between the concentrations of pH indicator in comparison to the concentration of the donor did not allow realising a sensor with satisfactorily stability properties. Therefore on the basis of enough overlapping of the absorption spectrum of deprotonated form of phenol red (i.e. in absence of CO_2) with luminescence of Ru(dpp) phenol red was chosen as an acceptor (Fig. 10) . This dye provides enough intensity when the energy transfer with Ru(dpp) takes place, has pK_a value in the physiological range (6.8) and expand the measurement range up to 20 % of CO_2 (corresponding to 147 mmHg).

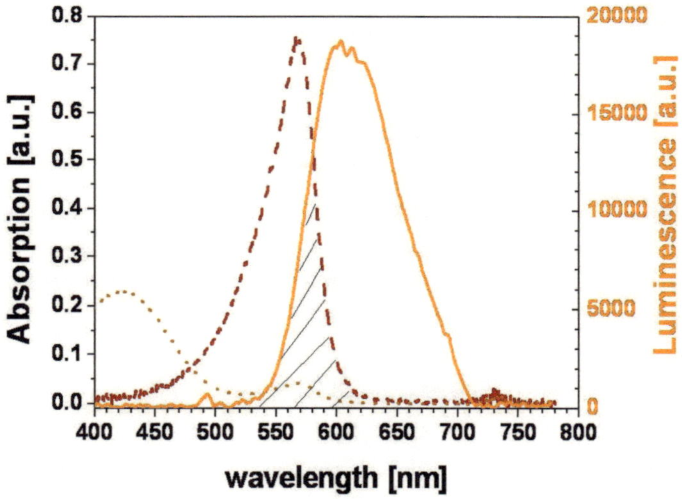

Fig. 10. Luminescence emission spectrum of $Ru(dpp)_3^{2+}$ (solid line) and absorption spectra of phenol red in the absence (dashed line) and presence (dotted line) of CO_2 (20 %). The hatched area indicates the spectral overlap of the donor emission and acceptor absorption.

The CO_2 sensor is solid type sensor, what means that no buffer was used. Instead this the pH indicator (A) forms a hydrated ion-pair ($A^-Q^+ \cdot xH_2O$) by an addition of quaternary ammonium base (Q^+OH^-) (Mills et al., 1992). To stabilize the deprotonated form of the indicator dye in the matrix and to form a lipophilic bicarbonate buffer system in the polymer, tetraoctyl ammonium hydroxide was used. The excess amount of TOA-OH is neutralised by atmospheric carbon dioxide to form a lipophilic hydrogen carbonate buffer in the solid matrix, and this can tune the sensitivity and improve the storage stability of the sensor. The sensing mechanism of an acceptor-Q^+OH ion-pair is described by (3):

$$TOA^+A^-.xH_2O + CO_2 \leftrightarrow TOA^+HCO_3^-.\left(x-1\right) H_2O + AH \tag{3}$$

where [TOA^+A^-. xH_2O] is the complex of the deprotonated pH indicator dye ion paired with the quaternery ammonium cation TOA^+ and a number x means the hydrophilic sites present in the polymer matrix.

In the case of Resonance Energy Transfer (RET) as sensing scheme the average lifetime is converted into a parameter (phase angle) that reflects the concentration of deprotonated pH-indicator (acceptor). The CO_2 dependent lifetime information is detected using the technique of phase fluorometry. In this technique, the lifetime data (τ) are converted to phase shift data (ϕ) as it was described by equation (1).

Efficiency of energy transfer E is expressed as:

$$E = \frac{k_{RET}}{k_{RET} + k_{other}} = 1 - \frac{\tau_{\max}}{\tau} \tag{4}$$

where k_{RET} is rate of resonance energy transfer and k_{other} is the sum of the rates of all other deactivation processes including luminescence. τ_{max} reflects the lifetime of donor (complete protonation of acceptor at very high pCO_2, when no energy transfer occurs) and τ is a lifetime in the absence of carbon dioxide. By rearrangement of the equation 4 the ratio τ_{max}/τ can be expressed as:

$$\frac{\tau_{\max}}{\tau} = \frac{k_{other} + k_{RET}}{k_{RET}} = 1 + \frac{k_T}{k_F} \tag{5}$$

From Förster theory it follows, that the rate of energy transfer k_T is proportional to the spectral overlap integral J between donor emission and acceptor absorption ($k_T \propto J$). When the pH-indicator is completely completely protonated, there is no spectral overlap with the donor and overlap integral J is proportional to concentration of deprotonated dye D^-:

$$\frac{\tau_{\max}}{\tau} - 1 = \frac{k_{RET}}{k_{other}} \cong J \cong \left[D^-\right] \tag{6}$$

This model corresponds to the situation of dynamic quenching of luminescence described by the Stern-Volmer equation. Therefore pCO_2 was calculated from the measured phase shift of the luminescence using as described by equation (7):

$$\phi = \phi_0 \left[\frac{f_0}{1 + K \cdot pCO_2} + 1 - f_0\right] \quad \phi = \phi_0 \left[\frac{f_0}{1+K \cdot pCO_2} + 1 - f_0\right] \tag{7}$$

3.6 Laboratory calibration protocol

Before of the *in vivo* measurements, the fiber optical pO_2 and pCO_2 sensors have been calibrated in laboratory. In the case of pO_2 sensors, a glass vessel filled with Ringer solution and kept in a water bath (Lauda RC 6 CP thermostat) at the constant temperature of 37 °C has been equilibrated with a gas mixture of nitrogen/oxygen at defined concentrations (typically 0, 5, 10 and 21 % O_2) using a mass flow controller (Bronkhorst, The Netherlands). Similarly, in the case of the pCO_2 sensors the Ringer solution has been equilibrated with a gas mixture of nitrogen/carbon dioxide at defined concentrations (typically 0, 5, 10, 15 and 20 % CO_2). A gas purity analyzer (Servomex) with both a paramagnetic O_2 sensor and a CO_2 infrared sensor has been used as reference to analyse the composition of the gas mixture dissolved in the Ringer solution. The fiber optic sensors have been put in contact with the Ringer solution and the phase shift at each defined gas concentration has been recorded. An example of the phase shift response at exposed gas concentration is shown in Fig. 11 for pO_2 sensors and in Fig. 12 for pCO_2 sensors, respectively. It can be seen that for both sensors usually two cycles of dissolved O_2 and CO_2 concentrations have been measured. The first cycle has been used for sensor calibration purpose, whereas the following cycle is used to check the specifications of the oxygen and carbon dioxide measurements calculated with calibration parameters obtained from equations 2 and 7, respectively.

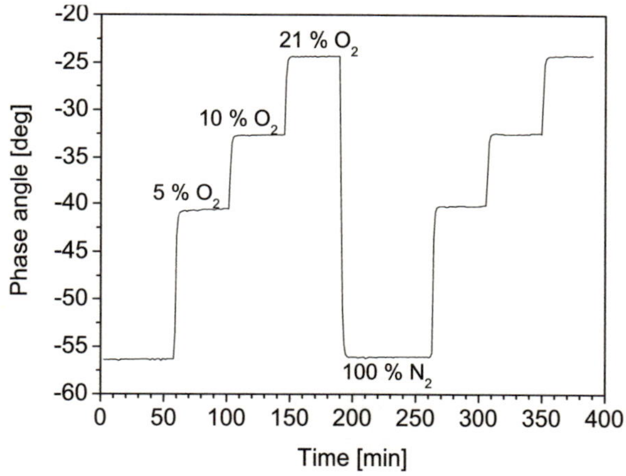

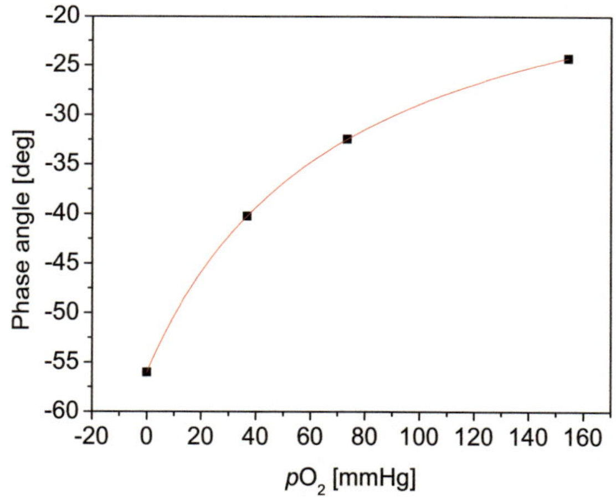

Fig. 11. Oxygen sensor response curve, phase angle as a function of different O_2 concentrations; below: calibration curve obtained by the first cycle of dissolved oxygen concentrations.

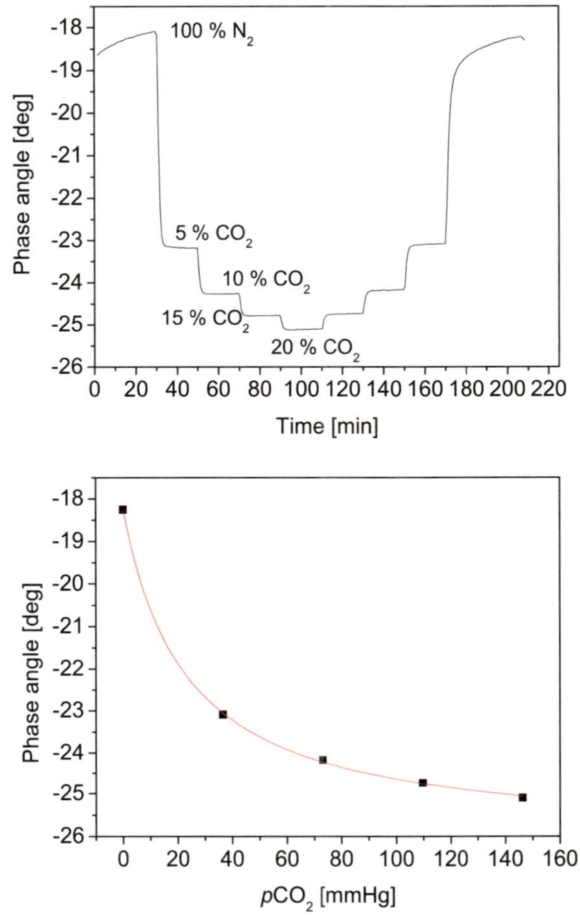

Fig. 12. Carbon dioxide response curve, phase angle as a function of different CO_2 concentrations; below: calibration curve obtained by the first cycle of dissolved CO_2 concentrations.

3.7 *In vivo* pO_2 and pCO_2 measurements on animal by use of fiber optical sensors

In vivo experiments on test animals have been performed at the Medical Univeristy Graz, Clinical Department of Heart Surgery. Fourteen pigs (30.7 $\pm$ 2.5 kg) were investigated by monitoring the pO_2 and pCO_2 content in tissue of the three vital organs: brain, heart and liver in order to examine the oxygen delivery to the tissue of these organs during simulated heart surgery operations. Heart surgery intervention is necessary to heal and treat cardiac defect or valvular heart diseases. During this kind of surgeries an extracorporeal blood circulation can be carried out using the so-called "heart lung machine" (HLM) which temporarily takes over the functions of the heart and lungs during surgeries, maintaining the circulation of blood in the body. A heart lung machine maintains perfusion to other body organs and tissues while the surgeon works in a bloodless surgical field.

The aim of test on animals was the simultaneously evaluation of pO_2/pCO_2 exchanges in three vital organs: brain, heart and liver using two different methods of extra corporal circulation: conventionally cardiopulmonary bypass (CPB) and Minimal Extra Corporal Circulation (MECC) (see Fig. 13). The questions that appeared were: are the optical fiber sensors suitable for the evaluation of metabolic variables and organ specific O_2/CO_2 exchanges during heart surgery and furthermore, is it on the basis of the obtained results capable to estimate difference between CPB and MECC?

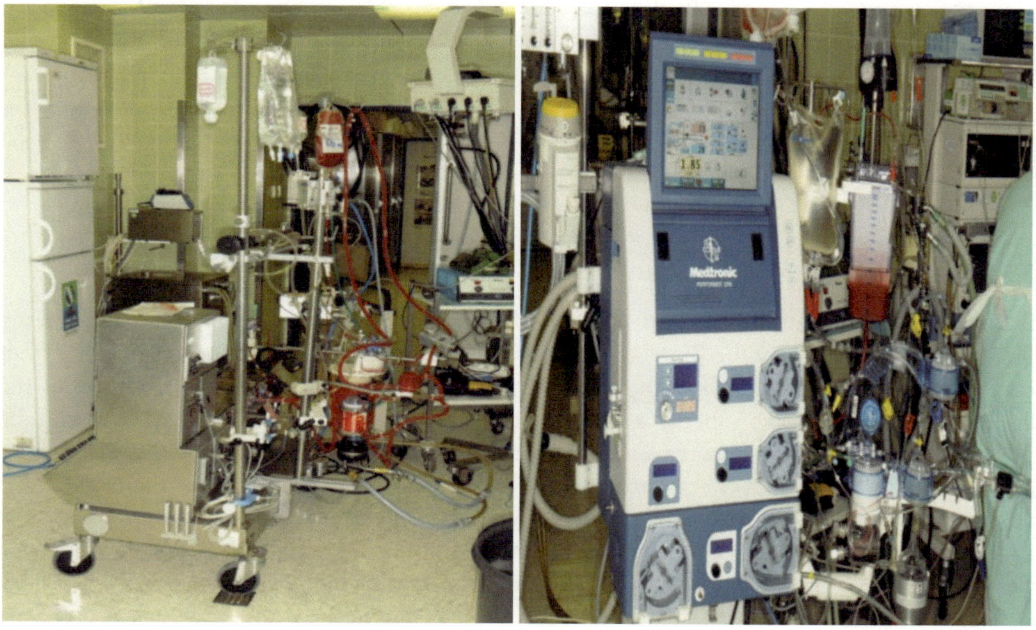

Fig. 13. CPB (left) and MECC (right)

3.7.1 Experimental protocol

pO_2 and pCO_2 have been measured continuously with fiber optical sensors previously calibrated in laboratory and checked again before the *in vivo* measurements, directly at the place of the in vivo measurements. In order to protect the fragile tip of the optical fibers and to position the two type of sensors as near as possible to each other at the local place of the measurement, each sensor (one pO_2 and one pCO_2) was inserted in the needle (inner diameter of 500 mm) and then implanted into the tissue of the organ. Firstly, a drill hole has been made in temporal region of the skull above ear in order to insert a needle containing the optical pO_2 and pCO_2 sensors deeply into the brain tissue. In this region is the skull of pigs thinnest due to the thickness of bones. Afterwards was the thorax and abdomen opened to insert the optical sensors into the liver and finally into heart tissue. After baseline measurements (recording of the first blood values and the first values obtained with the optical sensors) the aorta was cross clamped for a period of 60 min and the HLM was put into operation. The blood values and the tissue values of pO_2 and pCO_2 have been taken at intervals of time of 15 minutes. After one hour of operation the clamps of the HLM have been removed and the heart has been reanimated for 30 minutes. The physical state of the animal has been monitored another half of hour before the experiment ended. The exact study protocol is shown in Table 2:

Time	Event
T_1	Baseline before connection of HLM under physiologic conditions
T_2	During cooling on HLM after 15 min (then cross clamping of aorta)
T_3	15 min after cross clamping the aorta
T_4	30 min after cross clamping the aorta
T_5	45 min after cross clamping the aorta (then rewarming)
T_6	60 min after cross clamping the aorta (then de clamping of aorta)
T_7	15 min after normothermic reperfusion
T_8	30 min after normothermic reperfusion (off-HLM)
T_9	15 min after end of using of HLM
T_{10}	30 min after end of HLM (end of experiment)

Table 2. Protocol of *in vivo* experiments

3.7.2 Results of *in vivo* experiments

Since the brain tissue perfusion is one of the most important substrate for energy metabolism delivered by the blood, an example of pO_2 and pCO_2 profiles measured by fiber optical sensor in this organ are shown in Fig. 14. Despite the very low $pO2$ values at the beginning of the measurement, a slightly decrease in pO_2 value during HLM can be observed, followed by an increase up to 55 mmHg during recirculation. On the contrary, pCO_2 increases up to 40 mmHg during HLM and after recirculation reaches the same value as at the beginning (10 mmHg). These measurements confirmed a heterogeneous distribution of local pO_2 and pCO_2 values in brain. The tissue concentration is determined by the balance between supply and utilisation and the anatomical distribution of concentration and within the brain was reported to be wide, depending on the depth of penetration of the sensor into the tissue (Luebbers & Baumgaertl, 1997). The pO_2 and pCO_2 responses as shown in Fig. 14 have an opposite trend which is in agreement with medical point of view that anaerobic condition of tissue correlate with increase of CO_2 content.

The supply of brain with oxygen depends on the cerebral perfusion and on difference between arterial and venous O_2 partial pressures. CO_2 partial pressure has a significant influence on the blood circulation in the brain. When the arterial CO_2 partial pressure p_aCO_2 increases, pCO_2 in brain increases proportional with this increment. If on the contrary Hyperventilation causes a decrease of pCO_2 down to 20 mmHg, than a cerebral Hypoxia and increased cerebral Lactacidose is provoked due to the vasoconstriction.

Despite the fact that brain are very active organ with higher oxygen and energy demand, low pO_2 values below 10 mmHg has been often measured. Such sensor behavior could be obtained due to the influence of the narcosis on the brain activity and due to the position of the sensor tip in brain. Latter can result in holes by implantation of the optical sensor probes into brain region since the brain consists of the tissue and the hollow spaces. As a consequence of the positioning of the optical fiber in hollow spaces, lower pO_2 values can be obtained during measurements in brain.

Fig. 15 and 16 show oxygen and carbon dioxide profile in liver and heart, respectively. In vivo measurements in liver reflected lower pO_2 values before HLM was put into an operation (pO_2 values app. 5 mmHg) and afterwards a decrease of pO_2 vales in the range of 1.5 mmHg. After corporal recovery, the pO_2 values reach the value at the beginning of the measurement. Similarly, pCO_2 values in liver are at beginning of 8 - 10 mmHg, and during HLM operation an increase in pCO_2 has been noticed (app. 13 mmHg). After recirculation pCO_2 reaches the values similar at the beginning of the experiment (8 mm Hg).

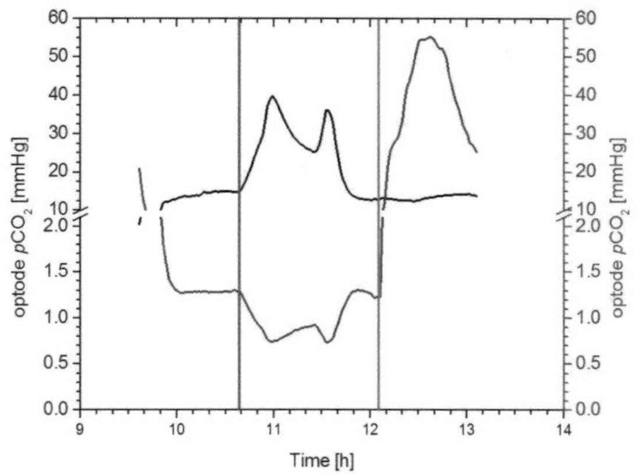

Fig. 14. *In vivo* measurement in brain obtained with inserted pO_2 and pCO_2 fiber optical micro sensors: red line indicates the start of functionality of HLM, while green line indicates the return to the normal intra-corporeal circulation. The x scale shows the real time of the measurement, so it can be seen that the experiment has started at about 9:30 am and ended few minutes after 13:00 for an overall duration of less than 4 hours.

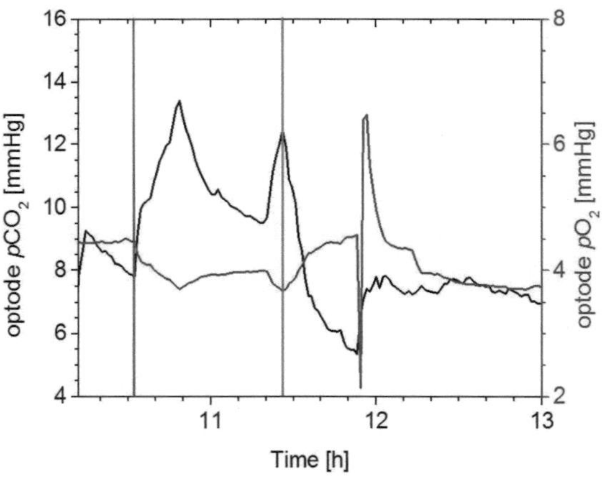

Fig. 15. *In vivo* measurement in liver obtained with inserted pO_2 and pCO_2 fiber optical micro sensors: red line indicates the start of functionality of HLM, while green line indicates the return to the normal intra-corporeal circulation.

For the comparison of all performed *in vivo* experiments on animals using either CPB or MECC as extracorporeal circulation system the measured pO_2 and pCO_2 values were statistically arranged with a two way analysis of variance (ANOVA) model. The measured parameters were normalized to the initial values at a level of 100 % due to the high variances of initial values of those variables measured with optical sensors (pO_2, pCO_2). Those results are shown in Fig. 17 to 19 for the brain, liver and heart, respectively.

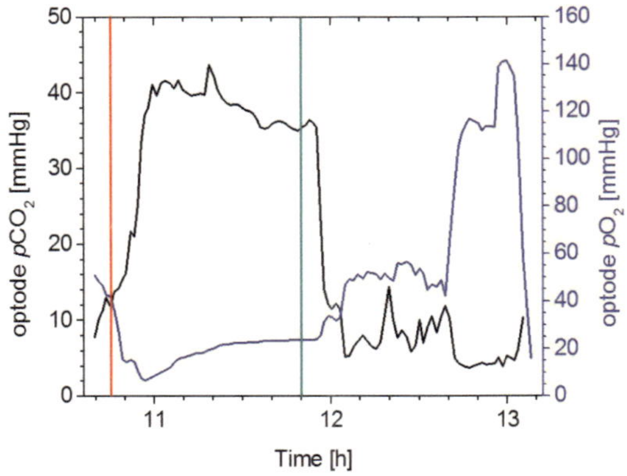

Fig. 16. *In vivo* measurement in heart obtained with inserted pO_2 and pCO_2 fiber optical micro sensors: red line indicates the start of functionality of HLM, while green line indicates the return to the normal intra-corporeal circulation.

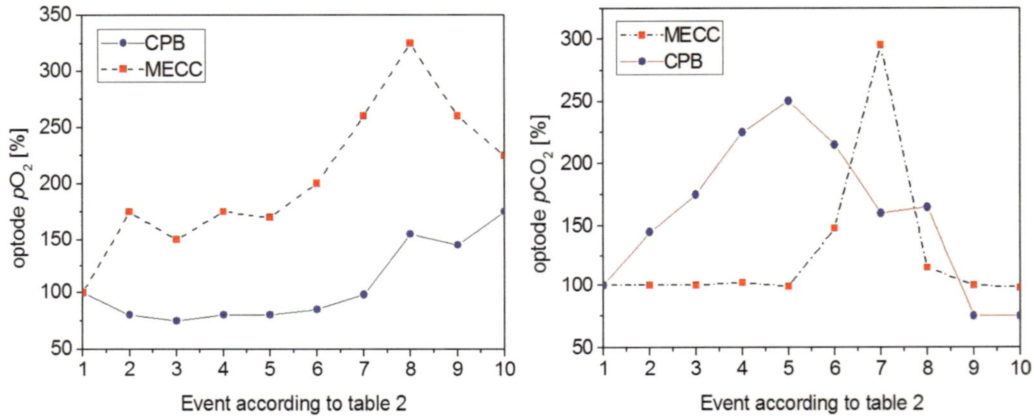

Fig. 17. Comparison of the values measured with optical fibers and normalized to the initial values at a level of 100 % for *in vivo* measurements in brain.

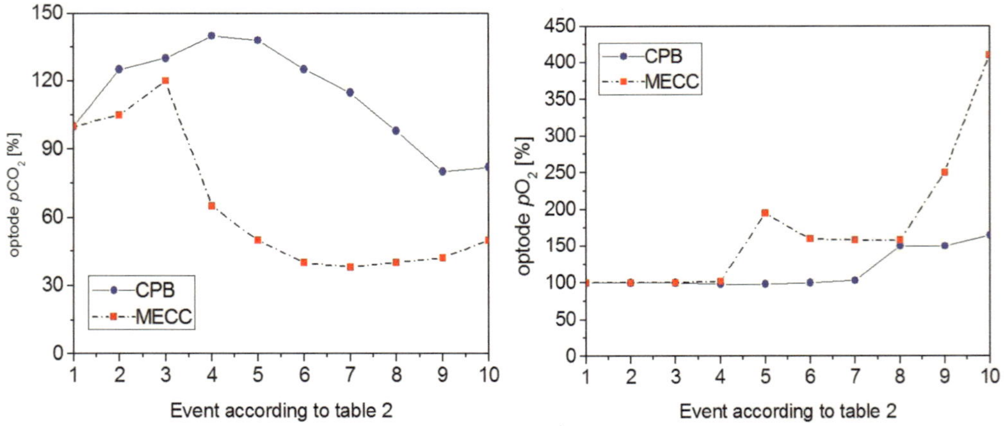

Fig. 18. Comparison of the values measured with optical fibers and normalized to the initial values at a level of 100 % for *in vivo* measurements in liver.

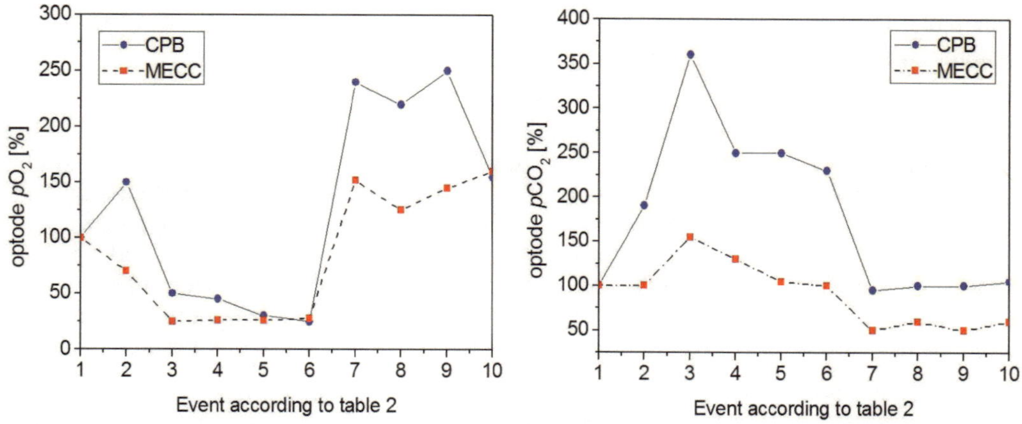

Fig. 19. Comparison of the values measured with optical fibers and normalized to the initial values at a level of 100 % for *in vivo* measurements in heart.

The above Figures indicate on the similar trends of pO_2 and pCO_2 curves for both extracorporeal circulation systems. It is shown that under standardised conditions pO_2 curves in brain were absolutely higher in the case of mini-ECC. In the case of liver it was found that under standardised conditions intrahepatic parenchymal pCO_2 curves were absolutely higher and they are significantly correlated to arterial and central venous lactate in CPB.

3.8 Conclusions

In the presented work a novel combined optochemical measurement system which allows a simultaneously measurement of dissolved oxygen and carbon dioxide in tissue of vital organs is described. The tissue oxygenation is frequently monitored with electrochemical micro sensors (e.g. polarographic needle electrodes) which one limitation is that due to the oxygen consumption during pO_2 measurements, the probe needs to be moved through the tissue (Lowry & Fillenz, 2001; Luebbers, 1996, Nair, 1987). On the contrary fiber optic probe can be left in the same position in the tissue for a prolonged period of time because is non-invasive. This makes it possible to directly temporal changes in pO_2 as well as the oxygenation-modifying interventions over time in a tissue. Furthermore, optochemical sensors based on luminescence intensity show higher sensitivity and the use of phase modulation fluorometry as measuring techniques for both oxygen and carbon dioxide sensors enable the realisation of optical sensors with excellent characteristics of reliability and measurement stability possible superior to those of electrodes. Further advantage of optical sensors over the electrodes is ease of miniaturisation. In presented work tapered optical fiber tips obtained through chemically etching improved the signal-to-noise ratio of about one order of magnitude. Based on this fact further miniaturisation can be achieved, using a fiber optical cables with smaller diameter (e.g. 50 μm). This type of the optical sensors also offers a possibility to use them for endoscopy. Furthermore, the described optical sensors are almost calibration free using only one point recalibration.

Despite the fact, that the last years are witnessed to an increased number of optochemical oxygen sensors based on phase modulation fluorometry in field of medical applications, especially for analysing the spatial heterogeneity in oxygen tension in tumor tissue using optical oxygen probes OxyLite (Oxford Optronix, Oxford, UK), at present and up to our knowledge there has been no reports about fiber optical CO_2 sensors for measuring pCO_2 content in tissue of organs.

The described optochemical instrumentation has been used in in vivo tests for simultaneously detection of oxygen and carbon dioxide exchange in brain, heart and liver using two different methods as extra corporal circulation: Minimal Extra corporal Circulation (MECC) and Cardiopulmonary bypass (CPB). It has been turned out that a significantly higher demand of blood transfusion is required and higher lactate values (correlated to higher pCO_2 values) are observed when using classical method CPB. The continuous acquisition of metabolic parameters (pO_2 and pCO_2) has showed that in the case of using MECC as extra corporal circulation, significantly higher pO_2 values are obtained, while the use CPB resulted in significantly higher pCO_2 in liver and heart tissue.

It is demonstrated that with a help of miniaturised non-invasive fiber optical sensors for continuously measurement of tissue pO_2 and pCO_2 a significant better O_2/CO_2 metabolism is achieved in the case of MECC as extra corporal circulation. Furthermore, MECC produced less lactate and showed significantly favourable results concerning O_2/CO_2 metabolism and transfusion demand.

4. Acknowledgement

The research work was part of the Zukunftsfonds funded by the Styria Country Government A3 Science and Research, Reference A3-22M1/206-4. The aim of this project was to develop a combine optochemical measurement system for determination of oxygen and carbon dioxide for medical applications.

The tests on animals have been made in cooperation with Univ. Prof. Dr. Igor Knez and Dr. med. univ. Elisabeth Beran at the Medical Univeristy of Graz, Clinical Department for Heart Surgery (Austria).

5. References

Astrup, P. & J. Severinghaus, W. (1986). *The Hystory of blood gases, Acids and Bases*, Munksgaard, Copenhagen, 1st ed.

Bizzarri, A.; Koehler, H.; Cajlakovic, M.; Pasic, A.; Schaupp, L.; Klimant, I. & Ribitsch, V. (2006). Continuous oxygen monitoring in subcutaneous adipose tissue using microdialysis. *Analytica Chimica Acta*, Vol. 573-574, (July 2006) pp. 48-56.

Bizzarri, A.; Konrad, C.; Čajlaković, M. & Ribitsch, V. (2007). New developments of an optochemical measurement system for the continuous monitoring in subcutaneous tissue by microdialysis. *Proceedings IEEE Sensors 2007*, (October 2007), Atlanta, USA.

Cajlakovic, M.; Bizzarri, A. & Ribitsch, V. (2006). Luminescence lifetime-based carbon dioxide optical sensor for optical applications. *Analitica Chimica Acta*, Vol. 573-574, (July 2006) pp. 57-64.

Cajlakovic, M.; Bizzarri, A.; Suppan, M.; Konrad, C. & Ribitsch, V. Continuous monitoring of pO_2 and pCO_2 by microdialysis indicates physiologic status of the critically ill patients. (2009). *Sensors and Actuators B*, Vol. 139, Issue 1 (May 2009) pp. 181-186.

Chang, P. P.; Sussman, M. S.; Conte, J. V.; Grega, M. A.; Schulman, S. P.; Gerstenblith, G.; Wang, N. Y.; Capriotti, A.; Weiss, J. L. (2002) Post-operative ventricular function, and cardiac enzymes after on-pump versus off-pump CABG surgery. *American Journal Cardiology*, Vol. 89 (2002) pp.1107–1110.

Cooney, C. G. & Towe, B. C. (1997). Intravascular carbon dioxide monitoring using micro-flow colorimetry. *Biosensors and Bioelectronics*, Vol. 12, No.1, (January 1997) pp. 11.

Cooney C. G.; Towe, B. C. & Eyster, C. R. (2000). Optical pH, oxygen and carbon dioxide monitoring using a microdialysis approach. *Sensors & Actuators* , Vol. 69 (September 2000), pp. 183.

Goossens, G. H.; Bizzarri, A.; Venteclef, N.; Essers, Y.; Cleutjens, J. P.; Konings, E.; Jocken, J. W. E.; Cajlakovic, M.; Ribitsch, V.; Clément, K & Blaak E. E. (2011). Adipose tissue hyperoxia is accompanied by insulin resistance, impaired adipose tissue capillarisation and inflammation in obese men. *Circulation*, (July 2011) pp. 1-10.

Goossens, G. H; Saris W. H. M. & Blaak E.E. Microdialysis on Adipose Tissue. In: Tsai, T-H, ed. Applications of Microdialysis in Pharmaceutical Science. Hoboken, NJ: John Wiley & Sons, Inc., *In press*..

Hahn, C. E. W. (1998). Electrochemical Analysis of Clinical Blood-gases, Gases and Vapours. *Analyst*, Vol. 123, (June 1998), pp. 57R-86R.

Harsten, A.; Berg, B.; Inerot, S. & Muth, L (1988). Importance of correct handling of samples for the results of blood gas analysis. *Acta Anaesthesiologica Scand*inavica, Vol. 32 (1988), pp. 365-368.

Hou, H.; Grinberg, O. Y.; Taie, S.; Leichtweis, S.; Miyake, M.; Grinberg, S.; Xie, H.; Csete, M. & Swartz, H. M. (2003). Electron Paramagnetic Resonance Assessment of Brain Tissue Oxygen Tension in Anesthetized Rats. *Anesthesia & Analgesia*, Vol. 96, (May 2003), pp. 1467–1472.

Immer, F. F.; Pirovino, C.; Gygax, E.; Englberger, L.; Tevaearai, H.; Carrel, T. P. (2005). *Minimal versus Conventional cardiopulmonary bypass: assessment of intraoperative myocardial damage in coronary bypass surgery*, Vol. 28, Nu. 5 (November 2005) pp. 701-704.

Kamiya, H.; Kofidis, T.; Haverich, A. & Klima. (2006). Preliminary experience with the mini-extracorporeal circulation system (Medtronic resting heart system). Interactive Cardio Vascular and Thoracic Surgery, Vol. 5, No. 6 (August 2006) pp. 680-682.

Kofidis, T.; Baraki, H.; Singh, H.; Kamiya, H.; Winterhalter, M. & Didilis. (2008). The minimized extracorporeal circulation system causes less inflammation and organ damage. *Perfusion*, Vol. 23, No. 3 (May 2008) pp. 147-151.

Lakowicz, J.R. (1983). Principles of Fluorescence Spectroscopy, Plenum Press, New York, London, 1983, pp. 257.

Levy, J.H.; Tanaka, K. A. (2003) Inflammatory response to cardiopulmonary bypass. *Annual Thoracic Surgery*, Vol. 75 (2003) pp. 715—720.

Lowry, J. P. & Fillenz, M. (2001). Real-time monitoring of brain energy metabolism in vivo using microelectrochemical sensors: the effects of anesthesia. *Bioelectrochemistry*, Vol. 54, (August 2001) pp. 39-47.

Luebbers, D. W. (1996). Oxygen electrodes and optodes and their application in vivo. *Advances in Experimental Medicine and Biology*, Vol. 388, (October 1996) pp.13-34.

Luebbers, D. W. & Baumgaertl, H. (1997). Heterogenities and profiles of oxygen pressure in brain and as Examples of the pO_2 distribution in the living tissue. *Kidney International*, Vol. 51, (February 1997) pp. 372-380.

Maldonado, F. A; Weil, M. H.; Tang, W.; Bisera, J.; Gazmuri, R.; Johnson, B. & D´Alessio A. (1993). Myocardial Hypercarbic Acidosis educes Cardiac Resuscitability. *Anesthesiology*, Vol. 78, No.2, (February 1993), pp. 343-352.

Mills, Q.; Chang, A. & McMurray, N. (1992). Equilibrium study on colorimetric plastic film sensors for carbon dioxide. *Analytical Chemistry*, Vol. 64, No. 13 (July 1992) pp. 1383-1389.

Nair, P. K.; Buerk, D. G. & Halsey, J. H. (1987). Comparison of oxygen metabolism and tissue pO_2 in cortex and hippocampus. *Stroke*, Vol. 18, No. 3 (May-June 1987) pp. 616-622.

Pasic, A.; Koehler, H., Schaupp, L.; Pieber, T. R. & Klimant, I. Fiber-optic flow-through sensor for online monitoring of glucose. (2006). *Analytical and Bioanalytical Chem*istry, Vol. 386, No. 5, (October 2006) pp. 1293.

Pentila,¨ H. J; Lepoja¨rvi, M. V. K.; Kiviluoma, K. T.; Kaukoranta, P. K.; Hassinen, E.; Peuhkurinen, K.J. (2001) Myocardial preservation during coronary surgery with and without cardiopulmonary bypass. *Annual Thoracic Surgery*, Vol. 71 (June 2001) pp. 565–71.

Rosser, D. M.; Stidwill, R. P.; Jacobson, D. & Singer, M. (1995). Oxygen Tension in the Bladder Epithelium Rises in Both High and Low Cardiac Output Endotoxemic sepsis. *Journal of Applied Physiology*, Vol. 79 (December 1995), pp. 1878-1882.

Rosser, D. M.; Stidwill, R. P.; Jacobson, D. & Singer M. (1996). Cardiorespiratory and Tissue Oxygen Dose Response to Rat Endotoxemia. *Heart and Circulatory Physiology*, Vol. 271, (September 1996), pp. H891- H895.

Suzuki, H.; Arakawa, H.; Sasaki, S. & Karube, I. (1999). Micromachined Severinghaus-type carbon dioxide electrode. *Analytical Chemistry*, Vol.71 (May 1999), pp. 1737-1743.

Suzuki, H., Hirakawa, T.; Hoshi, T. & Toyooka, H. (2001). Micromachined sensding module for pO_2, pCO_2 and pH and its design optimization for practical use. *Sensors and Actuators B*, Vol. 76 (June 2001), pp. 565-572.

Tang, W.; Weil, M.; Gazmuri, R.; Bisera, J. & Rackow (1991). Reversible Impairment of Myocardial Contractility due To Hypercarbic Acidosis in the Isolated Perused Rat Heart. *Critical Care Medicine*, Vol. 19, No.2, pp. 218.

Tsukada, K.; Sakai, S.; Hase, K. & Minamitani, H. 82003). Development of catheter-type optical oxygen sensor and applications to bioinstrumentation. *Biosensors and Bioelectronics*, Vol. 18 (October 2003), pp. 1439-1445.

Tusa, J. K. & Leiner, M. J. (2001) Fluorescent optical sensor for critical care analytes. *Annals de Biologie Clinique*, Vol. 61, (September 2001), pp. 183-191.

Ungerstedt, U. (1991). Microdialysis-principles and applications for studies in animals and man. *Journal of Internal Medicine*, Vol. 230, No. 4, (October 1991) pp. 365.

Vollmar, B.; Conzen, P. F.; Kerner, T.; Habazettl, H.; Vierl, M.; Waldner, H. & Peter, K. (1992). Blood Flow and Tissue Oxygen Pressures of Liver and Pancreas in Rats: Effects of Volatile Anesthetics and of Hemorrhage. *Anesthesia & Analgesia*, Vol. 75, (September 1992), pp. 421-430.

Whitehouse, T.; Stotz, M.; Taylor, V.; Stidwill, R.& Singer, M. (2006). Tissue Oxygen and Hemodynamics in Renal Medulla, Cortex and Corticomedullary Junction During Hemorrhage-Reperfusion. *American Journal of Physiology - Renal Physiology*, Vol. 291, (March 2006), pp. F647-F653.

Wolfbeis, O. S.; (1991). From fluorescent probes to optical sensors, *Analytical Proceeding*, Vol. 28 (1991), pp. 357.

Ye J.; Gao Z.; Yin J.; He Q.(2007). Hypoxia is a potential risk factor for chronic inflammation and adiponectin reduction in adipose tissue of dietary obese mice. *American Journal of Physiology*, Vol. 293, No. 4, E1118-1128.

5

Chemical Sensors Based on Photonic Structures

Vittorio M. N. Passaro, Benedetto Troia,
Mario La Notte and Francesco De Leonardis
*Photonics Research Group, Dipartimento di Elettrotecnica ed Elettronica,
Politecnico di Bari,
Italy*

1. Introduction

Photonic sensors have been the subject of intensive research over the last two decades for use in civil and military environments, especially for detection of a wide variety of biological, chemical and nuclear agents. Photonic sensor technologies involve a lot of application fields like chemical, temperature, strain, biomedical, electrical, magnetic, rotation, pressure, position, acoustic and vibration sensors. Important efforts have been carried out by the international scientific community (academia, industry-R&D and all interested parties), to develop and improve the know-how and the state-of-the-art of photonic sensing. In this context, optical Lab-on-a-chip systems, based on chemical and biochemical sensors, represent the state of the art of photonic sensing, since they are expected to exhibit higher sensitivity and selectivity as well as high stability, immunity to electromagnetic interference, and product improvements such as smaller integration sizes and lower cost (De Leonardis et al., 2007).

In recent years, rapid advancements in photonic technologies have significantly enhanced the photonic biochemical sensor performance, particularly in the areas of light - analyte interaction, device miniaturization and multiplexing, and fluidic design and integration. This has led to drastic improvements in sensor sensitivity, limit of detection, advanced fluidic handling capability, lower sample consumption, faster detection time, and lower overall detection cost per measurement. This trend is not a casual phenomenon, indeed it justifies the economic interest that many industries reveal to photonic chemical and biochemical sensors. With future commercialization of photonic biosensors in lab-on-a-chip systems, next generation biosensors are expected to be reliable and portable, able to be fabricated with mass production techniques to reduce the cost as well as to do multi-parameter analysis, enabling fast and real-time measurements of a large amount of biologic parameters within a single, compact sensor chip. For example, needs are expected to be boosted by healthcare, such as the increasing prevalence of diabetes in the population and the growing demand for home and point-of-care testing and monitoring tools.

In conclusion, optical biosensors (Passaro et al., 2007a) have reached a high degree of maturity in crucial areas of application such as environmental monitoring, biotechnology, medical diagnostics, drug screening, food safety (Arshak et al., 2009), and security (Leheny & McCants, 2009).

2. Photonic bio-chemical sensors: Sensing principles and architectures

A chemical sensor can be defined as an analytical device that coverts chemical or bio-chemical information (e.g., concentration, composition analysis), into a quantifiable and processable signal.

One of the possible chemical sensor classifications, concerns with the principle of the transducer, i.e. the device that transforms the chemical information about the sample (chemical analytes, molecules, cells or gases), into an analytical signal. To this purpose, electrochemical, electrical, mass sensitive, magnetic, thermometric and optical sensors can be classified as different types of chemical sensors.

A photonic chemical sensor is characterized by an optical-based transduction. This one includes a large number of sensing principles based on absorption, reflection, refraction, dispersion, Raman effect, chemiluminescence, fluorescence and phosphorescence, to name a few. The general scheme of a photonic chemical sensor can be represented as in Fig. 1 below.

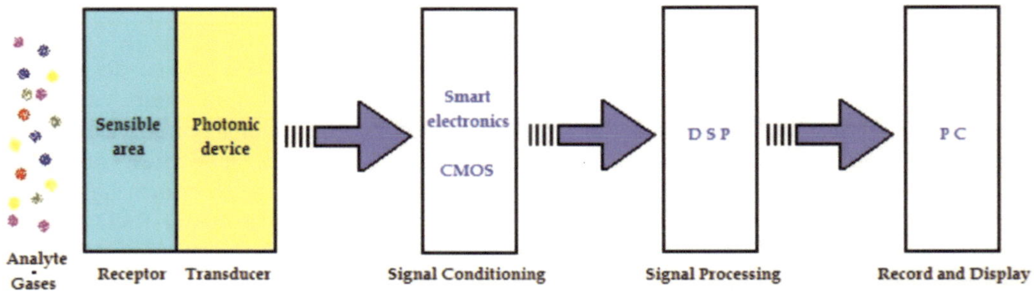

Fig. 1. Schematic diagram of a photonic bio-chemical sensor.

Colored blocks named "Sensible area" and "Photonic device", represent the receptor and the transducer of the photonic biochemical sensor, respectively.

The receptor is a fundamental part of the sensor. In fact, it has to be designed in order to catalyze the specific chemical reaction or biochemical process, exhibiting a high selectivity for the analyte or the chemical specie to be detected. Generally, the receptor can be realized by covering the sensor surface (sensible area), with polymeric layers (~ nm-thin). These ones are characterized by selective receptors, able to capture and immobilize only one type of analyte in a complex chemical sample (e.g., glucose in blood, metallic particles in water, antigen/antibody in organic solution). By this way, the sensible area is the direct maker of the sensor selectivity and can be defined as a chemical prerogative.

In an intrinsic photonic bio-chemical sensor, the transducer is generally represented by an integrated optical waveguide (e.g., optical fiber, slot-waveguide, photonic crystal waveguide, photonic wire waveguide). By this way, this photonic device has to satisfy different parallel functions. The first one consists in guiding the photonic signal from the optical source (e.g., led, laser), to the sensible area. Consequently, the photonic waveguide has the rule of transducer, enhancing the interaction between the chemical/biochemical process and the optical signal. By this way, the chemical information can be properly transduced into an optical one and, finally, guided to the photo-detector to be transformed into an analytically useful electrical signal. In this context, the transducer can be considered as an engineering prerogative, because several technical solutions have to be implemented in order to maximize the confinement of the optical field in the sensible area.

The remaining blocks labeled as "signal conditioning" and "signal processing", represent the sensor front-end, generally identified by integrated digital/analog CMOS electronics. The electrical signal obtained by the conversion of the photonic one, has to be restored by amplification and filtering, then processed in order to decode the analyte (chemical) information to be recorded and displayed ("record and display").

Intrinsic photonic biosensors are generally integrated chips, characterized by small footprints (~ mm^2, ~ μm^2). On the contrary, in extrinsic photonic biochemical sensors the waveguide does not directly interact with the chemical/biochemical process. In fact, the photonic waveguide cannot represent the transducer because it is designed only to able the optical signal propagation. Extrinsic photonic sensors can be classified as "benchtop" systems. They are expensive and not suitable for mass-scale production. In addition, they are characterized by large dimensions and low portability.

Nowadays, integrated photonic chemical sensors represent the state-of-the-art of photonic sensing. In fact, these sensing systems exhibit ultra high performance and can be realized by standard fabrication processes available in Microelectronics industry (e.g., photolithography and ICP etching). Consequently, additional start-up costs to be invested in clean room equipments updating are not needed, and mass-scale production can represent a concrete business.

The most popular technological platform adopted for the fabrication of integrated photonic sensors is called silicon-on-insulator (SOI). This one works by placing a thin, insulating layer, such as silicon dioxide or glass, between a thin layer of silicon and the silicon substrate. In Fig. 2 a typical SOI rib optical waveguide designed at 1.55 µm (operative wavelength, λ_{op}) is shown. Both quasi-TE and quasi-TM optical field distributions are plotted.

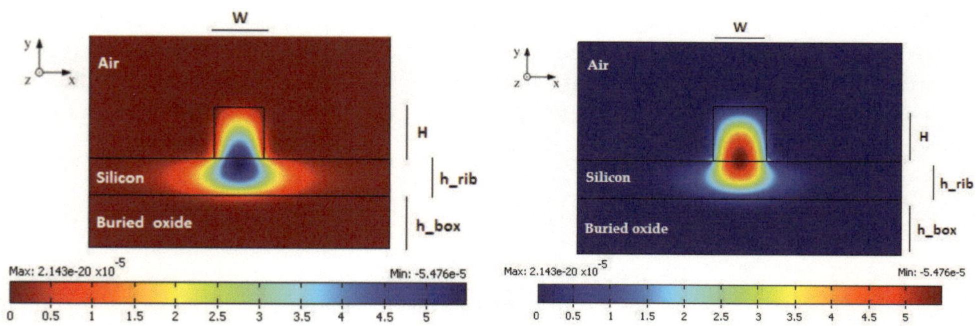

(a) (b)

Fig. 2. Optical (a) E_x-field (quasi-TE) and (b) E_y-field (quasi-TM) distributions in a SOI rib waveguide (λ_{op} = 1.55µm, W = 1µm, H = 1µm, h_rib = 700nm, h_box = 1µm).

In particular, it is possible to recognize the insulating layer (SiO_2) characterized by a refractive index n_{SiO2} = 1.45, and built on the silicon substrate (bulk).

A thin silicon layer (n_{Si} = 3.45 @ 1.55 µm) is properly etched in order to realize the rib waveguide, covered by air (n_{air} = 1). The architecture presented above exhibits a good optical transverse electric (TE) and transverse magnetic (TM) field confinement in the rib structure, directly exposed to the cover medium and characterized by the highest refractive index n_{Si}. However, in both quasi-TE and quasi-TM modal distributions, the optical field is quasi entirely confined in the high index region, reducing the interaction with analytes or gases in the cover medium. To this purpose, other SOI-based waveguide architectures, such as the SOI photonic wire waveguide sketched in Fig. 3, can be adopted in order to improve the field confinement in the cover medium, as well as the interaction between the propagating optical field and the chemical/biochemical species, then resulting in an optimized transduction process.

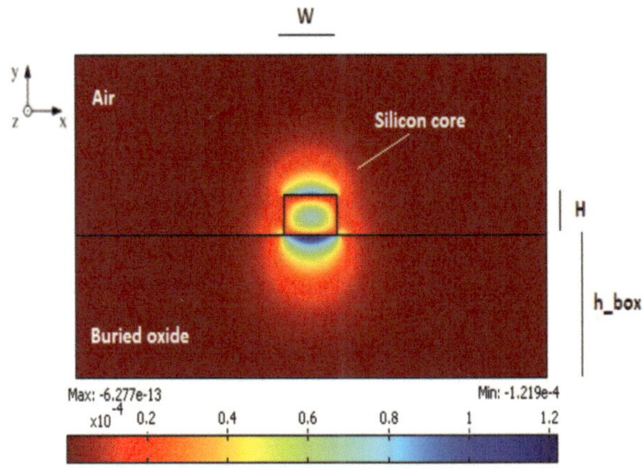

Fig. 3. Optical E_y-field distribution in a silicon-wire waveguide (quasi-TM), $\lambda_{op} = 1.55\mu m$.

In this specific waveguide configuration (h_box = 2μm, W = 450nm, H = 250nm), the optical transverse magnetic (TM) field is concentrated at both interfaces between the high refractive index region (silicon core) and the low refractive index ones (air at the upper side and SiO_2 at the bottom one). By this way, photonic wire waveguides represent a better choice if compared with previously analyzed rib ones. For an electromagnetic wave propagating in the z direction (Fig. 3), the major E-field component of the quasi-TM eigenmode undergoes a discontinuity at the horizontal silicon wire interfaces that, according to Maxwell's equations, is determined by the following relation:

$$\left|\frac{E_L}{E_H}\right| = \left(\frac{n_H}{n_L}\right)^2 \quad (1)$$

In Eq. 1 E_L is the component of the E-field, evaluated in the low refractive index (n_L) region, while E_H is the same component in the high refractive index (n_H) region. In particular, the greater the refractive index contrast $\Delta n = n_H - n_L$, the higher the E_L-field confinement. In this context, SOI technology represents a suitable solution for photonic sensing because it is possible to obtain a high refractive index contrast Δn ($n_{Si} - n_{SiO2} \approx 2$ @ 1.55 μm).

By using one of the two SOI-based waveguides described in the section above, it is possible to realize an intrinsic photonic biosensor. In particular, two different sensing principles characterizing the operating regime of a photonic chemical sensor can be used, i.e. surface and homogeneous sensing (Dell'Olio & Passaro, 2007).

In surface sensing a change of the modal effective index n_{eff} of the propagating optical signal is due to a change of thickness of an ultra-thin layer of selective receptor molecules which are immobilized on the functionalized waveguide surface. In particular, when the photonic sensor surface is exposed to a complex chemical solution, only target molecules will be recognized by selective receptors and will contribute to the bio-chemical process. By this way, if the thickness of the sensor surface layer was ρ_0 before the exposure to the chemical sample, the thickness of the same layer will be greater, for example $\rho_0 + \rho_1$, after the selective analyte adsorption. The thickness change $\Delta\rho$ can be transduced in an effective index change Δn_{eff}.

In homogenous sensing, the modal effective index change is produced by a change of cover medium refractive index, n_c. For example, by assuming that the photonic sensor is initially exposed to air ($n_{air} = 1$ @ 1.55μm), when a specific gas (e.g., He, CO_2, Ar, N_2, C_2H_2) will cover the sensor surface, a sensible cover refractive index change Δn_c will be induced. In addition, the same analysis can be carried out if the sensor is initially covered by water ($n_{water} = 1.33$ @ 1.55μm) and subsequently exposed to another liquid solution (e.g., NaCl). In Table 1 it is possible to appreciate the relative effective index changes under different operative conditions (De Leonardis et al., 2011).

RI Cover medium	RI Gas/Liquid	Δn_C (%)
$n_{air} = 1$	$n_{He} = 1.000035$	0.0035
$n_{air} = 1$	$n_{CO2} = 1.000059$	0.0059
$n_{air} = 1$	$n_{Ar} = 1.000278$	0.0278
$n_{air} = 1$	$n_{N2} = 1.000294$	0.0294
$n_{air} = 1$	$n_{C2H2} = 1.000593$	0.0593
$n_{water} = 1.33$	$n_{NaCl} \approx 1.33$	0.0018[1]

Table 1. Refractive index (RI) changes in homogeneous sensing for different gas and aqueous solution concentrations. All data are referred to $\lambda_{op} = 1.55$μm.

Integrated silicon wire waveguides define a class of photonic chemical sensors called "photonic wire evanescent field" (PWEF), because of the interaction between the evanescent optical field and the chemical/biochemical process to be detected (see Fig. 4).

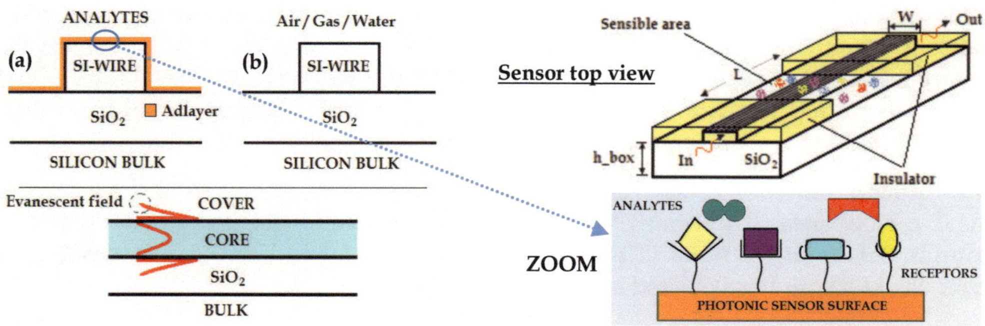

Fig. 4. Schematic representation of surface (a) and homogeneous (b) sensing in a SOI-based wire waveguide.

In both homogeneous and surface sensing it is possible to define an important performance parameter of the photonic chemical /biochemical sensor: the sensitivity.

[1] The refractive index of a NaCl aqueous solution changes 0.0018 RIU per 1% mass concentration (Sun et al., 2009).

$$S_h = \frac{\partial n_{eff}}{\partial n_c}, S_s = \frac{\partial n_{eff}}{\partial \rho} \tag{2}$$

In homogeneous sensing, according with variational theorem for dielectric waveguides, it is possible to write:

$$S_h = \left.\frac{\partial n_{eff}}{\partial n_c}\right|_{n_C = n_C^0} = \frac{2n_C^0}{\eta_0 P} \iint_C \left|\vec{E}(x,y)\right|^2 dxdy = \frac{2n_C^0 \Gamma_C}{\eta_0 P} \iint_\infty \left|\vec{E}(x,y)\right|^2 dxdy \tag{3}$$

where

$$P = \iint_\infty \left[\left(\vec{E} \times \vec{H}^* + \vec{E}^* \times \vec{H}\right) \bullet \hat{z}\right] dxdy \tag{4}$$

η_0 is the free space impedance (377Ω), $\vec{E}$ and $\vec{H}$ are the electric and magnetic field vectors, respectively, n_c^0 is the unperturbed value of cover medium refractive index, $\hat{z}$ indicates the unit vector along z direction (propagation direction) and Γ_C is the confinement factor in the cover medium (Dell'Olio & Passaro, 2007).

In surface sensing, shift of n_{eff} (Δn_{eff}) due to a change $\Delta\rho$ of ad-layer thickness can be calculated by using a perturbation approach as:

$$\Delta n_{eff} = \frac{n_m^2 - \left(n_C^0\right)^2}{\eta_0 P} \iint_M \left|\vec{E}(x,y)\right|^2 dxdy \tag{5}$$

where n_m is the molecular ad-layer refractive index and M is the region in which the ad-layer increases. Surface sensitivity S_s, can be calculated by using the definition in Eq. 2.

In conclusion, there is a last consideration concerning with units of measurement of both analyzed sensitivities, S_h and S_s. In the first case, the performance parameter S_h is dimensionless, because it is the ratio of two dimensionless physical quantities. On the contrary, surface sensitivity is generally measured in nm^{-1}, because the ad-layer due to analyte or molecules adsorption is characterized by a nanometer-scale thickness (2-5 nm).

Sensor limit of detection (LOD) represents another important parameter to characterize the sensor performance. It indicates the minimum resolvable signal and can be defined by taking into account the noise in the transduction signal (σ) and the sensor sensitivity (S). Generally, LOD can be calculated as the ratio σ/S.

Photonic chemical sensors based on homogeneous and surface sensing are generally defined RI-based sensors, because the transduction process consists always in a modal effective index change, as previously described. However, there is another class of chemical sensors based on optical absorption detection. In particular, the absorption coefficient, usually indicated with the variable α (cm^{-1}), depends on the operative wavelength of the photonic signal and material (e.g., gas, solid, liquid solutions) electronic and optical properties. For example, fundamental vibrational and rotational modes associated with most inorganic and organic molecules are spectroscopically accessible within the mid-infrared range (mid-IR). By this way, the interaction between mid-IR photons and organic molecules provides particularly sharp transitions, which, despite of the wide variety of organic molecules, provide unique mid-IR absorption spectra. Therefore, it is possible to reflect the molecularly characteristic arrangement of chemical bonds within the probed molecules via the frequency position of the associated vibrations and mixed rotational-vibrational transitions. In addition to this, a lot of environmental harmful gases like carbon dioxide (CO_2), methane (CH_4) and sulfur dioxide (SO_2), to name a few, are characterized by absorption spectra in mid-IR, as sketched in Fig. 5.

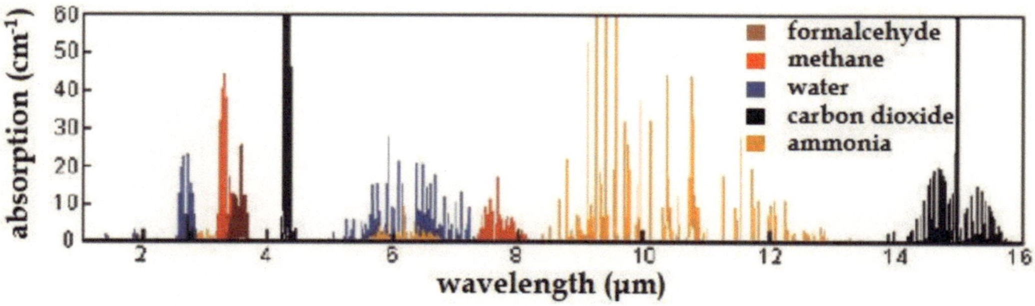

Fig. 5. Absorption spectra of several gases and liquid solutions.

It is possible to detect specific gas species or chemical analytes with high sensitivity and selectivity, by analyzing the transmission spectra of the light intensity in a suitable wavelength range. In particular, the photonic signal intensity is linked to the gas or analyte concentration C via the Beer-Lambert law:

$$I = I_0 \exp(-\alpha L) \tag{6}$$

where I is the light intensity at the end section of the whole path length (L), α linearly depends on C and I_0 is the light intensity at the initial section of the sensible area of photonic sensor. By this way, the sensor readout is an optical intensity one, thus it will be possible to register steep peaks in the transmission spectra corresponding to specific operative wavelengths, properly selected to be the absorption wavelengths of gas or molecule to be detected. Optical absorption detection allows to design photonic chemical sensors characterized by ultra low LOD (e.g., pg/mm^2, ng/mL^1).

In typical bio-chemical sensors, detection of specific pathogens or proteins require transduction labeling elements, such as fluorescent dyes or radioactive isotope, in order to generate a physically useful signal from a recognition event. For example, in fluorescence-based detection the intensity of the fluorescence indicates the presence of the target molecules as well as their concentration. Although label detection exhibits high sensitivity and ultra low LOD down to a single molecule, labeling chemistry is expensive and time-consuming, and may interfere with the function of bio-molecules. In this context, label-free photonic bio-sensing described in this paragraph, allows to preserve the natural form of target molecules and their natural interaction with selective receptors. In conclusion, label-free photonic sensors nowadays represent a universal platform for biochemical assays.

2.1 Slot-waveguides for sensing applications

Slot waveguides represent a very interesting and promising architecture for photonic chemical and bio-chemical ultra-high performance sensing. In fact, by using slot waveguides it is possible to confine an extremely high optical field in the low refractive index region called "slot region", where the chemical solution or gas will be detected (Almeida et al., 2004; Iqbal et al., 2008). In Fig. 6, it is possible to appreciate the optical field distribution in a silicon slot waveguide and a relative geometrical definition of the slab slot architecture.

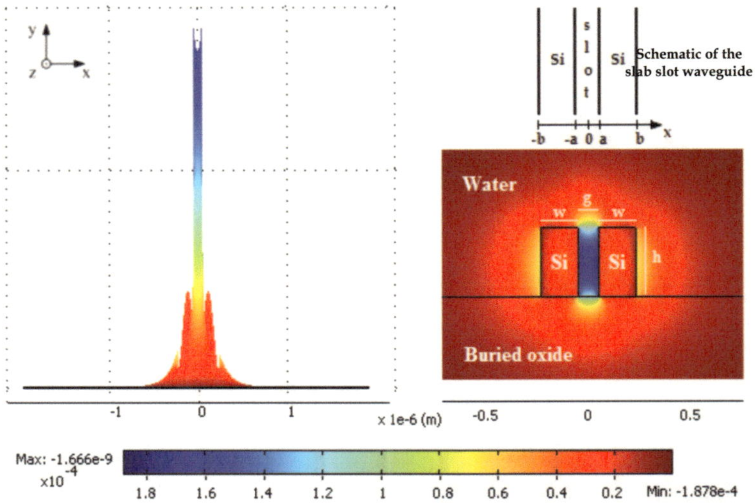

Fig. 6. 2D-views of the E_x-field spatial distribution (quasi-TE) solved for a Si-wire slot waveguide optimized @ 1.55μm; h = 324nm, g = 100nm, w = 180nm.

The mathematical expression of the E_x-field of TM mode in a slab slot waveguide (Fig. 6) is as follows:

$$E_x(x) = A\begin{cases} \frac{1}{n_L^2}\cosh(\gamma_S x) & ;|x| < a \\ \frac{1}{n_H^2}\cosh(\gamma_S a)\cos\left[k_H(|x|-a)\right] + \frac{\gamma_s}{n_L^2 k_H}\sinh\left[k(|x|-a)\right] & ;a < |x| < b \\ \frac{1}{n_L^2}\left\{\cosh(\gamma_S a)\cos\left[k_H(b-a)\right] + \frac{n_H^2\gamma_S}{n_L^2 k_H}\sinh(\gamma_S a)\sin\left[k_H(b-a)\right]\right\}\exp\left[-\gamma_C(|x|-b)\right] & ;|x| > b \end{cases} \tag{7}$$

where

$$A = A_0\frac{\sqrt{k_0^2 n_H^2 - k_H^2}}{k_0} \tag{8}$$

A_0 is an arbitrary constant, k_0 ($2\pi/\lambda_{op}$) is the vacuum wave number, k_H is the transverse wave number in the high refractive index (n_H) region, γ_C is the field decay coefficient in the cover medium and γ_S is the field decay coefficient in the slot region (low RI region, n_L).

By using slot waveguides instead of PWEF or rib ones, it is possible to enhance the interaction between the propagating optical field and the chemical test sample. In fact, slot-based photonic sensors can exhibit ultra high performance (e.g., $S_h > 1$, LOD ≈ 10^{-4} RIU).

To this purpose, interesting results have been presented in literature. For example the slot waveguide optimized for homogeneous sensing and presented in Fig. 6, is characterized by ultra high sensitivity, S_h = 1.0076 for quasi-TE propagation mode. A SOI slot waveguide, (w=135 nm, h=965 nm, g=100 nm) exhibiting ultra-high sensitivity of 1.1, has been also recently demonstrated by the authors. A value $S_h > 1$ implies that an effective index change $\Delta n_{eff} > \Delta n_C$ is induced by a cover index shift Δn_C.

Performance parameters such as sensitivity and confinement factors in cover medium and slot region, depend on waveguide geometrical parameters and materials selected for the fabrication. In fact, in Fig. 7 it is possible to note the sensitivity changes for homogenous sensing as a function of slot height "h", slot region width "g" and wire width "w", in a slot waveguide sensor. By this way, appropriate choice and design of photonic waveguide for chemical or biochemical sensing applications, have to be carried out in order to enhance all performance parameters as mentioned above (Passaro, 2009a).

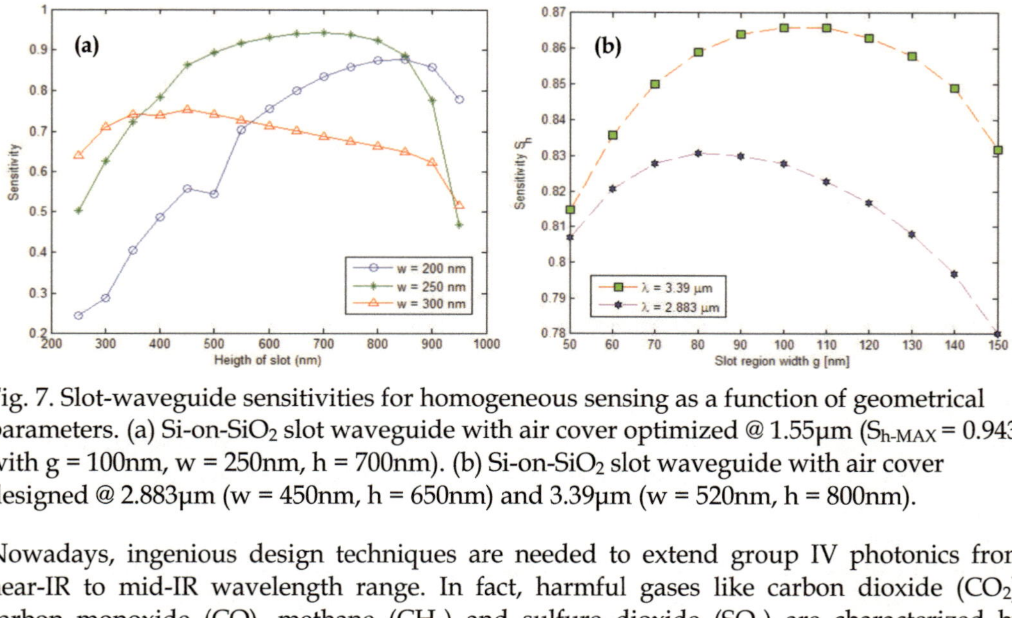

Fig. 7. Slot-waveguide sensitivities for homogeneous sensing as a function of geometrical parameters. (a) Si-on-SiO_2 slot waveguide with air cover optimized @ 1.55μm ($S_{h\text{-}MAX}$ = 0.943, with g = 100nm, w = 250nm, h = 700nm). (b) Si-on-SiO_2 slot waveguide with air cover designed @ 2.883μm (w = 450nm, h = 650nm) and 3.39μm (w = 520nm, h = 800nm).

Nowadays, ingenious design techniques are needed to extend group IV photonics from near-IR to mid-IR wavelength range. In fact, harmful gases like carbon dioxide (CO_2), carbon monoxide (CO), methane (CH_4) and sulfure dioxide (SO_2) are characterized by absorption spectra in mid-IR, specially in the range 2-8μm, according to Fig. 5. To this purpose, several group IV material systems (e.g., SiGe, SiGeSn, SiGeC, GeC, SiSn) have been investigated (Troia et al., 2011). These alloys have been proposed for the design of slot waveguide sensors for homogeneous sensing, working at 3.39μm and 2.883μm. Interesting theoretical performances have been obtained by optimizing SOI-based slot waveguides designed for optical transparency at working λ_{op} and ultra high refractive index contrast Δn. In particular, $S_h > 1$ has been demonstrated for two slot waveguides constituted by group IV material system layers, properly designed and stacked to form symmetrical wires, as sketched in Fig. 8 (Passaro et al., 2011). The most important advantages of these mid-IR "slot" sensors with respect to conventional slot sensors optimized at 1.55μm, concern with higher performance, relaxed fabrication tolerances, and presence of a second-order slot mode, able to be used for sensing. Similar investigations have been also proposed about novel optical slot waveguides working at 2.883μm ($Si_{0.15}Ge_{0.85}/Si/SiO_2$ and $Si_{0.15}Ge_{0.85}/SiO_2$).

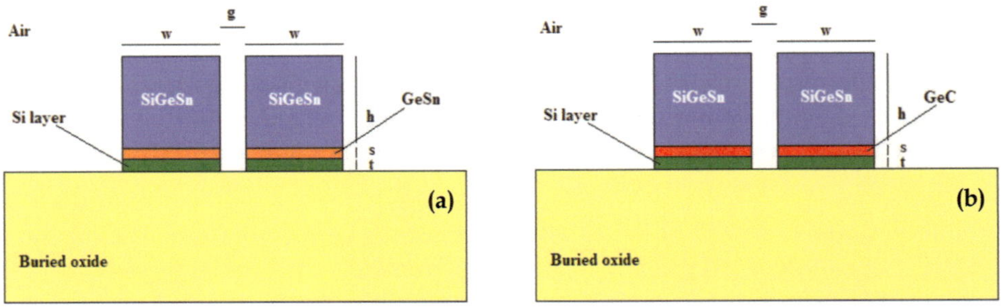

Fig. 8. Optimized novel slot photonic waveguides based on group IV alloys and material systems, designed at λ_{op} = 3.39μm. (a) $Si_{0.08}Ge_{0.78}Sn_{0.14}/Ge_{0.91}Sn_{0.09}/Si/SiO_2$ with w = 380nm, h = 520nm, t = 20nm, s = 50nm, g = 100nm; (b) $Si_{0.08}Ge_{0.78}Sn_{0.14}/Ge_{0.97}C_{0.03}/Si/SiO_2$ with w = 390nm, h = 560nm, t = 20nm, s = 50nm, g = 100nm.

In addition, these slot waveguides exhibit ultra-high performance (S_h > 1, LOD ~ 4×10^{-5}). The most important limitation that characterizes some of these intriguing alloys is the impossibility to grow the analyzed material systems selectively and directly on silicon dioxide. The effect of this limitation imposes some inevitable technological restrictions in the design of above described novel photonic sensors. For example, GeSn and SiGeSn alloys can be grown either on Si(100) or Ge/Si(100) wafers, using the SnD_4, $SiGeH_6$, Ge_2H_6 and Si_3H_8 hydride compounds as a source of Sn, Ge and Si constituent atoms, respectively. However, the Si-Ge-Sn class of materials, including all alloys mentioned in this section, constitutes a new paradigm in the integration of Si based electronics with optical components on a single chip. In fact, all fabrication techniques needed for new group IV alloys, are perfectly compatible with CMOS standard processes and facilities.

2.1.1 Fabrication tolerances

Slot waveguides and PWEF sensors presented until now, have been represented assuming an ideal geometry. In fact, vertical sidewalls (Fig. 4 a-b, Fig. 6, Fig. 8 a-b) are very difficult to be obtained by the state-of-the-art etching processes (e.g., inductively coupled plasma, ICP). Thus, deviations from ideal case have to be considered and one of the most important parameters quantifying this effect is the tilting angle θ, as shown in Fig. 9. By this way, it is possible to distinguish between vertical (θ=0°) and non vertical (θ≠0°) sidewalls.

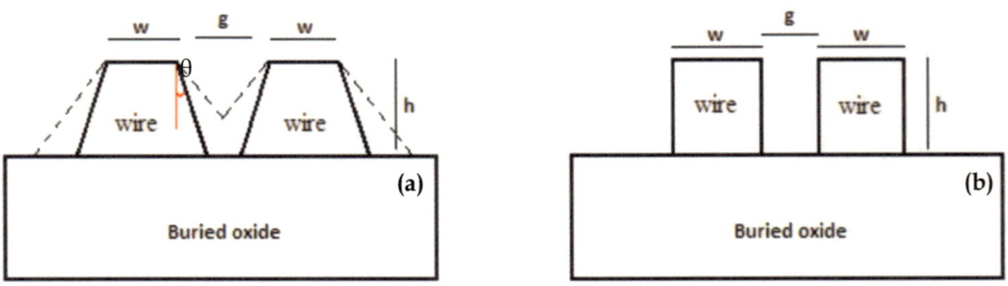

Fig. 9. Schematic view of slot waveguide. (a) real non vertical architecture for theoretical analysis of fabrication tolerances; (b) ideal slot waveguide with vertical sidewalls.

All well known performance parameters (S_h, Γ_C, Γ_S) have to be evaluated for different tilting angles θ° in the range 0°-10°. To this purpose, interesting results have been plotted in Fig. 10.

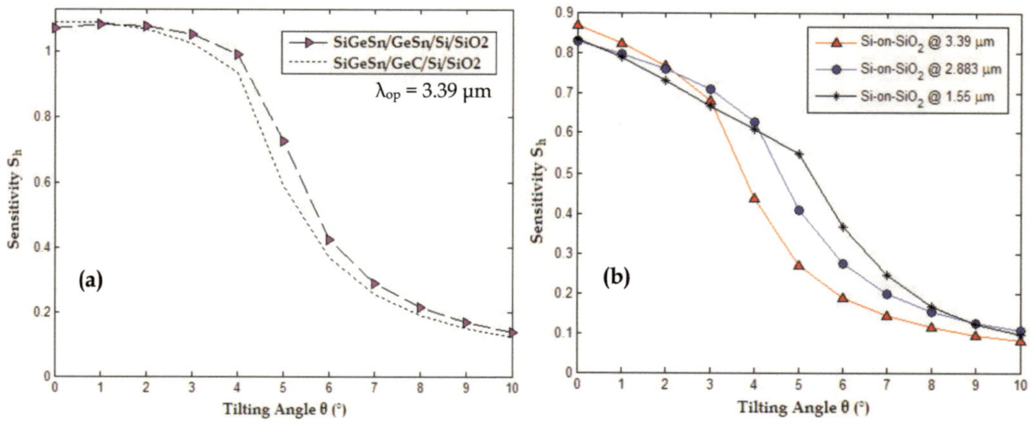

Fig. 10. Sensitivity of optimized slot waveguides as a function of the tilting angle θ with slot region width g fixed at 100nm in all cases. (a) Tolerance analysis of group IV-based optimized slot waveguides presented in Fig. 8; (b) Tolerances analysis of conventional SOI slot waveguides (@ 1.55µm, w = 230nm, h = 500nm, @ 3.39 & 2.883 µm, see also Fig. 7b).

All sensitivity values corresponding to θ = 0°, are referred to optimized slot waveguides characterized by vertical sidewalls. By observing Fig. 10a and 10b, it is evident that the greater the tilting angle, the lower the sensitivity. This effect is justified by the fact that increasing values of θ will produce a reduction of the slot region volume. Moreover, the slot region width g is not constant along y-direction. In conclusion, for θ > 4°, the slot mode cannot propagate because the optical field confinement in the slot region approaches zero.

Fabrication tolerance analysis represents an important aspect to estimate the real sensor performance. By this way, the solution to the technological problems presented above, is strictly related to the optimization of the etching process for a given combination of material systems. However, standard etching processes can assure tilting angles within 1°-2°. Generally, slot waveguides designed to operate in mid-IR wavelength range, exhibit greater tolerance margins because of their relaxed dimensions (Fig. 10a). To this purpose, θ < 3°-4° represents a good trade-off between technological constraints and device performance.

2.1.2 State of the art of slot waveguides designed for chemical sensing

All theoretical and experimental results demonstrate that slot waveguides are the best suitable photonic devices for sensing applications. In fact, if properly designed, they allow to concentrate a high optical field percentage in the cover medium, improving the interaction between the optical field and the chemical process, thus the limit of detection.

To this purpose, alternative solutions have been proposed in literature with the aim to enhance the efficiency and performance of photonic chemical sensors. For example, slot waveguides based on polymeric materials can ensure high performance, although characterized by low refractive index contrast Δn (Bettotti et al., 2011). In addition, multiple-slot waveguides represent an intriguing solution. By using multiple slot waveguides such as those presented in Fig. 11a-b-c, it is possible to significantly increase the sensitivity for both homogeneous ($\Delta n_{eff}/n_c \sim 0.2$) and surface sensing ($\Delta n_{eff}/\Delta\rho \sim 10^{-4}$ RIU/nm) (Kargar & Lee, 2009; Sun et al. 2009).

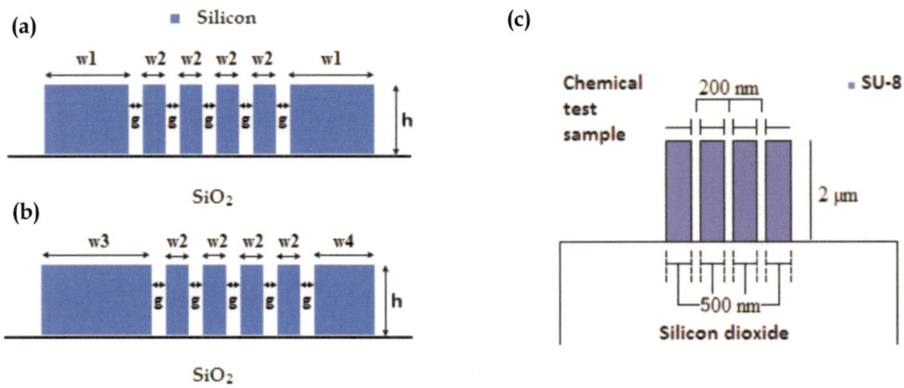

Fig. 11. Cross section of a symmetric (w1 = 133nm, w2 = 33nm, g = 20nm, h = 250nm) (a), and asymmetric (w3 = 200nm, w2 = 33nm, g = 20nm, w4 = 67nm, h = 250nm) (b), silicon quintuple-slot waveguides. Cross section of SU-8 (n_{SU-8} = 1.565 @ 1.55µm) multiple slot waveguide (c).

This improvement is justified by the fact that, in multiple-slot waveguides, the interaction between the optical field and the chemical test sample can be distributed in a larger area, in comparison to single slot waveguides. However, technological problems and reduced mode excitation usually limit the practical number of slots to three.

Another slot waveguide architecture evolution is represented by horizontal slot waveguide, as sketched in Fig. 12a. By this way, the optical propagating slot mode is quasi-TM, because the E_y-field is the only E-field component that undergoes discontinuities.

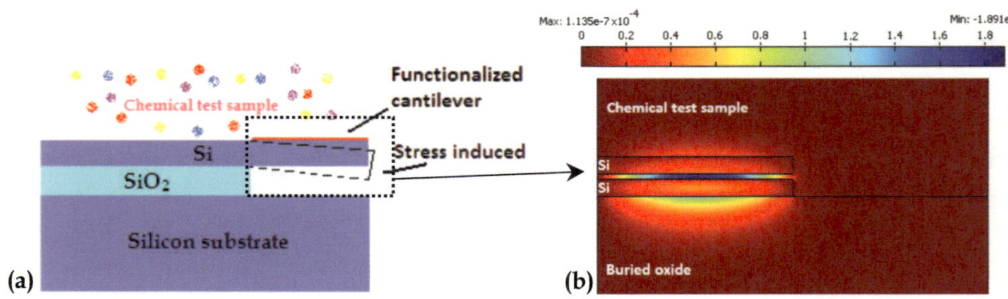

Fig. 12. Cross sectional view of the horizontal slot chemical sensor (a) with the spatial distribution of the fundamental TM optical mode (b).

A vertical slot fabrication can be realized by ICP etching, for example. However, etching in a very narrow region (slot region width g ~ 20-100nm) can produce large roughness in the vertical interface. By this way, in vertical slot waveguides, where the E_x-field is concentrated at the interface between the high and low refractive index region, propagation losses can be very high, such as 11.6 ± 3.6 dB/cm in a single slot of 50nm (Baehr-Jones et al., 2005). This problem can be partially solved by using horizontal slot waveguides (Fig. 12a-b). In fact, the latter can be fabricated by using thermal oxidation or deposition of different layers. By this way, the quasi TM-mode is not affected by surface roughness and propagation losses can be significantly reduced, down to 6.3 ± 0.2 dB/cm, 7.0 ± 0.2 dB/cm for a single and multiple slot waveguides, respectively (Sun et al., 2007). In Fig. 12a it is possible to see how the adsorption of target molecules on the cantilever surface, causes a small variation in the position of the functionalized cantilever. Consequently, an effective index change Δn_{eff} is caused by strain optic effect. Normally, the stress induced by proteins lies in a range between 3000 and 15000 $\mu\varepsilon$ (ε is the strain induced on the cantilever). Thus, the sensor response and the sensitivity are quite high. On the contrary, there are some aspects that limit the mass-scale production of these particular photonic sensors. First of all, the mechanical deformation, directly responsible of the sensing principle described above, has to be ensured also by very accurate technological steps, strongly influencing the sensor lifetime. Another important limitation concerns with the reliability of the sensing procedure. In fact, if the cantilever is corrupted by other molecules or particles not involved in the sensing detection, the calibration of the same sensor could be wrong, thus procedure has to be repeated more than one time.

In conclusion, both vertical and horizontal slot waveguides represent very promising photonic devices for chemical sensing applications. The high refractive index contrast between silicon and its oxide provides excellent optical confinement. By this way, it is possible to guide high optical intensity in the cover medium, preventing non linear effects in silicon, such as two-photon absorption (TPA), that can lead to high optical losses.

3. Advanced platforms for chemical photonic sensing

Generally, the choice of the photonic device suitable for label-free sensing applications (rib waveguide, photonic wire and slot waveguide) concerns with the possibility to ensure the best overlapping between optical field and chemical/biochemical sample. As analyzed until now, in both homogeneous and surface sensing the optical transduction leads to the effective index change Δn_{eff}. However, this optical variable has to be transformed into an analytically useful signal. To this purpose, several photonic integrated platforms have been proposed in literature. In particular, it is possible to distinguish between two different types of optical readout: the intensity and wavelength readout (power and wavelength interrogation, respectively). For example, photonic chemical sensors based on optical absorption are characterized by an intrinsic intensity readout, because the change of the absorption coefficient α is directly linked to the output optical intensity, according to the Beer-Lambert law. By this way, it is possible to register the chemical-to-optical transduction by using a photodiode as a transducer (Fig. 1).

In this context, advanced optical platforms for chemical sensing will be investigated in this Section.

3.1 Mach-Zehnder interferometer

The operation principle of the integrated Mach-Zehnder interferometer (MZI) is illustrated in Fig. 13a. Light is introduced through a photonic waveguide (e.g., slot waveguide) in the Y-branch on the scheme left side. By this way, the input optical power I_{IN} is split into two optical beams with an half power $I_{IN}/2$ in the upper and lower arms, called "sensing" and "reference" arm, respectively. If any optical phase delay is applied to the guided mode in the sensing arm ($\Delta\varphi = 0$), light will be combined at the output Y-branch exhibiting an output optical power $I_{OUT} = I_{IN}$. In all different cases ($\Delta\varphi \neq 0$ in sensing arm), the optical output power will be different from the input one ($I_{OUT} \neq I_{IN}$), according to Eq. 9.

$$I_{OUT} = I_{IN} \cos^2\left(\frac{\Delta\varphi}{2}\right) = I_{IN} \cos^2\left(\frac{\pi \Delta n_{eff} L}{\lambda}\right) \tag{9}$$

In Eq. 9, L is the interaction length, thus the guiding path in which the overlapping between the optical field and the chemical test sample occurs, and λ is the operative wavelength.

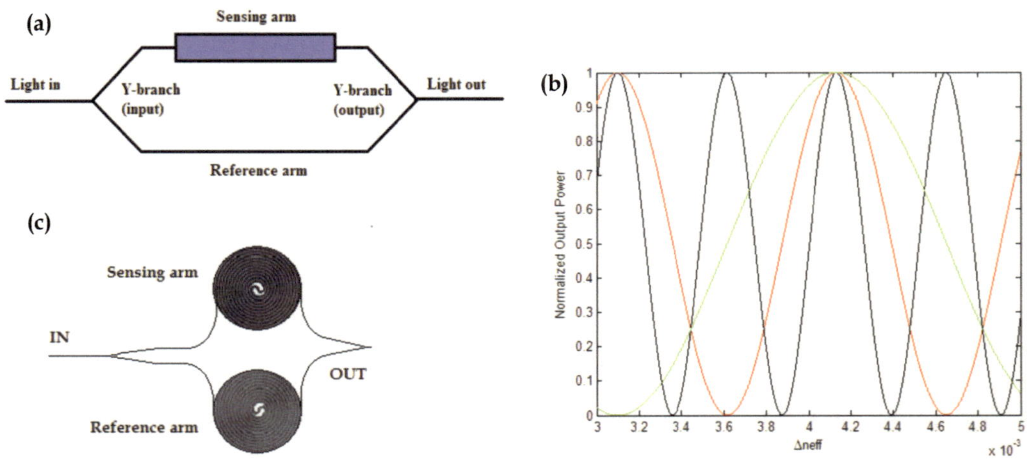

Fig. 13. (a) Schematic view of Mach-Zehnder interferometer; (b) Normalized output power as a function of different values of L (black = 3mm, red = 1.5mm, green = 750nm, λ_{op} = 1.55µm); (c) Image of a balanced MZI characterized by spiral arms.

In Fig. 13b, it is possible to see the effect of different values of L on sensing efficiency. In fact, sensor resolution is the capability to detect the minimum effective index perturbation. By this way, it is possible to conclude that the longer the interaction length, the higher the sensor resolution. Graphically, this concept can be observed by analyzing the steepness and width of each $\cos^2$ lobe. In conclusion, the closer the transmission spectra lobes, the smaller the detectable effective index change. Obviously, it is not convenient to realize cm-long arms because photonic chemical sensors have to be characterized by very small footprints. To this purpose, interesting technological solutions have been presented in literature. In Fig. 13c, a spiral path configuration is presented. By this way, it is possible to improve sensor performance and efficiency, without compromising the sensor dimensions. In fact, mm-long arms (~ 2-3mm) can be realized and concentrated in spiral architecture characterized by µm-long diameter.

Very high integration with spiral-path configurations allows to design biosensor arrays for multiplexed real-time and label-free molecular detection. Ultra high performance can be achieved in terms of sensitivity and LOD. For example, the measured sensitivity of the 2.1 mm-long interferometer fabricated on SOI wafer with 2µm silica under 260nm of silicon, is $d\varphi/dn_C$ = 4930 rad/RIU, while the calculated change of polydimethylsiloxane (PDMS) refractive index upon xylene is 8.7×10^{-7} RIU/ppm (Saunders et al., 2010). PDMS has been used as sensor cladding in order to enable sensing of BTEX (Benzene, Toluene, Ethylbenzene, Xylenes) volatile organic compound. In addition, a surface mass coverage of 0.3 pg/mm^2 can be detected with a silicon photonic wire MZI sensor array, with integrated SU-8 micro-fluidic channel (Densmore et al., 2009). In conclusion, the principal penalty of using MZI photonic sensors is represented by the cosine-dependent intensity function.

In fact, the signal change is not easily resolvable near the maximum and minimum of the cosine function.

3.2 Surface plasmon resonance biosensors

A surface plasmon is a localized electromagnetic wave that propagates along the metal-dielectric interface and exponentially decays into both media. Surface plasmons can be excited due to the resonant transfer of the incident photon energy and momentum to collectively oscillating electrons in a noble metal. The plasmon resonance condition is given by the following expression (Jha, R. & Sharma, 2010):

$$k_0 n_C \sin(\theta) = k_0 \left(\frac{\varepsilon_{mr} n_S^2}{\varepsilon_{mr} + n_S^2} \right)^{\frac{1}{2}} \tag{10}$$

In Eq. 10, ε_{mr} is the real part of the metal dielectric function, n_C is the refractive index of light coupling glass, n_S is the refractive index of sensing medium (dielectric), λ is the wavelength of incident light and k_0 is the wave number ($2\pi/\lambda$). The resonance condition is satisfied when β (propagation constant of the light beam coupled at the angle θ) is equal to β_{SPW} (propagation constant of the surface plasmon polariton). The resonant condition is expressed by Eq. 10 when $\theta = \theta_{SPR}$. At the resonant condition, the reflected light intensity encounters a sharp dip due to optical absorption by surface plasmon wave (SPW). This dip can be easily detected by the photo-detector, as reported in the scheme proposed in Fig. 14.

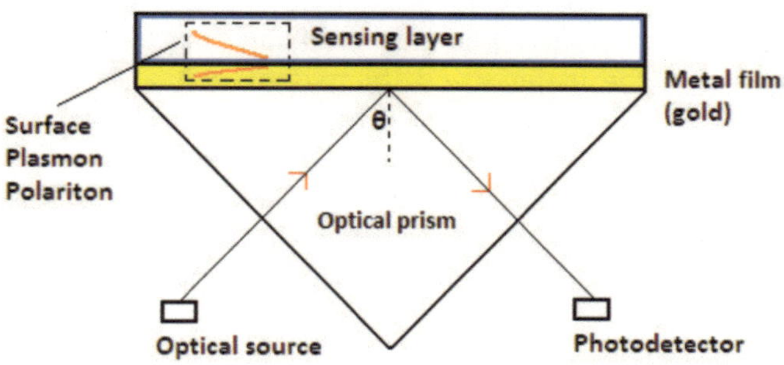

Fig. 14. Schematic view of a SPR biosensor characterized by prism coupling configuration.

Bio-molecular recognition elements at the metal surface selectively capture target analytes in a chemical liquid test sample, the reflectance dip makes a shift due to the change of local refractive index near the surface. In particular, plasmonic related sensing procedure has been used for label-free detection of biological analytes at pico/femto-molar level, pesticides, immunoassays, DNA with a refractive index resolution of 1.4×10^{-7} (Le et al. 2011), RNA, allergens and human blood-group identification (Jha & Sharma, 2010). For example, a SPR mid-infrared immunosensor used in a Fourier-transform infrared (FTIR) spectroscopy platform, exhibits a sensitivity of 3022 nm/RIU and a limit of detection of ~70 pg/mm^2 (DiPippo et al., 2010). Generally, the detection of the SPR change due to the adsorption of target molecules on the sensing surface, can be quantified by monitoring the resonant intensity, wavelength or angle change. Although SPR chemical sensors exhibit high sensing performance, they are characterized by some operative constrains. In particular, surface plasmon evanescent field penetrates in the sensing layer only for ~100nm. In this way, it is not possible to detect large target molecules like cells and bacteria. Another limitation concerns with the impossibility of the SPR photonic sensor, to distinguish between the refractive index surface change and the bulk solution refractive index change. Then, if the sensor is covered by a complex chemical solution (e.g., blood, bacteria-solution), it is very difficult to ensure high detection resolution. To this purpose, interesting architectures have been proposed in literature such as a novel optical structure, in which the metal layer is sandwiched by two dielectric layers with similar refractive index. In this way, both long (LSPR) and short (SSPR) surface plasmon modes can be generated at both metal-dielectric interfaces, allowing to differentiate the background and surface bound refractive index change (Li et al., 2010). In conclusion, all commercialized SPR biosensors can be characterized by different coupling methods. In addition to prism coupling configuration sketched in Fig. 14, there are guided-wave, fiber optic and grating coupling configurations (Marrocco et al., 2010). Slot plasmonic waveguides have been also proposed as an integrated chemical sensor with high sensitivity (Hu et al., 2010).

3.3 Chemical sensors based on photonic crystals

Photonic crystals (PhCs) are periodic systems characterized by symmetrical separation between high dielectric and low dielectric regions. Generally, it is possible to distinguish between three different geometrical configurations as a function of the spatial periodicity. In fact, there are one-dimensional PhCs where the dielectric function periodicity is distributed only to one dimension (1D), thus bi-dimensional (2D) and three-dimensional (3D) PhCs. By using photonic crystals it is possible to guide and trap the light. In addition, photonic crystals exhibit a very large selective photonic band gap that prevents photons in the band gap from propagating in the material. By this way, it is possible to design a photonic waveguide by creating a line defect in the structure. The basic element of a photonic crystal is called "elemental" or "primary cell" and it can be represented by a line, circle, hexagonal or tetrahedral geometry, to name a few. For example, the periodical repetition of the hexagonal geometry produces a photonic crystal with a particular structure defined as "honeycomb". Photonic crystals exhibit interesting and promising features, suitable for high performance chemical sensing.

3.3.1 Integrated optical Bragg-grating-based chemical sensors

Photonic Bragg gratings are characterized by a periodic spatial distribution of the dielectric constant, thus of the refractive index, along the propagation direction. Consequently they are photonic crystals and exhibit a selective range of propagating wavelengths.

It is possible to realize an integrated optical sensor based on Bragg gratings in SOI technology, as sketched in Fig. 15 (Passaro et al., 2008). A Bragg grating is characterized by a periodic perturbation of the waveguide refractive index along the propagating direction z:

$$n(x,y,z) = n_0(x,y) + \Delta n(x,y,z) \tag{11}$$

In particular, $n_0(x,y)$ is the waveguide refractive index distribution, $\Delta n(x,y,z)$ represents the periodic refractive index change characterized by a period Λ (Fig. 15a). Λ represents a physical parameter that has to be designed in order to ensure a precise center wavelength of the Bragg grating spectral response (Fig. 15b), following the relation as in Eq. 12. In particular, m is the grating order, n_{eff} is the effective index of the propagating mode and λ_C is the central wavelength, previously defined.

$$\Lambda = \frac{m}{2n_{eff}}\lambda_C \tag{12}$$

In conclusion, the grating reflectivity can be expressed by the following expression:

$$R(\delta) = \frac{|k|^2 \tanh^2(\gamma L)}{\gamma^2 + \delta^2 \tanh^2(\gamma L)} \tag{13}$$

where k is the coupling coefficient between backward and forward propagating optical modes, L is the grating length, γ and δ are two optical parameters defined as:

$$\delta = \frac{2\pi}{\lambda} n_{eff} - \frac{\pi}{\Lambda} \; ; \gamma = \sqrt{|k|^2 - \delta^2} \tag{14}$$

By this way, it is possible to explain the sensor device operative regime. In particular, all geometrical parameters (h, H, D, W, d, Λ) and the index contrast Δn, have to be designed in order to determinate the central operative wavelength λ_C. In Fig. 15b, it corresponds to 1550 nm. By observing Eq. 12, it is possible to see that λ_C only depends on the effective index of the mode propagating, because the grating period Λ is fixed after the device fabrication. Consequently, any effective index change Δn_{eff} induced by homogeneous or surface sensing, causes a central wavelength shift, $\Delta\lambda_C$.

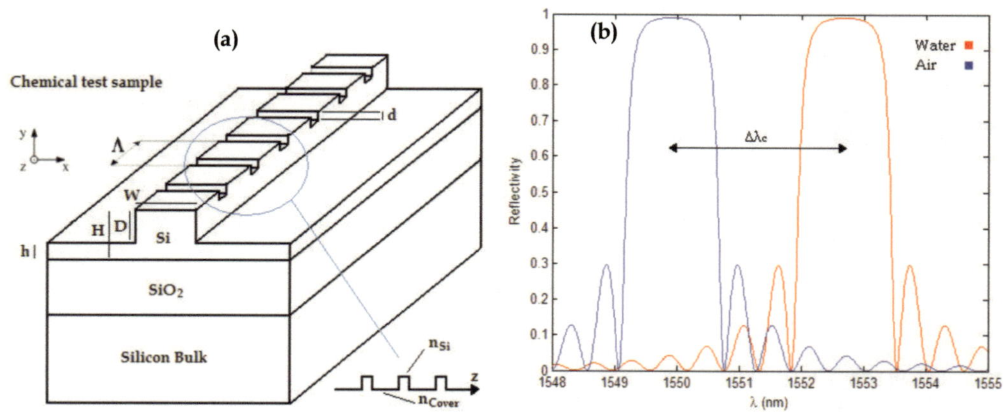

Fig. 15. (a) Schematic view of a grating-based sensor device. (b) Spectral response of the first-order grating for different values of cover refractive index (H = 1μm, d = 30nm, W = 0.56μm, D = 0.62μm, Λ = 240nm, $n_{air} = 1$, $n_{water} = 1.33$).

In the graph reported in Fig. 15b, the sensor operative regime is characterized by λ_C = 1550nm with air cover. When the sensor is exposed to aqueous solution (e.g., water), λ_C = 1552.7nm. By this way, if the selected operative wavelength λ_{op} is 1550nm, than it is possible to detect a change in the solution composition by measuring the new center wavelength shift, $\Delta\lambda_C$ = 2.69nm. By using third-order grating based sensor devices (m = 3), it is possible to maximize the device efficiency and employ shorter length chips. In conclusion, grating sensitivity can be defined as following:

$$S = \frac{\partial \lambda_C}{\partial n_{clad}} = \frac{\partial \lambda_C}{\partial n_{eff}} \frac{\partial n_{eff}}{\partial n_{clad}} = \frac{\Lambda}{m} S_w = \frac{\lambda_C}{2n_{eff}} S_w \tag{15}$$

In Eq. 15, n_{clad} is the cladding refractive index and S_w is the waveguide sensitivity, whose definition depends on sensing mechanism (homogeneous or surface sensing). A device sensitivity S around 120nm/RIU is practically obtainable. In conclusion, Bragg gratings analyzed above can be adopted in different waveguide architectures such fiber optics (fiber Bragg gratings, FBG). Sensitivity of 92 nm/RIU has been demonstrated for refractive index change of water in a FBG chemical sensor (Lee et al., 2010).

The sensing principle described above is based on the wavelength interrogation of the Bragg grating. However, another type of interrogation can be used, based on the incident angle θ (°) that the propagating light forms with the Bragg grating surface. The device operation can be expressed by the following relation:

$$m\lambda = 2\Lambda \sin\theta \tag{16}$$

Coupling wavelength λ depends on the grating period and interrogation angle. By this way, by monitoring changes in coupling angle, molecular binding at the waveguide surface can be detected and measured by using SPR sensors, as in Fig 16a-b. In particular, the reflected intensity spectrum exhibits a sharp resonance peak at the coupling wavelength.

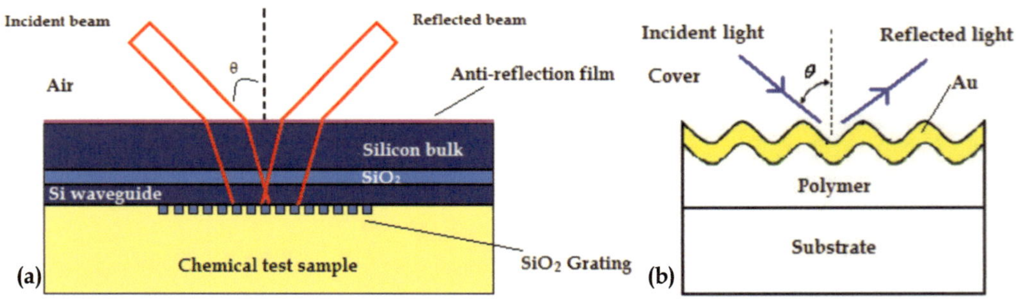

Fig. 16. (a) Schematic view of a photonic silicon waveguide sensor based on surface grating interrogation. (b) Chemical grating-coupled surface plasmon resonance sensor.

The sensing principle based on angle interrogation can be used also in surface plasmon resonance sensor characterized by a grating coupling. In particular, a refractive index resolution of 1.5×10^{-6} RIU has been demonstrated by using a sensor as that sketched in Fig. 16b (Kuo & Chang, 2010).

3.3.2 State of the art of integrated chemical sensors based on photonic crystals

Nowadays, one and two dimensional photonic crystals represent an interesting and promising research field for sensing applications, because of their compactness and high resolution in the detection process. Recently, a photonic crystal slot waveguide has been fabricated and used as infrared (IR) absorption spectrometer for methane monitoring. The principle of IR absorption spectroscopy is based on the Beer-Lambert law, described in the previous section (Lai et al., 2011).

The photonic device shown in Fig. 17a, has been realized by using a circular geometry as primary cell.

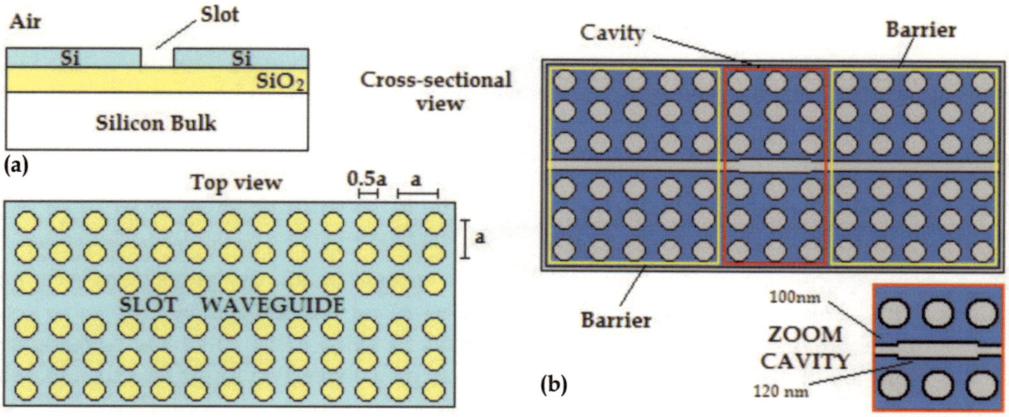

Fig. 17. (a) Photonic crystal slot waveguide built on SOI platform. (b) Photonic crystal resonant cavity based on slot waveguide for sensing application.

Consequently, a periodic 3D spatial distribution of the refractive index is realized with a lattice constant a. In fact, it is possible to observe that circular holes are characterized by low refractive index (n_{SiO2}), while the reciprocal space represents the high refractive index region (n_{Si}). A slot waveguide has been realized by introducing a line defect. By this way, a high electric field intensity is localized in a low-index (n_{air}) 90nm wide slot. The silicon photonic crystal slot waveguide device is characterized by a total length of 300µm in order to increase the optical absorption path length. A methane concentration of 100ppm in nitrogen has been measured.

In Fig. 17b a photonic crystal air-slot cavity has been presented (Jagerska et al., 2010). The structure is similar to the previously analyzed slot waveguide. In fact, the proposed air-slot cavity has been processed on a 220nm thick silicon-on-insulator wafer with a 2µm buried oxide layer. The cavity has been realized by reducing the line defect width by 20nm, thus unchanging the photonic crystal lattice constant. Reduced-slit width results in the formation of reflective barriers for the cavity mode. By this way, only resonant wavelengths can propagate inside the photonic cavity. The sensor has been tested by exposing the chip to gases of different refractive index such as N_2 (RI = 1.00270), He (1.00032) and CO_2 (1.00407) at the operative wavelength, λ_{op} = 1570nm. The refractive index change (Δn) due to the sensor exposition to different gases, causes a resonant wavelength shift ($\Delta\lambda$). Consequently, it is possible to detect different gas concentrations by monitoring the cavity transmission spectrum.

The sensor sensitivity can be defined as follows:

$$S = \frac{\Delta \lambda}{\Delta n} = \Gamma \frac{\lambda}{n_{eff}} \tag{17}$$

In Eq. 17, λ is the resonance wavelength of the cavity, Γ is the mode field overlap and n_{eff} the modal effective index. The high quality factor of the cavity (Q ~ 2.6×10^4) allows to achieve an experimental sensitivity S = 510 nm/RIU and a detection limit below 1×10^{-5} RIU.

In conclusion, other intriguing solutions have been proposed such as a single-defect silicon 2D photonic crystal nano-cavity for strain sensing (Tung et al., 2011). The sensing principle is the same as the gas sensor described above, but the resonant wavelength shift is induced by geometrical longitudinal and transverse strain. The theoretical minimum detectable strain for the photonic crystal cavity has been estimated to be 8.5×10^{-9} ε (8.5nε). By this way, advanced chemical sensing applications can be contemplated, based on cantilever sensing architectures.

3.4 Integrated optical directional coupler biosensors

The use of optical directional couplers for the measurement of interactions between biological molecules and the detection of target chemical and biochemical species, is an interesting research field for sensing applications. In fact, these photonic sensors are characterized by high performance and real-time monitoring of multiple outputs. The scheme of a directional coupler is presented in Fig. 18.

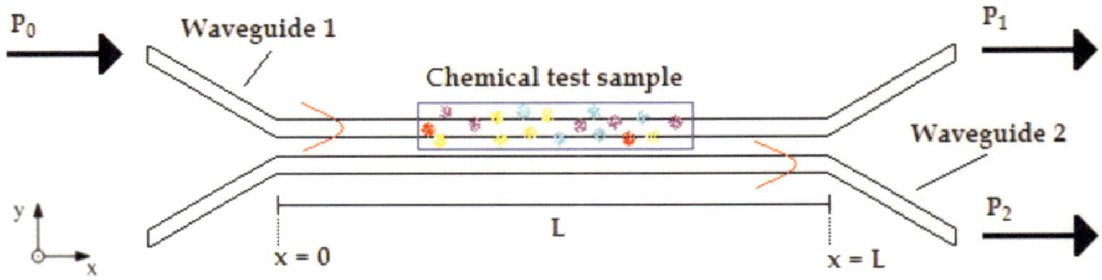

Fig. 18. Schematic representation of an integrated direction coupler.

The sensing principle is based on the super-mode theory. In Fig. 18 it is possible to see that only one waveguide (in this case the waveguide 1), is directly exposed to the chemical test sample. The other one has to be isolated by using appropriate coatings (e.g., polymers, Teflon). Generally, both waveguides have to be designed as symmetrical (e.g., identical geometries and contrast index Δn), in order to exhibit the same propagation features. In this way, β_1 and β_2, the propagation constants of the optical modes propagating in the waveguide 1 and 2, respectively, can be considered equal ($\beta_1 = \beta_2$) in a "synchronous" regime, where the propagation constant mismatch is equal to zero ($\Delta\beta = 0$). By assuming that an optical power P_0 is introduced in the waveguide 1 at the section x = 0, optical powers (P_1, P_2) at the output section (x = L) can be calculated by the following expressions:

$$P_2 = \frac{\sin^2\left\{\left[(kL)^2 + \left(\frac{\Delta\beta L}{2}\right)^2\right]^{\frac{1}{2}}\right\}}{1+\left(\frac{\Delta\beta}{2k}\right)^2} P_0, \; P_1 = P_0 - P_2 \tag{18}$$

where k is the coupling coefficient and depends on the overlap between the two propagating modal fields, and L is the device length. In synchronism condition ($\Delta\beta = 0$), complete power transfer from the waveguide 1 to the waveguide 2 is achieved when kL = π/2, 3π/2, 5π/2, etc. In this way, it will results $P_2 = P_0$ and $P_1 = 0$, at the end section x = L.

When a chemical test sample, gas or liquid solution, covers the waveguide 1, the refractive index of the cover medium will change, causing a modal effective index change of the propagating mode n_{eff1}. Consequently, the synchronous coupling condition will not be ensured because $\beta_1 \neq \beta_2$, thus $\Delta\beta \neq 0$. According to Eq. 18, P_1 will be different to zero because a complete optical power transfer now cannot occur. By monitoring the change of coupler outputs P_1 and P_2, it is possible to detect a particular chemical target and estimate its concentration with high limit of detection. The sensitivity S can be calculated as follows:

$$S = n_{clad(REF)} \left.\frac{\partial T}{\partial n_{clad}}\right|_{nclad=nclad(REF)} \tag{19}$$

where T is the transmittance registered at the output of the reference waveguides, n_{clad} is the chemical analyte refractive index and $n_{clad(REF)}$ is the reference analyte refractive index value for which the derivative is calculated. The highest sensitivity can be achieved at the point of maximum slope of the main peak of the transmittance curve. Consequently, this condition indicates the operating point of the sensor. Obviously, the overall sensitivity depends on the waveguide sensitivity, S_W. In this context, the use of slot waveguides represents the best solution. To this purpose, interesting results have been obtained by using coupled slot SOI waveguides sketched in Fig. 19a. In particular, a sensitivity S = 215.29 was achieved for quasi-TE mode with a coupler length L = 400μm (Passaro et al., 2009b).

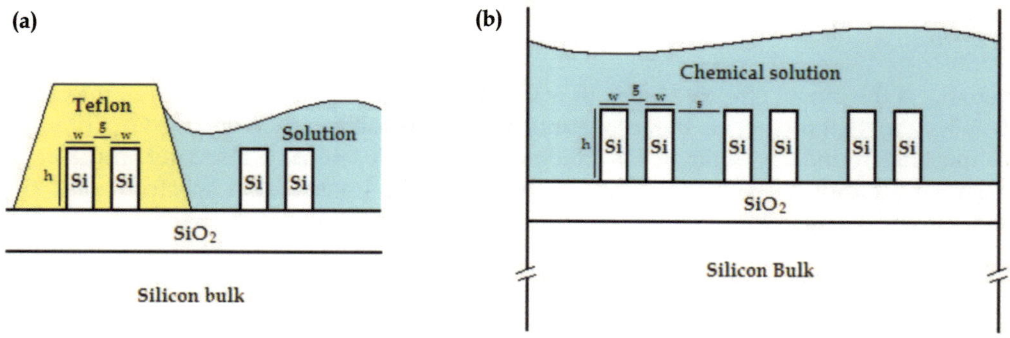

Fig. 19. (a) Architecture of coupled SOI slot waveguides for chemical sensing (h = 324nm, w = 180nm, g = 100nm). (b) Multi-channel directional coupler with slot waveguides (h = 300nm, w = 180nm, g = 100nm, s = 500nm).

The sensor sketched in Fig. 19a allows to estimate an analyte concentration in the aqueous solution. In particular, the photonic sensor integrated with a footprint of ~ 1mm^2, can detect a minimum refractive index change of ~10^{-5} and it can be adopted to estimate the glucose concentration in aqueous solution with a theoretical resolution of 0.1 g/L.

In Fig. 19b, a multi-channel directional coupler with slot waveguides is shown (McCosker & Town, 2010a). The sensing principle adopted in this device is different with respect to the previous one. In particular, it is possible to observe that there is not an isolated waveguide, because all silicon rails are simultaneously covered with the chemical solution. In this way, it is impossible to register a propagation constant mismatch change, because it will be constant. Thus, the sensing principle consists in the coupling coefficient change Δk, as a function of the cover refractive index change Δn_{clad}. In particular, the coupling coefficient k has an exponential dependence on analyte refractive index (McCosker & Town, 2010b). Interesting results have been theoretically demonstrated with a 1,607μm-long device. For example, a concentration of glucose or ethanole dispersed in deionised water could be detected with sensitivities of -172 (negative T slope in Eq.19) and +155 (positive T slope), respectively.

Photonic sensors based on directional couplers, exhibit very high performance. However, it is necessary to achieve a good trade-off between sensitivity and device length. In fact, in both sensing principles as analyzed above, mm-long interaction lengths allow to appreciate transmittance changes at the device output.

3.5 Chemical sensors based on integrated optical resonant microcavities

Photonic devices exhibiting best performance for chemical sensing are those based on resonant microcavities (Passaro et al., 2007b). In fact, they allow to achieve ultra low detection limits and ultra high sensitivities due to their wavelength interrogation, poorly influenced by optical noise. Optical resonant microcavities can be designed with different paths, for example ring, race-track or spiral-path ring. Generally, a resonance structure consists in a waveguide closed into a loop and coupled with one or two input/output bus waveguides (Fig. 20a). Resonance theory imposes the propagation of precise guided stationary modes inside the resonance cavity, in particular those modes whose wavelengths satisfy the well known resonance condition:

$$\lambda = \frac{2\pi n_{eff} R}{m} \tag{20}$$

where n_{eff} is the modal effective index, R is the ring resonator radius and m is an integer (m = 1, 2, 3,..., n), that represents the resonant order. This result has been obtained by imposing that the total round trip shift of the guided mode inside the ring resonator, must be an integer multiple of 2π. The analytical expression of the transmission spectrum of a ring resonator with a single bus (Fig. 20a), is given by the following equation:

$$T_{ring}(\lambda) = \frac{\alpha^2 + t^2 - 2\alpha t \cos\left(\frac{2\pi L n_{eff}}{\lambda}\right)}{1 + \alpha^2 t^2 - 2\alpha t \cos\left(\frac{2\pi L n_{eff}}{\lambda}\right)} \tag{21}$$

In Eq. 21, α is the loss factor that gives the field attenuation after one round trip through the ring cavity waveguide (i.e., $\alpha = 1$ in lossless case). The cosine argument represents the round trip phase, $L = 2\pi R$ is the cavity length depending on the ring radius, and t is the coupling ratio between the bus waveguide and the ring cavity.

In this context, the sensing principle consists in a resonant wavelength shift $\Delta\lambda$ due to the modal effective index change Δn_{eff} caused by the presence of the chemical analyte in the sensing area (Fig. 20b).

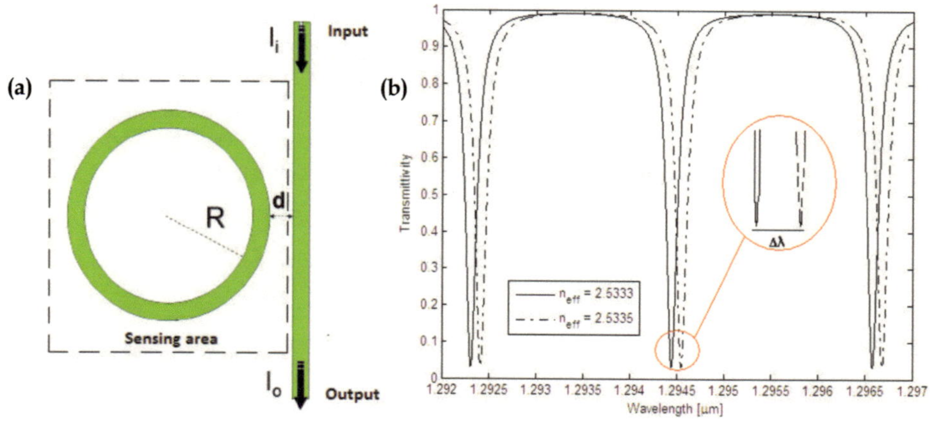

Fig. 20. (a) Schematic top view diagram of a silicon photonic ring resonator with a single bus waveguide. (b) Transmission spectrum of the ring resonator before (n_{eff} = 2.5333) and after (n_{eff} = 2.5335) the exposure to the chemical analyte.

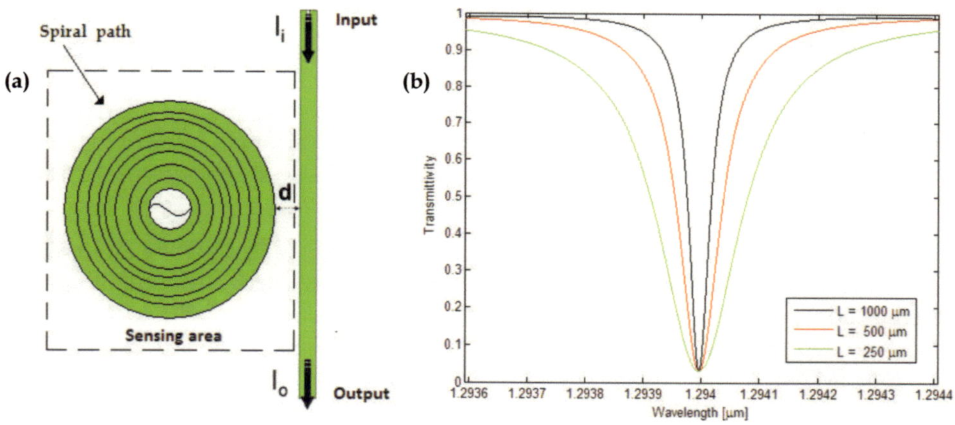

Fig. 21. (a) Top view of the spiral-path folded resonant cavity coupled to a bus waveguide. (b) Parametric analysis of the transmission spectrum (Lorentzian), windowed to a single resonant wavelength (λ_{op} = 1.294μm), as a function of three different interaction lengths.

The transmission spectrum shown in Fig. 20b has been calculated for a 250μm-long cavity characterized by α = 0.8875, t = 0.9198. In particular, it is possible to appreciate that a modal effective index change $\Delta n \sim 10^{-4}$ theoretically induces a resonant wavelength shift $\Delta\lambda$ of about 0.1nm. As analyzed about the MZI configuration (see. Para. 3.1), in chemical sensors based on single ring resonator as sketched in Fig. 20a, the resonant cavity length assumes a key role in the photonic sensor optimization. In fact, in Fig. 21b it is possible to see that the longer the interaction length "L" between the optical field and the chemical analyte, the higher the detectable resolution.

The most important parameter that allows to appreciate the ring-based sensor resolution is the linewidth of its transmission spectrum. In fact, in Fig. 21b it is possible to see that the longer the interaction length, the sharper the resonant peak. By this way, it is possible to see that a small linewidth corresponds to a high slope of the resonant peak, resulting in a signal change easily resolvable near the maximum and minimum of the transmission function. However, very long interaction lengths will result in large device areas that are not suitable for high scale integration. To this purpose, intriguing folded spiral–path cavities have been proposed (Fig. 21a). In fact, they allow to achieve mm-long interaction length concentrated in very small footprint (~μm^2). The sharpness of the resonant peak in a ring resonator transmission spectrum is usually expressed in terms of its quality factor Q:

$$Q = \frac{\lambda}{\Delta\lambda_{FWHM}} = \frac{\pi n_g L}{\lambda}\left[\arccos\left(\frac{2\alpha t}{1+\alpha^2 t^2}\right)\right]^{-1} \tag{22}$$

In Eq. 22, λ is the resonant wavelength, $\Delta\lambda_{FWHM}$ is the resonance full-width-at-half-maximum (*FWHM*) and n_g is the ring resonator group index. It is evident that the smaller the linewidth, the higher the resonant cavity quality factor.

In conclusion, it is possible to revisit the concept of sensitivity and limit of detection (LOD):

$$S = S_W \frac{\lambda}{n_{eff}}, LOD = \frac{\Delta\lambda}{S} \tag{23}$$

The sensitivity *S* (nm/RIU) depends on the waveguide sensitivity S_W and other well known parameters (resonant operative wavelength λ and modal effective index n_{eff}). The limit of detection *LOD* (RIU), depends on the minimum detectable resonant wavelength shift $\Delta\lambda$. By this way, ultra high performance for chemical, biochemical and gas sensing (S ~ 2000 nm/RIU, LOD ~ 3.8×10^{-5} RIU) can be achieved by using slot waveguides (Passaro et al., 2011).

Particular research efforts are still doing today in this field. In fact, the state-of-the-art of photonic sensors for bio-sensing applications is mainly dedicated to the optimization and characterization of novel resonant microcavities based on slot waveguides. In particular, a novel integrated SOI optic racetrack resonator has been proposed for bio-sensing applications (Malathi et al., 2010). The device is capable of distinguishing compressive and tensile stresses on a cantilever due to conformational changes of protein, Bovine Serum Albumin (BSA) and Immunoglobulin G (IgG). The change of surface stress upon adsorption of IgG is compressive, while for BSA it is tensile. The sensing principle of this specific photonic sensor configuration can be expressed by the following equation:

$$\frac{\Delta\lambda}{\lambda} = \frac{\Delta L}{L} + \frac{\Delta n_{eff}}{n_{eff}} \tag{24}$$

When the bio-molecules adhere to the device surface, both effective refractive index and racetrack length change due to photo elastic effect. When the cantilever bends, one arm of the racetrack experiences a small variation in L by ΔL and n_{eff} changes by Δn_{eff}, due to the strain optic effect. Bio-molecules stress on the cantilever has been simulated and the wavelength shift from the resonance has been found to be 0.3196×10^3 nm/με, where ε is the strain induced on the cantilever. Normally, the stress induced by proteins lies in a range between 3000με and 15000με, thus the sensor response and sensitivity obtained are quite high.

In this context, an ultrasensitive nano-mechanical photonic sensor based on horizontal slot waveguide resonator, has been also proposed on SOI platform (Barrios, 2006). The effective index variation can be read out as a resonant wavelength shift of the disk resonator, obtaining interesting results. In particular, a deflection sensitivity of 33 nm^{-1}, a detection limit of the slot waveguide disk resonator of 3×10^{-5}nm, an estimated cantilever stress sensitivity of 1.76 nm/(mJ/m^2), that means that the minimum detectable surface stress is $\Delta\sigma_{min}= 1.7\times10^{-5}$ mJ/m^2, have been achieved.

One of the most important aspects for photonic bio-sensing concerns with temperature effects affecting final measurements. The temperature effect is directly linked to very small refractive index changes of the adopted material system. Consequently, the global modal effective index perturbation can be written as:

$$\Delta n_{eff} = \Delta n_{eff,SENSING} \pm \Delta n_{eff,TEMP.} \tag{25}$$

By this way, the resonance wavelength of the ring resonator can change not only due to sensing principles (e.g., homogenous or surface sensing), but also due to temperature changes. The final result is a corrupted measure of biological analytes and molecules to be detected. Three approaches have been used for thermal noise reduction: active temperature control, a-thermal waveguide design, and temperature drift compensation by on chip referencing.

In this context, an on-chip temperature compensation in an integrated slot waveguide ring resonator refractive index sensor array has been investigated (Gylfanson et al., 2010). Experimental study has demonstrated a low temperature dependence of -16.6 pm/K while, at the same time, a large refractive index sensitivity of 240±10 nm/RIU. Furthermore, by using on chip temperature referencing, a differential temperature sensitivity of only 0.3 pm/K has been obtained, without any individual sensor calibration. This low value indicates good sensor-to-sensor repeatability, thus enabling use in highly parallel chemical assays. The detection limit has been demonstrated to be 8.8×10^{-6} RIU in a 7K temperature operating windows. Another interesting solution to the problem of temperature effects in SOI wire waveguide ring resonator label-free biosensor arrays, has been proposed (Xu et al., 2010). The device is constituted by four ring resonators with a reference ring for tracking sensor temperature changes. The reference ring is protected by a 2μm-thick upper cladding layer and optically isolated from the sensing medium. By this way, real-time measurements have shown that the reference resonator resonances are linked to the temperature changes without any noticeable time delay, enabling an effective cancellation of temperature-induced shifts. A concentration of 20pM has been demonstrated by monitoring the binding between complementary IgG protein pairs. In particular, the sensor is able to detect a fluid refractive index fluctuation of $\pm 4\times10^{-6}$. Better results have been achieved by using an array of 32 silicon ring resonator sensors integrated on the same platform.

With a bulk refractive index sensitivity of 7.6×10^{-7} it is possible to detect a concentration of ~ 60fM, by using immobilized biotin to capture streptavidin diluted in bovine serum albumin solution (Iqbal et al., 2010).

In conclusion, a folded cavity SOI micro-ring sensor characterized by a folded spiral path geometry with a 1.2mm long ring waveguide, enclosed in a 150μm diameter sensor area, has been designed (Xu et al., 2008). The spiral cavity resonator is used to monitor the streptavidin protein binding with a detection limit of ~3 pg/mm^2, or a total mass of ~5fg.

The sensor presented above is characterized by Q-factor of 90,650 and an extinction ratio of 8.60 dB.

3.5.1 Resonant architectures for high performance photonic chemical sensing

Due to the increasing demand of high sensitivity and high limit of detection in biological and chemical detections, many kinds of sophisticated photonic sensors have been proposed. In particular, the minimum detectable wavelength shift for a traditional ring sensor is given by the resonance linewidth and resolution of the optical spectra analysis (OSA). In order to achieve ultra high limit of detection, very high resolution optical spectrum analysis and very stable tunable lasers have to be usually used for the whole measurement setup, usually not convenient and very expensive. An innovative approach consists in the joint use of two resonance cavities, for example a ring resonator or a racetrack integrated inside a Fabry-Perot cavity or two ring resonators integrated in the same photonic chip (Fig. 22a). By this way, interesting ultra high performance have been demonstrated.

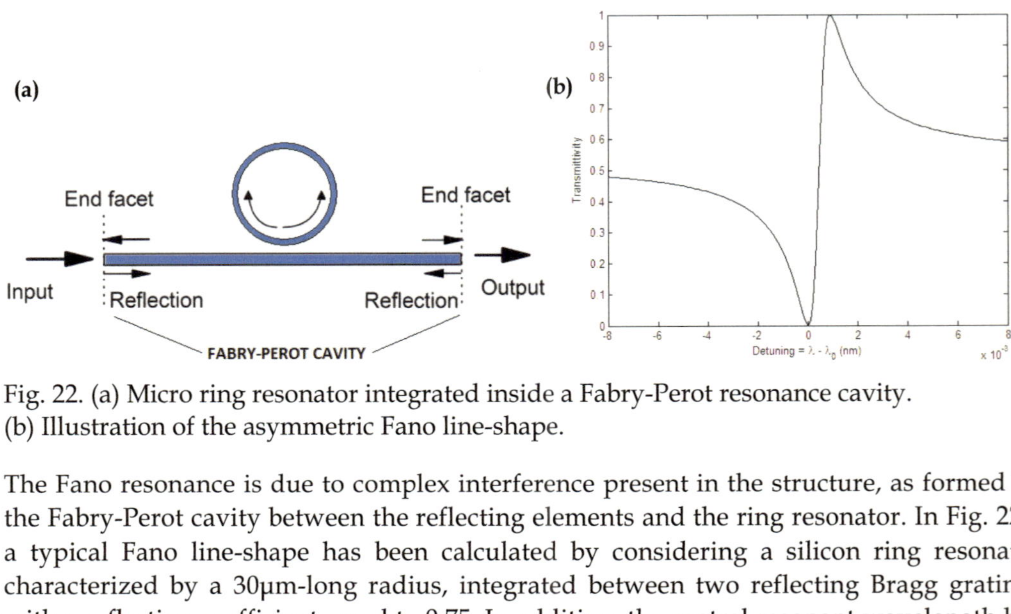

Fig. 22. (a) Micro ring resonator integrated inside a Fabry-Perot resonance cavity.
(b) Illustration of the asymmetric Fano line-shape.

The Fano resonance is due to complex interference present in the structure, as formed by the Fabry-Perot cavity between the reflecting elements and the ring resonator. In Fig. 22b, a typical Fano line-shape has been calculated by considering a silicon ring resonator characterized by a 30µm-long radius, integrated between two reflecting Bragg gratings with a reflection coefficient equal to 0.75. In addition, the central resonant wavelength has been imposed to be λ_0 = 1.535µm. Fano line-shape is characterized by a resonant peak steeper than the Lorentzian's one, that can be obtained with a standard ring resonator (Fig. 21b). By this way, the resolution of the photonic device and the limit of detection are improved. The improvement in terms of sensor performance is obtained because in Fano resonance the slope of the line-shape is greater than that obtained with Lorentzian resonance. To this purpose, a highly sensitive silicon micro-ring sensor with sharp asymmetrical resonance has been presented in literature (Yi et al., 2010). Coupled waveguides and micro-ring resonator have been fabricated using a SOI wafer which has a 1µm buffered oxide layer topped with 230nm of Si. Using a Fabry-Perot cavity characterized by a cavity length of 10mm, a quality factor Q = 3.8×10^4 and a limit of detection of $\sim10^{-8}$ have been measured.

The second approach allowing to ensure an ultra-sensitivity and the reduction of the detector complexity, consists in the integration of two cascaded ring resonators in the same chip. The principle used in this configuration is called "Vernier-scale" or "Vernier-effect". It consists of two wavelength scales with different periods, of which one slides the other one. The overlap between lines of two scales is used to perform the measurements.

In Fig. 23, two resonators with different optical roundtrip lengths ($R_2 < R_1$) are cascaded, so that drop signal of the first ring resonator serves as the input for the second one.

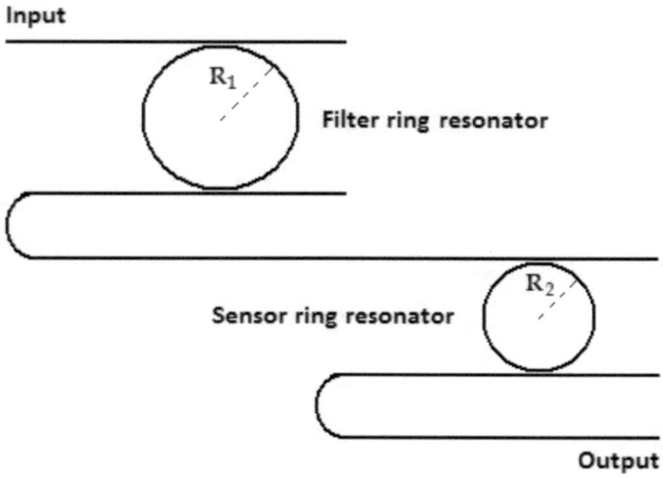

Fig. 23. Schematic view of a cascaded double ring photonic sensor based on Vernier effect.

Each individual ring resonator has a comb-like transmission spectrum with peaks at its resonance wavelengths, as already seen. The transmission spectrum of the cascade of two ring resonators is the product of the transmission spectra of the isolated individual resonators. Consequently, it will only exhibit peaks at wavelengths for which two resonance peaks of the respective ring resonators overlap, and the height of each of these peaks will be determined by the amount of this overlap. The sensor ring resonator will act as the sliding part of the Vernier-scale, as its evanescent field can interact with the refractive index in the environment of the sensor, where a change will cause a shift of the resonance wavelengths. The filter ring resonator is shielded from these refractive index changes by the cladding and will act as the constant part of the Vernier-scale. By this way, the whole architecture can be designed such that a small shift of the resonant wavelengths of the sensor ring resonator will result in a much larger shift of the transmission spectrum of the cascade. To this purpose, very interesting results have been demonstrated by using a silicon photonic biosensor whom architecture is the same as that sketched in Fig. 23 (Claes et al., 2010). The sensor has been tested with three aqueous solutions of NaCl with different concentrations and the operative wavelength range is from 1.52μm to 1.54μm. A sensitivity of 2169 nm/RIU has been experimentally determined in aqueous environment. In addition, the minimum detectable wavelength shift $\Delta\lambda$ is 18pm and the LOD is 8.3×10^{-6} RIU.

In conclusion, both integrated architectures analyzed above, can be designed by adopting slot waveguides in order to further improve their performance.

4. Conclusions

Photonic sensors represent a real revolution in chemical sensing technologies. In fact, their most important features are:

- Extremely high selectivity and sensitivity;
- Multi-variable and parallel processing in a chip;
- Wavelength readout (noise and interference immunity);
- Low-cost and high integration with front-end and support electronic systems (silicon, CMOS-compatible processing);
- Real-time processing.

In addition, the silicon-on-insulator technology platform allows to integrate photonics (laser, sensing architecture, photo-detector) and electronics (CMOS data processing) on the same chip in partially monolithic form to constitute the so called Lab-on-a-chip photonic system. Moreover, the refractive index contrast of the Si/SiO_2 material system, enables record reduction in photonic device footprint with ultra-high performance and portability. Undoubtedly, photonic Lab-on-a-chip represents the innovative approach to the main mission of the modern age: the improvement of the human life quality.

Finally, a comparative Table is presented below in order to appreciate all features and performance of actual different photonic sensor configurations.

Architecture	Technology	Performance	Size	Analyte	Author
MZI	SOI	8.7×10^{-7} RIU/ppm	2.1mm-long	BTEX	Saunders et al., 2010
MZI	CMOS-compatible	0.3 pg/mm^2	1.8mm-long (×9 - array)	IgG goat ,rabbit	Densmore et al., 2009
SPR	CMOS-compatible	3022nm/RIU 70pg/mm^2	~ 800μm^2	Molecules	DiPippo et al., 2010
Grating	SOI	~ 120nm ~ 10^{-4}RIU	173μm-long	Biological reactions	Passaro et al., 2008
PhC-slot	SOI	100ppm	300μm-long	Methane	Lai et al., 2011
PhC-slot	SOI	510nm/RIU 1×10^{-5}RIU	2μm-cavity length	Gases N_2, He, CO_2	Jagerska et al., 2010
Directional coupler	SOI	0.1 g/L	~ 1mm^2 (footprint)	Glucose	Passaro et al., 2009
MMI	SOI	+152, -172	1.607μm-long	Glucose, etanole	McCosker & Town, 2010b
Slot-ring resonator	SOI	2000nm/RIU 3.8×10^{-5}RIU	~ 1mm^2 (footprint)	Molecules, Gases	Passaro et al., 2011
Ring resonator	SOI	60fM	175×500μm^2 (×32- array)	DNA	Iqbal et al., 2010
Cascaded resonators	SOI	2169nm/RIU 8.3×10^{-6}RIU	200×70μm^2 (2x- array)	NaCl, molecules	Claes et al., 2010

Table 2. Comparative analysis of several silicon photonic platforms optimized for chemical and biochemical sensing.

5. Acknowledgments

This work has been supported by Fondazione della Cassa di Risparmio di Puglia, Bari, Italy.

6. References

Almeida, V.R.; Xu, Q.; Barrios, C.A. & Lipson, M. (2004). Guiding and confining light in void nanostructure. *Optics Letters,* Vol.29, No.11, (June 2004), pp. 1209-1211

Arshak, K.; Velusamy, V.; Korostynska, O.; Oliwa-Stasiak, K. & Adley, C. (2009). Conducting Polymers and Their Applications to Biosensors: Emphasizing on Foodborne Pathogen Detection. *IEEE Sensors Journal,* Vol.9, No.12, (December 2009), pp. 1942-1951

Baehr-Jones, T.; Hochberg, M.; Walker, C. & Scherer, A. (2005). *Applied Physics Letters,* Vol.86, No.8, (February 2005), art. 081101

Barrios, C.A. (2006). Ultrasensitive Nanomechanical Photonic Sensor Based on Horizontal Slot-Waveguide Resonator", *IEEE Photonic Technology Letters,* Vol.18, No.22, (November 2006), pp. 2419-2421

Bettotti, P.; Pitanti, A.; Rigo, E.; De Leonardis, F.; Passaro, V.M.N. & Pavesi, L. (2011). Modeling of Slot Waveguide Sensors Based on Polymeric Materials. *Sensors,* Vol.11, No.8, (August 2011), pp. 7327-7340

Claes, T.; Bogaerts, W. & Bienstam, P. (2010). Experimental characterization of a silicon photonic sensor consisting of two cascaded ring resonators based on the Vernier-effect and introduction of a curve fitting method for an improved detection limit. *Optics Express,* Vol.18, No.22, pp. 22747-22761

De Leonardis, F. & Passaro, V.M.N. (2007). Modeling and Performance of a Guided-Wave Optical Angular Velocity Sensor based on Raman Effect in SOI, *IEEE J. Lightwave Technol.,* Vol.25, No.9, (September 2007), pp. 2352-2366

De Leonardis, F.; Giannoccaro, G.; Troia, B.; Passaro, V.M.N. & Perri, A.G. (2011). Design of Optimized SOI Slot Waveguides for Homogeneous Optical Sensing in Near Infrared. *Int. Workshop on Sensors and Interfaces (IWASI 2011),* ISBN: 978-1-4577-0624-0, pp. 142-147, June 28-29, 2011

Dell'Olio, F. & Passaro, V.M.N. (2007). Optical sensing by optimized silicon slot waveguides. *Optics Express,* Vol.15, No.8, (April 2007), pp. 4977-4993

Densmore, A.; Vachon, M.; Xu, D.-X.; Janz, S.; Ma, R.; Li, Y.-H.; Lopinski, G.; Delage, A.; Lapointe, J.; Luebbert C.C.; Liu, Q.Y.; Cheben, P. & Schmid J.H. (2009). Silicon photonic wire biosensor array for multiplexed real-time and label free molecular detection. *Optics Letters,* Vol.34, No.23, (November 2009), pp. 3598-3600

DiPippo, W.; Jae Lee, B. &Park, K. (2010). Design analysis of doped-silicon Plasmon resonance immunosensors in mid-infrared range. *Optics Express,* Vol.18, No.18, (August 2010), pp. 19396-19406

Gylfason, K.B.; Carlborg, C.F.; Kazmierczak, A.; Dortu, F.; Sohlstrom, H.; Vivien, L.; Barrios, C.A.; Van Der Wijngaart, W. & Stemme, G. (2010). On-chip temperature compensation in an integrated slot waveguide ring resonator refractive index sensor array. *Optics Express,* Vol.18, No.4, (February 2010), pp. 3226-3237

Hu, S.; Liu, F.; Wan, R. & Huang, Y. (2010). Coupling characteristics between slot plasmonic mode and dielectric waveguide mode. *Communications and Photonics Conference and Exhibition (ACP)*, ISBN: 978-1-4244-7111-9, December 8-12, 2010

Iqbal, M.; Zheng, Z. & Jiansheng, L. (2008). Microwave and Millimeter Wave Technology, *Proceedings of ICMMT*, pp. 878-881, ISBN 978-1-4244-1879-4, April 21-24, 2008

Iqbal, M.; Gleeson, M.A.; Spaugh, B.; Tybor, F.; Gunn, W.G.; Hochberg, M.; Baehr-Jones, T.; Bailey, R.C. & Gunn, L.C. (2010). Label-Free Biosensor Arrays Based on Silicon Ring Resonators and High-Speed Optical Scanning Instrumentation. *IEEE Journal of Selected Topics in Quantum Electronics*. Vol.16, No.3, (May-June 2010), pp. 654-661

Jagerska, J.; Zhang, H.; Diao, Z.; Le Thomas, N. & Houdrè, R. (2010). Refractive index sensing with an air-slot photonic crystal nanocavity. *Optics Letters*, Vol.35, No.15, (August 2010), pp. 2523-2525

Jha, R. & Sharma, A.K. (2010). Design of a silicon-based plasmonic biosensor chip for human blood-group identification. *Sensors and Actuators B: Chemical*, Vol.145, No.1, (March 2010), pp. 200-204

Kargar A. & Lee C. (2009). Optical sensing by multiple-slot waveguide microring resonators. *9th IEEE Conference on Nanotechnology*. ISBN: 978-1-4244-4832-6, July 26-30, 2009

Kuo, W.-K. & Chang C.-H. (2010). Phase detection properties of grating-coupled surface Plasmon resonance sensors. *Optics Express*, Vol.18, No.19, (September 2010), pp. 19656-19664

Lai, W.-C.; Chakravarty, S.; Wang, X.; Lin, C. & Chen, R.T. (2011). On-chip methane sensing by near-IR absorption signatures in a photonic crystal slot waveguide. *Optics Letters*, Vol.36, No.6, (March 2011), pp. 984-986

Le, L.; Suihua, M.; Yanhong, J.; Xinyuan, C.; Zhiyi, L.; Yonghong, H. & Jihua, G. (2011). A two-dimensional polarization interferometry based parallel scan angular surface plasmon resonance biosensor. *Review of Scientific Instruments*, Vol.82, No.2, (February 2011), art. 023109

Lee, S.M.; Saini, S.S. & Jeong, M.Y. (2010). Parametric Discrimination between Refractive Index, Temperature and Strain using Etched-Core FBG. *23rd Annual Meeting of the IEEE Photonics Society*, ISBN: 978-1-4244-5368-9, November 7-11, 2010

Leheny, R.F. & McCants, C.E. (2009). Technologies for Photonic Sensor Systems, *Proceedings of the IEEE*, Vol.97, No.6, (June 2009), pp. 957-970

Li, S.; Yin, S.; Jiang, Y.; Yin, C; Deng, Q. & Du, C. (2010). Specific Protein Detection in Multiprotein Coexisting Environment by Using LSPR Biosensor. *IEEE Transactions on Nanotechnology*, Vol.9, No.5, (September 2010), pp. 554-557

Malathi, S.; Rani, K.E. & Srinivas, T. (2010). Design and analysis of SOI based biosensor with integrated optic readout. *Computing Communication and Networking Technologies*, ISBN: 978-1-4244-6591-0, July 29-31, 2010

Marrocco, V.; Vincenti, M.A.; Grande, M.; Calò, G.; Petruzzelli, V.; Prudenzano F. & D'Orazio, A. (2010). Field localization in Bragg waveguide assisted by metal layers. *Journal of the Optical Society of America B*, Vol.27, No.4, (March 2010), pp. 703-707

McCosker, R.J. & Town, G.E. (2010a). Optical chemical sensor using a multi-channel directional coupler with slot waveguides. *International Conference on Photonics (ICP)*, ISBN: 978-1-4244-7186-7, July 5-7, 2010

McCosker, R.J. & Town, G.E. (2010b). Multi-channel directional coupler as an evanescent field sensor. (2010). *Sensors and Actuators B: Chemical*, Vol.150, No.1, (September 2010), pp. 417-424

Passaro, V.M.N.; Dell'Olio, F.; Casamassima, B. & De Leonardis, F. (2007a). Guided-Wave Optical Biosensors, *Sensors*, Vol.7, No.4, (April 2007), pp. 508-536

Passaro, V. M. N.; Dell'Olio, F. & De Leonardis, F. (2007b). Ammonia Optical Sensing by Microring Resonators. *Sensors*, Vol.7, No.11, (November 2007), pp. 2741-2749

Passaro, V.M.N.; Loiacono, R.; D'Amico, G. & De Leonardis, F. (2008). Design of Bragg Grating Sensors Based on Submicrometer Optical Rib Waveguides in SOI. *IEEE Sensors Journal*, Vol.8, No.9, (September 2008), pp. 1603-1611

Passaro, V.M.N. (2009a). *Modeling of Photonic Devices*, NOVA Science Publ., ISBN: 978-1-60456-980-3, New York, USA

Passaro, V.M.N.; Dell'Olio F.; Ciminelli C. & Armenise M.N. (2009b). Efficient Chemical Sensing by Coupled Slot SOI Waveguides. *Sensors*, Vol.9, No.2, (February 2009), pp. 1012-1032

Passaro, V.M.N.; Troia, B.; De Leonardis, F. (2011). Group IV Photonic Slot Structures for Highly Efficient Gas Sensing in mid-IR, *IARIA Int. Conference on Sensor Device Technologies and Applications (SENSORDEVICES 2011)*, Nizza (FR), ISBN: 978-1-61208-145-8, August 21-27, 2011

Saunders, J.; Dreher, M.A.; Barnes, J.A.; Crudden, C.M.; Du, J.; Loock, H.; Dan-Xia Xu; Densmore, A.; Ma, R.; Janz, S.; Vachon, M.; Lapointe, J.; Delage, A.; Schmid, J. & Cheben, P. (2010). Detection of lead contamination of water and VOC contamination of air using SOI micro-optical devices. *7th IEEE International Conference on Group IV Photonics*. ISBN: 978-1-4244-6344-2, 1-3 September 2010

Sun, R.; Dong, P.; Feng, N.; Hong, C.; Michel, J.; Lipson, M. & Kimerling, L. (2007). Horizontal single and multiple slot waveguides: optical transmission at $\lambda = 1550$nm. *Optics Express*, Vol.15, No.26, (December 2007), pp. 17967-17972

Sun, H.; Chen, A. & Dalton, L.R. (2009). Enhanced Evanescent Confinement in Multiple-Slot Waveguides and Its Application in Biochemical Sensing. *IEEE Photonics Journal*, Vol.1, No.1, (June 2009), pp. 48-57

Troia, B.; Passaro, V. M. N. & De Leonardis, F. (2011). Bandgap Engineering and Optical Properties of Group IV Material Compounds, *Proceedings of 13° Nat. Conf. of Photonic Technologies (Fotonica 2011)*, ISBN: 978-8-8872-3712-2, Genova, Italy, May 9-11, 2011

Tung, B.T.; Dao, V.D.; Ikeda, T.; Kanamori, Y.; Hane, K. & Sugiyama S. (2011). Investigation of strain sensing effect in modified single-defect photonic crystal nanocavity. *Optics Express*, Vol.19, No.9, (April 2011), pp. 8821-8829

Xu, D.-X.; Densmore, A.; Delage, A.; Waldron, P.; McKinnon, R.; Janz, S.; Lapointe, J.; Lopinski, G.; Mischki, T.; Post, E; Cheben, P. & Schmid, J.H. (2008). Folded cavity SOI microring sensors for high sensitivity and real time measurement of biomolecular binding. *Optics Express*, Vol.16, No.19, (September 2008), pp. 15137-15148

Xu, D.-X.; Vachon, M.; Densmore, A.; Ma, R.; Janz, S.; Delage, A.; Lapointe, J; Cheben, P.; Schmid, J.H.; Post, E.; Messaoudene, S. & Fedeli, J.-M. (2010). Real-time cancellation of temperature induced resonance shifts in SOI wire waveguide ring resonator label-free biosensor arrays. *Optics Express*, Vol.18, No.22, (October 2010), pp. 22867-22879

Yi, H.; Citrin, D.S. & Zhou, Z. (2010). Highly sensitive silicon microring sensor with sharp asymmetrical resonance. *Optics Express*, Vol.18, No.3, (February 2010), pp. 2967-2972

6

Coumarin-Derived Fluorescent Chemosensors

Hongqi Li, Li Cai and Zhen Chen
College of Chemistry,
Chemical Engineering and Biotechnology,
Donghua University, Shanghai,
China

1. Introduction

Fluorescent chemosensors are highly valuable in a variety of fields including environmental chemistry, analytical chemistry, and bio-medicinal science. They have provided accurate, on-line, and low-cost detection of toxic heavy metal ions, anions, and enzymes with high selectivity and sensitivity. Coumarins, with the structure of benzopyrone, have many advantages including high fluorescence quantum yield, large Stokes shift, excellent light stability, and less toxicity. Therefore coumarins have been widely used in the fields of biology, medicine, perfumes, cosmetics, and fluorescent dyes. By far coumarin derivatives have been used as fluorescent probes of pH, for detection of nitric oxide, nitroxide, and hydrogen peroxide. Moreover, coumarin derivatives have served as good chemosensors of anions including cyanide, fluoride, pyrophosphate, acetate, benzoate, and dihydrogenphosphate as well as various metal ions comprised of Hg(II), Cu(II), Zn(II), Ni(II), Ca(II), Pb(II), Mg(II), Fe(III), Al(III), Cr(III), and Ag(I). Several systems containing coumarin exhibited simultaneous sensitivity toward two or more different metal ions, e.g. Ca(II) and Mg(II), Ni(II) and Co(II), Cu(II) and Hg(II), Na(I) and K(I), Cu(II) and Ni(II), Hg(II) and Ag(I), Cu(II)/Ni(II)/Cd(II), Zn(II)/Cd(II)/Pb(II), or Ni(II)/Pd(II)/Ag(I). Herein a brief review of fluorescent chemical sensors derived from coumarins is presented.

2. Structural characteristics of coumarin

The fusion of a pyrone ring with a benzene ring gives rise to a class of heterocyclic compounds known as benzopyrones, of which two distinct types are recognized, namely benzo-α-pyrones, commonly called coumarins, and benzo-γ-pyrones, called chromones, the latter differing from the former only in the position of the carbonyl group in the heterocyclic ring as shown in Fig. 1 (Sethna & Shah, 1945). It is well known that stilbene with a *trans* conformation is highly fluorescent. From the viewpoint of molecular structure, coumarins bear a carbon-carbon double bond which is fixed as *trans* conformation as in *trans*-stilbene through a lactone structure. This can help to avoid the *trans-cis* transformation of the double bond under ultraviolet (UV) irradiation as observed in stilbene compounds and results in strong fluorescence and high fluorescence quantum yield and photostability in most of coumarin derivatives.

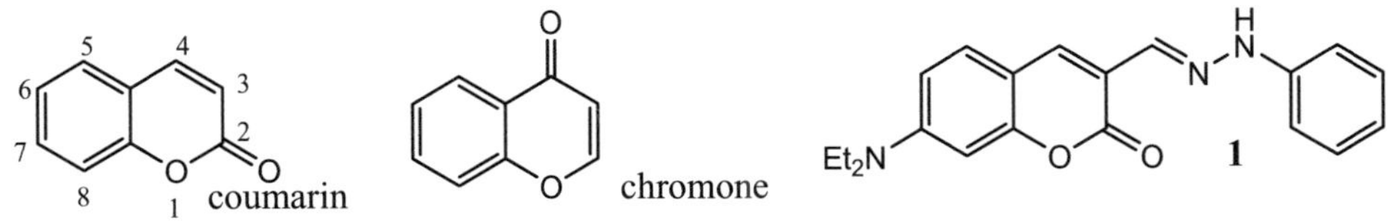

Fig. 1. Structures and numbering scheme of coumarin and related compounds

It was showed in the late 1950s that substitutions on the coumarin structure shifted the fluorescence band. For instance, adding a methyl group to the 4-position of 7-hydroxy- or 7-methoxycoumarin red shifts the fluorescence spectra. Addition of electron-repelling groups in the 4-, 6-, or 7-position or electron-attracting groups in the 3-position all shifts the fluorescence band to longer wavelengths. When the carbonyl is substituted with a thione, the absorbance was red shifted and the fluorescence was quenched (Trenor et al., 2004). The reduction of the acceptor strength of the substituent at the 3-position, as in a 7-diethylaminocoumarin dye **1** (the structure of which is shown in Fig. 1) does not always result in sustained fluorescence in polar, aprotic solvents. Thus 7-diethylamino-3-styrylcoumarin dyes are not technically important as laser dyes; however, the extreme sensitivity of coumarin **1** to the medium polarity could provide an opportunity to probe the microenvironment experienced by the molecule (Bangar Raju & Varadarajan, 1995).

Changing the solvent or the solution pH also affected the fluorescence spectra. Study on the effect of solution pH on 7-hydroxy-4-methylcoumarin showed that increasing the solution pH raised the fluorescence intensity. Studies on the effect of changing the solvent polarity on 13 coumarin derivatives revealed that increasing solvent polarity red shifted the absorbance as well as red shifted and broadened the emission of the coumarins due to increased hydrogen bonding. Studies on the excited-state properties of 4- and 7-substituted coumarin derivatives revealed that solvent polarity shifted both the emission and absorption peaks, with a greater shift observed in the emission spectra, indicating that the excited- state dipole moment of the solute molecule was greater than the ground-state dipole moment (Trenor et al., 2004).

3. Coumarin-derived fluorescent chemosensors

3.1 Coumarin-derived fluorescent chemosensors for metal ions

Considering the threat of mercury to the environment and human health, a great effort has been devoted to the utilization of fluorescent methods for detection of Hg^{2+} ions. More than ten coumarin-derived fluorescent chemosensors for Hg^{2+} ions have been reported. The structures and references of these chemical sensors are listed in Table 1.

The recognition mechanisms of these chemosensors mainly involve photoinduced electron transfer (PET), intramolecular charge transfer (ICT), fluorescence resonance energy transfer (FRET), coordination, and desulfurization. For instance, the fluorescence detection of **2** upon Hg^{2+} addition is promoted by a Hg^{2+}-induced desulfurization of the thiourea moiety, leading to a decrease in an ICT character of the excited-state coumarin moiety (Shiraishi et al., 2010). Coumarin-thiazolobenzo-crown ether based chemosensor **3** has been developed for Hg(II) ions that utilizes the strong coordination of Hg(II) ions on the crown oxygen and thiazole nitrogen. The complexation of Hg(II) disrupts the ICT from the oxygen donor to the

Structure of sensor	Reference	Structure of sensor	Reference
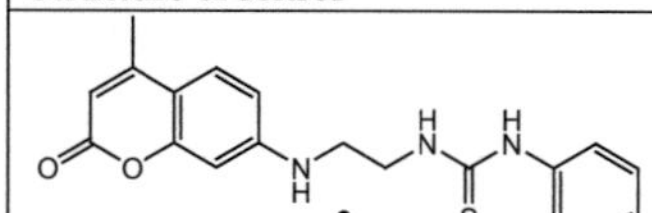 **2**	Shiraishi et al., 2010	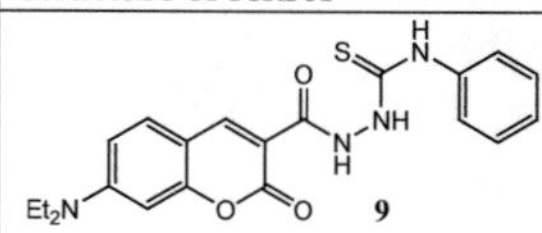 **9**	W. Ma et al., 2010
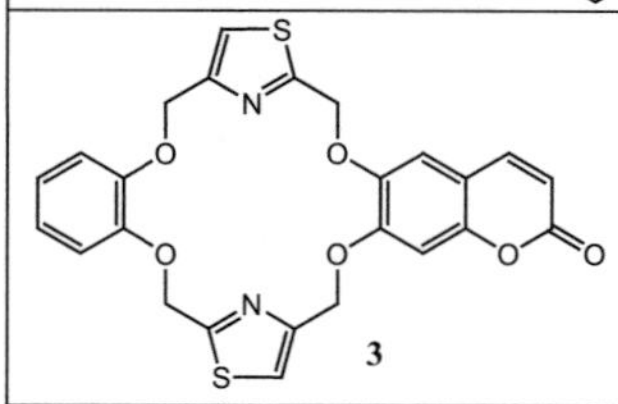 **3**	S. H. Lee et al., 2010	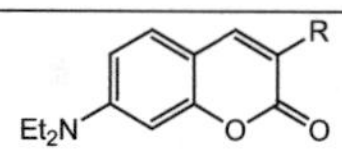 **10**: R = acetylene  **11**: R = 1,3-dithiane-2-yl 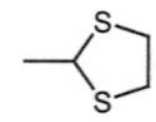	D.-N. Lee et al., 2009 J. H. Kim et al., 2009
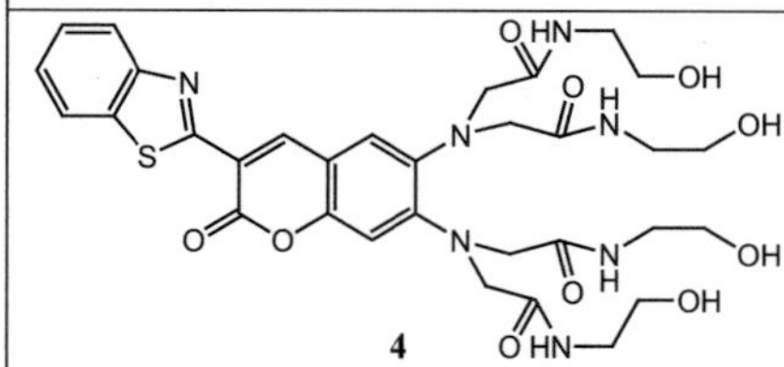 **4**	Wang et al., 2006	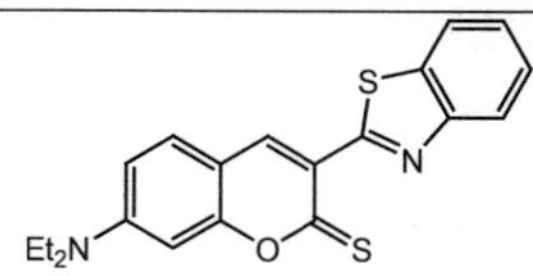 **12**	M. G. Choi et al., 2009
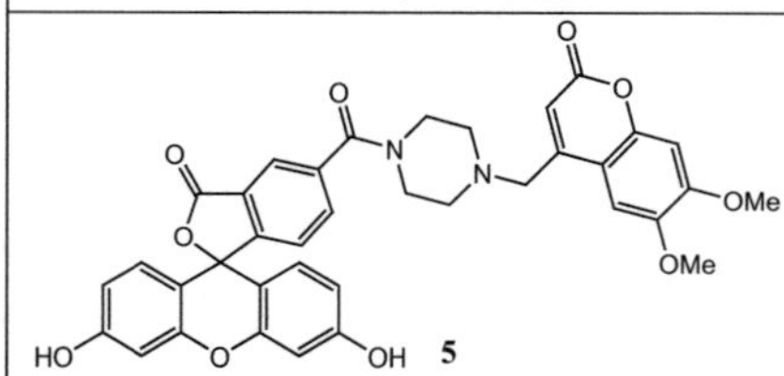 **5**	Ryu et al., 2010	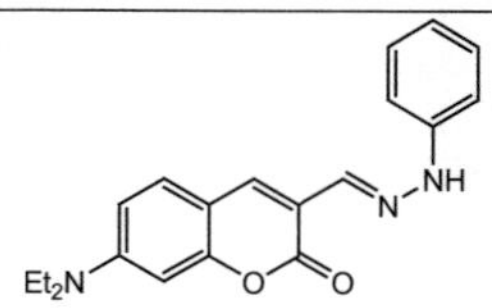 **13**	H. J. Kim et al., 2010
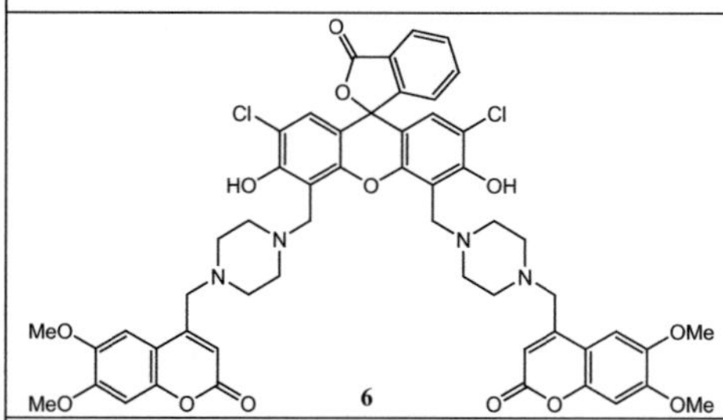 **6**	H. J. Kim et al., 2009a	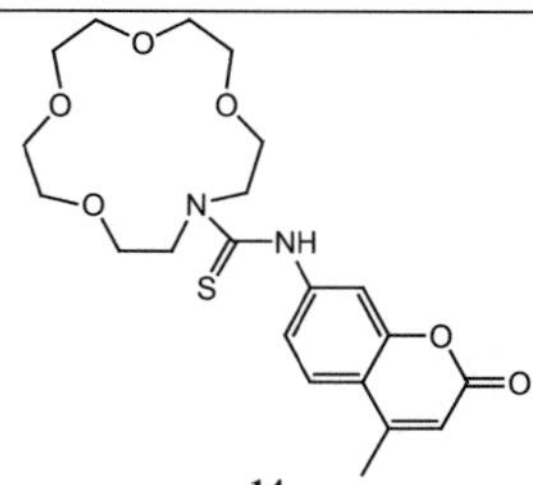 **14**	Voutsadaki et al., 2010
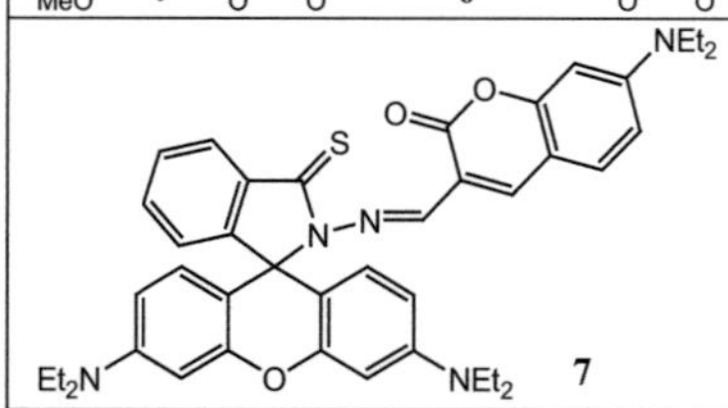 **7**	Q.-J. Ma et al., 2010	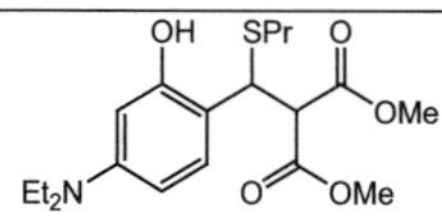 **15**	Jiang & Wang, 2009
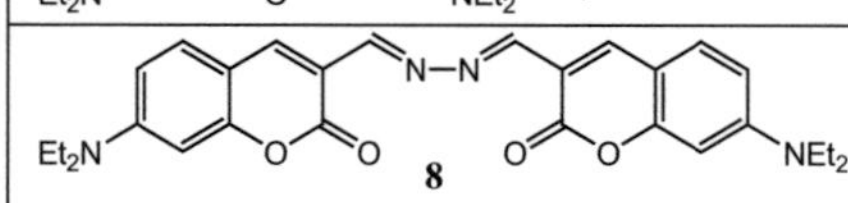 **8**	R. Sheng et al., 2008	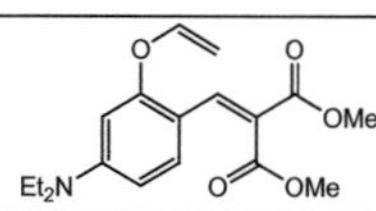 **16**	Cho & Ahn, 2010

Table 1. Structures and references of coumarin-derived chemosensors for Hg(II) ions

coumarin fluorophore resulting in blue-shift in absorption and quenching of the fluorescence (S. H. Lee et al., 2010). The fluorescent ratiometric Hg^{2+} ion sensor **4,** based on a coumarin platform coupled with a tetraamide receptor, can specifically detect Hg^{2+} ions through the ICT mechanism (Wang et al., 2006). Fluorescein-coumarin chemodosimeter **5** for signaling Hg^{2+} ions is designed based on FRET arising from the interaction between a pair of fluorophores (Ryu et al., 2010). Rhodamine-coumarin conjugate **7** was developed as a probe for Hg(II) ions. The fluorescence response to Hg(II) ions is attributed to the 1:1 complex formation between probe **7** and Hg^{2+} (Q.-J. Ma et al., 2010). Chemosensor **9** based on the coumarin thiosemicarbazide displays a selective fluorescence enhancement for Hg^{2+}, which is attributed to the transformation of thiosemicarbazide unit to 1,3,4-oxadiazoles via Hg^{2+}-induced desulfurization reaction in aqueous media (W. Ma et al., 2010). Coumarinyldithiane **11** and thiocoumarin **12** selectively sense Hg^{2+} also due to the Hg^{2+}-induced desulfurization reaction. Probe **14** belongs to the turn-on class of sensors, functioning via a PET process (Voutsadaki et al., 2010).

Chemosensors **15** and **16** are not coumarin derivatives but in the presence of Hg(II) ions the weakly fluorescent precursor **15** can be transformed to strongly fluorescent coumarin **17** via a desulfurization-lactonization cascade reaction as shown in Fig. 2 (Jiang & Wang, 2009). Similarly, **16** selectively senses inorganic mercury in the turn-on mode through a Hg(II) ion-promoted hydrolysis-cyclization reaction that leads to coumarin **17** as shown in Fig. 3 (Cho & Ahn, 2010).

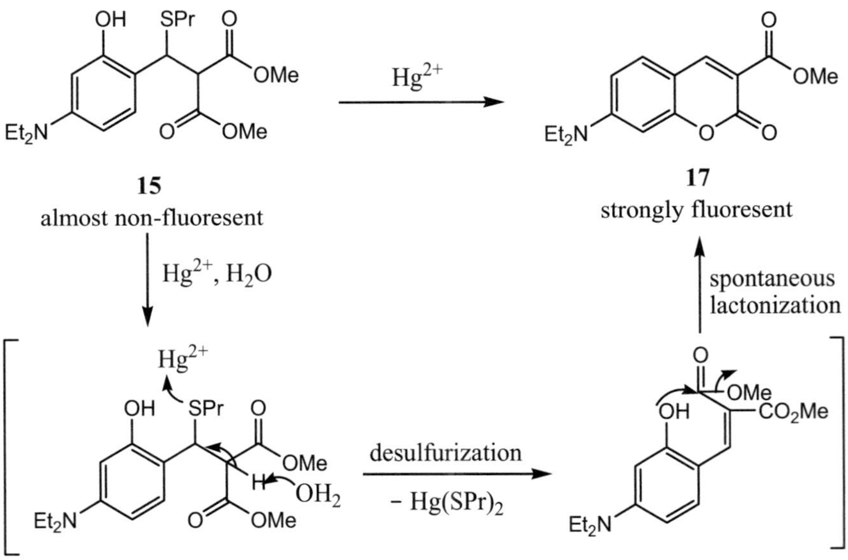

Fig. 2. Transformation of sensor **14** to fluorescent coumarin **16** via Hg(II)-induced desulfurization-lactonization

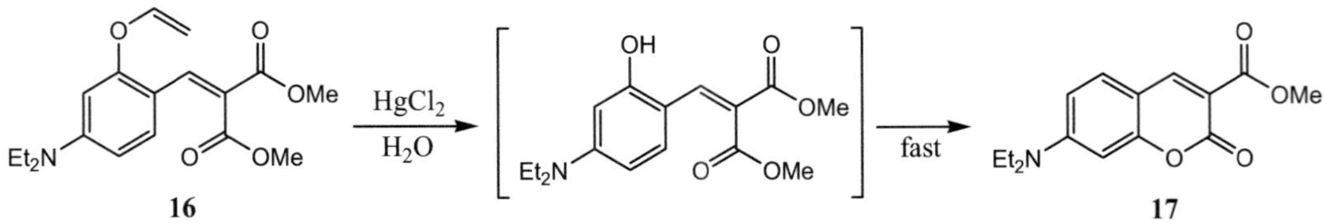

Fig. 3. The reaction of probe **16** with $HgCl_2$

Detection of trace amount of Cu^{2+} is important not only for environmental applications, but also for toxicity determination in living organs. Following the report that a new cavitand bearing four coumarin groups acts as fluorescent chemosensor for Cu^{2+} (Jang et al., 2006), over ten more coumarin-derived fluorescent chemosensors for Cu(II) ions have been envisaged. The structures and references of these chemical sensors are listed in Table 2.

Structure of sensor	Reference	Structure of sensor	Reference
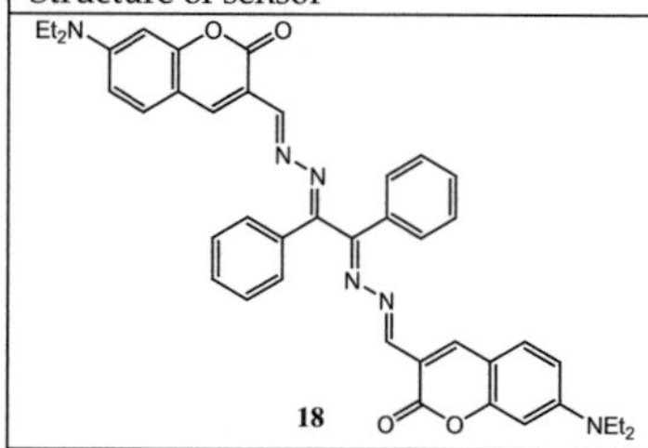 18	G. He et al., 2010a	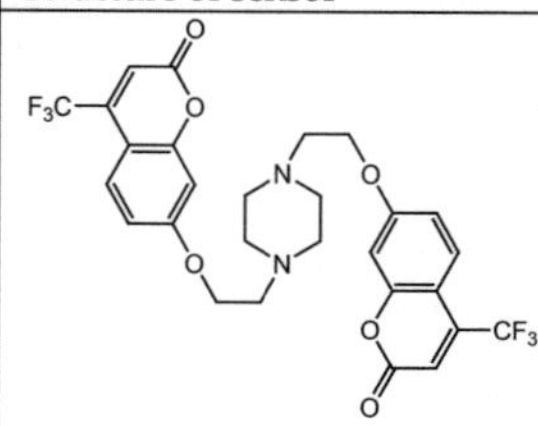 24	X. Chen et al., 2011
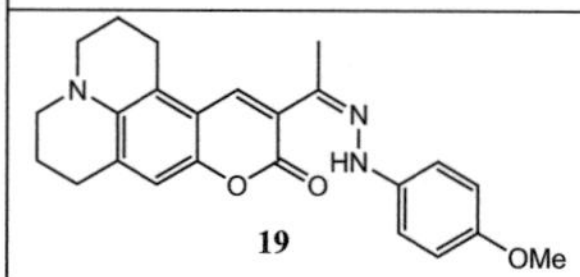 19	M. H. Kim et al., 2009	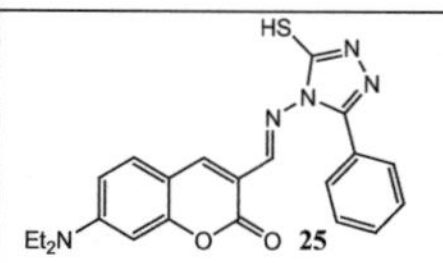 25	Ko et al., 2011
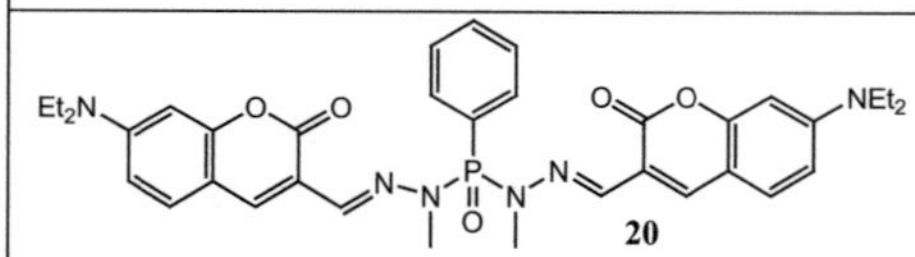 20	Chandrasekhar et al., 2009	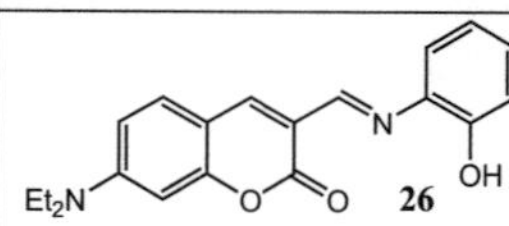 26	N. Li et al., 2010
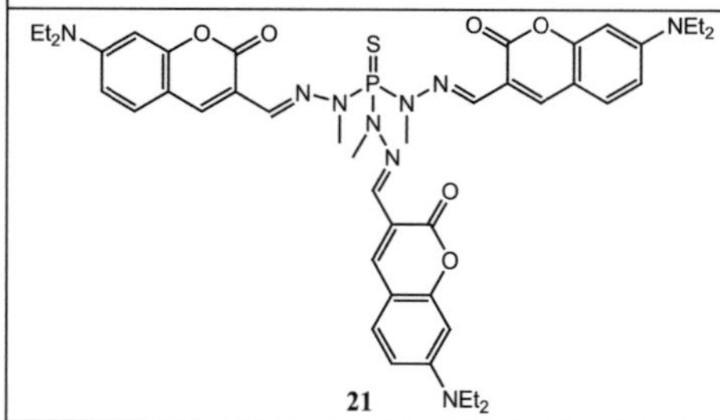 21	Chandrasekhar et al., 2009	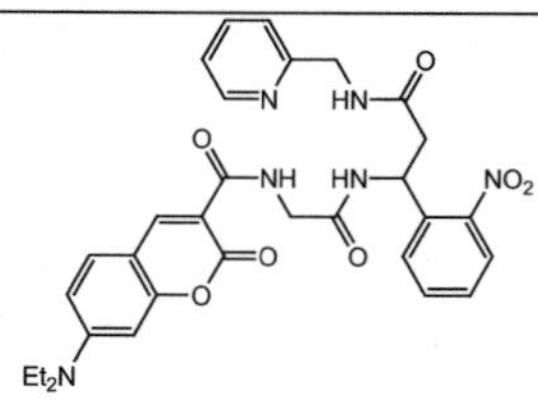 27	Ciesienski et al., 2010
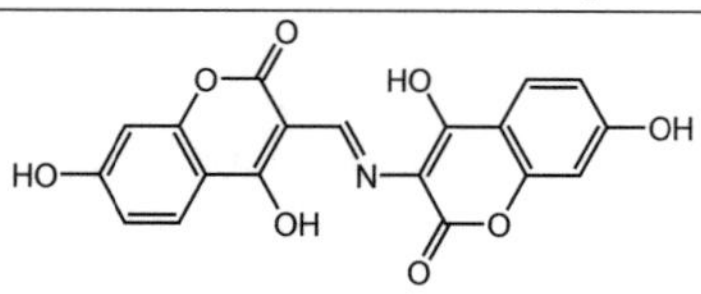 22	J. R. Sheng et al., 2008	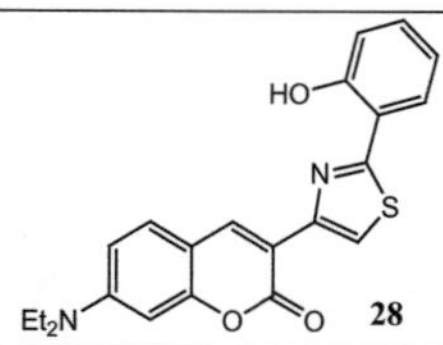 28	Helal et al., 2011
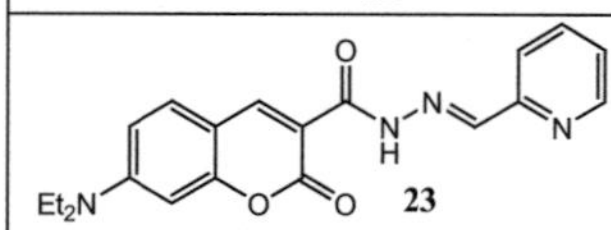 23	Zhao et al., 2010	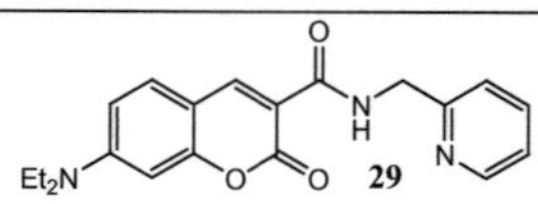 29	Jung et al., 2009

Table 2. Structures and references of coumarin-derived chemosensors for Cu(II) ions

A turn-on fluorescent probe **18** for Cu^{2+} is presented by incorporating coumarin fluorophores within the benzyl dihydrazone moiety. It is among the brightest Cu^{2+} binding sensors in aqueous media reported to date (G. He et al., 2010a). Coumarin **19** is a highly effective turn-on fluorescent sensor that is catalytically hydrolyzed by Cu^{2+} leading to a large increase in the fluorescence intensity (M. H. Kim et al., 2009). Studies on interaction of phosphorus-supported multidentate coumarin-containing fluorescent sensors **20** and **21** with various transition metal ions reveal substantial fluorescence enhancement upon interaction with Cu^{2+} enabling a selective detection mechanism for Cu^{2+} (Chandrasekhar et al., 2009). Biscoumarin **22** linked by a C=N double bond is highly sensitive and selective to Cu^{2+} and the fluorescent sensing mechanism is based on C=N isomerization (J. R. Sheng et al., 2008). Coumarin probe **23** is highly selective for Cu^{2+} over biologically relevant alkali metals, alkaline earth matals and the first row transition metals due to the formation of a 1:2 complex between Cu^{2+} and **23** (Zhao et al., 2010). Fluorescent biscoumarin **24** linked by a piperazine unit shows high selectivity towards Cu^{2+} (Chen et al., 2011). Rationally designed iminocoumarin fluorescent sensor **25** displays high selectivity for Cu^{2+} over a variety of competing metal ions in aqueous solution with a significant fluorescence increase (Ko et al., 2011). Nonfluorescent coumarin derivative **26** is synthesized as an efficient turn-on fluorescent chemodosimeter for Cu^{2+} in water. Mechanism studies suggest that **26** forms a complex with Cu^{2+} in a 1:2 metal-to-ligand ratio, and a 50-fold fluorescence enhancement is observed when the complex simultaneously undergoes Cu^{2+}-promoted hydrolysis (N. Li et al., 2010). Sensor **27** relies on a coumarin-tagged ligand that selectively binds Cu^{2+} over other biometals to induce fluorescence quenching, which is subsequently relieved upon UV irradiation to provide the turn-on response (Ciesienski et al., 2010). For chemosensor **28** the mechanism of fluorescence is based on ICT, which is modified by the introduction of an electron-donating diethylamino group making it chromogenic and increasing the binding affinity (Helal et al., 2011). Coumarin **29** appending 2-picolylamide enables efficient tridentate complexation for Cu(II) in preference to a variety of other common heavy and toxic metal ions (Jung et al., 2009).

Owing to the important role of zinc, the second most abundant transition metal in the human body, more and more attention has been paid to development of Zn^{2+}-specific chemosensors including coumarin-derived fluorescent chemosensors for Zn(II) ions, the structures and references of which are listed in Table 3.

Study on the sensory capabilities of two novel di(2-picolyl)amine (DPA)-substituted coumarins **30** and **31** shows that the variation of the point of attachment of the DPA group to the coumarin framework controls their sensing behavior: the 4-subsituted system **30** is a chelation-enhanced fluorescence (CHEF)-type sensor which shows a significant increase in fluorescence intensity upon Zn^{2+} binding, whereas the 3-substituted coumarin **31** is a ratiometric sensor (N. C. Lim & Brückner, 2004). Coumarin Schiff-base **32** acts as a turn-on fluorescent chemosensor for Zn(II) ions (H. Li et al., 2009) and **33** does not show any two-photon activity in the wavelength range 760-860 nm but in the presence of Zn(II) **33** exhibits large two-photon absorption as well as emission in the same wavelength range (Ray et al., 2010). Another coumarin Schiff-base **43** is a highly sensitive and selective fluorescent probe for Zn^{2+} in tetrahydrofuran (THF) (Yan et al., 2011). In coumarin Schiff-base **34** the fluorescence quenching is dominant because of the nitrogen lone pair orbital contribution to the excitation. Upon Zn^{2+} coordination **34** shows a significant fluorescence enhancement due to the blocking of the nitrogen lone pair orbital by metal coordination (Jung et al., 2010). A series of coumarin-based fluorescent probes **35-38** for detecting Zn^{2+} with high affinities show the ratiometric fluorescent properties (Mizukami et al., 2009). Another series of coumarin-derived chemosensors **39-42** belong to the CHEF-type and have been showed to

Structure of sensor	Reference	Structure of sensor	Reference
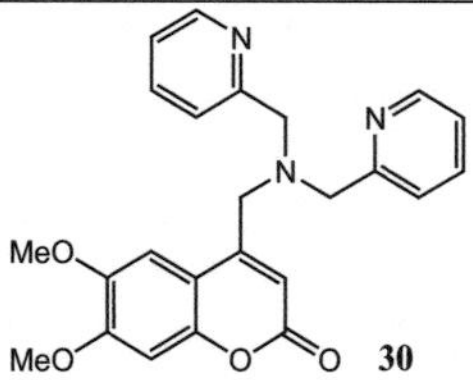 30	N. C. Lim & Brückner, 2004	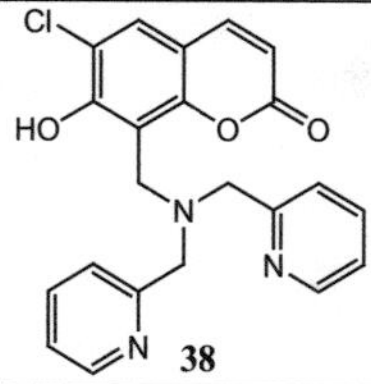 38	Mizukami et al., 2009
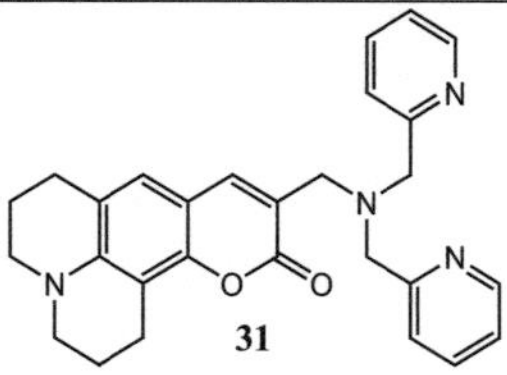 31	N. C. Lim & Brückner, 2004	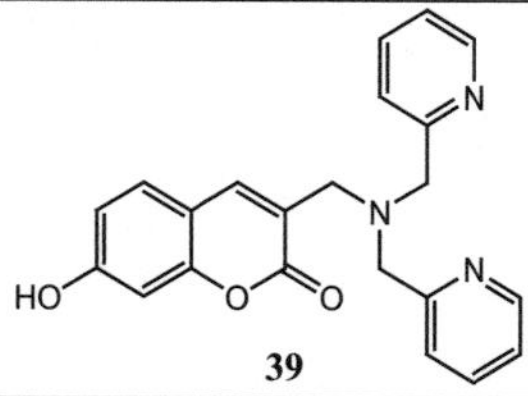 39	N. C. Lim et al., 2005
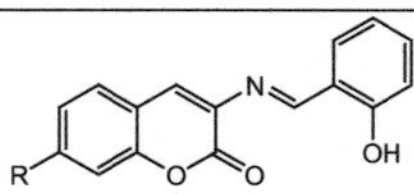 **32**: R = OH **33**: R = NEt$_2$	H. Li et al., 2009 Ray et al., 2010	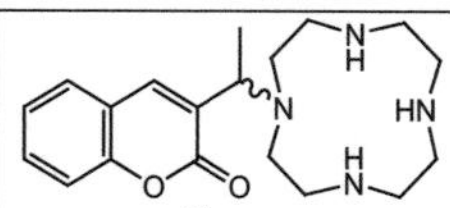 40 • 4 HCl	N. C. Lim et al., 2005
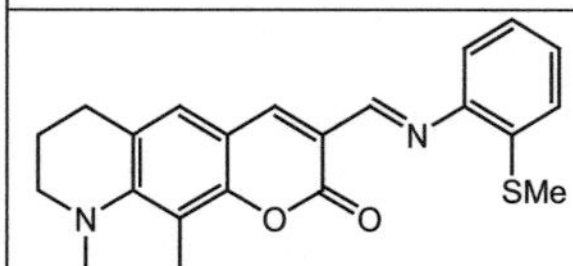 34	Jung et al., 2010	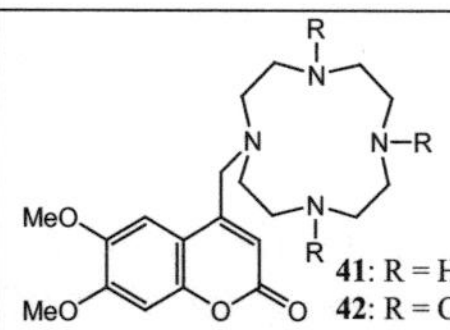 **41**: R = H **42**: R = CH$_2$CO$_2$Et	N. C. Lim et al., 2005
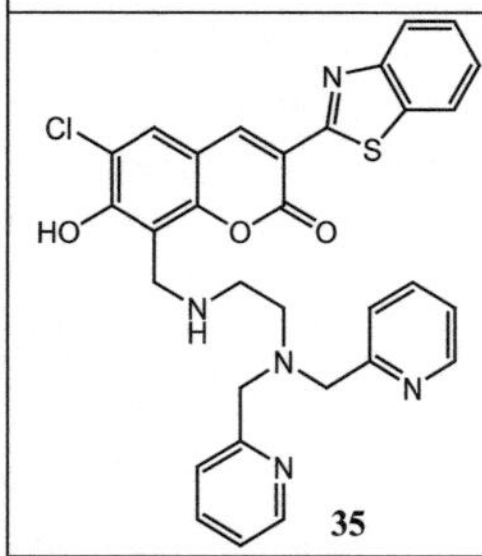 35	Mizukami et al., 2009	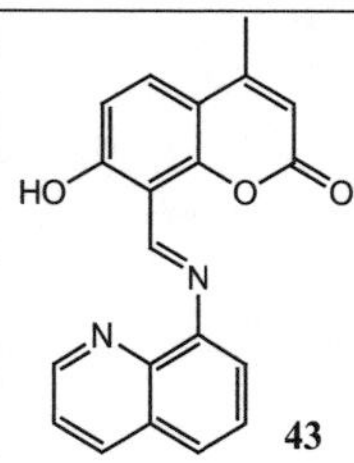 43	Yan et al., 2011
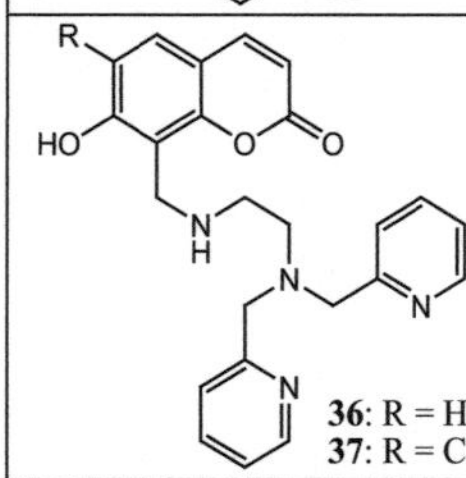 **36**: R = H **37**: R = Cl	Mizukami et al., 2009	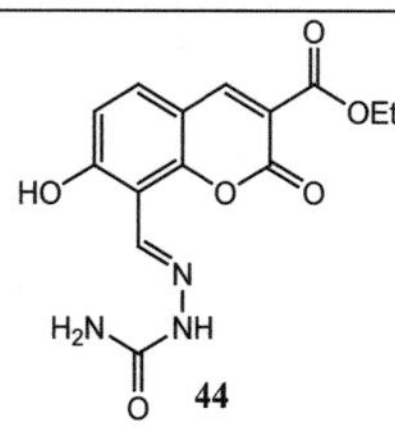 44	Su et al., 2010

Table 3. Structures and references of coumarin-derived chemosensors for Zn(II) ions

be competent for detecting zinc pools in cultured rat pituitary (GH3) and hepatoma (H4IIE) cell lines (N. C. Lim et al., 2005). Coumarin **44** is a fluorescent sensor for Zn^{2+} and exhibits lower background fluorescence due to intramolecular PET but upon mixing with Zn^{2+} in aqueous ethanol, a turn-on fluorescence emission is observed (Su et al., 2010). Recently it was reported that a biscoumarin linked by bi-thiazole acted as a colorimetric receptor selectively for Zn^{2+} (Upadhyay & Mishra, 2010).

Structures of several coumarin-derived fluorescent chemosensors for iron(III) are shown in Fig. 4. Squarate hydroxamate-coumarin conjugate **45** is designed as a CHEF-type sensor for Fe(III). Due to a PET process, **45** possesses a low fluorescence yield but upon exposure of **45** to Fe(III), an irreversible 9-fold fluorescence intensity increase is observed as the result of an oxidation/hydrolysis reaction (N. C. Lim et al., 2009). Coumarin derivative **46** exhibits high selectivity for Fe^{3+} and the selectivity is not affected by the presence of representative alkali metals, alkali earth metals and other transition metal salts (Yao t al., 2009). Mugineic acid-coumarin derivative **47** synthesized by click chemistry acts as a fluorescent probe for Fe^{3+} (Namba et al., 2010). Coumarin-based hexadentate fluorescent probes for selective quantification of iron(III) have also been designed and synthesized (Y. M. Ma & Hider, 2009).

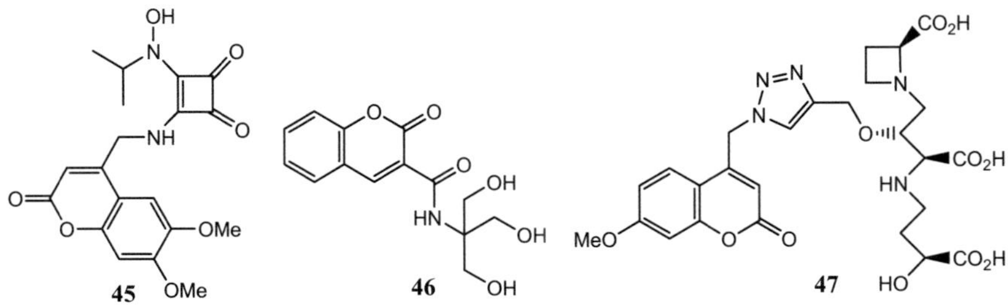

Fig. 4. Structures of coumarin-derived fluorescent chemosensors for iron(III) ion

Structures of coumarin-derived fluorescent chemosensors for Mg(II) are shown in Fig. 5. Coumarin-based two-photon probe **48** is developed for the detection of free Mg^{2+} ions in living cells and living tissues. The probe can be excited by 880 nm laser photons, emits strong two-photon excited fluorescence in response to Mg^{2+} ions (H. M. Kim et al., 2007). Coumarin Schiff-base **49**, without two-photon activity in the wavelength range 760-860 nm, exhibits large two-photon absorption as well as emission in the presence of Mg^{2+} (Ray et al., 2010). Coumarin-based chromoionophore **50** implemented in a transparent membrane can be used as an optical one-shot sensor for Mg^{2+} (Capitán-Vallvey, 2006). Two coumarin salen-based sensors **51** and **52** exhibit a pronounced fluorescence enhancement response toward Mg^{2+} in the presence of Na^{+} as a synergic trigger (Dong et al., 2011). Coumarin-derived fluorescent molecular probes **53** and **54** can be used for highly selective detection of Mg^{2+} versus Ca^{2+} by means of monitoring the absorption and fluorescence spectral change (Suzuki et al., 2002).

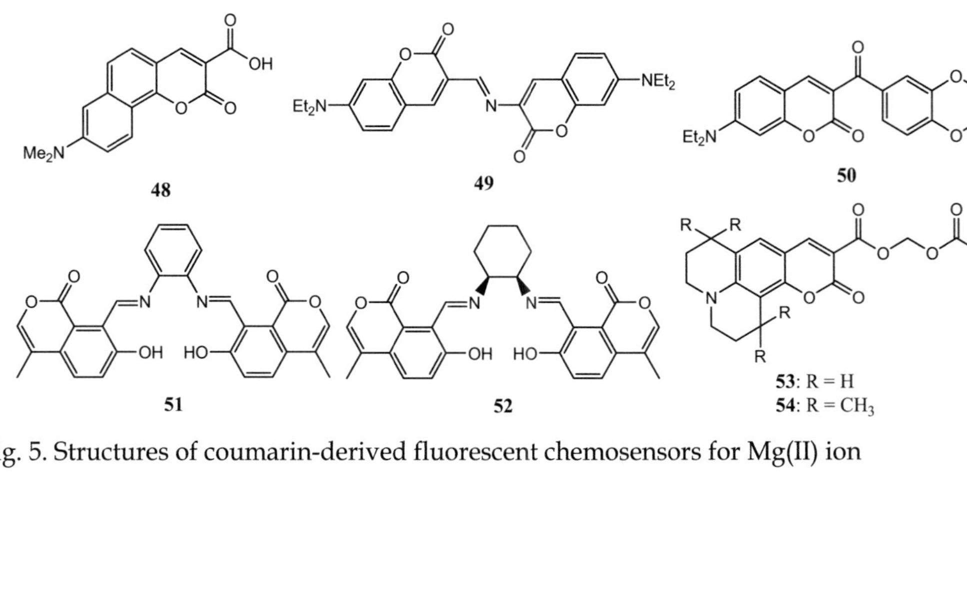

Fig. 5. Structures of coumarin-derived fluorescent chemosensors for Mg(II) ion

Fig. 6 shows the structures of coumarin-derived fluorescent chemosensors for Ag(I). A highly sensitive and selective fluorescent chemosensor **55** for Ag^+ based on a coumarin-Se_2N chelating conjugate is developed. Due to inhibiting a PET quenching pathway, a fluorescent enhancement of 4-fold is observed under the binding of the Ag^+ cation to **55** with a detection limit down to the 10^{-8} M range (S. Huang et al., 2011). Coumarin **56** and **57** are highly silver ion selective fluorescence ionophores (Sakamoto et al., 2010). By a microwave-assisted dual click reaction, fluorogenic 3-azidocoumarin can be rapidly introduced onto a 3,4-dipropargylgluco- or galactosyl scaffold with restored fluorescence. Subsequent desilylation leads to water soluble sugar-bis-triazolocoumarin conjugates which are applicable toward selective Ag^+ detection in aqueous media via fluorescence spectroscopy (X.-P. He et al., 2011).

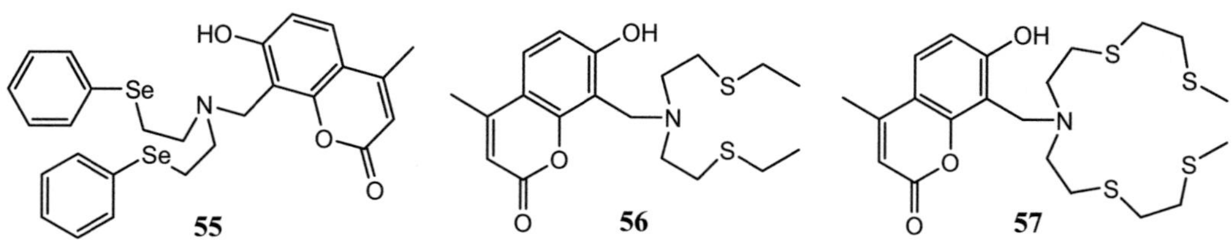

Fig. 6. Structures of coumarin-derived fluorescent chemosensors for silver(I) ion

Structures of coumarin-derived fluorescent chemosensors for lead(II) are shown in Fig. 7. Fluorescent chemosensor **58** based on a coumarin-crown ether conjugate exhibits a high affinity and selectivity for Pb^{2+} (C.-T. Chen & W.-P. Huang, 2002). Coumarin dyes **59-61** seem to fulfill most of the criteria required for intracellular lead indicators, as they exhibit high selectivity for Pb^{2+} (Roussakis et al., 2008). Ion competition studies and fluorescence experiments show that a fullerene-coumarin dyad is selective for Pb^{2+} complexation (Pagona et al., 2010).

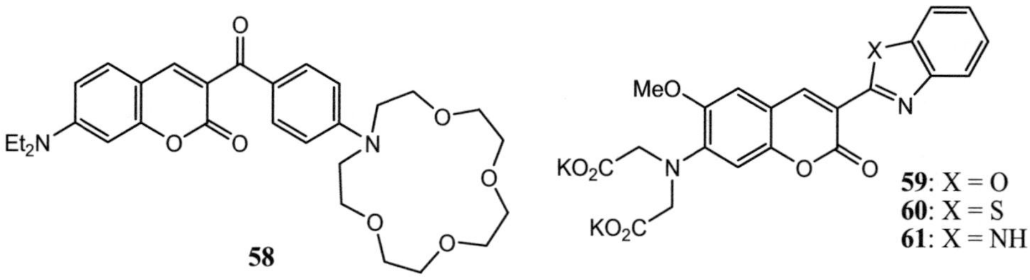

Fig. 7. Structures of coumarin-derived fluorescent chemosensors for lead(II) ion

Apart from the above-mentioned metal ions, coumarin-derived fluorescent chemosensors have been used for detection of other metal ions including Cd^{2+} (Taki et al., 2008), Al^{3+} (Maity & Govindaraju, 2010), and Cr^{3+} (Hu et al., 2011). Some coumarin-derived fluorescent chemosensors exhibit simultaneous sensitivity toward two or more different metal ions, e.g. Cu(II) and Hg(II) based on FRET mechanism (G. He et al., 2010b) or via selective anion-induced demetallation (Lau et al., 2011), Cu^{2+} and Ni^{2+} (H. Li et al., 2011), Ni^{2+} and Zn^{2+} (Chattopadhyay et al., 2006), Ni^{2+} and Co^{2+} (Lin et al., 2009), Ca^{2+} and Mg^{2+} (Suresh & Das, 2009), Na^+ and K^+ (Ast et al., 2011), Hg^{2+} and Ag^+ (Tsukamoto et al., 2011), $Cu^{2+}/Ni^{2+}/Cd^{2+}$ (Lin et al., 2008), $Zn^{2+}/Cd^{2+}/Pb^{2+}$ (Kulatilleke et al., 2006), or $Ni^{2+}/Pd^{2+}/Ag^+$ (Santos et al., 2009).

3.2 Coumarin-derived fluorescent chemosensors for anions

Development of highly efficient chemosensors for cyanide is of extreme significance due to the detrimental aspect of cyanide. Much attention has been paid to the utilization of fluorescent methods for the detection of cyanide. Several coumarin-derived fluorescent chemosensors for CN^- have been reported, the structures of which are shown in Fig. 8.

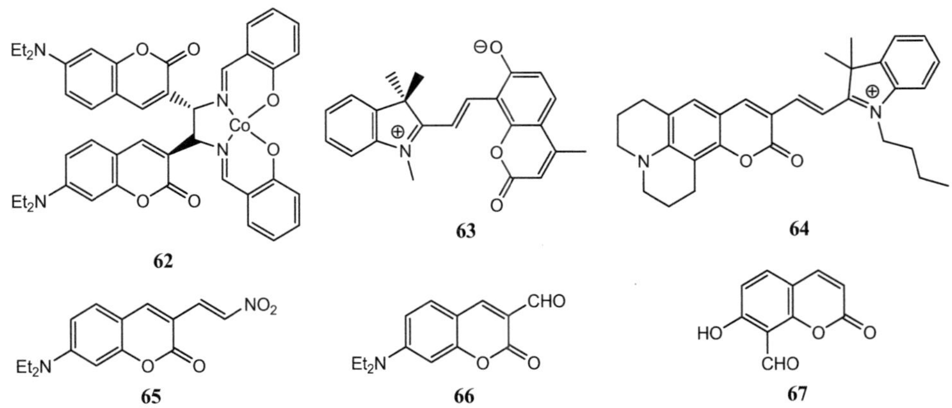

Fig. 8. Structures of coumarin-derived fluorescent chemosensors for cyanide

Cobalt(II)-coumarinylsalen complex **62** exhibits selective and tight binding to a cyanide anion and displays a significant fluorescence enhancement upon the addition of cyanide owing to the interruption of PET from the coumarin fluorophore to the cobalt(II)-salen moiety (J. H. Lee et al., 2010). Coumarin-spiropyran conjugate **63** is a highly sensitive chemosensor for CN^- and shows a CN^--selective fluorescence enhancement under UV irradiation (Shiraishi et al., 2011). An indole conjugated coumarin **64** for KCN chemodosimeter displays considerable dual changes in both absorption (blue-shift) and emission (turn-on) bands exclusively for KCN. The fluorescence enhancement of the **64**-KCN is mainly due to blocking of the ICT process (H. J. Kim et al., 2011). Doubly activatived coumarin **65** acts as a colorimetric and fluorescent chemodosimeter for cyanide (G.-J. Kim & H.-J. Kim, 2010a). A simple aldehyde-functionalized coumarin **66** has been utilized as a doubly activated Michael acceptor for cyanide (G.-J. Kim & H.-J. Kim, 2010b). Coumarin-based fluorescent chemodosimeter **67** with a salicylaldehyde functionality as a binding site has shown selectivity for cyanide anions over other anions in water at biological pH (K.-S. Lee et al., 2008a).

Recognition and detection of fluoride, the smallest anion with unique chemical properties is of growing interest. Several coumarin-derived fluorescent chemosensors for F^- have been developed, the structures of which are shown in Fig. 9.

Coumarin derivative **68** has been developed as a fluorescent probe for detection of F^- ion in water and bioimaging in A549 human lung carcinoma cells (S. Y. Kim et al., 2009). Coumarin **69** is a simple, highly selective, and sensitive chemosensor for fluoride anions in organic and aqueous media based on the specific affinity of fluoride anion to silicon (Sokkalingam & Lee, 2011). Coumarin-derived chemosensor **70** shows an obvious color change from yellow to blue upon addition of F^- ion with a large red shift of 145 nm in acetonitrile (Zhuang et al., 2011). Coumarin-based hydrazone **71** is an ICT probe for fluoride in aqueous medium (Upadhyay et al., 2010a). Coumarin-based system **72** has been developed as a novel turn-on fluorescent and colorimetric sensor for fluoride anions (J. Li et al., 2009).

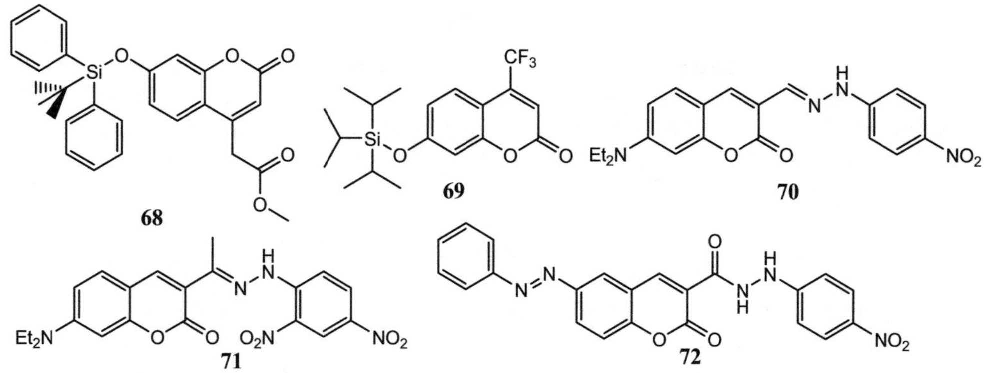

Fig. 9. Structures of coumarin-derived fluorescent chemosensors for F^- ion

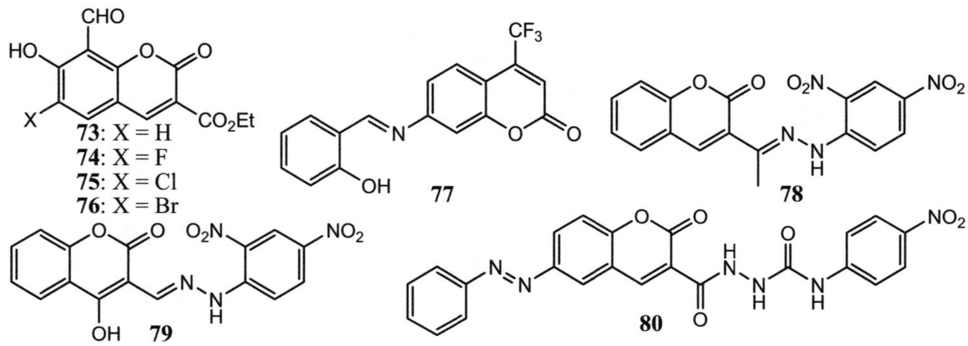

Fig. 10. Structures of coumarin-derived fluorescent chemosensors for anions other than CN^- and F^-

Structures of coumarin-derived fluorescent chemosensors for detection of anions other than CN^- and F^- have been shown in Fig. 10. Coumarin-based fluorescent probes **73-76** with salicylaldehyde functinoality as recognition unit have been developed for selective detection of bisulfite anions in water (K. Chen et al., 2010). Coumarin Schiff-base **77** is a highly selective and sensitive turn-on fluorogenic probe for detection of HSO_4^- ions in aqueous solution (H. J. Kim et al., 2009b). Coumarin-based hydrazone **78** acts as an ICT probe for detection of acetate, benzoate, and dihydrogenphosphate (Upadhyay et al., 2010b). Another coumarin-based hydrazone **79** has been utilized as both a colorimetric and ratiometric chemosensor for acetate and a selective fluorescence turn-on probe for iodide (Mahapatra et al., 2011). Compound **80** is a colorimetric and fluorescence anion sensor with the urea group as binding site and the coumarin moiety as signal unit. The sensor displays significant fluorescence enhancement response to anions such as acetate, because of complex formation (Shao, 2010). Coumarin-derived fluorescent chemosensors have also been used for detection of pyrophosphate (S. K. Kim et al., 2009), $H_2PO_4^-$ and $PhPO_3H^-$ (K. Choi & Hamilton, 2001), and multiple anions including pyrophosphate, citrate, ATP and ADP (Mizukami et al., 2002). A new Zn^{II}-2,2':6',2''-terpyridine complex, derivatized with a coumarin moiety, acts as a fluorescent chemosensor for different biologically important phosphates like PPi, AMP and ADP in mixed aqueous media (Das et al., 2011). A strapped calix[4]pyrrole-coumarin conjugate has been developed as a fluorogenic anions (Cl^-, Br^- and AcO^-) receptor modulated by cation and anion binding (Miyaji et al., 2005).

3.3 Coumarin derivatives as fluorescent probes of pH

Structures of coumarin-derived fluorescent chemosensors for sensing pH are shown in Fig. 11. Several iminocoumarin (**81**) derivatives **82-84** have been synthesized as a new type of fluorescent pH indicator. The indicators possess moderate to high brightness, excellent photostability and compatibility with light-emitting diodes. The indicators can be covalently immobilized on the surface of amino-modified polymer microbeads which in turn are incorporated into a hydrogel matrix to afford novel pH-sensitive materials. When a mixture of two different microbeads is used, the membranes are capable of optical pH sensing over a very wide range comparable to the dynamic range of the glass electrode (pH 1-11) (Vasylevska et al., 2007). Coumarin derivative **85** containing piperazine and imidazole moieties has been developed as a fluorophore for hydrogen ions sensing. The fluorescence enhancement of the sensor with an increase in hydrogen ions concentration is based on the hindering of PET from the piperazinyl amine and the imidazolyl amine to the coumarin fluorophore by protonation. The sensor **85** has a novel molecular structure design of fluorophore-spacer-receptor(1)-receptor(2) format and therefore is proposed to sense two range of pH from 2.5 to 5.5 and from 10 to 12 instead of sensing one pH range (Saleh et al., 2008). By using rational molecular design, two molecular functions, the transport by vesicular monoamine transporter (VMAT) and ratiometric optical pH sensing, have been integrated to develop ratiometric pH-responsive fluorescent false neurotansmitter (FFN) probes (M. Lee et al., 2010). A FRET sensor with a donor and an acceptor attached to each end of pH-sensitive polysulfoamides exhibits an instantaneous conformation change from coil to globule at a specific pH, which results in the drastic on-and-off FRET efficiency. To detect a specific pH region, sulfadimethoxine and sulfamethizole are selected among various sulfonamides since their pK_a values are in the physiological pH. For tuning the emission color arising from FRET, 7-hydroxy-4-bromomethylcoumarin and coumarin 343 are used as a FRET donor and an acceptor, respectively, for a blue-to-green FRET sensor (Hong & Jo, 2008).

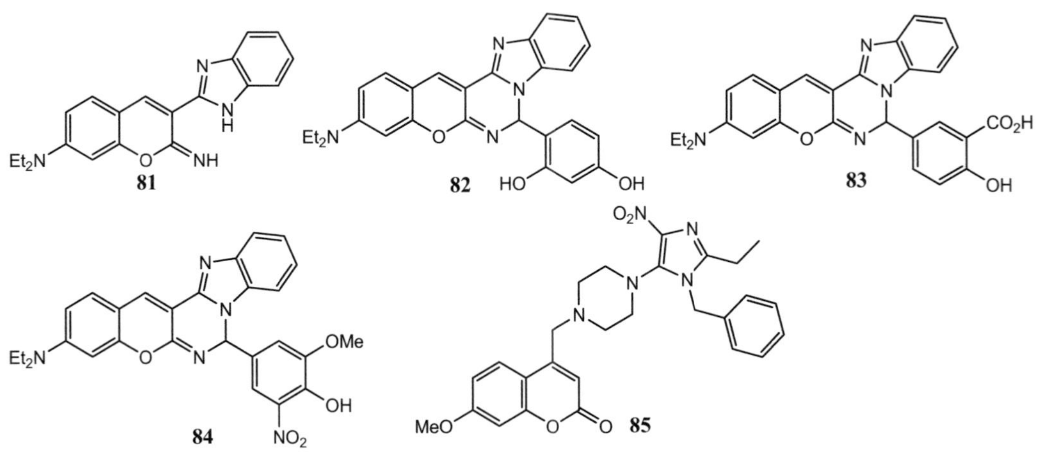

Fig. 11. Structures of coumarin-derived fluorescent chemosensors for sensing pH

3.4 Coumarin-derived fluorescent chemosensors for thiols

Biothiols such as cysteine (Cys), homocysteine (Hcy), and glutathione (GSH) are of great significance because they are involved in myriad vital cellular processes including redox homeostasis and cellular growth. Alteration of the cellular biothiols is also implicated in cancer and AIDS. Study on fluorescent and colorimetric probes for detection of thiols has received much attention and many coumarin-derived fluorescent chemosensors for detection of thiols have been reported. The structures of these chemical sensors are shown in Fig. 12.

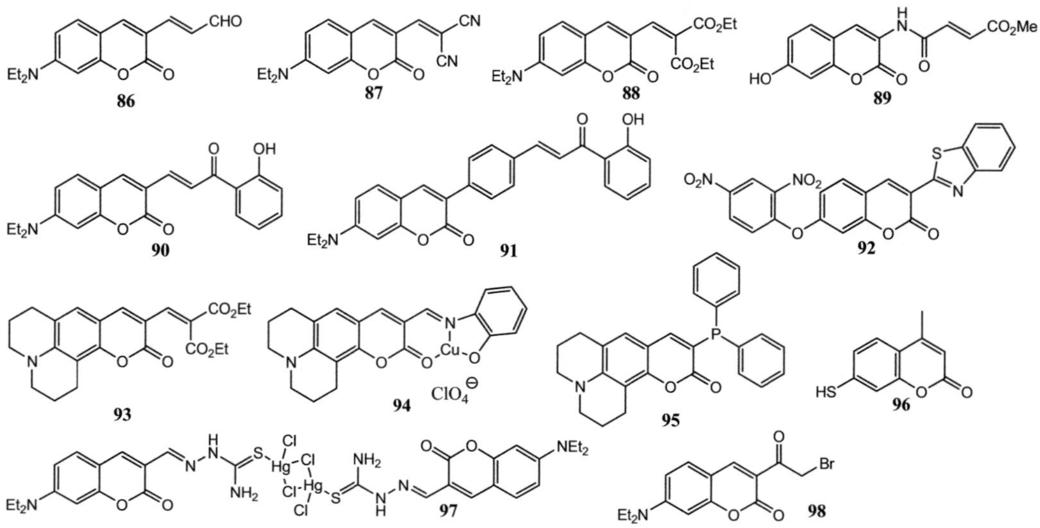

Fig. 12. Structures of coumarin-derived chemosensors for detection of thiols

Generally detection of thiols by optical probes is based on two features of thiols, their strong nucleophilicity and high binding affinity toward metal ions. Accordingly, most of the fluorescent probes for thiols are in fact chemodosimeters, which involve specific reactions between probes and thiols, such as Michael addition, cyclization with aldehyde (or ketone), cleavage of disulfide by thiols, metal complexes-displace coordination, demetalization from Cu-complex, thiolysis of dinitrophenyl ether, and Staudinger ligation. For instance, detection of thiols by chemosensors **87-91** and **93** involves Michael addition between the probes and thiols. For sensors **67** (Fig. 8), **86** and **98**, cyclization reactions occur between the sensors and thiols in the detection process. Coumarin **86** is a ratiometric fluorescent probe for specific detection of Cys over Hcy and GSH based on the drstic distinction in the kinetic profiles (Yuan et al., 2011). Nonfluorescent coumarin-malonitrile conjugate **87** can be transformed into a strongly fluorescent molecule through the Michael addition and thus exhibits a highly selective fluorescence response toward biothiols including Cys, Hcy and GSH with micromolar sensitivity (Kwon et al., 2011). Similarly, nonfluorescent **88** displays a highly selective fluorescence enhancement with thiols and has been successfully applied to thiols determination in intracellular, in human urine and blood samples (Zuo et al., 2010). Coumarin **89** has been developed as a water-soluble, fast-response, highly sensitive and selective fluorescence thiol quantification probe (Yi et al., 2009). Compound **90** (G.-J. Kim et al., 2011) and **91** (S. Y. Lim et al., 2011) with a hydrogen bond act as highly selective ratiometric fluorescence turn-on probes for GSH. Structure **92** has been judiciously designed and synthesized as a new type of selective benzenethiol fluorescent probe based on the thiolysis of dinitrophenyl ether (Lin et al., 2010a). Coumarin-based chemodosimeter **93** effectively and selectively recognizes thiols based on a Michael type reaction, showing a preference for Cys over other biological materials including Hcy and GSH (Jung et al.,

2011a). Iminocoumarin-Cu(II) ensemble-based chemodosimeter **94** sensitively senses thiols followed by hydrolysis to give a marked fluorescence enhancement over other amino acids based on demetalization from Cu-complex (Jung et al., 2011b). Nonfluorescent coumarin-phosphine dye **95** reacts with S-nitrosothiols (RSNOs) to form a fluorescent coumarin derivative and thus may be used as a tool in the detection of RSNOs. The reaction mechanism is similar to the well-known Staudinger ligation (Pan et al., 2009). 7-Mercapto-4-methylcoumarin **96** is a reporter of thiol binding to the CdSe quantum dot surface (González-Béjar et al., 2009). Coumarin-derived complex **97** has been developed as a reversible fluorescent probe for highly selective and sensitive detection of mercapto biomolecules such as Cys, Hcy and GSH (J. Wu et al., 2011). A simple coumarin derivative **98** is the first fluorescence turn-on probe for thioureas by the double functional group transformation strategy. The probe exhibits high sensitivity and selectivity for thioureas over other structurally and chemically related species including urea and thiophenol (Lin et al., 2010b). The simple coumarin sensor **67** (Fig. 8) has shown fluorescence selectivity for not only cyanide anions but also Hcy and Cys in water (K.-S. Lee et al., 2008b). A new coumarin-containing zinc complex has been developed as a colorimetric turn-on and fluorescence turn-off sensor which shows high selectivity for hydrogen sulfide in the presence of additional thiols like Cys or GSH (Galardon et al., 2009).

3.5 Coumarin-derived fluorescent chemosensors for H_2O_2, O_2, hydroxyl radicals or chemical warfare agents

Structures of coumarin-derived fluorescent chemosensors for hydrogen peroxide, oxygen, hydroxyl radicals or chemical warfare agents are shown in Fig. 13. Water-soluble umbelliferone-based fluorescent probe **99** shows very large increase (up to 100-fold) in fluorescent intensity upon reaction with hydrogen peroxide, and good selectivity over other reactive oxygen species (Du et al., 2008). Another water-soluble fluorescent hydrogen peroxide probe **100** based on a 'click' modified coumarin fluorophore shows significant intensity increases (up to fivefold) in near-green fluorescence upon reaction with H_2O_2, and good selectivity over other reactive oxygen species (Du Ý et al., 2010). More recently a simple and highly sensitive fluorometric method was proposed for the determination of H_2O_2 in milk samples. In this method, nonfluorescent coumarin was oxidized to highly fluorescent 7-hydroxycoumarin by hydroxyl radicals generated in a Fenton reaction, and the oxidation product had strong fluorescence with a maximum intensity at 456 nm and could be used as a fluorescent probe for H_2O_2 (Abbas et al., 2010). Thiazo-coumarin ligand directly cyclometallated Pt(II) complex **101** has been used for luminescent O_2 sensing (W. Wu et al., 2011). A hybrid coumarin-cyanine platform **102** has been developed as the first ratiometric fluorescent probe for detection of intracellular hydroxyl radicals (Yuan et al., 2010). More recently a coumarin-neutral red (CONER) nanoprobe was developed for detection of hydroxyl radical based on the ratiometric fluorescence signal between 7-hydroxy coumarin 3-carboxylic acid and neutral red dyes. Biocompatible poly lactide-*co*-glycolide nanoparticles containing encapsulated neutral red were produced using a coumarin 3-carboxylic acid conjugated poly(sodium *N*-undecylenyl-Nε-lysinate) as moiety reactive to hydroxyl radicals. The response of the CONER nanoprobe was dependent on various parameters such as reaction time and nanoparticle concentration. The probe was selective for hydroxyl radicals as compared with other reactive oxygen species including $O_2^{\bullet -}$, H_2O_2, $^1O_2$ and OCl^- (Ganea et al., 2011).

The current rise in international concern over criminal terorist attacks using chemical warfare agents has brought about the need for reliable and affordable detection methods of toxic gases. One of the applicable technologies is the design of fluorogenic chemosensors for the specific detection of nerve agents (Royo et al., 2007). A coumarin oximate **103** has been developed for detection of chemical warfare simulants based on the PET mechanism that gives an "off-on" fluorescent response with a half-time of approximately 50 ms upon phosphorylation of a reactive oximate functionality (Wallace et al., 2006). Coumarin-derived hydroxy oxime **104** serves as a nerve agent sensor based on the reaction of β-hydroxy oxime with organophosphorus agent mimics (Dale & Rebek, 2009). A FRET approach towards potential detection of phosgene has been developed as shown in Fig. 14. When both coumarins **105** and **106** are mixed together with triphosgene in the presence of Et_3N in $CHCl_3$, hybrid urea **107** forms in a statistical yield. Significant fluorescence enhancement is detected which is particularly important since the acceptor unit alone does not emit under the same condition. The fluorescence increase is obviously due to the formation of urea **107**. Simultaneously, the fluorescence from the donor unit decreases due to the quenching, indicating that efficient energy transfer takes place from the donor to the acceptor. This system is selective, since other gases/agents rarely can serve for cross-linking (Zhang & Rudkevich, 2007).

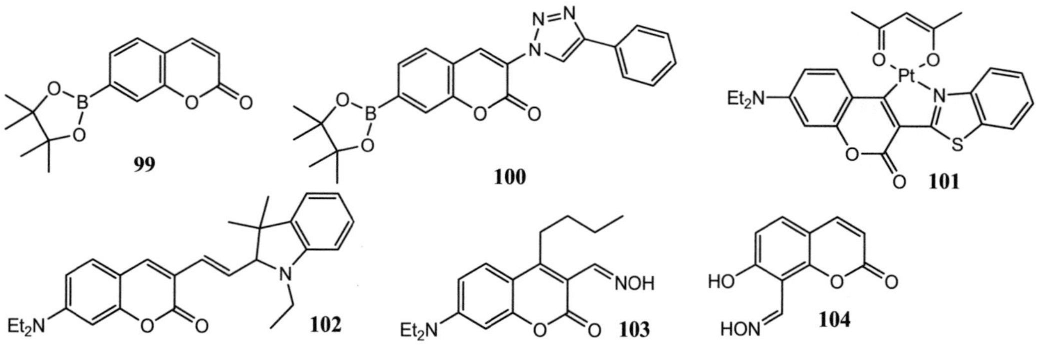

Fig. 13. Structures of coumarin-derived fluorescent sensors for H_2O_2, O_2, hydroxyl radicals or chemical warfare agents

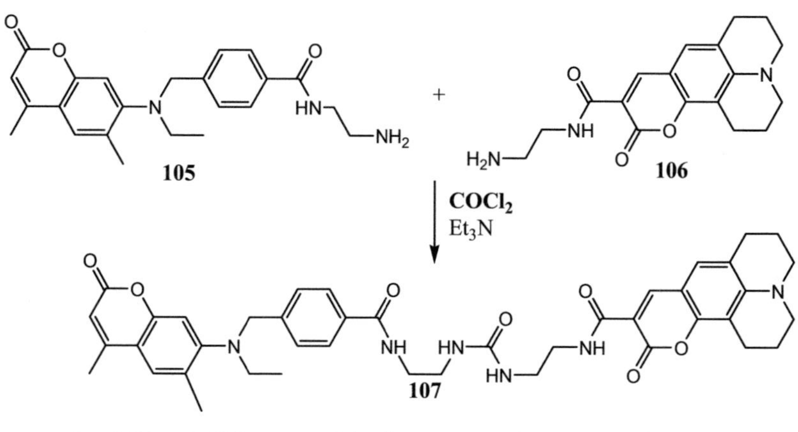

Fig. 14. Coumarins **105** and **106** react with phosgene to form urea **107**

3.6 Coumarin-derived fluorescent chemosensors for amines, amino acids or other organic compounds

Structures of coumarin-derived fluorescent chemosensors for amines, amino acids or other organic compounds are shown in Fig. 15. Butyl-subsituted coumarin aldehyde **108** is an excellent chemosensor for detection of amines and unprotected amino acids in aqueous conditions by formation of highly fluorescent iminium ions (Feuster & Glass, 2003). Boronic acid-containing coumarin aldehyde **109** binds to primary catecholamines with good affinity and acts as an effective colorimetric sensor for dopamine and norepinephrine with excellent selectivity over epinephrine, amino acids, and glucose. In the fluorescence manifold, sensor **109** responds differentially to catechol amines over simple amines, giving a fluorescence decrease in response to catechol-containing compounds and a fluorescence increase with other amines (Secor & Glass, 2004). Coumarin-based fluorescent functional monomers containing a carboxylic acid functionality, **110** and **111** have been synthesized, which allow for the preparation of fluorescent imprinted polymer sensors for chiral amines (Nguyen & Ansell, 2009). Coumarin aldehyde **66** (Fig. 8) can be utilized as not only a doubly activated Michael acceptor for cyanide but also a highly selective and sensitive fluorescence turn-on probe for proline (G.-J. Kim & H.-J. Kim, 2010c). Coumarin-azacrown ether conjugate **112** has been developed as a fluorescent probe for identifying melamine (Xiong et al., 2010). Thiocoumarin **113** can be efficiently desulfurized to its corresponding coumarin by the reaction with mCPBA, and results in a pronounced fluorescence turn-on type signaling. The conversion also provides a significant change in absorption behavior which allows a ratiometric analysis, providing a convient detection method for mCPBA in aqueous environment (Cha et al., 2010). Polymers containing 4,8-dimethylcoumarin have been developed for detection of 2,4-dinitrotoluene (DNT) and 2,4,6-trinitrotoluene (TNT).

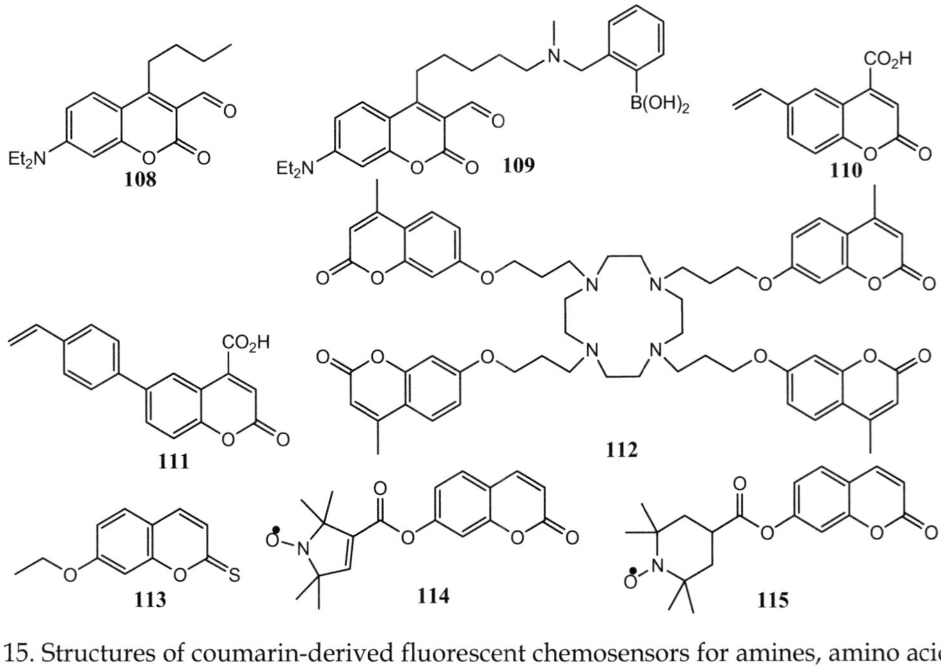

Fig. 15. Structures of coumarin-derived fluorescent chemosensors for amines, amino acids or other organic compounds

The fluorescence quenching of these copolymers in solution can be attributed to the collisional quenching. The response of these polymeric sensors is promising and can easily detect DNT and TNT at few parts per billion levels (Kumar et al., 2010). A novel kind of luminescent vesicular chemosensors for the recognition of biologically important ions and molecules such as imidazoles has been developed by the self-assembly of lipids, amphiphilic binding sites, and fluorescent coumarin reporter dyes that are sensitive to their environment (Gruber et al., 2010). Two hybrid compounds **114** and **115**, linked via an ester-bond between the 7-hydroxyl residue of an umbelliferone and a carboxylic acid residue of two nitroxide radicals, act as fluorescence and spin-label probes. The ESR intensities of **114** and **115** are proportionally reduced after the addition of ascorbic acid sodium salt, and their fluorescence intensities are increased maximally by eight- and nine-fold, respectively (Sato et al., 2008).

3.7 Coumarin-derived fluorescent chemosensors for TiO_2, monolayer, polymerization or polymeric micelles

Structures of coumarin-derived fluorescent chemosensors for detection of TiO_2, monolayer, or photopolymerization are shown in Fig. 16. A novel acac-coumarin chromophore linker **116** for robust sensitization of TiO_2 has been developed to find molecular chromophores with suitable properties for solar energy conversion. The synthesis and spectroscopic characterization confirms that **116** yields improved sensitization to solar light and provides robust attachment to TiO_2 even in aqueous conditions (Xiao et al., 2011). A new amphiphilic coumarin dye, 7-aminocoumarin-4-acetic acid octadecylamide (**117**) forms a stable monolayer at the air-water interface and may be utilized as an efficient fluorescent probe for monolayer studies (Kele et al., 2001). Performance of amidocoumarins **118-120** as probes for monitoring of cationic photopolymerization of monomers by fluorescence probe technology has been investigated. 7-Diethylamino-4-methylcoumarin **118** can be used for monitoring cationic photopolymerization of monomers using the fluorescence intensity ratio as an indicator of the polymerization process. The replacement of diethylamino group in **118** with benzamido or acetamido groups eliminates the effect of the probe protonation on kinetics of cationic photopolymerization. 7-Benzamido-4-methylcoumarin **119** and 7-acetamido-4-methylcoumarin **120** can be used as fluorescent probes for monitoring progress of cationic polymerization of vinyl ethers under stationary measurement conditions, using normalized fluorescence intensity as an indicator of the polymerization progress (Ortyl et al., 2010). Coumarin 153 has been used as a fluorescent probe molecule to monitor the possible micellization of several amphiphilic block copolymers (Basu et al., 2009).

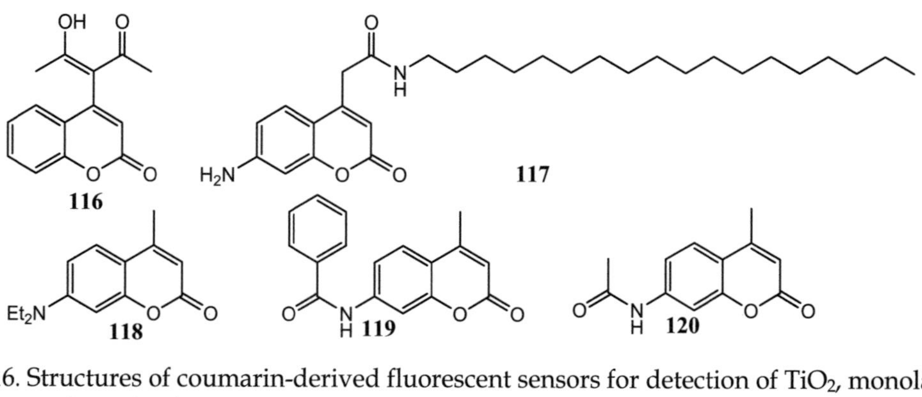

Fig. 16. Structures of coumarin-derived fluorescent sensors for detection of TiO_2, monolayer, or photopolymerization

3.8 Coumarin-derived fluorescent chemosensors for enzymes

Structures of coumarin-derived fluorescent chemosensors for enzymes are shown in Fig. 17. Hemicyanine-coumarin hybrid **121** represents a new class of far-red emitting fluorogenic dyes whose fluorescence is unveiled through an enzyme-initiated domino reaction and thus acts as a fluorogenic probe for penicillin G acylase (Richard et al., 2008). Similarly novel self-immolative spacer systems **122** and **123** have been developed and are utilized as fluorogenic probes for sensing penicillin amidase (Meyer et al., 2008). A library of 6-arylcoumarins has been developed as candidate fluorescent sensors of which **124** has the strongest fluorescence intensity, whose quantum yield is similar to that of ethyl 7-diethylaminocoumarin-3-carboxylate, a well-known fluorophore as labeling or sensing biomolecule. The transormation of the methoxy group (**125**) to a hydroxyl group (**126**) induces a change of fluorescence intensity, which suggests that **125** may be useful as a fluorescent sensor for dealkylating enzymes such as glycosidase. Coumarin **127** shows 50% decrease of the fluorescence intensity at pH 8.0 compared with that at pH 6.0 and this decrease may be derived from the deprotonation of the triazole ring. Thus **127** may be used as a fluorescent sensor for nitric oxide (Hirano et al., 2007). Histone deacetylases are intimately involved in epigenetic regulation and, thus, are one of the key therapeutic targets for cancer. Coumarin-suberoylanilide hydroxamic acid **128** is a fluorescent probe for determining binding affinities and off-rates of histone deacetylase inhibitors (Singh et al., 2011). A quinone-methide-rearrangement reaction as the off-on optical switch has been successfully implemented into the design of the first long-wavelength latent fluorogenic substrate **129** which is a sensitive fluorimetric indicator for analyte determination in salicylate hydroxylase-coupled dehydrogenase assay (S.-T. Huang et al., 2010). Another switch-on long-wavelength latent fluorogenic substrate **130** is a fluorescent probe for nitroreductase (H.-C. Huang et al., 2011).

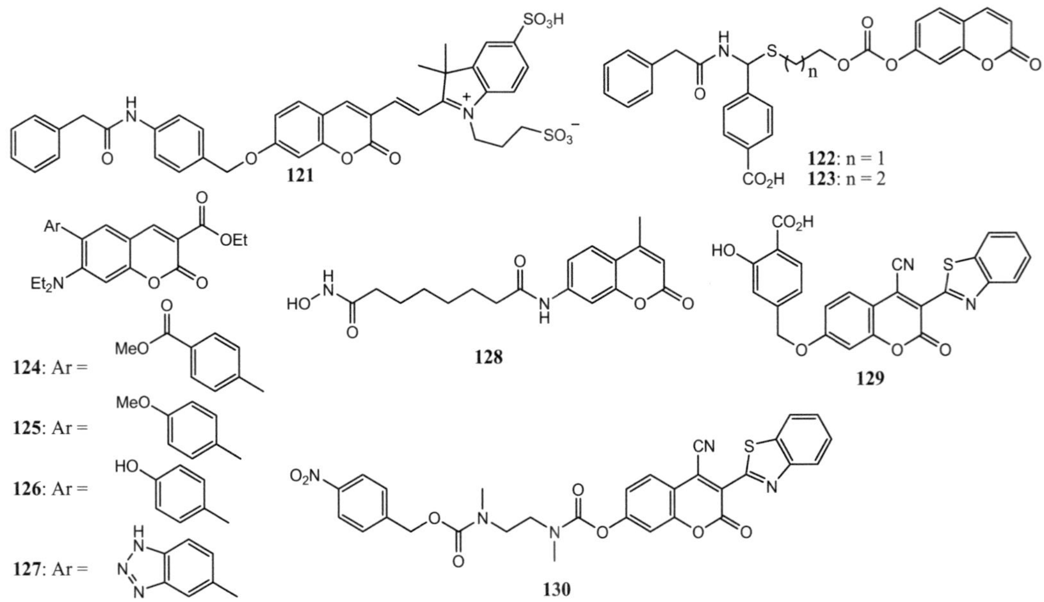

Fig. 17. Structures of coumarin-derived fluorescent chemosensors for enzymes

Apart from the above-mentioned coumarin-derived fluorescent chemosensors for enzymes, clikable biocompatible nanoparticles have been prepared in a one-pot process by microemulsion polymerization, which are then readily modified by the Huisgen Cu(I)-catalyzed azide-alkyne cycloaddition reaction to afford a coumarin-containing subtilisin responsive nanosensor (Welser et al., 2009). A coumarin-containing time-resolved fluorescence probe for dipeptidyl peptidase 4 has also been reported (Kawaguchi et al., 2010). A coumarin-derived triple-signaling fluorescent probe has been successfully applied for intracellular measurement of different enzyme activity (Y. Li et al., 2011). As shown in Fig. 18, a sensitive, selective, and fluorogenic probe **131** has been developed for monoamine oxidases (MAO A and B). Nonfluorescent aminocoumarin **131** can be converted to fluorescent pyrrolocoumarin **132** in the presence of MAO A and B (G. Chen et al., 2005). A new fluorogenic transformation based on a quinone reduction/lactonization sequence as shown in Fig. 19 has been developed and evaluated as a tool for probing redox phenomena in a biochemical context.

Fig. 18. Systematic mapping of aminocoumarin **131** and the corresponding pyrrolocoumarin **132**

3.9 Coumarin-derived fluorescent chemosensors for proteins, DNA, RNA and other uses

Structures of coumarin-derived fluorescent chemosensors for DNA, RNA, nitroxyl and proteins are shown in Fig. 20. A novel coumarin C-riboside **133** is designed and synthesized based on the well-known photoprobe Coumarin 102. The coumarin C-glycoside **133** has been incorporated synthetically into DNA oligomers, and has been used to probe ultrafast dynamics of duplex DNA using time-resolved Stokes shift methods (Coleman et al., 2007). Coumarin-triazole **134** reacts with $CuCl_2$ to form a chelated Cu(II)-**134** complex which shows highly selective turn-on type fluorogenic behavior upon addition of Angeli's salt ($Na_2N_2O_3$) and can be used for detection of nitroxyl in living cells (Zhou et al., 2011). A simple coumarin derivative 7-diethylaminocoumarin-3- carboxylic acid **135** has been used as an acceptor to construct a useful and effective FRET system for detection of RNA-small molecule binding (Xie et al., 2009). Coumarin dye bearing an indolenine substituent **136** displays high emission and bright fluorescence and offers promise as an fluorescent chemosensor for protein detection (Kovalska et al., 2010). Coumarin-containing trifunctional probe **137**, assembled using a cleavable linker, is useful for efficient enrichment and detection of glycoproteins (Tsai et al., 2010). Coumarin 6 has been used as a fluorescent probe to monitor protein aggregation and can distinguish between both amorphors and fibrillar aggregates (Makwana et al., 2011). An S_NAr reaction-triggered fluorescence probe is developed using a new fluorogenic compound derivatized from 7-aminocoumarin for oligonucleotides detection (Shibata et al., 2009). Coumarin C343 has been conjugated to silica nanoparticles and entrapped in a sol-gel matrix to produce a nanosensor capable of monitoring lipid peroxidation (Baker et al., 2007). Coumarin-containing dual-emission chemosensors for nucleoside polyphosphates have been developed based on a new mechanism involving binding-induced recovery of FRET. These sensors demonstrate that binding-induced modulation of spectral overlap is a powerful strategy for the rational design of FRET-based chemosensors (Kurishita et al., 2010).

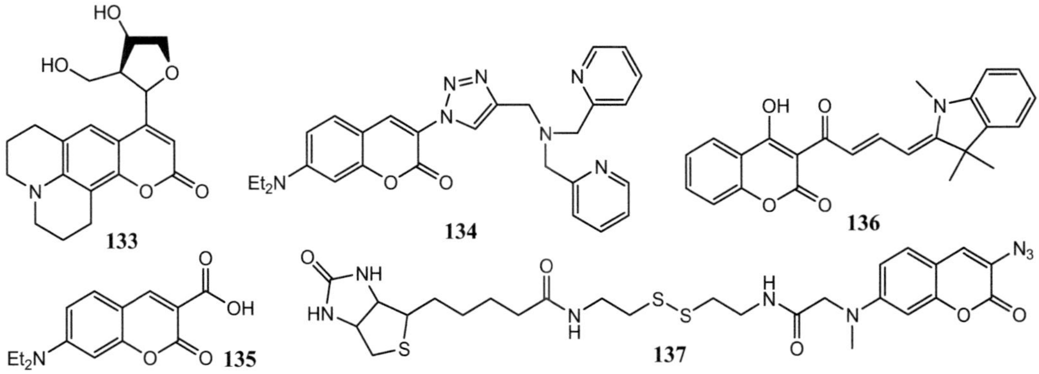

Fig. 19. Structures of coumarin-derived fluorescent chemosensors for DNA, RNA, nitroxyl and proteins

4. Conclusions

Coumarin-derived fluorescent chemosensors have been extensively applied in a variety of fields. Though these sensors are effective for detection of many species, their performance toward different species might decrease in the following order: metal ions, anions, biothiols, enzymes, pH, amines and amino acids, chemical warfare agents, proteins, hydrogen peroxide, hydroxyl radicals, polymerization and polymeric micelles, DNA and RNA, oxygen, titania. Continuous efforts will be devoted to development of fluorescent chemosensors with higher selectivity and sensitivity for more single target or simultaneously for multiple targets, thus providing practical fluorescent chemosensors for application in environmental chemistry, analytical chemistry, and bio-medicinal science.

5. References

Abbas, M. E.; Luo, W.; Zhu, L.; Zou, J. & Tang, H. (2010). Fluorometric determination of hydrogen peroxide in milk by using a Fenton reaction system. *Food Chem.*, vol.120, No.1, (May 2010), pp. 327-331, ISSN 0308-8146

Ast, S.; Müller, H.; Flehr, R.; Klamroth, T.; Walz, B. & Holdt, H.-J. (2011). High Na^+ and K^+-induced fluorescence enhancement of a π-conjugated phenylaza-18-crown-6-triazol-substituted coumarin fluoroionophore. *Chem. Commun.*, vol.47, No.16, (April 2011), pp. 4685-4687, ISSN 1359-7345

Baker, N.; Greenway, G. M.; Wheatley, R. A. & Wiles, C. (2007). A chemiluminescence nanosensor to monitor lipid peroxidation. *Analyst*, vol.132, No.2, (January 2007), pp. 104-106, ISSN 0003-2654

Bangar Raju, B. & Varadarajan, T. S. (1995). Spectroscopic studies of 7-diethylamino-3-styrylcoumarins. *J. Photochem. Photobio. A*, vol.85, No.3, (January 1995), pp. 263-267, ISSN 1010-6030

Basu, S.; Chatterjee, D. P.; Chatterjee, U.; Mondal, S. & Mandal, D. (2009). Fluorescence probing of block copolymeric micelles using Coumarin 153. *Colloids Surf. A*, vol.341, No.1-3, (June 2009), pp. 13-20, ISSN 0927-7757

Capitán-Vallvey, L. F.; Fernández-Ramos, M. D.; Lapresta-Fernández, A.; Brunet, E.; Rodríguez-Ubis, J. C. & Juanes, O. (2006). Magnesium optical one-shot sensor based on a coumarin chromoionophore. *Talanta*, vol.68, No.5, (February 2006), pp. 1663-1670, ISSN 0039-9140

Cha, S.; Hwang, J.; Choi, M. G. & Chang, S.-K. (2010). Dual signaling of *m*-chloroperbenzoic acid by desulfurization of thiocoumarin. *Tetrahedron Lett.*, vol.51, No.50, (December 2010), pp. 6663-6665, ISSN 0040-4039

Chandrasekhar, V.; Bag, P. & Pandey, M. D. (2009). Phosphorus-supported multidentate coumarin-containing fluorescence sensors for Cu^{2+}. *Tetrahedron*, vol.65, No.47, (November 2009), pp. 9876-9883, ISSN 0040-4020

Chattopadhyay, N.; Mallick, A. & Sengupta, S. (2006). Photophysical studies of 7-hydroxy-4-methyl-8-(4'-methylpiperazin-1'-yl) methylcoumarin: a new fluorescent chemosensor for zinc and nickel ions in water. *J. Photochem. Photobio. A*, vol.177, No.1, (January 2006), pp. 55-60, ISSN 1010-6030

Chen, C.-T. & Huang, W.-P. (2002). A highly selective fluorescent chemosensor for lead ions. *J. Am. Chem. Soc.*, vol.124, No.22, (June 2002), pp. 6246-6247, ISSN 0002-7863

Chen, G.; Yee, D. J.; Gubernator, N. G. & Sames, D. (2005). Design of optical switches as metabolic indicators: new fluorogenic probes for monoamine oxidases (MAO A and B). *J. Am. Chem. Soc.*, vol.127, No.13, (April 2005), pp. 4544-4545, ISSN 0002-7863

Chen, K.; Guo, Y.; Lu, Z.; Yang, B. & Shi, Z. (2010). Novel coumarin-based fluorescent probe for selective detection of bisulfite anion in water. *Chin. J. Chem.*, vol.28, No.1, (January 2010), pp. 55-60, ISSN 1001-604X

Chen, X.; Xi, H.; Sun, X.; Zhao, T.; Meng, Q. & Jiang, Y. (2011). Synthesis and fluorescent probes properties of a coumarin-based piperazine containing fluorine. *Chin. J. Org. Chem.*, vol.31, No.4, (April 2011), pp. 544-547, ISSN 0253-2786

Cho, Y.-S. & Ahn, K. H. (2010). A 'turn-on' fluorescent probe that selectively responds to inorganic mercury species. *Tetrahedron Lett.*, vol.51, No.29, (July 2010), pp. 3852-3854, ISSN 0040-4039

Choi, K. & Hamilton, A. D. (2001). A dual channel fluorescence chemosensor for anions involving intermolecular excited state proton transfer. *Angew. Chem. Int. Ed.*, vol.40, No.20, (October 2001), pp. 3912-3915, ISSN 1433-7851

Choi, M. G.; Kim, Y. H.; Namgoong, J. E. & Chang, S.-K. (2009). Hg^{2+}-selective chromogenic and fluorogenic chemodosimeter based on thiocoumarins. *Chem. Commun.*, No.24, (June 2009), pp. 3560-3562, ISSN 1359-7345

Ciesienski, K. L.; Hyman, L. M.; Derisavifard, S. & Franz, K. J. (2010). Toward the detection of cellular copper(II) by a light-activated fluorescence increase. *Inorg. Chem.*, vol.49, No.15, (August 2010), pp. 6808-6810, ISSN 0020-1669

Coleman, R. S.; Berg, M. A. & Murphy, C. J. (2007). Coumarin base-pair replacement as a fluorescent probe of ultrafast DNA dynamics. *Tetrahedron*, vol.63, No.17, (April 2007), pp. 3450-3456, ISSN 0040-4020

Dale, T. J. & Rebek, J., Jr. (2009). Hydroxy oximes as organophosphorus nerve agent sensors. *Angew. Chem. Int. Ed.*, vol.48, No.42, (October 2009), pp. 7850-7852, ISSN 1433-7851

Das, P.; Ghosh, A.; Kesharwani, M. K.; Ramu, V.; Ganguly, B. & Das, A. (2011). Zn^{II}-2,2′:6′,2′′-Terpyridine-based complex as fluorescent chemosensor for PPi, AMP and ADP. *Eur. J. Inorg. Chem.*, No.20, (July 2011), pp. 3050-3058, ISSN 1434-1948

Dong, Y.; Li, J.; Jiang, X.; Song, F.; Cheng, Y. & Zhu, C. (2011). Na^+ triggered fluorescence sensors for Mg^{2+} detection based on a coumarin salen moiety. *Org. Lett.*, vol.13, No.9, (May 2011), pp. 2252-2255, ISSN 1523-7060

Du, L.; Li, M.; Zheng, S. & Wang, B. (2008). Rational design of a fluorescent hydrogen peroxide probe based on the umbelliferone fluorophore. *Tetrahedron Lett.*, vol.49, No.19, (May 2008), pp. 3045-3048, ISSN 0040-4039

Du Ý, L.; Ni Ý, N.; Li, M. & Wang, B. (2010). A fluorescent hydrogen peroxide probe based on a 'click' modified coumarin fluorophore. *Tetrahedron Lett.*, vol.51, No.8, (February 2010), pp. 1152-1154, ISSN 0040-4039

Feuster, E. K. & Glass, T. E. (2003). Detection of amines and unprotected amino acids in aqueous conditions by formation of highly fluorescent iminium ions. *J. Am. Chem. Soc.*, vol.125, No.52, (December 2003), pp. 16174-16175, ISSN 0002-7863

Galardon, E.; Tomas, A.; Roussel, P. & Artaud, I. (2009). New fluorescent zinc complexes: towards specific sensors for hydrogen sulfide in solution. *Dalton Trans.*, No.42, (November 2009), pp. 9126-9130, ISSN 1477-9226

Ganea, G. M.; Kolic, P. E.; El-Zahab, B. & Warner, I. M. (2011). Ratiometric coumarin-neutral red (CONER) nanoprobe for detection of hydroxyl radicals. *Anal. Chem.*, vol.83, No.7, (April 2011), pp. 2576-2581, ISSN 0003-2700

González-Béjar, M.; Frenette, M.; Jorge, L. & Scaiano, J. C. (2009). 7-Mercapto-4-methylcoumarin as a reporter of thiol binding to the CdSe quantum dot surface. *Chem. Commun.*, No.22, (June 2009), pp. 3202-3204, ISSN 1359-7345

Gruber, B.; Stadlbauer, S.; Späth, A.; Weiss, S.; Kalinina, M. & König, B. (2010). Modular chemosensors from self-assembled vesicle membranes with amphiphilic binding sites and reporter dyes. *Angew. Chem. Int. Ed.*, vol.49, No.39, (September 2010), pp. 7125-7128, ISSN 1433-7851

He, G.; Zhang, X.; He, C.; Zhao, X. & Duan, C. (2010b). Ratiometric fluorescence chemosensors for copper(II) and mercury(II) based on FRET systems. *Tetrahedron*, vol.66, No.51, (December 2010), pp. 9762-9768, ISSN 0040-4020

He, G.; Zhao, X.; Zhang, X.; Fan, H.; Wu, S.; Li, H.; He, C. & Duan, C. (2010a). A turn-on PET fluorescence sensor for imaging Cu^{2+} in living cells. *New J. Chem.*, vol.34, No.6, (June 2010), pp. 1055-1058, ISSN 1144-0546

He, X.-P.; Song, Z.; Wang, Z.-Z.; Shi, X.-X. & Chen, K. (2011). Creation of 3,4-bis-triazolocoumarin-sugar conjugates via fluorogenic dual click chemistry and their quenching specificity with silver(I) in aqueous media. *Tetrahedron*, vol.67, No.19, (May 2011), pp. 3343-3347, ISSN 0040-4020

Helal, A.; Rashid, M. H. O.; Choi, C.-H. & Kim, H.-S. (2011). Chromogenic and fluorogenic sensing of Cu^{2+} based on coumarin. *Tetrahedron*, vol.67, No.15, (April 2011), pp. 2794-2802, ISSN 0040-4020

Hirano, T.; Hiromoto, K. & Kagechika, H. (2007). Development of a library of 6-arylcoumarins as candidate fluorescent sensors. *Org. Lett.*, vol.9, No.7, (March 2007), pp. 1315-1318, ISSN 1523-7060

Hong, S. W. & Jo, W. H. (2008). A fluorescence resonance energy transfer probe for sensing pH in aqueous solution. *Polymer*, vol.49, No. 19, (September 2008), pp. 4180-4187, ISSN 0032-3861

Hu, X.; Zhang, X.; He, G.; He, C. & Duan, C. (2011). A FRET approach for luminescence sensing Cr^{3+} in aqueous solution and living cells through functionalizing glutathione and glucose moieties. *Tetrahedron*, vol.67, No.6, (February 2011), pp. 1091-1095, ISSN 0040-4020

Huang, H.-C.; Wang, K.-L.; Huang, S.-T.; Lin, H.-Y. & Lin, C.-M. (2011). Development of a sensitive long-wavelength fluorogenic probe for nitroreductase: a new fluorimetric indicator for analyte determination by dehydrogenase-coupled biosensors. *Biosensors Bioelectronics*, vol.26, No.8, (April 2011), pp. 3511-3516, ISSN 0956-5663

Huang, S.; He, S.; Lu, Y.; Wei, F.; Zeng, X. & Zhao, L. (2011). Highly sensitive and selective fluorescent chemosensor for Ag^+ based on a coumarin-Se_2N chelating conjugate. *Chem. Commun.*, vol.47, No.8, (February 2011), pp. 2408-2410, ISSN 1359-7345

Huang, S.-T.; Teng, C.-J.; Lee, Y.-H.; Wu, J.-Y.; Wang, K.-L. & Lin, C.-M. (2010). Design and synthesis of a long-wavelength latent fluorogenic substrate for salicylate hydroxylase: a useful fluorimetric indicator for analyte determination by dehydrogenase- coupled biosensors. *Anal. Chem.*, vol.82, No.17, (September 2010), pp. 7329-7334, ISSN 0003-2700

Jang, Y. J.; Moon, B.-S.; Park, M. S.; Kang, B.-G.; Kwon, J. Y.; Hong J. S. J.; Yoon, Y. J.; Lee, K. D. & Yoon, J. (2006). New cavitand derivatives bearing four coumarin groups as fluorescent chemosensors for Cu^{2+} and recognition of dicarboxylates utilizing Cu^{2+} complex. *Tetrahedron Lett.*, vol.47, No.16, (April 2006), pp. 2707-2710, ISSN 0040-4039

Jiang, W. & Wang, W. (2009). A selective and sensitive "turn-on" fluorescent chemodosimeter for Hg^{2+} in aqueous media via Hg^{2+} promoted facile desulfurization-lactonization reaction. *Chem. Commun.*, No.26, (July 2009), pp. 3913-3915, ISSN 1359-7345

Jung, H. S.; Han, J. H.; Habata, Y.; Kang, C. & Kim, J. S. (2011b). An iminocoumarin-Cu(II) ensemble-based chemodosimeter toward thiols. *Chem. Commun.*, vol.47, No.18, (May 2011), pp. 5142-5144, ISSN 1359-7345

Jung, H. S.; Ko, K. C.; Kim, G.-H.; Lee, A.-R.; Na, Y.-C.; Kang, C.; Lee, J. Y. & Kim, J. S. (2011a). Coumarin-based thiol chemosensor: synthesis, turn-on mechanism, and its biological application. *Org. Lett.*, vol.13, No.6, (March 2011), pp. 1498-1501, ISSN 1523-7060

Jung, H. S.; Ko, K. C.; Lee, J. H.; Kim, S. H.; Bhuniya, S.; Lee, J. Y.; Kim, Y.; Kim, S. J. & Kim, J. S. (2010). Rationally designed fluorescence trun-on sensors: a new design strategy based on orbital control. *Inorg. Chem.*, vol.49, No.18, (September 2010), pp. 8552-8557, ISSN 0020-1669

Jung, H. S.; Kwon, P. S.; Lee, J. W.; Kim, J. I.; Hong, C. S.; Kim, J. W.; Yan, S.; Lee, J. Y.; Lee, J. H.; Joo, T. & Kim, J. S. (2009). Coumarin-derived Cu^{2+}-selective fluorescence sensor: synthesis, mechanisms, and applications in living cells. *J. Am. Chem. Soc.*, vol.131, No.5, (February 2009), pp. 2008-2012, ISSN 0002-7863

Kawaguchi, M.; Okabe, T.; Terai, T.; Hanaoka, K.; Kojima, H.; Minegishi, I. & Nagano, T. (2010). A time-resolved fluorescence probe for dipeptidyl peptidase 4 and its application in inhibitor screening. *Chem. Eur. J.*, vol.16, No.45, (December 2010), pp. 13479-13486, ISSN 0947-6539

Kele, P.; Orbulescu, J.; Mello, S. V.; Mabrouki, M. & Leblanc, R. M. (2001). Langmuir-Blodgett film characterization of a new amphiphilic coumarin derivative. *Langmuir*, vol.17, No.23, (November 2001), pp. 7286-7290, ISSN 0743-7463.

Kim, G.-J. & Kim, H.-J. (2010a). Doubly activated coumarin as a colorimetric and fluorescent chemodosimeter for cyanide. *Tetrahedron Lett.*, vol.51, No.1, (January 2010), pp. 185-187, ISSN 0040-4039

Kim, G.-J. & Kim, H.-J. (2010b). Coumarinyl aldehyde as a Michael acceptor type of colorimetric and fluorescent probe for cyanide in water. *Tetrahedron Lett.*, vol.51, No.21, (May 2010), pp. 2914-2916, ISSN 0040-4039

Kim, G.-J. & Kim, H.-J. (2010c). Highly selective and sensitive fluorescence turn-on probe for proline. *Tetrahedron Lett.*, vol.51, No.35, (September 2010), pp. 4670-4672, ISSN 0040-4039

Kim, G.-J.; Lee, K.; Kwon, H. & Kim, H.-J. (2011). Ratiometric fluorescence imaging of cellular glutathione. *Org. Lett.*, vol.13, No.11, (June 2011), pp. 2799-2801, ISSN 1523-7060

Kim, H. J.; Bhuniya, S.; Mahajan, R. K.; Puri, R.; Liu, H.; Ko, K. C.; Lee, J. Y. & Kim, J. S. (2009b). Fluorescence turn-on sensors for HSO_4^-. *Chem. Commun.*, No.46, (December 2009), pp. 7128-7130, ISSN 1359-7345

Kim, H. J.; Kim, Y.; Kim, S. J.; Park, S. Y.; Lee, S. Y.; Kim, J. H.; No, K. & Kim, J. S. (2010). Iminocoumarin-based Hg(II) ion probe. *Bull. Korean Chem. Soc.*, vol.31, No.1, (January 2010), pp. 230-233, ISSN 0253-2964

Kim, H. J.; Ko, K. C.; Lee, J. H.; Lee, J. Y. & Kim, J. S. (2011). KCN sensor: unique chromogenic and 'turn-on' fluorescent chemodosimeter: rapid response and high selectivity. *Chem. Commun.*, vol.47, No.10, (March 2011), pp. 2886-2888, ISSN 1359-7345

Kim, H. J.; Park, J. E.; Choi, M. G.; Ahn, S. & Chang, S.-K. (2009a). Selective chromogenic and fluorogenic signalling of Hg^{2+} ions using a fluorescein-coumarin conjugate. *Dyes Pigments*, vol.84, No.1, (January 2010), pp. 54-58, ISSN 0143-7208

Kim, H. M.; Yang, P. R.; Seo, M. S.; Yi, J.-S.; Hong, J. H.; Jeon, S.-J.; Ko, Y.-G.; Lee, K. J. & Cho, B. R. (2007). Magnesium ion selective two-photon fluorescent probe based on a benzo[h]chromene derivative for in vivo imaging. *J. Org. Chem.*, vol.72, No.6, (March 2007), pp. 2088-2096, ISSN 0022-3263

Kim, J. H.; Kim, H. J.; Kim, S. H.; Lee, J. H.; Do, J. H.; Kim, H.-J.; Lee, J. H. & Kim, J. S. (2009). Fluorescent coumarinyldithiane as a selective chemodosimeter for mercury(II) ion in aqueous solution. *Tetrahedron Lett.*, vol.50, No.43, (October 2009), pp. 5958-5961, ISSN 0040-4039

Kim, M. H.; Jang, H. H.; Yi, S.; Chang, S.-K. & Han, M. S. (2009). Coumarin-derivative-based off-on catalytic chemodosimeter for Cu^{2+} ions. *Chem. Commun.*, No.32, (August 2009), pp. 4838-4840, ISSN 1359-7345

Kim, S. K.; Lee, D. H.; Hong, J.-I. & Yoon, J. (2009). Chemosensors for pyrophosphate. *Acc. Chem. Res.*, vol.42, No.1, (January 2009), pp. 23-31, ISSN 0001-4842

Kim, S. Y.; Park, J.; Koh, M.; Park, S. B. & Hong, J.-I. (2009). Fluorescent probe for detection of fluoride in water and bioimaging in A549 human lung carcinoma cells. *Chem. Commun.*, No.31, (August 2009), pp. 4735-4737, ISSN 1359-7345

Ko, K. C.; Wu, J.-S.; Kim, H. J.; Kwon, P. S.; Kim, J. W.; Bartsch, R. A.; Lee, J. Y. & Kim, J. S. (2011). Rationally designed fluorescence 'turn-on' sensor for Cu^{2+}. *Chem. Commun.*, vol.47, No.11, (March 2011), pp. 3165-3167, ISSN 1359-7345

Kovalska, V. B.; Volkova, K. D.; Manaev, A. V.; Losytskyy, M. Y.; Okhrimenko, I. N.; Traven, V. F. & Yarmoluk, S. M. (2010). 2-Quinolone and coumarin polymethines for the detection of proteins using fluorescence. *Dyes Pigments*, vol.84, No.2, (February 2010), pp. 159-164, ISSN 0143-7208

Kulatilleke, C. P.; de Silva, S. A. & Eliav, Y. (2006). A coumarin based fluorescent photoinduced electron transfer cation sensor. *Polyhedron*, vol.25, No.13, (September 2006), pp. 2593-2596, ISSN 0277-5387

Kumar, A.; Pandey, M. K.; Anandakathir, R.; Mosurkal, R.; Parmar, V. S.; Watterson, A. C. & Kumar, J. (2010). Sensory response of pegylated and siloxanated 4,8-dimethylcoumarins: a fluorescence quenching study by nitro aromatics. *Sensors Actuators B*, vol.147, No.1, (May 2010), pp. 105-110, ISSN 0925-4005

Kurishita, Y.; Kohira, T.; Ojida, A. & Hamachi, I. (2010). Rational design of FRET-based ratiometric chemosensors for in vitro and in cell fluorescence analyses of nucleoside polyphosphates. *J. Am. Chem. Soc.*, vol.132, No.38, (September 2010), pp. 13290-13299, ISSN 0002-7863

Kwon, H.; Lee, K. & Kim, H.-J. (2011). Coumarin-malonitrile conjugate as a fluorescent turn-on probe for biothiols and its cellular expression. *Chem. Commun.*, vol.47, No.6, (February 2011), pp. 1773-1775, ISSN 1359-7345

Lau, Y. H.; Price, J. R.; Todd, M. H. & Rutledge, P. J. (2011). A click fluorophore sensor that can distinguish Cu^{II} and Hg^{II} via selective anion-induced demetallation. *Chem. Eur. J.*, vol.17, No.10, (March 2011), pp. 2850-2858, ISSN 0947-6539

Lee, D.-N.; Kim, G.-J. & Kim, H.-J. (2009). A fluorescent coumarinylalkyne probe for the selective detection of mercury(II) ion in water. *Tetrahedron Lett.*, vol.50, No.33, (August 2009), pp. 4766-4768, ISSN 0040-4039

Lee, J. H.; Jeong, A. R.; Shin, I.-S.; Kim, H.-J. & Hong, J.-I. (2010). Fluorescence turn-on sensor for cyanide based on a cobalt(II)-coumarinylsalen complex. *Org. Lett.*, vol.12, No.4, (February 2010), pp. 764-767, ISSN 1523-7060

Lee, K.-S.; Kim, H.-J.; Kim, G.-H.; Shin, I. & Hong, J.-I. (2008a). Fluorescent chemodosimeter for selective detection of cyanide in water. *Org. Lett.*, vol.10, No.1, (January 2008), pp. 49-51, ISSN 1523-7060

Lee, K.-S.; Kim, T.-K.; Lee, J. H.; Kim, H.-J. & Hong, J.-I. (2008b). Fluorescence turn-on probe for homocysteine and cysteine in water. *Chem. Commun.*, No.46, (December 2008), pp. 6173-6175, ISSN 1359-7345

Lee, M.; Gubernator, N. G.; Sulzer, D. & Sames, D. (2010). Development of pH-responsive fluorescent false neurotransmitters. *J. Am. Chem. Soc.*, vol.132, No.26, (July 2010), pp. 8828-8830, ISSN 0002-7863

Lee, S. H.; Helal, A. & Kim, H.-S. (2010). Fluorescence sensing properties of thiazolobenzo-crown ether incorporating coumarin. *Bull. Korean Chem. Soc.*, vol.31, No.3, (March 2010), pp. 615-619, ISSN 0253-2964

Li, H.; Cai, L.; Li, J.; Hu, Y.; Zhou, P. & Zhang, J. (2011). Novel coumarin fluorescent dyes: synthesis, structural characterization and recognition behavior towards Cu(II) and Ni(II). *Dyes Pigments*, vol.91, No.3, (December 2011), pp. 309-316, ISSN 0143-7208

Li, H.; Gao, S. & Xi, Z. (2009). A colorimetric and "turn-on" fluorescent chemosensor for Zn(II) based on coumarin Schiff-base derivative. *Inorg. Chem. Commun.*, vol.12, No.4, (April 2009), pp. 300-303, ISSN 1387-7003

Li, J.; Lin, H.; Cai, Z. & Lin, H. (2009). A novel coumarin-based switching-on fluorescent and colorimetric sensor for F^-. *J. Lumin.*, vol.129, No.5, (May 2009), pp. 501-505, ISSN 0022-2313

Li, N.; Xiang, Y. & Tong, A. (2010). Highly sensitive and selective "turn-on" fluorescent chemodosimeter for Cu^{2+} in water via Cu^{2+}-promoted hydrolysis of lactone moiety in coumarin. *Chem. Commun.*, vol.46, No.19, (May 2010), pp. 3363-3365, ISSN 1359-7345

Li, Y.; Wang, H.; Li, J.; Zheng, J.; Xu, X. & Yang, R. (2011). Simultaneous intracellular β-D-glucosidase and phosphodiesterase I activities measurements based on a triple-signaling fluorescent probe. *Anal. Chem.*, vol.83, No.4, (February 2011), pp. 1268-1274, ISSN 0003-2700

Lim, N. C. & Brückner, C. (2004). DPA-substituted coumarins as chemosensors for zinc(II): modulation of the chemosensory characteristics by variation of the position of the chelate on the coumarin. *Chem. Commun.*, No.9, (May 2004), pp. 1094-1095, ISSN 1359-7345

Lim, N. C.; Pavlova, S. V. & Brückner, C. (2009). Squaramide hydroxamate-based chemidosimeter responding to iron(III) with a fluorescence intensity increase. *Inorg. Chem.*, vol.48, No.3, (February 2009), pp. 1173-1182, ISSN 0020-1669

Lim, N. C.; Schuster, J. V.; Porto, M. C.; Tanudra, M. A.; Yao, L.; Freake, H. C. & Brückner, C. (2005). Coumarin-based chemosensors for zinc(II): toward the determination of the design algorithm for CHEF-type and ratiometric probes. *Inorg. Chem.*, vol.44, No.6, (March 2005), pp. 2018-2030, ISSN 0020-1669

Lim, S.-Y.; Lee, S.; Park, S. B. & Kim, H.-J. (2011). Highly selective fluorescence turn-on probe for glutathione. *Tetrahedron Lett.*, vol.52, No.30, (July 2011), pp. 3902-3904, ISSN 0040-4039

Lin, W.; Cao, X.; Yuan, L. & Ding, Y. (2010b). Double functional group transformations for fluorescent probe construction: a fluorescence turn-on probe for thioureas. *Chem. Eur. J.*, vol.16, No.22, (June 2010), pp. 6454-6457, ISSN 0947-6539

Lin, W.; Long, L. Tan, W. (2010a). A highly sensitive fluorescent probe for detection of benzenethiols in environmental samples and living cells. *Chem. Commun.*, vol.46, No.9, (March 2010), pp. 1503-1505, ISSN 1359-7345

Lin, W.; Yuan, L.; Cao, X.; Tan, W. & Feng, Y. (2008). A coumarin-based chromogenic sensor for transition-metal ions showing ion-dependent bathochromic shift. *Eur. J. Org. Chem.*, No.29, (October 2008), pp. 4981-4987, ISSN 1434-193X

Lin, W.; Yuan, L.; Cao, Z.; Feng, J. & Feng, Y. (2009). Fluorescence enhancement of coumarin-quinoline by transition metal ions: detection of paramagnetic Ni^{2+} and Co^{2+}. *Dyes Pigments*, vol.83, No.1, (October 2009), pp. 14-20, ISSN 0143-7208

Ma, Q.-J.; Zhang, X.-B.; Zhao, X.-H.; Jin, Z.; Mao, G.-J.; Shen, G.-L. & Yu, R.-Q. (2010). A highly selective fluorescent probe for Hg^{2+} based on a rhodamine-coumarin conjugate. *Anal. Chim. Acta*, vol.663, No.1, (March 2010), pp. 85-90, ISSN 0003-2670

Ma, W.; Xu, Q.; Du, J.; Song, B.; Peng, X.; Wang, Z.; Li, G. & Wang, X. (2010). A Hg^{2+}-selective chemodosimeter based on desulfurization of coumarin thiosemicarbazide in aqueous media. *Spectrochim. Acta A*, vol.76, No.2, (July 2010), pp. 248-252, ISSN 1386-1425

Ma, Y. M. & Hider, R. C. (2009). The selective quantification of iron by hexadentate fluorescent probes. *Bioorg. Med. Chem.*, vol.17, No.23, (December 2009), pp. 8093-8101, ISSN 0960-894X

Mahapatra, A. K.; Hazra, G.; Roy, J. & Sahoo, P. (2011). A simple coumarin-based colorimetric and ratiometric chemosensor for acetate and a selective fluorescence turn-on probe for iodide. *J. Lumin.*, vol.131, No.7, (July 2011), pp. 1255-1259, ISSN 0022-2313

Maity, D. & Govindaraju, T. (2010). Conformationally constrained (cumarin-triazolyl-bipyridyl) click fluoroionophore as a selective Al^{3+} sensor. *Inorg. Chem.*, vol.49, No.16, (August 2010), pp. 7229-7231, ISSN 0020-1669

Makwana, P. K.; Jethva, P. N. & Roy, I. (2011). Coumarin 6 and 1,6-diphenyl-1,3,5-hexatriene (DPH) as fluorescent probes to monitor protein aggregation. *Analyst*, vol.136, No.10, (May 2011), pp. 2161-2167, ISSN 0003-2654

Meyer, Y.; Richard, J.-A.; Massonneau, M.; Renard, P.-Y. & Romieu, A. (2008). Development of a new nonpeptidic self-immolative spacer. Application to the design of protease sensing fluorogenic probes. *Org. Lett.*, vol.10, No.8, (April 2008), pp. 1517-1520, ISSN 1523-7060

Miyaji, H.; Kim, H.-K.; Sim, E.-K.; Lee, C.-K.; Cho, W.-S.; Sessler, J. L. & Lee, C.-H. (2005). Coumarin-strapped calyx[4]pyrrole: a fluorogenic anion receptor modulated by cation and anion binding. *J. Am. Chem. Soc.*, vol.127, No.36, (September 2005), pp. 12510-12512, ISSN 0002-7863

Mizukami, S.; Nagano, T.; Urano, Y.; Odani, A. & Kikuchi, K. (2002). A fluorescent anion sensor that works in neutral aqueous solution for bioanalytical application. *J. Am. Chem. Soc.*, vol.124, No.15, (April 2002), pp. 3920-3925, ISSN 0002-7863

Mizukami, S.; Okada, S.; Kimura, S. & Kikuchi, K. (2009). Design and synthesis of coumarin-based Zn^{2+} probes for ratiometric fluorescence imaging. *Inorg. Chem.*, vol.48, No.16, (August 2009), pp. 7630-7638, ISSN 0020-1669

Namba, K.; Kobayashi, K.; Murata, Y.; Hirakawa, H.; Yamagaki, T.; Iwashita, T.; Nishizawa, M.; Kusumoto, S. & Tanino, K. (2010). Mugineic acid derivatives as molecular probes for the mechanistic elucidation of iron acquisition in barley. *Angew. Chem. Int. Ed.*, vol.49, No.51, (December 2010), pp. 9956-9959, ISSN 1433-7851

Nguyen, T. H. & Ansell, R. J. (2009). Fluorescent imprinted polymer sensors for chiral amines. *Org. Biomol. Chem.*, vol.7, No.6, (March 2009), pp. 1211-1220, ISSN 1477-0520

Ortyl, J.; Sawicz, K. & Popielarz, R. (2010). Performance of amidocoumarins as probes for monitoring of cationic photopolymerization of monomers by fluorescence probe technology. *J. Polym. Sci. A*, vol.48, No.20, (October 2010), pp. 4522-4528, ISSN 1099-0518

Pagona, G.; Economopoulos, S. P.; Tsikalas, G. K.; Katerionpoulos, H. E. & Tagmatarchis, N. (2010). Fullerene-coumarin dyad as a selective metal receptor: synthesis, photophysical properties, electrochemistry and ion binding studies. *Chem. Eur. J.*, vol.16, No.39, (October 2010), pp. 11969-11976, ISSN 0947-6539

Pan, J.; Downing, J. A.; McHale, J. L. & Xian, M. (2009). A fluorogenic dye activated by S-nitrosothiols. *Mol. BioSyst.*, vol.5, No.9, (September 2009), pp. 918-920, ISSN 1742-206X

Ray, D.; Nag, A.; Jana, A.; Goswami, D. & Bharadwaj, P. K. (2010). Coumarin derived chromophores in the donor-acceptor-donor format that give fluorescence enhancement and large two-photon activity in presence of specific metal ions. *Inorg. Chim. Acta*, vol.363, No.12, (October 2010), pp. 2824-2832, ISSN 0020-1693

Richard, J.-A.; Massonneau, M.; Renard, P.-Y. & Romieu, A. (2008). 7-Hydroxycoumarin—hemicyanine hybrids: a new class of far-red emitting fluorogenic dyes. *Org. Lett.*, vol.10, No.19, (October 2008), pp. 4175-4178, ISSN 1523-7060

Roussakis, E.; Pergantis, S. A. & Katerinopoulos, H. E. (2008). Coumarin-based ratiometric fluorescent indicators with high specificity for lead ions. *Chem. Commun.*, No.46, (December 2008), pp. 6221-6223, ISSN 1359-7345

Royo, S.; Martínez-Máñez, R.; Sancenón, F.; Costero, A. M.; Parra, M. & Gil, S. (2007). Chromogenic and fluorogenic reagents for chemical warfare nerve agents' detection. *Chem. Commun.*, No.46, (JDecember 2007), pp. 4839-4847, ISSN 1359-7345

Ryu, D. H.; Noh, J. H. & Chang, S.-K. (2010). Selective ratiometric signaling of Hg^{2+} ions by a fluorescein-coumarin chemodosimeter. *Bull. Korean Chem. Soc.*, vol.31, No.1, (January 2010), pp. 246-249, ISSN 0253-2964

Sakamoto, H.; Ishikawa, J.; Osuga, H.; Doi, K. & Wada, H. (2010). Highly silver ion selective fluorescent ionophore: fluorescent properties of polythiazaalkane derivatives bearing 8-(7-hydroxy-4-methyl)coumarinyl moiety in aqueous solution and in liquid-liquid extraction systems. *Analyst*, vol.135, No.3, (February 2010), pp. 550-558, ISSN 0003-2654

Saleh, N.; Al-Soud, Y. A. & Nau, W. M. (2008). Novel fluorescent pH sensor based on coumarin with piperazine and imidazole substituents. *Spectrochim. Acta A*, vol.71, No.3, (December 2008), pp. 818-822, ISSN 1386-1425

Santos, H. M.; Pedras, B.; Tamayo, A.; Casabó, J.; Escriche, L.; Covelo, B.; Capelo, J. L. & Lodeiro, C. (2009). New chemosensors based on thiomacrocycle-containing coumarin-343 fluoroionophor: X-ray structures and previous results on the effect of cation binding on the photophysical properties. *Inorg. Chem. Commun.*, vol.12, No.11, (November 2009), pp. 1128-1134, ISSN 1387-7003

Sato, S.; Suzuki, M.; Soma, T. & Tsunoda, M. (2008). Synthesis and properties of umbelliferone-nitroxide radical hybrid compounds as fluorescence and spin-label probes. *Spectrochim. Acta A*, vol.70, No.4, (September 2008), pp. 799-804, ISSN 1386-1425

Secor, K. E. & Glass, T. E. (2004). Selective amine recognition: development of a chemosensor for dopamine and norepinephrine. *Org. Lett.*, vol.6, No.21, (October 2004), pp. 3727-3730, ISSN 1523-7060

Sethna, S. M. & Shah, N. M. (1945). The chemistry of coumarins. *Chem. Rev.*, vol.36, No.1, (February 1945), pp. 1-62, ISSN 0009-2665

Shao, J. (2010). A novel colorimetric and fluorescence anion sensor with a urea group as binding site and a coumarin group as signal unit. *Dyes Pigments*, vol.87, No.3, (November 2010), pp. 272-276, ISSN 0143-7208

Sheng, J. R.; Feng, F.; Qiang, Y.; Liang, F. G.; Sen, L. & Wei, F.-H. (2008). A coumarin-derived fluorescence chemosensors selective for copper(II). *Anal. Lett.*, vol.41, No.12, (August 2008), pp. 2203-2213, ISSN 0003-2719

Sheng, R.; Wang, P.; Liu, W.; Wu, X. & Wu, S. (2008). A new colorimetric chemosensor for Hg^{2+} based on coumarin azine derivatives. *Sensors Actuators B*, vol.128, No.2, (January 2008), pp. 507-511, ISSN 0925-4005

Shibata, A.; Abe, H.; Ito, M.; Kondo, Y.; Shimizu, S.; Aikawa, K. & Ito, Y. (2009). DNA templated nucleophilic aromatic substitution reactions for fluorogenic sensing of oligonucleotides. *Chem. Commun.*, No.43, (November 2009), pp. 6586-6588, ISSN 1359-7345

Shiraishi, Y.; Sumiya, S. & Hirai, T. (2010). A coumarin-thiourea conjugate as a fluorescent probe for Hg(II) in aqueous media with a broad pH range 2-12. *Org. Biomol. Chem.*, vol.8, No.6, (March 2010), pp. 1310-1314, ISSN 1477-0520

Shiraishi, Y.; Sumiya, S. & Hirai, T. (2011). Highly sensitive cyanide anion detection with a coumarin-spiropyran conjugate as a fluorescent receptor. *Chem. Commun.*, vol.47, No.17, (May 2011), pp. 4953-4955, ISSN 1359-7345

Singh, R. K.; Mandal, T.; Balasubramanian, N.; Cook, G. & Srivastava, D. K. (2011). Coumarin-suberoylanilide hydroxamic acid as a fluorescent probe for determining binding affinities and off-rates of histone deacetylase inhibitors. *Anal. Biochem.*, vol.408, No.2, (January 2011), pp. 309-315, ISSN 0003-2697

Sokkalingam, P. & Lee, C.-H. (2011). Highly sensitive fluorescence "turn-on" indicator for fluoride anion with remarkable selectivity in organic and aqueous media. *J. Org. Chem.*, vol.76, No.10, (May 2011), pp. 3820-3828, ISSN 0022-3263

Su, Z.; Chen, K.; Guo, Y.; Qi, H.; Yang, X.-F. & Zhao, M.. (2010). A coumarin-based fluorescent chemosensor for Zn^{2+} in aqueous ethanol media. *J. Fluoresc.*, vol.20, No.4, (July 2010), pp. 851-856, ISSN 1053-0509

Suresh, M. & Das, A. (2009). New coumarin-based sensor molecule for magnesium and calcium ions. *Tetrahedron Lett.*, vol.50, No.42, (October 2009), pp. 5808-5812, ISSN 0040-4039

Suzuki, Y.; Komatsu, H.; Ikeda, T.; Saito, N.; Araki, S.; Citterio, D.; Hisamoto, H.; Kitamura, Y.; Kubota, T.; Nakagawa, J.; Oka, K. & Suzuki, K. (2002). Design and synthesis of Mg^{2+}-selective fluoroionophores based on a coumarin derivative and application for Mg^{2+} measurement in a living cell. *Anal. Chem.*, vol.74, No.6, (March 2002), pp. 1423-1428, ISSN 0003-2700

Taki, M.; Desaki, M.; Ojida, A.; Iyoshi, S.; Hirayama, T.; Hamachi, I. & Yamamoto, Y. (2008). Fluorescence imaging of intracellular cadmium using a dual-excitation ratiometric chemosensor. *J. Am. Chem. Soc.*, vol.130, No.38, (September 2008), pp. 12564-12565, ISSN 0002-7863

Trenor, S. R.; Shultz, A. R.; Love, B. J. & Long, T. E. (2004). Coumarins in polymers: from light harvesting to photo-cross-linkable tissue scaffolds. *Chem. Rev.*, vol.104, No.6, (June 2004), pp. 3059-3077, ISSN 0009-2665

Tsai, C.-S.; Liu, P.-Y.; Yen, H.-Y.; Hsu, T.-L. & Wong, C.-H. (2010). Development of trifunctional probes for glycoproteomic analysis. *Chem. Commun.*, vol.46, No.30, (August 2010), pp. 5575-5577, ISSN 1359-7345

Tsukamoto, K.; Shinohara, Y.; Iwasaki, S. & Maeda, H. (2011). A coumarin-based fluorescent probe for Hg^{2+} and Ag^{+} with an *N'*-acetylthioureido group as a fluorescence switch. *Chem. Commun.*, vol.47, No.17, (May 2011), pp. 5073-5075, ISSN 1359-7345

Upadhyay, K. K. & Mishra, R. K. (2010). Zn^{2+} specific colorimetric receptor based on coumarin. *Bull. Chem. Soc. Jpn.*, vol.83, No.10, (October 2010), pp. 1211-1215, ISSN 0009-2673

Upadhyay, K. K.; Mishra, R. K.; Kumar, V. & Chowdhury, P. K. R. (2010a). A coumarin based ICT probe for fluoride in aqueous medium with its real application. *Talanta*, vol.82, No.1, (June 2010), pp. 312-318, ISSN 0039-9140

Upadhyay, K. K.; Mishra, R. K.; Kumar, A.; Zhao, J. & Prasad, R. (2010b). Self assembled pseudo double helix architecture and anion sensing behavior of a coumarin based ICT probe. *J. Mol. Struct.*, vol.963, No.2-3, (January 2010), pp. 228-233, ISSN 0022-2860

Vasylevska, A. S.; Karasyov, A. A.; Borisov, S. M. & Krause, C. (2007). Novel coumarin-based fluorescent pH indicators, probes and membranes covering a broad pH range. *Anal. Bioanal. Chem.*, vol.387, No.6, (March 2007), pp. 2131-2141, ISSN 1618-2642

Voutsadaki, S.; Tsikalas, G. K.; Klontzas, E.; Froudakis, G. E. & Katerinopoulos, H. E. (2010). A "turn-on" coumarin-based fluorescent sensor with high selectivity for mercury ions in aqueous media. *Chem. Commun.*, vol.46, No.19, (May 2010), pp. 3292-3294, ISSN 1359-7345

Wallace, K. J.; Fagbemi, R. I.; Folmer-Andersen, F. J.; Morey, J.; Lynth, V. M. & Anslyn, E. V. (2006). Detection of chemical warfare simulants by phosphorylation of a coumarin oximate. *Chem. Commun.*, No.37, (October 2006), pp. 3886-3888, ISSN 1359-7345

Wang, J.; Qian, X. & Cui, J. (2006). Detecting Hg^{2+} ions with an ICT fluorescent sensor molecule: remarkable emission spectra shift and unique selectivity. *J. Org. Chem.*, vol.71, No.11, (May 2006), pp. 4308-4311, ISSN 0022-3263

Welser, K.; Perera, M. D. A.; Aylott, J. W. & Chan, W. C. (2009). A facile method to clickable sensing polymeric nanoparticles. *Chem. Commun.*, No.43, (November 2009), pp. 6601-6603, ISSN 1359-7345

Wu, J.; Sheng, R.; Liu, W.; Wang, P.; Ma, J.; Zhang, H. & Zhuang, X. (2011). Reversible fluorescent probe for highly selective detection of mercapto biomolecules. *Inorg. Chem.*, vol.50, No.14, (July 2011), pp. 6543-6551, ISSN 0020-1669

Wu, W.; Wu, W.; Ji, S.; Guo, H. & Zhao, J. (2011). Accessing the long-lived emissive ^{3}IL triplet excited states of coumarin fluorophores by direct cyclometallation and its application for oxygen sensing and upconversion. *Dalton Trans.*, vol.40, No.22, (June 2011), pp. 5953-5963, ISSN 1477-9226

Xiao, D.; Martini, L. A.; Snoeberger III, R. C.; Crabtree, R. H. & Batista, V. S. (2011). Inverse design and synthesis of acac-coumarin anchors for robust TiO_2 sensitization. *J. Am. Chem. Soc.*, vol.133, No.23, (June 2011), pp. 9014-9022, ISSN 0002-7863

Xie, Y.; Dix, A. V. & Tor, Y. (2009). FRET enabled real time detection of RNA-small molecule binding. *J. Am. Chem. Soc.*, vol.131, No.48, (December 2009), pp. 17605-17614, ISSN 0002-7863

Xiong, M.; Xi, H.; Fu, Y. & Sun, X. (2010). Synthesis and properties of novel coumarin-based fluorescent probes for identifying melamine. *Chin. J. Org. Chem.*, vol.30, No.6, (June 2010), pp. 908-911, ISSN 0253-2786

Yan, M.; Li, T. & Yang, Z. (2011). A novel coumarin Schiff-base as a Zn(II) ion fluorescent sensor. *Inorg. Chem. Commun.*, vol.14, No.3, (March 2011), pp. 463-465, ISSN 1387-7003

Yao, J.; Dou, W.; Qin, W. & Liu, W. (2009). A new coumarin-based chemosensor for Fe^{3+} in water. *Inorg. Chem. Commun.*, vol.12, No.2, (February 2009), pp. 116-118, ISSN 1387-7003

Yi, L.; Li, H.; Sun, L.; Liu, L.; Zhang, C. & Xi, Z. (2009). A highly sensitive fluorescence probe for fast thiol-quantification assay of glutathione reductase. *Angew. Chem. Int. Ed.*, vol.48, No.22, (May 2009), pp. 4034-4037, ISSN 1433-7851

Yuan, L.; Lin, W. & Song, J. (2010). Ratiometric fluorescent detection of intracellular hydroxyl radicals based on a hybrid coumarin-cyanine platform. *Chem. Commun.*, vol.46, No.42, (November 2010), pp. 7930-7932, ISSN 1359-7345

Yuan, L.; Lin, W. & Yang, Y. (2011). A ratiometric fluorescent probe for specific detection of cysteine over homocysteine and glutathione based on the drastic distinction in the kinetic profiles. *Chem. Commun.*, vol.47, No.22, (June 2011), pp. 6275-6277, ISSN 1359-7345

Zhang, H. & Rudkevich, D. M. (2007). A FRET approach to phosgene detection. *Chem. Commun.*, No.12, (March 2007), pp. 1238-1239, ISSN 1359-7345

Zhao, X.; Zhang, Y.; He, G. & Zhou, P. (2010). Highly sensitive fluorescent coumarin-based probes for selective detection of copper ion. *Chin. J. Lumin.*, vol.31, No.3, (June 2010), pp. 433-438, ISSN 1000-7032

Zhou, Y.; Liu, K.; Li, J.-Y.; Fang, Y.; Zhao, T.-C. & Yao, C. (2011). Visualization of nitroxyl in living cells by a chelated copper(II) coumarin complex. *Org. Lett.*, vol.13, No.6, (March 2011), pp. 1290-1293, ISSN 1523-7060

Zhuang, X.; Liu, W.; Wu, J.; Zhang, H. & Wang, P. (2011). A novel fluoride ion colorimetric chemosensor based on coumarin. *Spectrochim. Acta A*, vol.79, No.5, (September 2011), pp. 1352-1355, ISSN 1386-1425

Zuo, Q.-P.; Li, B.; Pei, Q.; Li, Z. & Liu, S.-K. (2010). A highly selective fluorescent probe for detection of biological samples thiol and its application in living cells. *J. Fluoresc.*, vol.20, No.6, (November 2010), pp. 1307-1313, ISSN 1053-0509

Part 2

Chemical Sensor with Nanostructure

7

Surface-Functionalized Porous Silicon Wafers: Synthesis and Applications

Fahlman Bradley D.[1] and Arturo Ramírez-Porras[2]
[1]*Department of Chemistry, Central Michigan University, Mount Pleasant, MI,*
[2]*Escuela de Física and*
Centro de Investigación en Ciencia e Ingeniería de Materiales (CICIMA),
Universidad de Costa Rica, San Jose,
[1]*United States*
[2]*Costa Rica*

1. Introduction

Porous silicon (PS) can be defined as a semiconductor material resulting from the electrochemical attack of a strong acid (usually hydrofluoric acid, HF), to form a network of pores with typical diameters ranging from a few micrometers to nanometers. Sometimes this material is referred to be a *quantum sponge.* The high surface-to-volume ratio (typically in the order of 500 m^2/cm^3), and their inherent electronic and transport characteristics make this material suitable for development of photonic and sensing devices.

Although the attention focus driven by PS started in 1990, some previous works have to be mentioned here. Early works on electrochemical treatment of silicon surfaces dealt with problems of anodic oxidation, electropolishing and chemical etching as early as 1937 (Güntherschulze & Betz, 1937). A more detailed study was performed twenty years later (Schmidt & Michel, 1957). The first mention of PS material (without being named in that way) was reported in 1956, when A. Uhlir Jr. found unusual deposits on anodized silicon samples (Uhlir, 1956). He supposed that those deposits corresponded to oxide forms of silicon. Shortly after this, Turner reported a more detailed study of anodically formed films on silicon (Turner, 1958). Years later, in 1971, Watanabe and Sakai reported for the first time that the electrochemically formed films on silicon surfaces corresponded to a porous nature (Watanabe & Sakai, 1971). Theunissen modelled the "formation of etch channels which propagate in crystal-oriented directions in the monocrystal" of n-type silicon the following year (Theunissen, 1972). Subsequently, interest on porous silicon began to grow slowly, and important articles dealing with different aspects of the material were published.

One of the key points for which PS attracted much attention in the decade of 1990 is the capability of the material to show photoluminescence (PL). In 1984, PL was observed at a temperature of 4.2K (Pickering et al, 1984). In this work, the PL was attributed to amorphous phases present in the porous media, but no further studies could explain all features of the process. It was until 1990, when Canham´s paper explained PL in terms of *quantum confinement* effect (Canham, 1990). This study triggered a vigorous research in the scientific community, as illustrated in Fig. 1.

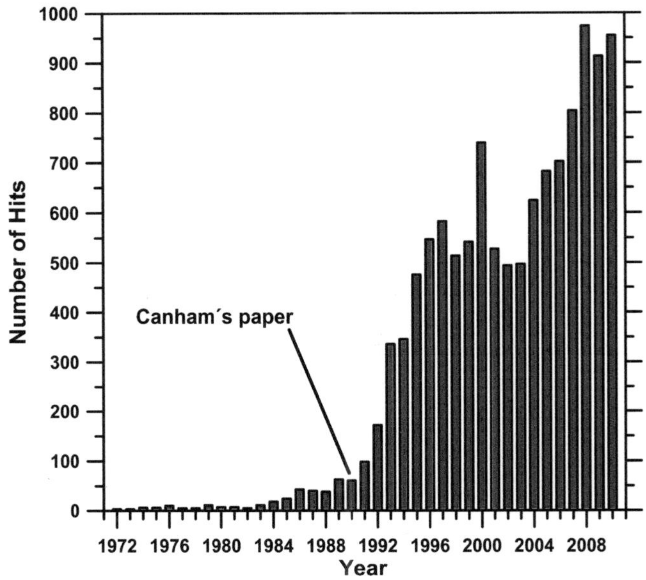

Fig. 1. Evolution of the number of publications that contain the words "porous silicon" as keywords in the period 1972 to 2010. The time location of the Canham seminal paper of 1990 is also shown.

In the early years after Canham´s report, besides the quantum confinement effect, many other models have been proposed to explain the PL emission. The most important to be mentioned here are: recombination via surface states on nanostructures (Koch et al, 1993), siloxene compounds (Brandt, 1992), amorphous phases (George, 1992) and SiH_X complexes (Prokes, 1992). A comprehensive model grouping the quantum confinement effect and the contribution of surface states as recombination centers was proposed by Wolkin and co-workers (Wolkin, 1999). This model was further generalized to consider stochastic variations of silicon nanocrystal sizes (Ramírez-Porras & Weisz, 2002).

The porous silicon era really started in 1990, as can be seen in Fig. 1. Three periods of time could be extracted from this. The first one goes from 1990 to 1997, where the rate of publications in the field exhibited a rapid growth, and the dominant research topic were the PL and electroluminescence (EL) of the material in view of a possible light emitting diode (LED) development fully compatible with the standard silicon electronics industry. Certainly, one of the key moments of this period was the publication of the porous silicon advancements in the front page of the *Physics Today* journal of January 1997 and its accompanying article therein (Collis, 1997). A second period characterized by a decrease in the field research interest went from 1998 to 2003, with a highest peak in 2000, coinciding with the 10th anniversary of Canham´s paper. This decrease was the result of the impossibility of producing a commercially available LED made out of porous silicon. Nevertheless, some very comprehensive reviews were published in this period of time, as the one from Bisi et al (Bisi et al, 2000) and from Lehmann (Lehmann, 2002). A renaissance period started in 2004, where the research interest shifted to other applications, generally dealing with the development of sensing devices (chemical and biological sensors mainly).

The electrochemical etch of silicon to produce porous silicon is a fairly inexpensive process, and therefore it can be implemented in a large group of countries. Fig. 2 shows the number of publications with affiliations belonging to the top 15 countries (Ramirez-Porras, 2011), taking into account the three periods of time already mentioned. It is noticeable the leader position earned by some European countries, United States and Japan in the first period, whereas in the last years China has undertaken a strong presence in the international community.

In this Chapter, we will briefly describe the synthetic procedure for porous silicon etching, followed by attempts to fabricate free-standing PS wafers. We will conclude with a discussion PS surface functionalization strategies for gas-sensing applications.

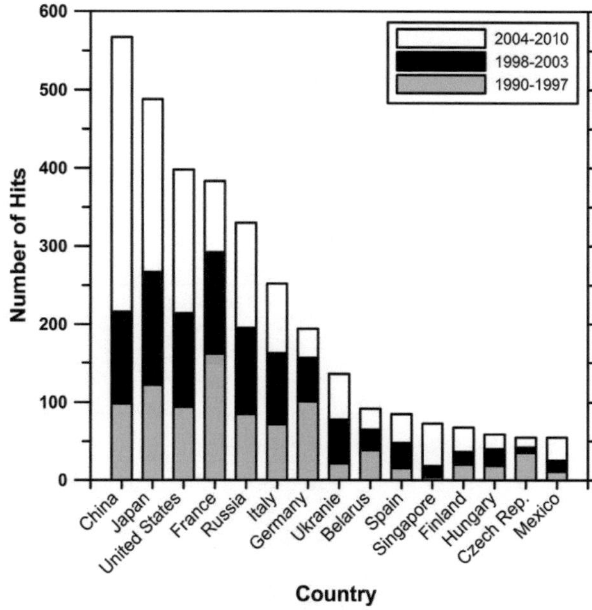

Fig. 2. Number of publications organized by affiliation that contain the words "porous silicon" as keywords. Data per country are stacked for three different periods of time: 1990-1997 (grey bars), 1998-2003 (black bars), and 2004-2010 (white bars).

2. Synthesis of porous silicon

The tunable porosity of porous silicon (p-Si) and concomitant physical properties such as photoluminescence and electrical conductivity have long been exploited for solar (Menna, 1995; Vitanov, 1997), biomedical (Low, 2009; MacInnes, 2009; Alvarez, 2009), and (bio)sensing (Razi, 2008; Palestino, 2008; Janshoff, 1998) applications. The most common method used to fabricate porous silicon is electrochemical etching (anodization) of a doped Si wafer in an aqueous/ethanolic HF electrolytic solution (Fahlman, 2011). The porosity of the resultant p-Si may be fine-tuned by altering a number of parameters such as the electrolyte concentration, etching current, anodization time, or n-/p- dopant concentrations (Table 1; Bisi, 2000). The diverse range of applications for p-Si are attributed to the morphological control of p-Si fabrication, exhibiting micro- to macroporosity (Fig. 3).

A variety of anodization cells may be employed for the electrochemical etching of silicon wafers (Fig. 4). These typically employ a platinum cathode and silicon wafer anode that is in contact with a HF-based electrolytic solution. The use of a constant DC source is usually used to ensure a constant concentration of electrolyte at the cathode tip, resulting in a more controllable corrosion rate. Although a number of hypotheses regarding the mechanism of silicon etching have been proposed (Smith, 1992), it is generally accepted that semiconductor carriers known as holes are important for the initial oxidation steps during pore formation. Hence, for n-type Si, an appreciable amount of anodization may only take place under light illumination, high fields, etc. The magnitude of illumination is dependent on the doping level of the native wafer. Equation 1 provides a plausible surface chemical reaction that occurs during the anodization of silicon:

$$Si + 6HF \rightarrow H_2SiF_6 + H_2 + 2H^+ + 2e^- \quad (1)$$

In order to offset the build-up of negative charge at the interface, holes must be injected from the substrate toward the Si/electrolyte interface during the anodization process.

Experimental Parameter[a]	**Resultant Porosity**	**Resultant Etching Rate**	**Resulting Critical Current**
HF concentration	Decrease	Decrease	Increase
Current density	Increase	Increase	No effect
Anodization time	Increase	Minimal effect	No effect
Temperature	No effect	No effect	Increase
Wafer doping (p-type)	Decrease	Increase	Increase
Wafer doping (n-type)	Increase	Increase	No effect

[a] Corresponds to an increase in each experimental parameter

Table 1. Effect of varying experimental parameters on the formation of porous silicon

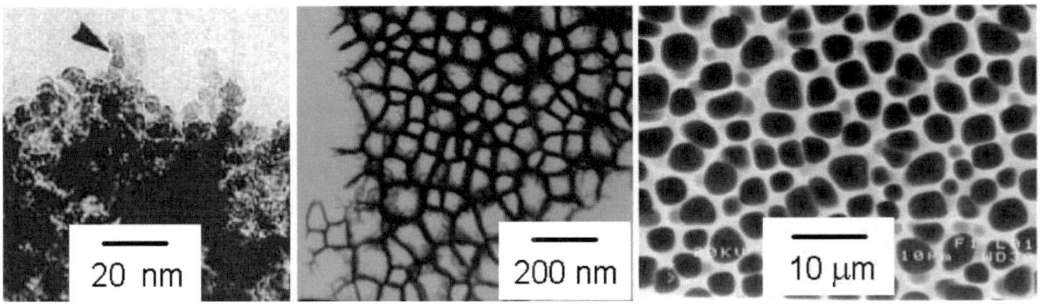

Fig. 3. TEM images of microporous (left), mesoporous (middle), and microporous (right) porous silicon. Images reproduced with permission from Gaburro, 2005.

Several techniques have been applied to characterize the interesting features that arise from the anodization process. One of these is the X-ray diffractometry (XRD), used to extract information about crystallinity. Figure 5 shows a XRD plot of both a crystalline silicon sample (c-Si) and a porous silicon sample (p-Si). The region in the vicinity of the (004) Bragg peak is shown. The X-Ray source yields two lines, the copper $K_{\alpha1}$ and $K_{\alpha2}$. The plots show the corresponding peaks for silicon and silicon dioxide (SiO_2). The ratio of the intensity lines between SiO_2 and Si is an estimation of the oxide cover on the surface. This ratio is notably higher in the p-Si sample, indicating the higher level of oxidation at the porous surface.

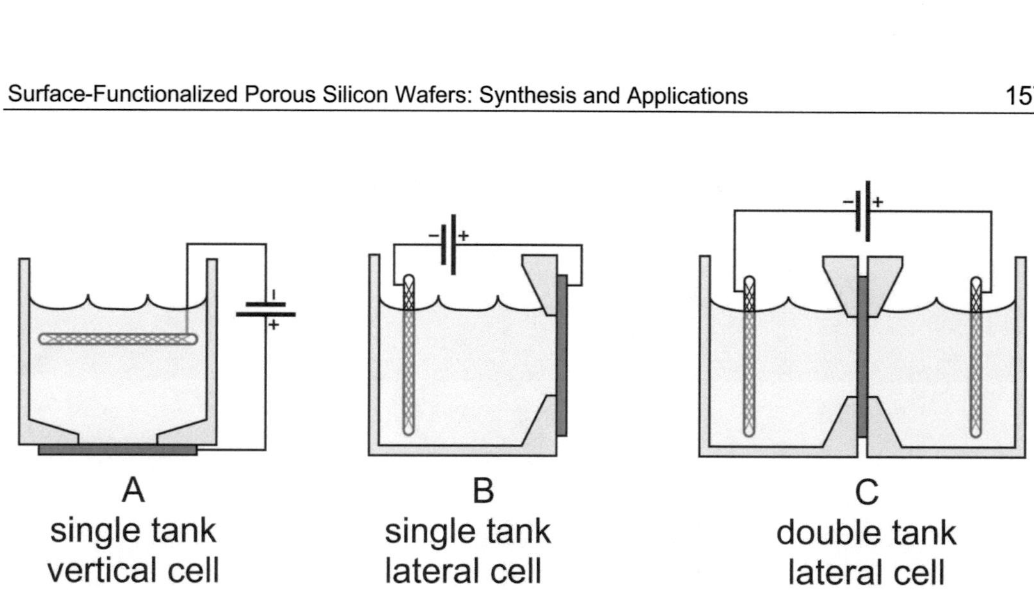

Fig. 4. Illustration of various types of anodization cells used to fabricate porous silicon wafers. Figure reproduced with permission from Gaburro, 2005.

The XRD peaks also contain information on the representative nanocrystal size. As seen in Figure 6, the peak can be deconvoluted into a set of peaks, one for the crystal phase and satellite peaks corresponding to nanocrystal phases. By the use of the Scherrer´s equation (Guinier, 1994), that relates the full width at half maximum (FWHM) of the satellite peaks with the mean structure diameter, one is able to find the typical nanocrystal size.

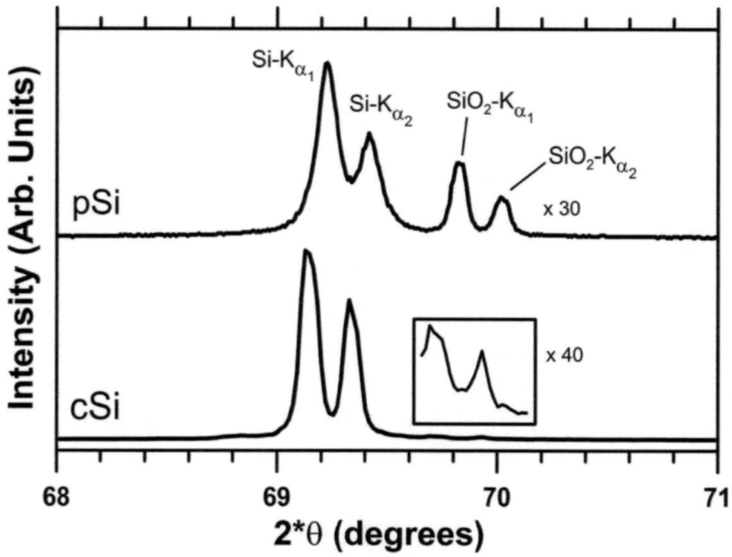

Fig. 5. XRD diffractometry plots of crystalline silicon (c-Si) and porous silicon (p-Si) surfaces near the (004) Bragg reflection zone. $K_{\alpha 1,2}$ lines for silicon and silicon dioxide are present in both cases. The intensity lines are not at the same scale (p-Si lines are increased 30 times with respect to the c-Si lines). A zoom of the silicon dioxide peaks in c-Si is also shown. (Unpublished results from the authors.)

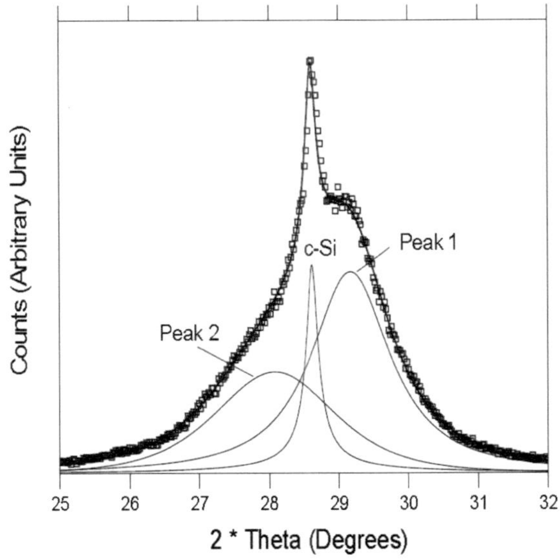

Fig. 6. XRD spectrum for a p-Si specimen near the (111) Bragg reflection plane. The satellite peaks surrounding the c-Si peak contain information on the size of the typical silicon nanocrystals, which in this case are in the 4 - 6 nm range. (Ramírez-Porras, 2002.)

3. Fabrication of free-standing porous silicon membranes

Whereas there are copious precedents of p-Si surfaces in contact with the native crystalline Si (c-Si) wafer, there are few reports of applications utilizing free-standing p-Si wafers (Liu, 2002; Bazrafkan, 2009). The most common method reported to fabricate free-standing p-Si has been electrochemical etching followed by a high-current electropolishing pulse that releases the p-Si as a free-standing flake (Papadimitriou, 2004; Lammel, 2000; Lo, 2009; Burstein, 1997). However, these methods do not yield reproducible p-Si membranes, with discrete flakes being comprised of varying thicknesses, porosities, and resultant physical/(opto)electronic properties. More recently, a sophisticated photolithographic technique has been utilized to yield free-standing p-Si microstructures (Garel, 2007). In this section, we will describe a much simpler electrochemical double-etching technique to generate free-standing p-Si wafers in a reproducible manner.

For electrochemical etching, we employed a current of 54 $mA.cm^{-2}$ for 20 min., using a p-type Si(100) substrate with a resistivity of 20 - 50 Ω.cm. This results in a macroporosity (70%), with average pores 2 - 3 μm in diameter and 40 - 50 μm depth.

In order to improve the etching uniformity over long times, the electrolyte was allowed to equilibrate in the pores during etching. This was accomplished by introducing a rest time (zero current) of 1 min. after every 10 min. of active etch time. In addition, an electrolyte stir-rate of 240 rpm was found to best control tghe reproducibility of our etching procedure. In order to maintain a sufficient current through the previously-etched p-Si layer and prevent subsequent leakage of the corrosive electrolyte solution, we introduced a sacrificial crystalline Si wafer prior to etching the second side (Fig. 7).

The optimum conditions for the fabrication of free-standing wafers consisted of 90 min. of first-side etching, followed by an additional 40 min. for the opposite side (Fig. 8).

Interestingly, it should be noted that attempts to etch for shorter times (20 min., 30 min., 40 min.) followed by longer increments (90 - 150 min.) did not yield yield suitable free-standing wafers. Instead, a large portion of un-etched crystalline silicon remained in the interior of the sample (Fig. 9). This indicates that after a period of time, the electrical current becomes more focused on the wafer edge than across the entire wafer surface. Although porous silicon is able to transport electrical current to etch the second side, it is not able to maintain this current uniformly across wafer surface for long periods of time due to increasing sample resistivity during the etching process.

Although Searson and coworkers have reported a simple one-step etching procedure (Searson, 1991), the reported porosities were non-uniform, with the formation of copious lateral pore channels. Our attempts to etch completely through the wafer using a similar one-step process were unsuccessful (Fig. 10). For reasons stated above, this process results in etching almost completely around the circumference of the p-Si wafer, which yields a back-side comprised of completely un-etched crystalline Si (Fig. 11). Further, partial etching followed by chemical removal of the remaining c-Si backside (using 1% KOH(aq); Navarro, 1997) resulted in a severely damaged p-Si surface due to the rapid etching of both c-Si and p-Si by the basic solution.

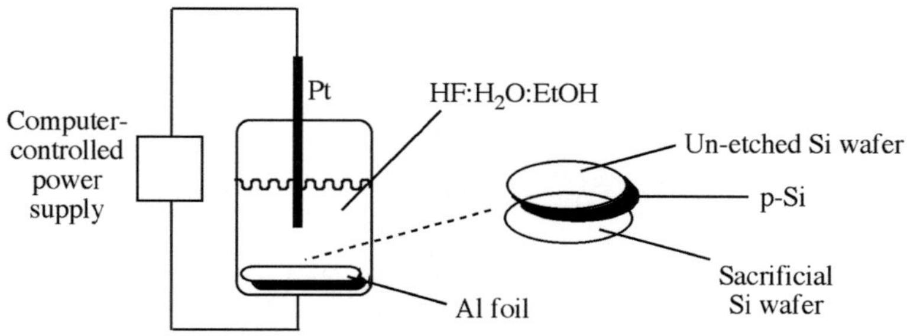

Fig. 7. Illustration of the cell used for the double electrochemical etching of silicon to yield a free-standing p-Si wafer.

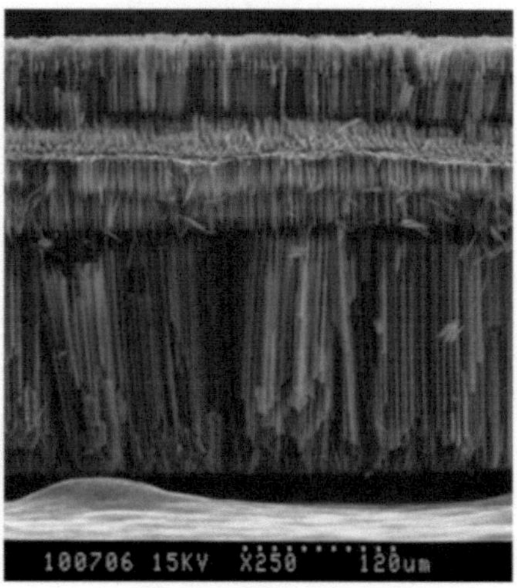

Fig. 8. Cross-section SEM image of a silicon wafer that has been electrochemically double-etched (90 min., first-side + 40 min., second side).

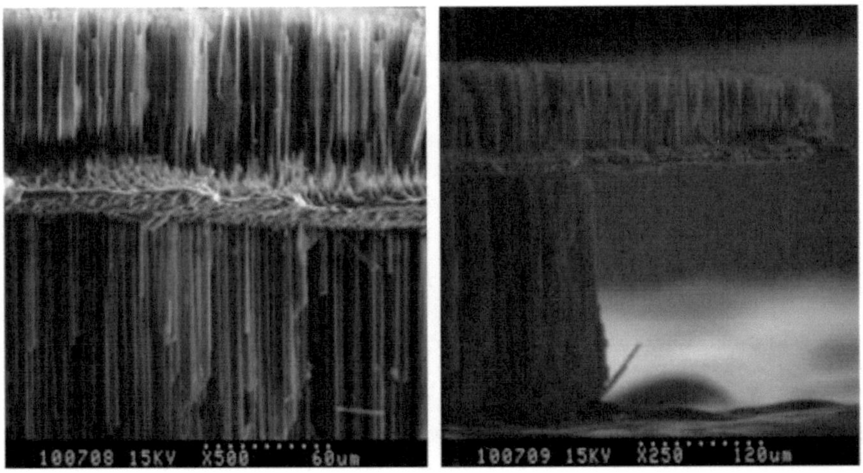

Fig. 9. Cross-section SEM images showing a region of un-etched crystalline silicon.

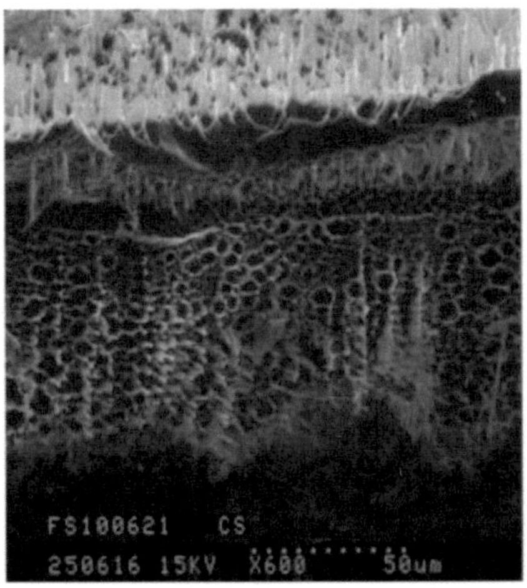

Fig. 10. Cross-section SEM image of a crystalline Si wafer following a one-step etching procedure, showing the formation of lateral pore channels.

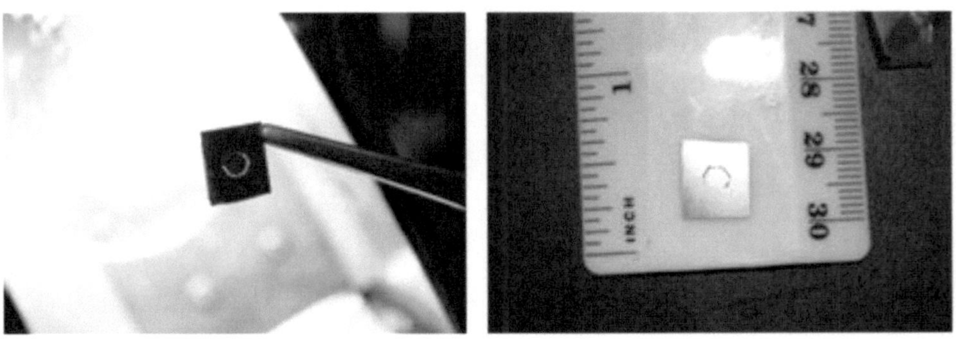

Fig. 11. Photograhs of the front (left) and backside (right) of a crystalline Si wafer, following a one-step etching procedure.

The overall thicknesses of our free-standing wafers were of the same magnitude as the starting crystalline wafer (*ca.* 200 μm). Similar pore dimensions were formed on both sides of the wafers, virtually identical to those observed in p-Si/c-Si supported films using the same etching parameters (Stefano, 2004). Interestingly, square porous arrays are observed on only one of the sides that was in contact with aluminum foil during etching of the first side (Fig. 12). These arrays duplicate the position of creases in the aluminum foil that was used as the anode during etching.

Fig. 12. SEM images illustrating the square arrays formed on the backside of the free-standing p-Si wafer, resulting from contact with the creases of aluminum foil used as the anode during etching.

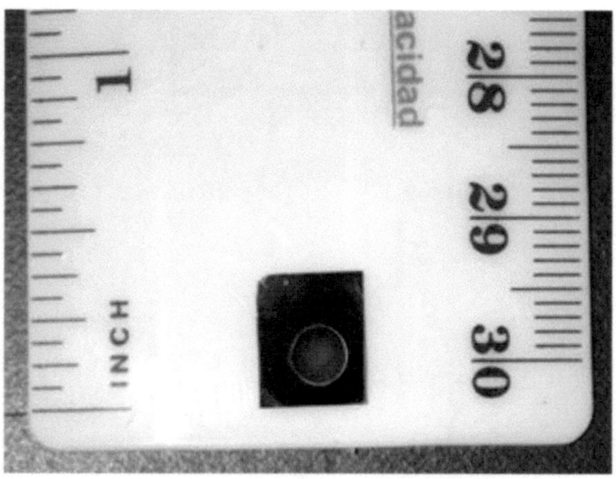

Fig. 13. Photograph illustrating uniform etching of the sacrificial crystalline Si substrate used in our double-etching procedure.

As illustrated in Figure 8, the porosity of the free-standing wafer exhibits a transitional region in its cross-section, at the point where the second-side etching is concluded. This is likely an artifact of induced diffusion currents of electrolyte at the interface of the two porous arrays. Nevertheless, the sacrificial crystalline Si was etched uniformly by the electrolytic solution (Fig. 13), indicating that the porous channels remain spatially homogenous throughout the entire thickness of the wafer.

4. Applications of surface-functionalized porous silicon

The surface of p-Si is terminated with Si-H groups immediately following electrochemical etching. However, these moieties are replaced by oxide groups over time due to exposure to ambient O_2/moisture. This will affect the electrical conductivity and optical properties of p-Si, which will alter its applicability for certain applications. Although a post-oxidation process has been used to yield a more reproducible and stable surface, it is often most desirable to introduce a surface-functionalization in order to effectively prevent post-oxidation following anodization.

A surface functionalizing group may serve a two-fold benefit - preventing post-oxidation, as well as providing a chemical handle for the selective binding of gaseous/liquid adsorbant species. The latter benefit is of interest for sensor applications (Archer, 2005; Schechter, 1995), wherein one is interested in the analyte of choice being bound selectively to a surface, which will alter a specific surface physical property (*e.g.*, conductivity, photoluminescence, *etc.*) that is required for analyte detection (Rossi, 2007).

Accordingly, we have assessed the utility of our free-standing p-Si wafers for gas-sensing applications. Indium contacts were soldered onto alternate sides of a p-Si wafer, and vapors of ethanol were generated by bubbling purified nitrogen gas (800 sccm) through the liquid at 20 °C (Fig. 14). The vapor was passed through a stainless steel chamber, which housed the porous silicon substrate in the absence of light during the analyses.

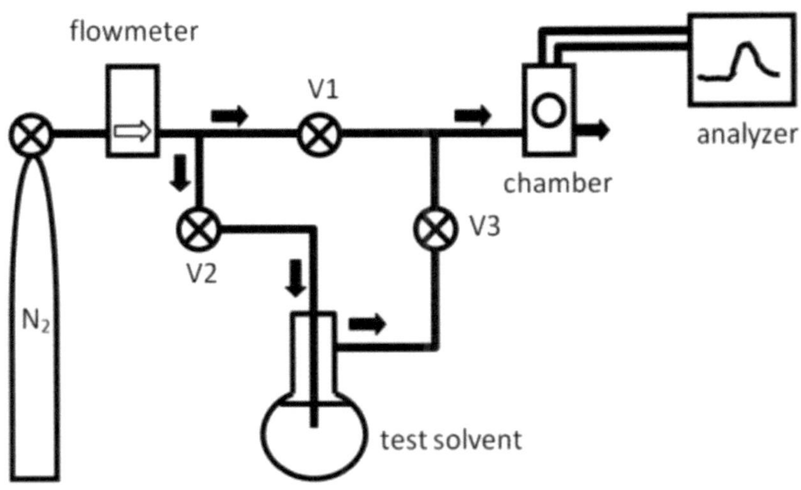

Fig. 14. Schematic of the experimental setup for vapor sensing. The concentration of the test solvent (ethanol) is controlled by varying the relative positions of valves V1 and V2/V3. The arrows indicate the direction of the gas and vapor flows.

We assessed both as-etched free standing p-Si wafers, as well as those functionalized with 4-aminopyridine. Our procedure for functionalization began by removing the native oxide layer by immersing the wafers for five minutes in a 2% ethanolic HF solution, followed by ethanol and THF rinses and drying under purified N_2. Chlorine gas was then added to the samples, and allowed to react for 30 minutes at 95 °C. Any unreacted Cl_2 was then removed from the system by purging with a flow of N_2 gas for 1 h. A 0.001 M solution of 4-aminopyridine in THF was introduced and allowed to react for two hours at 95 °C. After rinsing the samples with THF, the wafers were stored in vials under N_2 until use.

Whereas the contact angle of as-etched p-Si was 119° prior to functionalization, the contact angle decreased to 35° following functionalization with 4-aminopyridine. This may be compared to contact angles of p-Si following oxidation using ozone (ca. 0°) and O_2/heat (8-10°). The IR spectrum of functionalized p-Si (Fig. 13) is also in agreement with condensed 4-aminopyridine, exhibiting characteristic ν(N-H), ν(C-H), and ν(C-C_{arom}) resonances.

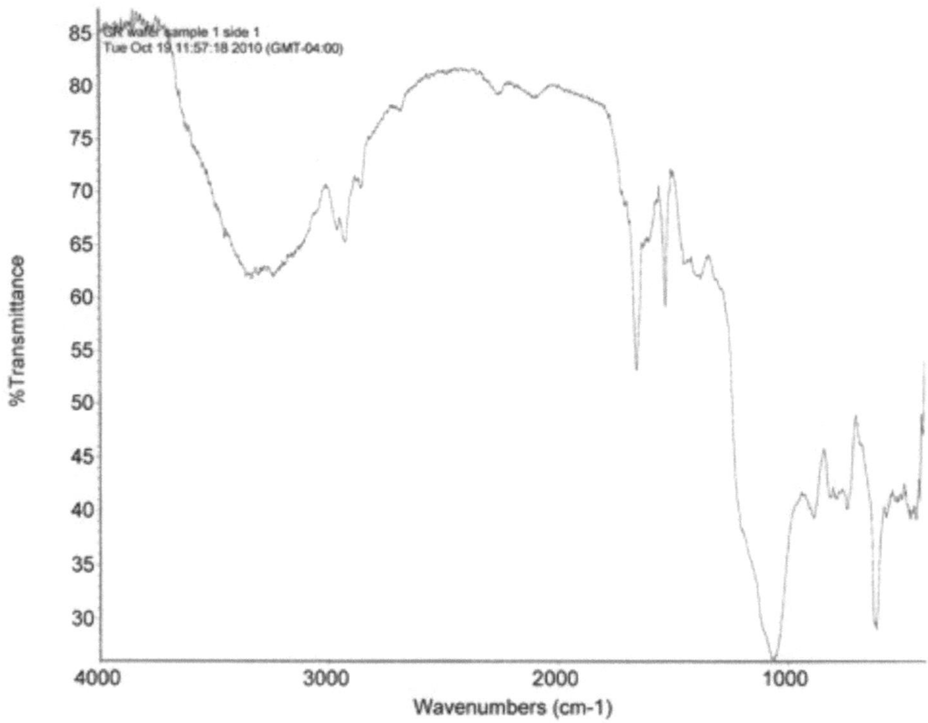

Fig. 15. DRIFT FTIR spectrum of free-standing porous silicon functionalized with 4-aminopyridine.

Figure 16 illustrates the electrical conductance response of the p-Si wafer toward a N_2 flux of ethanol of varying concentrations. As-etched free-standing wafers exhibit a strong conductive response. However, the recovery time is very slow due to the enhanced surface area of the substrate, concomitant with a slow release of condensed vapor from its pores (*vide infra*). In contrast, the amino-functionalized wafers result in a much faster response following exposure to ethanol vapor, while not significantly affecting the overall sensitivity of the wafer. Both as-etched and functionalized p-Si exhibited a detection limit of *ca.* 4 ppt toward ethanol vapor at 20 °C.

The enhanced binding reversibility of ethanol vapor onto functionalized p-Si is counter-intuitive based solely on the surface polarity following functionalization. That is, for a polar organic vapor, one would expect that the highly hydrophobic H-terminated p-Si to easily release the vapor following exposure. Our observance of slow conductive baseline recovery following an ethanol flux is in agreement with Rossi et al., who report vapor sensing by p-Si to be facilitated by condensation within the pore structure (Stefano, 2004). This mode of chemical binding would require long purging times in order to evaporatively sweep the molecules from the more pronounced pore volume of our free-standing wafers.

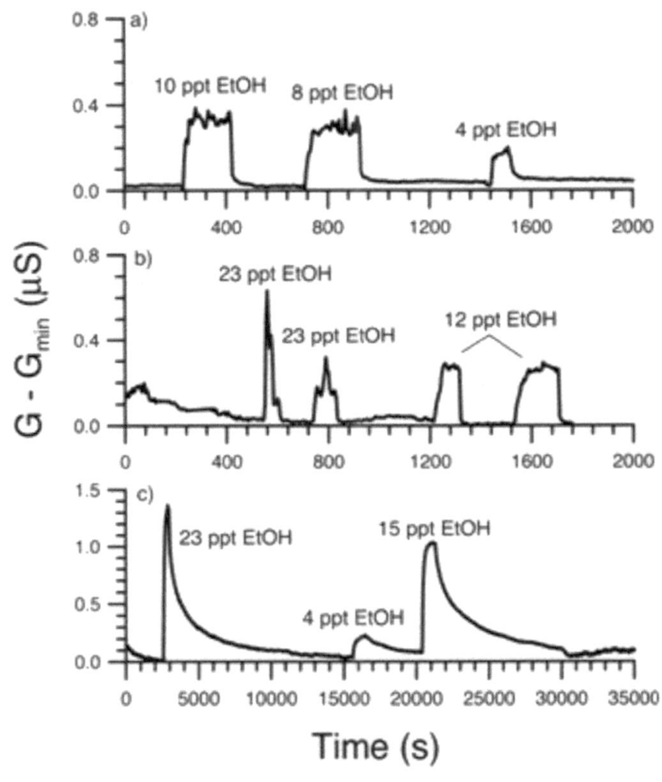

Fig. 16. Conductance response of free-standing p-Si toward ethanol vapor at varying concentrations. Both as-etched wafers (bottom) and 4-aminopyridine-functionalized wafers (top, middle) were used.

In contrast, surfaces functionalized with 4-aminopyridine may interact with a polar vapor via intermolecular interactions, that do not require an evaporative phase change in order to release the molecules from the surface/pore structure. It should be noted that we have also observed an analogous electrical behavior for c-Si supported functionalized p-Si films;

however, the sensitivity and concomitant detection limits were significantly inferior to the free-standing substrates reported herein.

5. Conclusions

In summary, we have demonstrated a simple electrochemical double-etching technique to generate free-standing p-Si wafers without lateral pore formation or residual bulk c-Si. Further, no secondary plasma or chemical treatment is required to remove the back-contact c-Si, which greatly simplifies the procedure making it widely amenable for industrial applications. As-etched and surface-functionalized wafers were shown to be effective to detect low concentrations of ethanol vapors using electrical impedance, with the latter exhibiting an extremely fast recovery time. We are working to improve the detection limits by morphological alterations to enhance the p-Si surface area, as well as expanding the sensing experiments to other organic vapors, both of which will be reported shortly.

6. Acknowledgements

The authors acknowledge the Vicerrectoría de Investigación of the Universidad de Costa Rica for providing funds by means of the project grant #816-B0-063. Partial support from the Consejo Nacional para Investigaciones Científicas y Tecnológicas (CONICIT) and the Ministerio de Ciencia y Tecnología (MICIT) of Costa Rica under the Fondo de Incentivos program is also acknowledged. BDF wishes to acknowledge the Department of Chemistry and College of Science and Technology (CST) at Central Michigan University, as well as the Universidad de Costa Rica (CICIMA, CELEQ and Escuela de Física) for hosting his sabbatical.

7. References

Alvarez, S. D.; Derfus, A. M.; Schwartz, M. P.; Bhatia, S. N.; Sailor, M. J. (2009). *Biomaterials,* Vol. 30, pp. 26-34, ISSN 0142-9612.

Arai, T. & Kragic, D. (1999). Variability of Wind and Wind Power, In: *Wind Power,* S.M. Muyeen, (Ed.), 289-321, Scyio, ISBN 978-953-7619-81-7, Vukovar, Croatia.

Archer, M.; Christophersen, M.; Fauchet, P. M. (2005). *Sensors and Actuators B,* Vol. 106, p. 347-357.

Bazrafkan, I.; Dariani, R. S. (2009). *Physica B,* Vol. 404, pp. 1638-1642.

Bisi, O. et al. (2000). Porous silicon : a quantum sponge structure for silicon based optoelectronics. *Surface Science Reports,* Vol. 38, pp. 1-136, ISSN 0167-5729.

Brandt, M.S. et al. (1992). The origin of visible luminescence from "porous silicon": A new interpretation. *Solid State Communications,* Vol. 81, No. 4, pp. 307-312, ISSN 0038-1098.

Burstein, L.; Shapira, Y.; Partee, J.; Shinar, J.; Lubianiker, Y.; Balberg, I. (1997). *Phys. Rev. B,* Vol. 55, p. R1930-R1933, ISSN 1098-0121.

Canham, L.T. (1990). Silicon quantum wire array fabrication by electrochemical and chemical dissolution of wafers. *Applied Physics Letters*, Vol. 57, pp. 1046-1048, ISSN 0003-6951.

Collis, R.T. et al. (1997). Porous Silicon: From Luminescence to LEDs. *Physics Today*, Vol. 50, No. 1, pp. 24-31, ISSN 0031-9228.

Fahlman, B. D. (2011). *Materials Chemistry* (2nd ed.), Springer, New York, ISBN 978-94-007-0692-7

Gaburro, Z.; Daldossoh, N.; Pavesi, L. (2005). *Encyclopedia of Condensed Matter Physics*, pp. 391-401.

Garel, O.; Breluzeau, C.; Dufour-Gergam, E.; Bosseboeuf, A.; Belier, B.; Mathet, V.; Verjus, F. (2007). *J. Micromech. Microeng.*, Vol. 17, p. S164-S167.

George, T. et al. (1992). Microstructural investigations of light-emitting porous Si layers. *Applied Physics Letters*, Vol. 60, No. 19, pp. 2359-2361, ISSN 0003-6951.

Guinier, A. (1994). *X-Ray Diffraction on Crystals, Imperfect Crystals, and Amorphous Bodies*, Dover Publications Inc., New York.

Güntherchulze. A. & Betz, H. (1937). *Elektrolyt-Kondensatoren* (Ed. Krayn, M.), Techn. Verlag H. Cram, Berlin.

Janshoff, A.; Dancil, K. -P. S.; Steinem, C.; Greiner, D. P.; Lin, V. S. -Y.; Gurtner, C.; Motesharei, K.; Sailor, M. J.; Ghadiri, M. R. J. (1998). *Am. Chem. Soc.*, Vol. 120, pp. 12108-12116.

Koch, F. et al (1993). The luminescence of porous Si: the case for the surface state mechanism. *Journal of Luminescence*, Vol. 57, No. 1-6, pp. 271-281, ISSN 0013-4651.

Lammel, G.; Renaud, P. (2000). *Sensors and Actuators B*, Vol. 85, pp. 356-360.

Lehmann, V. (2002). *Electrochemistry of Silicon: Instumentation, Science, Materials and Applications*. Wiley-VCH Verlag GmbH, Federal Republic of Germany, ISBN 3-527-29321-3.

Li, B.; Xu, Y. & Choi, J. (1996). Applying Machine Learning Techniques, *Proceedings of ASME 2010 4th International Conference on Energy Sustainability*, pp. 14-17, ISBN 842-6508-23-3, Phoenix, Arizona, USA, May 17-22, 2010.

Lima, P.; Bonarini, A. & Mataric, M. (2004). *Application of Machine Learning*, InTech, ISBN 978-953-7619-34-3, Vienna, Austria.

Liu, R.; Schmedake, T. A.; Li, Y. Y.; Sailor, M. J.; Fainman, Y. (2002). *Sensors and Actuators B*, Vol. 87, pp. 58-62.

Lo, S. -Z.; Rossi, A. M.; Murphy, T. E. (2009). *Physica Status Solidi A*, Vol. 206, pp. 1273-1277.

Low, S. P.; Voelcker, N. H.; Canham, L. T.; Williams, K. A. (2009). *Biomaterials*, Vol. 30, pp. 2873-2880, ISSN 0142-9612.

McInnes, S. J. P.; Thissen, H.; Choudhury, N. R.; Voelcker, N. H. J. (2009). *Coll. Interf. Sci.*, Vol. 32, pp. 336-344.

Menna, P.; Francia, G. D.; La Ferrara, V. Solar Energy Mater. Solar Cells, 1995, 37, 13 - 24.

Navarro, M.; Lopez-Villegas, J. M.; Samitier, J.; Morante, J. R.; Bausells, J.; Merlos, A. (1997). *J. Micromech. Microeng.* Vol. 7, p. 131-132.

Palestino, G.; Legros, R.; Agarwal, V.; Perez, E.; Gergely, C. (2008). *Sensors and Actuators B*, Vol. 135, pp. 27-34.

Papadimitriou, D.; Tsamis, C.; Nassiopoulou, A. G. (2004). *Sensors and Actuators B*, Vol. 103, pp. 356-361.

Pickering, C et al. (1984). Optical studies of the structure of porous silicon films formed in p-type degenerate and non-degenerate silicon. *Journal of Physics C: Solid State Physics*, Vol. 17, pp. 6535-6552, ISSN 0022-3719.

Prokes, S.M. et al. (1992). SiH_X excitation: An alternate mechanism for porous Si photoluminescence. *Physical Review B*, Vol. 45, No. 23, pp 13788-13791, ISSN 1098-0121.

Ramírez-Porras, A. (2002). Determination of nanometer-scale sizes in n^+-type porous silicon by the use of X-Ray and Raman spectroscopies. *Surface Review and Letters*, Vol. 9, Nos. 5-6, pp. 1769-1772.

Ramírez-Porras, A. & Weisz, S.Z. (2002). Stochastic approach to the smart quantum confinement model in porous silicon. *Surface Science Letters*, Vol. 515, No. 2-3, pp. L509-L513,

Ramírez-Porras, A. (2011). Search for publications in the 1972-2010 period for the keywords "porous silicon", in *Scopus*, 24.05.2011. Available from: http://www.scopus.com.

Razi, F.; Rahimi, F.; Irajizad A. (2008). *Sensors and Actuators B*, Vol. 132, pp. 40-44.

Rossi, A. M.; Wang, L.; Reipa, V.; Murphy, T. E. (2007). *Biosensors and Bioelectronics*, Vol. 23, pp. 741-745.

Schechter, I.; Ben-Chorin, M.; Kux, A. (1995). *Anal. Chem.*, Vol. 67, pp. 3727-3732.

Schmidt, P.F. & Michel, W. (1957). Anodic Formation of Oxide Films on Silicon. *Journal of The Electrochemical Society*, Vol. 104, No. 4, pp. 230-236, ISSN 0013-4651.

Searson, P. C. (1991). *Appl. Phys. Lett.*, Vol. 59, p. 832-833.

Siegwart, R. (2001). Indirect Manipulation of a Sphere on a Flat Disk Using Force Information. *International Journal of Advanced Robotic Systems,* Vol.6, No.4, (December 2009), pp. 12-16, ISSN 1729-8806

Smith, R. L.; Collins, S. D. J. (1992). *Appl. Phys.*, Vol. 71, p. R1.

Stefano, L. D.; Moretti, L.; Rendina, I.; Rossi, A. M. (2004.) *Sensors and Actuators B*, Vol. 100, p. 168-172.

Theunissen, M.J.J. (1972). Etch Channel Formation during Anodic Dissolution of N-Type Silicon in Aqueous Hydrofluoric Acid. *Journal of the Electrochemical Society*, Vol. 119, No. 3, pp. 351-360, ISSN 0013-4651.

Turner, D.R. (1958). Electropolishing Silicon in Hydrofluoric Acid Solutions. *Journal of The Electrochemical Society*, Vol. 105, No. 7, pp. 402-408, ISSN 0013-4651.

Uhlir Jr., A. (1956). Electrolytic Shaping of Germanium and Silicon. *Bell System Technical Journal*, Vol. 35, No. 2, pp. 333-347, ISSN 0005-8580.

Van der Linden, S. (2010). Integrating Wind Turbine Generators (WTG's) with Energy Storage, In: *Wind Power,* 17.06.2010, Available from http://sciyo.com/articles/show/title/wind-power-integrating-wind-turbine-generators-wtg-s-with-energy-storage

Vitanov, P.; Kamenova, M.; Tyutyundzhiev, N.; Delibasheva, M.; Goranova, E.; Peneva, M. (1997). *Thin Solid Films*, Vol. 1, pp. 299-303.

Watanabe, Y. & Sakai, T. (1971). Application of a thick anode film to semiconductor devices. *Reviews of the Electrical Communications Laboratories*, Vol. 19, No. 7-8, pp. 899, ISSN: 0029-067X.

Wolkin M.V. et al. (1999). Electronic States and Luminescence in Porous Silicon Quantum Dots: The Role of Oxygen. *Physical Review Letters*, Vol. 82, No. 1, pp. 197-200, ISSN 0031-9007.

8

Improvement of the Gas Sensing Properties in Nanostructured $Gd_{0.9}Sr_{0.1}CoO_3$

Carlos R. Michel[1], Narda L. López Contreras[1], Edgar R. López-Mena[1], Juan Carlos Ibarra[1], Arturo Chávez-Chávez[1] and Mauricio Ortiz-Gutiérrez[2]

[1]Universidad de Guadalajara, CUCEI,
[2]Universidad Michoacana de San Nicolás de Hidalgo,
México

1. Introduction

Global warming has become one of the most important issues worldwide (Kerr, 2007). The emission of large amounts of CO_2 has been identified as its main cause (Karl et al., 2003; Parry et al., 2008). In order to determine the concentration of this and other polluting gases, researchers around the world have developed solid state chemical sensors. Even though SnO_2, ZnO, TiO_2 and WO_3 have been some of the most studied gas sensor materials, other oxides, with unique physical and chemical properties are also appropriate for this application (Yamazoe, 2005). Ternary and quaternary oxides, whose crystal structures are the perovskite or spinel, have notable electrical and catalytical properties, useful for gas sensing purposes (Brosha et al., 2000; Dutta et al., 2004; Kong et al., 1996; Kosacki et al., 1998; Post et al., 1999; Suo et al., 1997).

$GdCoO_3$ is a semiconductor material, having the perovskite-type structure. It possess outstanding physical and physicochemical properties like magnetotransport, thermoelectricity, mixed ionic-electronic conductivity for solid oxide fuel cells, photocatalytical activity for the decomposition of dyes and phenols, and gas sensing activity (Rey-Cabezudo et al., 2002; Moon et al., 2000; Takeda et al., 1996; Wienhöfer et al., 2004; Mahata et al., 2007; Michel et al., 2009).

In this work, samples of $Gd_{1-x}Sr_xCoO_3$ (x = 0, 0.1) were prepared by two different methods. The main goals were to study the effect of chemical composition and microstructure on the gas sensing properties. The effect of the chemical composition was approached through strontium doping, and the effect of the microstructure was analyzed by preparing samples with different particle size. The preparation of $Gd_{1-x}Sr_xCoO_3$ powders was made in aqueous media, using the solution and solution-polymerization methods.

In recent years, the development of nanomaterials has attracted the attention of many research groups around the world. In order to decrease the particle size of ceramic materials, the use of polymerizing agents, such as polyvinyl alcohol and polyethylene glycol has been previously reported in the literature (Lee et al., 1998; Gülgün et al., 1999). A wide variety of nanostructured materials, such as cordierite, monoclinic yttrium aluminate, portland cement components, hydroxyapatite, and many more, have been synthesized by this method.

On the other hand, nanostructured materials provide large specific surface areas; which increases the interaction between the sensor surface and the surrounding gases. Therefore, in this work, the gas sensing properties of $Gd_{1-x}Sr_xCoO_3$ powders were evaluated. This characterization was done by using direct and alternating current on $Gd_{1-x}Sr_xCoO_3$ thick films.

2. Experimental

$GdCoO_3$ was synthesized by a solution-polymerization method, using stoichiometric amounts of $Gd(NO_3)_3 6H_2O$ (Alfa Aesar) and $Co(NO_3)_3 6H_2O$ (J.T. Baker). The reagents were dissolved in 100 ml of deionized water. To promote the polymerization effect, 0.44 g of polyvinyl alcohol (PVA, Alfa Aesar) and 3.2 g of sucrose (Alfa Aesar) were dissolved in 100 ml of deionized water. Due to the poor solubility of PVA in cool water, the water was preheated at 60°C. All the solutions were mixed under strong stirring for 2 h (60°C); Fig. 1A shows a typical solution obtained after this process. $Gd_{0.9}Sr_{0.1}CoO_3$ was prepared by the solution and solution-polymerization methods. Similarly to the previous synthesis, stoichiometric amounts of gadolinium, cobalt and strontium nitrates ($Sr(NO_3)_2$ (Alfa Aesar)) were dissolved in 100 ml of deionized water. By the solution method, 0.1 mol of citric acid (Alfa Aesar) was added to the nitrate solution. In the case of the solution-polymerization, the process previously described was used. Then, water evaporation was done by using microwave radiation; a domestic microwave oven (Panasonic) was used. The materials obtained after the microwave-assisted evaporation are shown in: Fig. 1B (solution method) and Fig. 1C (solution-polymerization). In the latter, a dry solid having extensive porosity was produced.

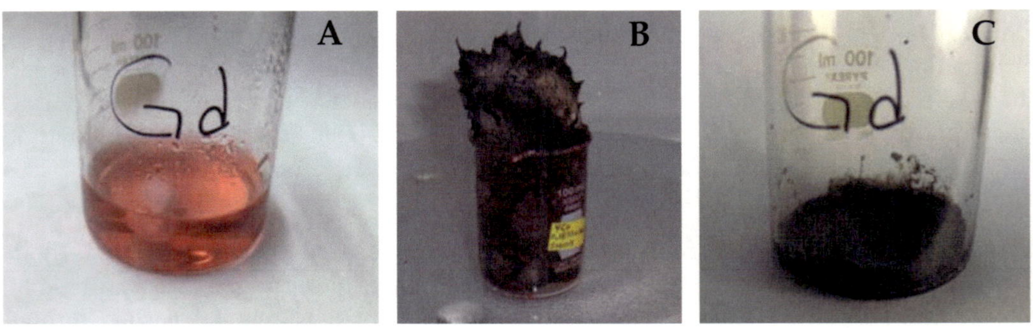

Fig. 1. (A) Aqueous solution obtained by solution-polymerization. Materials obtained after water evaporation: (B) solution and (C) solution-polymerization.

In order to obtain single-phase $GdCoO_3$ and $Gd_{0.9}Sr_{0.1}CoO_3$ powders, the precursors were calcined in the range of 550 to 800°C. The resulting powders were analyzed by X-ray powder diffraction (XRD), at room temperature, using a Rigaku Miniflex apparatus. Cu $K_{\alpha 1}$ radiation (1.5405 Å) was used. The surface morphology of the powders was observed by scanning electron microscopy (SEM), using a Jeol JSM-5400LV microscope, in secondary electron mode. For the identification of nanoparticles, transmission electron microscopy

(TEM, Jeol JEM-1010) was used. Thick films of $GdCoO_3$ and $Gd_{0.9}Sr_{0.1}CoO_3$ were prepared by depositing a suspension of the powders on alumina discs. In this process, ~0.5 g of each powder was dispersed in 2 ml of isopropyl alcohol, using ultrasonic vibration. Then, the suspension was poured on alumina substrates, producing films with 3 mm diameter and ~400 μm thickness. The electrical contacts were made with high-purity silver wires (diameter 0.2 mm). Fig. 2 shows a scheme of the gas sensor device.

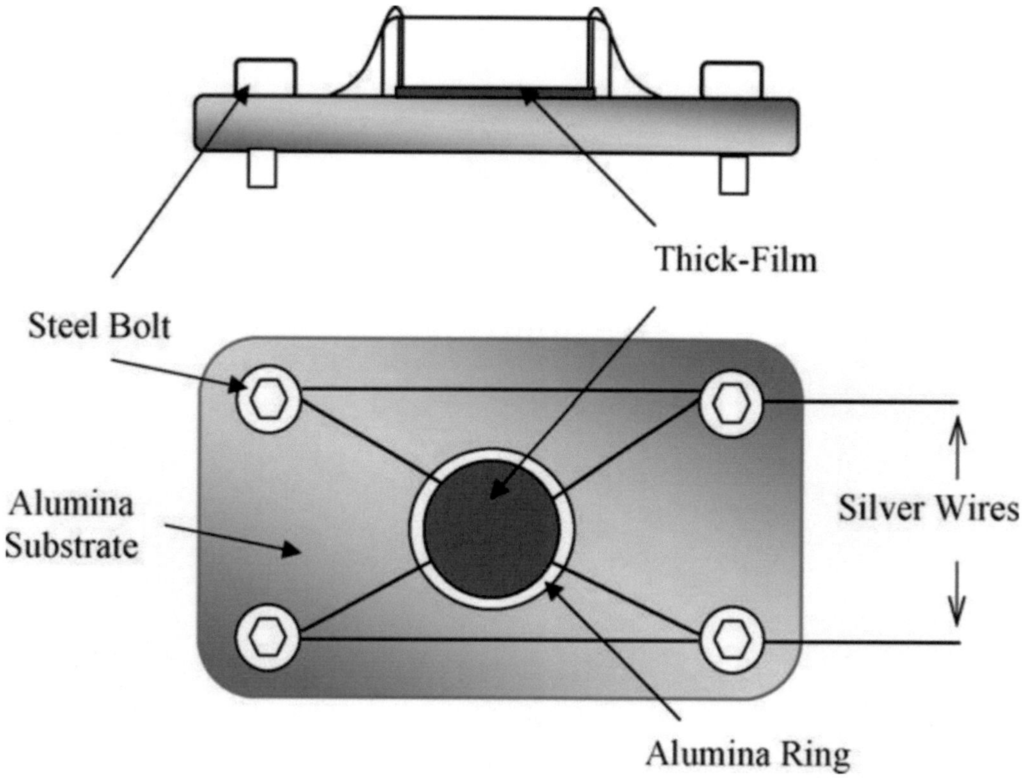

Fig. 2. Schematic illustration of the gas sensor device.

Electrical characterization was carried out by the two-point probe method, using a tube-type furnace having programmable temperature control. The direct current measurements (DC) were obtained by using a digital data acquisition unit (Agilent 34970A), having a multiplexer module. A digital voltmeter (Agilent 34401A) was also used. The polarization curves were recorded from -5V to 5V, using a Solartron 1285A potentiost/galvanostat. The alternating current characterization (AC) was performed by measuring the magnitude of the impedance (|Z|), with a LCR meter (Agilent 4263B). The graphs were obtained using the LabView 8.6 software (National Instruments).

The gases (CO_2, O_2 and extra dry compressed air) were supplied by using a mass flow controller (647C, MKS Instruments). Fig. 3 shows a diagram of the experimental setup and Fig. 4 shows an image of the actual instruments. It is important to mention that in this work the experimental setup was not pressurized; then, the pressure of the test chamber was slightly above 1013 milibars.

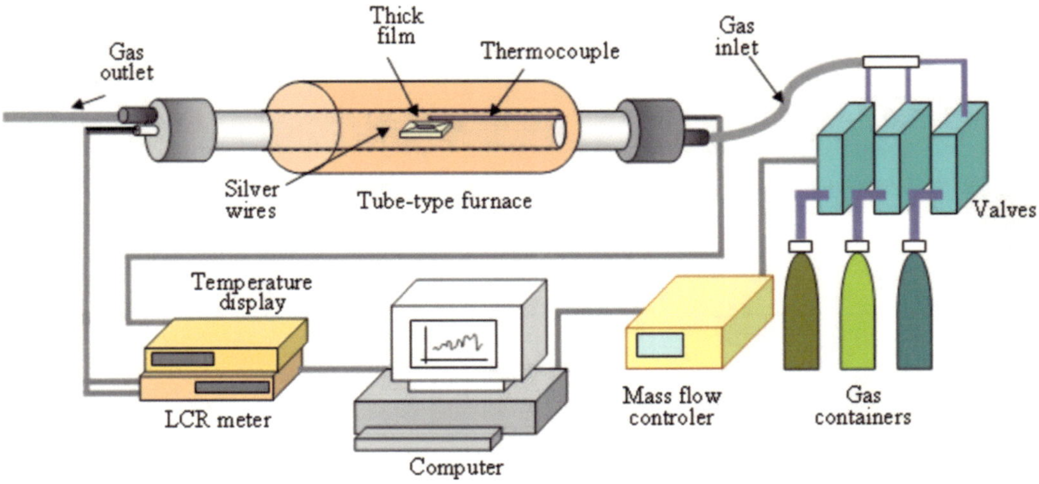

Fig. 3. Diagram of the experimental setup used for the gas sensing characterization.

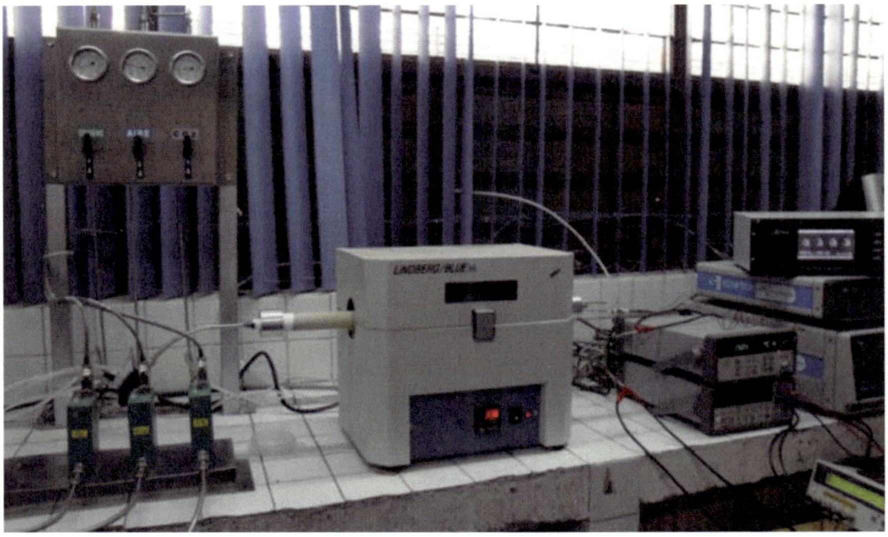

Fig. 4. Image of the instruments used for the gas sensing measurements.

3. Results

3.1 X-ray powder diffraction

Fig. 5 shows the XRD patterns of the precursor powder of $GdCoO_3$ calcined from 550 to 750°C in air. The calcination at 550°C produced no crystalline materials; however, at 600°C the main diffraction lines of Gd_2O_3 and Co_3O_4 were identified by using the JCPDF files 012-0797 and 043-1003 respectively. The calcination at 650 °C revealed that the diffraction lines of Gd_2O_3 and Co_3O_4 are still present; however, the main diffraction line of $Gd_{0.9}Sr_{0.1}CoO_3$ can be observe at 2θ = 33.8°. The identification of $Gd_{0.9}Sr_{0.1}CoO_3$ was made by using the JCPDF file 25-1057, which indeed corresponds to $GdCoO_3$; however, these oxides have the same crystal structure. However, it should be consider that the amount of gadolinium that can be replaced by strontium is small, because their ionic radii are 0.94Å (Gd) and 1.16Å (Sr). The calcination at 700°C increased the reaction rate between Gd_2O_3 and Co_3O_4, producing a larger amount of $Gd_{0.9}Sr_{0.1}CoO_3$. Firing at 750°C yielded near single-phase $Gd_{0.9}Sr_{0.1}CoO_3$, as it can be observed in the corresponding pattern. The Miller indices of each plane were assigned according to the JCPDF card.

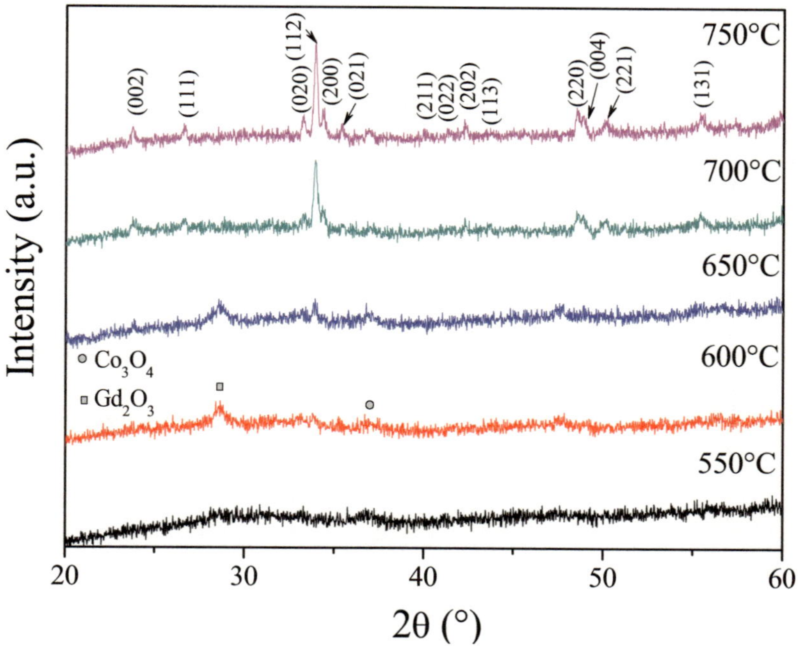

Fig. 5. X-ray powder diffraction patterns of $GdCoO_3$ obtained by solution-polymerization method, calcined at various temperatures.

Figure 6 shows the crystal structure evolution as temperature increased from 600 to 800°C. The calcination at 600°C produced a mixture Gd_2O_3, Co_3O_4 and $Gd_{0.9}Sr_{0.1}CoO_3$, being the latter the main crystalline phase. At 700°C, a significant increase of $Gd_{0.9}Sr_{0.1}CoO_3$ can be noticed; however, the main diffraction line of Gd_2O_3, placed at $2\theta = 28.6°$ is present. The calcination at 800°C produced almost single-phase $Gd_{0.9}Sr_{0.1}CoO_3$. Comparing these results with those obtained for $GdCoO_3$, $Gd_{0.9}Sr_{0.1}CoO_3$ appears at a lower calcination temperature;

Fig. 7 shows X-ray diffraction patterns obtained from precursor powders of $Gd_{0.9}Sr_{0.1}CoO_3$ (solution-polymerization), calcined at different temperatures. The calcination at 550°C produced an amorphous solid; however, at 600°C nearly single-phase $Gd_{0.9}Sr_{0.1}CoO_3$ was produced. Increasing the calcination temperature to 700°C produced no perceptible changes in the XRD patterns. Comparing the sequence of XRD patterns of Fig. 7, with those displayed in Fig. 6, the solution-polymerization method decreased the temperature at which $Gd_{0.9}Sr_{0.1}CoO_3$ can be formed. Besides, Fig. 7 does not display diffraction lines of Gd_2O_3 and Co_3O_4, indicating that the formation and further reaction of these oxides does not occur. The interaction, at molecular scale, of gadolinium, strontium and cobalt ions, could be an explanation of the formation of $Gd_{0.9}Sr_{0.1}CoO_3$ at lower temperature. Evidently the solution-polymerization method provides a notable save of energy in the preparation of ceramics, by using an inexpensive polymeric precursor.

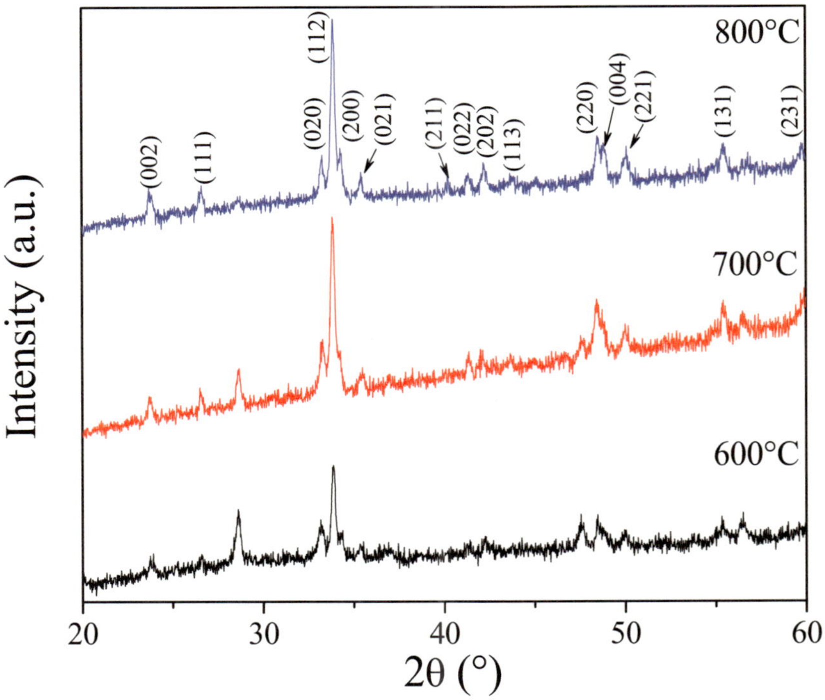

Fig. 6. Crystal structure evolution with temperature of $Gd_{0.9}Sr_{0.1}CoO_3$ prepared by the solution method.

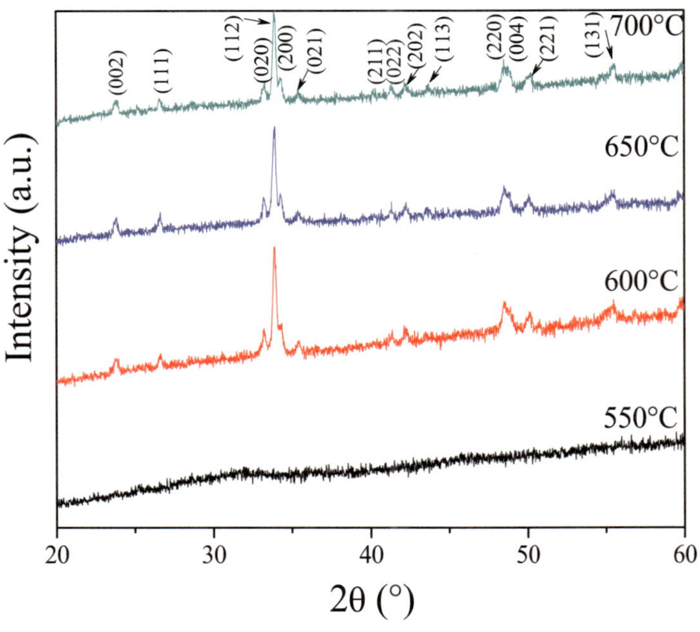

Fig. 7. X-ray powder diffraction patterns of $Gd_{0.9}Sr_{0.1}CoO_3$ obtained by solution-polymerization, calcined at various temperatures.

3.2 Scanning electron microscopy

Fig. 8 shows the typical surface morphology (SEM) of powders of $GdCoO_3$ calcined at: (A) 550, (B) 600, (C) 650, (D) 700 and (E) 750°C. The microstructure of samples calcined from 550 to 700°C was of thin laminas having smooth surface and semispherical cavities. The entire solids possess extensive porosity. The sample calcined at 750°C shows a slight different microstructure, because a granular shape can be noticed. This feature can be explained by the sintering process occurred at higher temperature. The observation of this sample by TEM will provide a better insight of the microstructure of this sample, as will be shown later in this chapter.

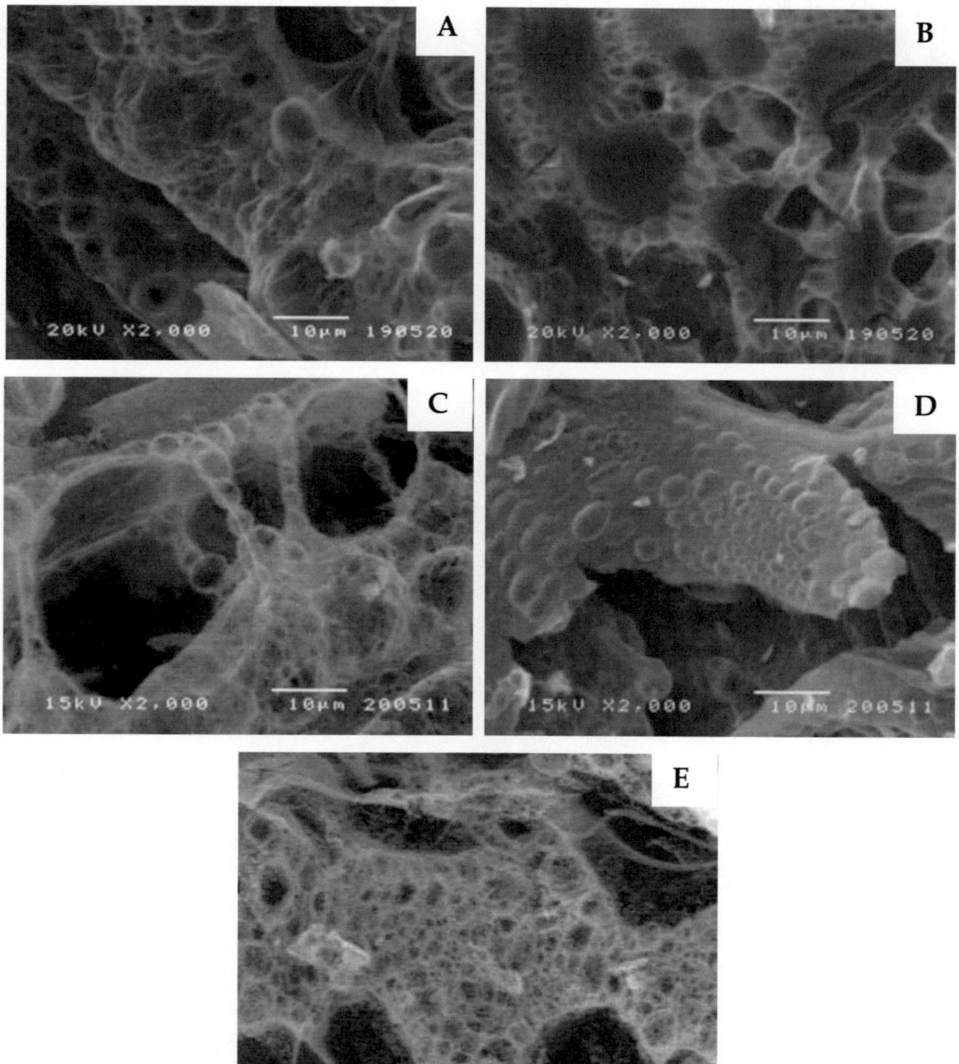

Fig. 8. SEM images of $GdCoO_3$ powder calcined at: A) 550°C, B) 600°C, C) 650°C, D) 700°C and E) 750°C, in air.

Fig. 9 shows the surface microstructure of $Gd_{0.9}Sr_{0.1}CoO_3$ samples (solution method), calcined at A) 600°C, B) 700°C and C) 800°C. These SEM images revealed that the increase in the calcination temperature had little effect on the surface microstructure. All the samples show no relevant features; however, in order to determine if this material is composed by nanoparticles, further observation by TEM was performed.

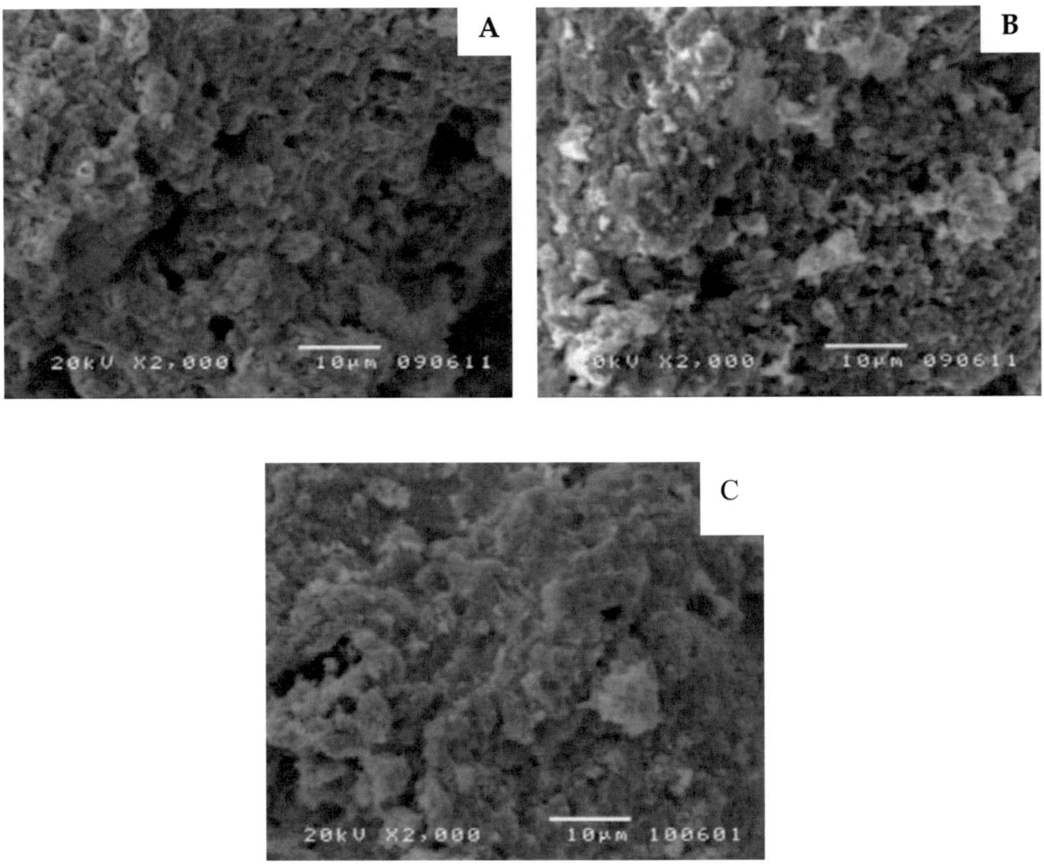

Fig. 9. SEM image of $Gd_{0.9}Sr_{0.1}CoO_3$ (solution method) calcined at: A) 600°C, B) 700°C and C) 800°C, in air.

Fig. 10 shows SEM photos of $Gd_{0.9}Sr_{0.1}CoO_3$ synthesized by the solution-polymerization route, calcined at: A) 550°C, B) 600°C, C) 650°C and D) 700°C. A highly porous material, having smooth surfaces was produced. By increasing the calcination temperature, the microstructure of the samples was not altered. Comparing these images with those obtained for $Gd_{0.9}Sr_{0.1}CoO_3$ (the solution method), significant differences can be noticed. The use of a polymerizing agent in aqueous solution produced at least a solid with high porosity. Due to $Gd_{0.9}Sr_{0.1}CoO_3$ prepared by solution-polymerization can be indeed formed by nanoparticles its observation by TEM was also done.

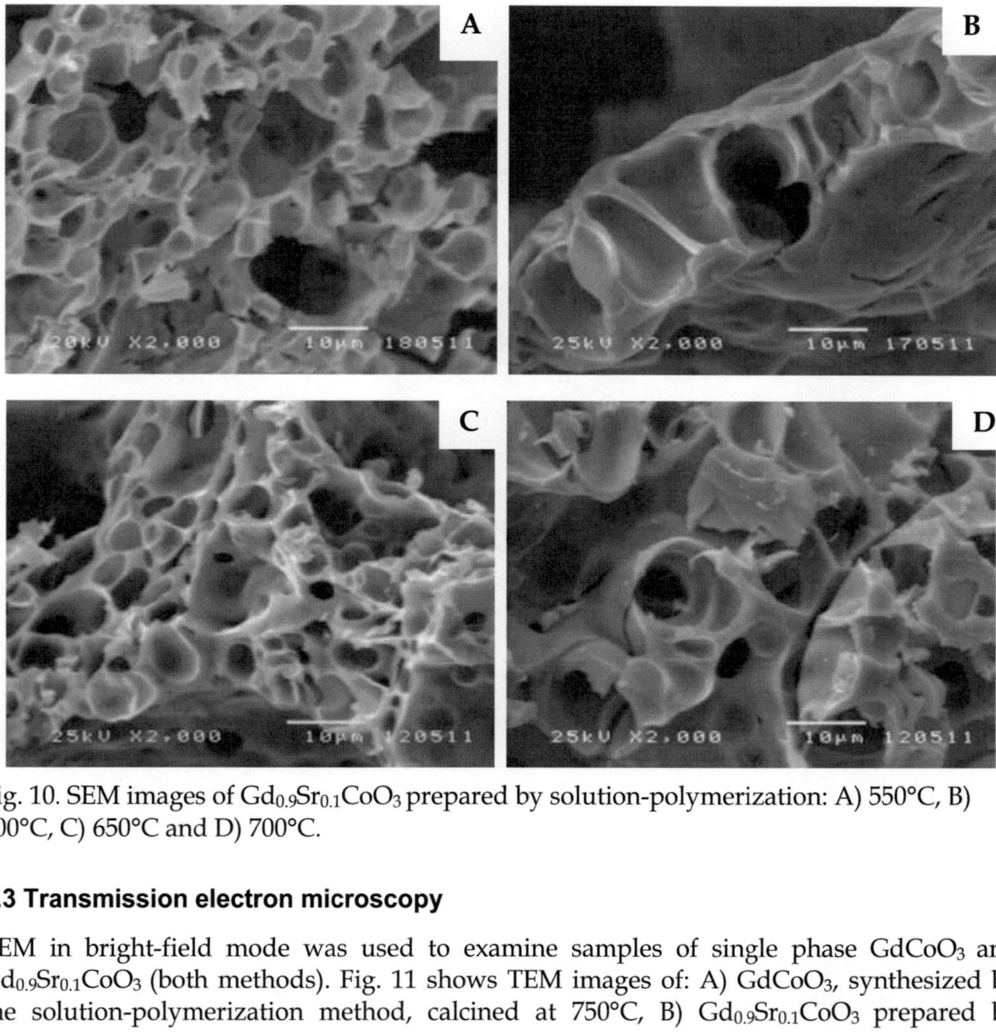

Fig. 10. SEM images of $Gd_{0.9}Sr_{0.1}CoO_3$ prepared by solution-polymerization: A) 550°C, B) 600°C, C) 650°C and D) 700°C.

3.3 Transmission electron microscopy

TEM in bright-field mode was used to examine samples of single phase $GdCoO_3$ and $Gd_{0.9}Sr_{0.1}CoO_3$ (both methods). Fig. 11 shows TEM images of: A) $GdCoO_3$, synthesized by the solution-polymerization method, calcined at 750°C, B) $Gd_{0.9}Sr_{0.1}CoO_3$ prepared by solution method, calcined at 800°C, and C) $Gd_{0.9}Sr_{0.1}CoO_3$ prepared by solution-polymerization (650°C). Fig. 11A shows interconnected, rounded or semispherical particles, with size larger than 100 nm. The formation of necks between these submicron particles was identified throughout the sample. Moreover, the interconnection between particles produced a rigid solid structure, with high porosity, which increases the contact surface between gaseous species and the perovskite.

Fig. 11B shows a typical TEM image of $Gd_{0.9}Sr_{0.1}CoO_3$ powder prepared by solution method. It exhibits a similar microstructure than that observed for $GdCoO_3$, having extensive connection among particles, forming a continuous solid. Fig. 11C displays the microstructure of $Gd_{0.9}Sr_{0.1}CoO_3$ made by solution-polymerization. Compared to previous samples, a notable smaller particle size can be noticed; an average particle size of 45 nm was measured. Even though, nanoparticle agglomeration is present, abundant nanoporosity can be observed. It can be concluded that the solution-polymerization method reduced the energy expenditure in the preparation of $Gd_{0.9}Sr_{0.1}CoO_3$, and produced a significant reduction in particle size. High porosity and high specific surface areas can be obtained by using this method.

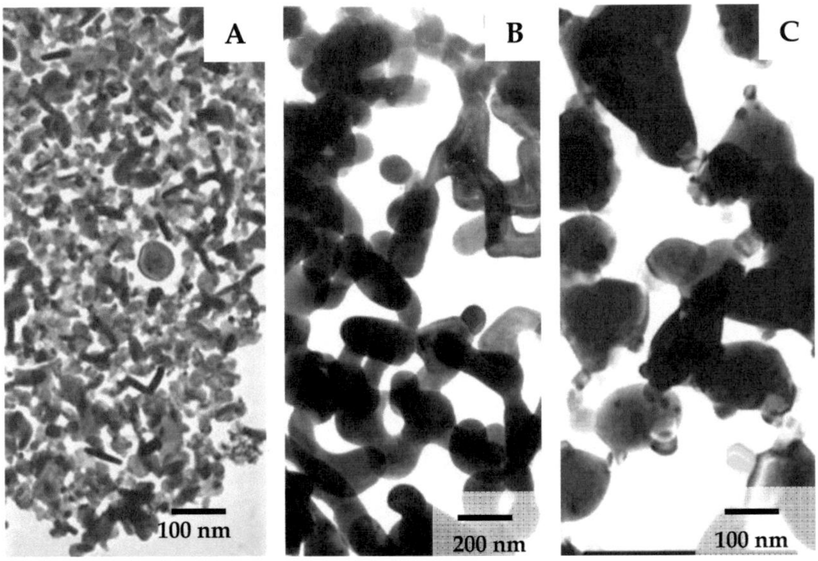

Fig. 11. TEM images of (A) $GdCoO_3$ (750°C), (B) $Gd_{0.9}Sr_{0.1}CoO_3$ prepared by solution method (800°C), and (C) $Gd_{0.9}Sr_{0.1}CoO_3$ prepared by solution-polymerization (650°C).

The abundant porosity of $Gd_{0.9}Sr_{0.1}CoO_3$ (solution-polymerization) is associated to the evolution of carbon dioxide and steam during the thermal decomposition of the polymerized metal ion-chelated complex (Das, 2001). A large amount of heat is also produced in this process. Theoretically, one mole of sucrose delivers 23 moles of gases, which avoid particle agglomeration and produce extensive porosity and fine particles in the final product. The thermal decomposition of sucrose occurs by the following reaction:

$$C_{12}H_{22}O_{11} + 12O_2 \rightarrow 12CO_2 + 11H_2O \quad (1)$$

Another role of sucrose in the preparation of $Gd_{0.9}Sr_{0.1}CoO_3$ (solution-polymerization) is the formation of sacaric acid; which is obtained after its contact with citric acid. Sacaric acid is known to be a good chelating agent for metal ions, which produces a uniform distribution of cations throughout the polymeric network.

3.4 Gas sensing characterization

3.4.1 Gas sensitivity curves

About the gas sensing properties of $GdCoO_3$ and $Gd_{0.9}Sr_{0.1}CoO_3$, Fig. 12 shows the sensitivity curves obtained in O_2 and CO_2. Fig. 12A displays the graphs obtained from $GdCoO_3$, whereas Fig. 12B corresponds to $Gd_{0.9}Sr_{0.1}CoO_3$ (solution), and Fig. 12C to $Gd_{0.9}Sr_{0.1}CoO_3$ (solution-polymerization). The sensitivity curves were obtained from the ratio: R_{gas} / R_{air}; where R_{gas} is the resistance measured in the test gas (O_2 or CO_2), and R_{air} the resistance measured in dry air. Table 1 shows the temperatures at which the maximum sensitivity values were detected. From these results, sensitivity maxima occur at lower temperature in $Gd_{0.9}Sr_{0.1}CoO_3$ (solution polymerization). Moreover, sensitivity maxima in CO_2 are sharper than those obtained in O_2; in the latter, broad curves were recorded instead. By comparing these results, with those reported from other materials such as ZnO and SnO_2, $Gd_{0.9}Sr_{0.1}CoO_3$ displays moderate gas sensitivity values.

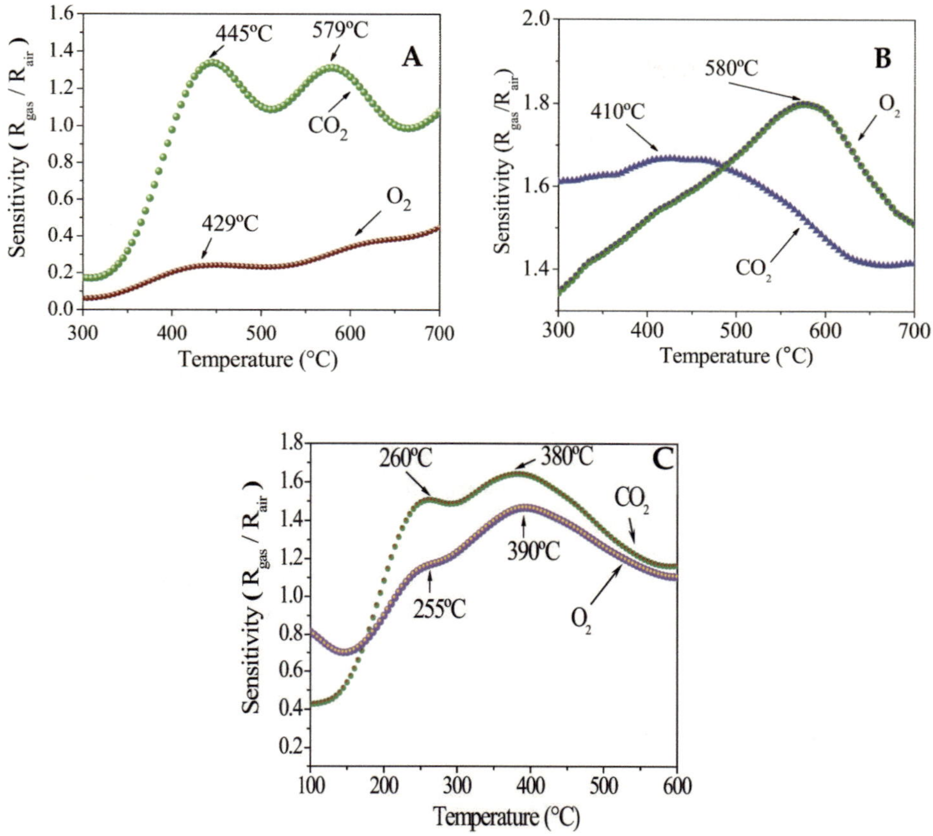

Fig. 12. Sensitivity *vs.* temperature plots recorded in O_2 and CO_2 for A) $GdCoO_3$, B) $Gd_{0.9}Sr_{0.1}CoO_3$ (solution method), and C) $Gd_{0.9}Sr_{0.1}CoO_3$ (solution-polymerization).

Sensitivity (R_{gas} / R_{air})	O_2		CO_2	
$GdCoO_3$	580°C		410°C	
$Gd_{0.9}Sr_{0.1}CoO_3$ (solution)	429°C	600°C	445°C	579°C
$Gd_{0.9}Sr_{0.1}CoO_3$ (solution-polymerization)	255°C	390°C	260°C	380°C

Table 1. Temperatures of maximum sensitivity values recorded in O_2 and CO_2 for $Gd_{1-x}Sr_xCoO_3$.

3.4.2 Dynamic gas sensing response

Fig. 13 shows resistance *vs.* time graphs recorded on $GdCoO_3$, in: (A) air/100 ppm of O_2 (in air), and (B) air/100 ppm of CO_2 (in air). These tests were done at a fixed temperature, which correspond to the maximum sensitivity values shown in Table 1. Fig. 13 shows a decrease of R after the introduction of O_2; in CO_2 the opposite behavior was registered. The variation of resistance in O_2 was -6 Ω; whereas, in CO_2 was ~40 Ω. However, in CO_2 the resistance did not stabilize.

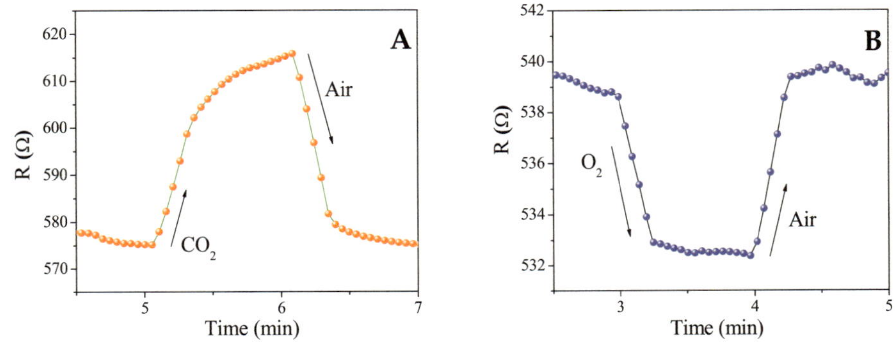

Fig. 13. Resistance *vs.* time graphs registered in air/O_2 (580°C), and air/CO_2 (410°C), for $GdCoO_3$.

Fig. 14 shows R *vs.* time graphs recorded in: (A) air/100 ppm of O_2 (in air), and (B) air/100 ppm of CO_2 (in air), for $Gd_{0.9}Sr_{0.1}CoO_3$ prepared by solution and solution-polymerization. These measurements were performed at 380°C, which is an intermediate sensitivity value. This temperature was chosen in order to compare the results obtained from different $Gd_{0.9}Sr_{0.1}CoO_3$ samples. Compared to the graphs of Fig. 13, the resistance decreased about one order of magnitude. $Gd_{0.9}Sr_{0.1}CoO_3$ (solution-polymerization) displayed the largest variation of resistance; however, the stabilization of resistance with time was not observed. $Gd_{0.9}Sr_{0.1}CoO_3$ (solution method), displays stabilization of resistance in both gases; however, further experiments demonstrated that $Gd_{0.9}Sr_{0.1}CoO_3$ (solution-polymerization) exhibited a better gas sensing performance.

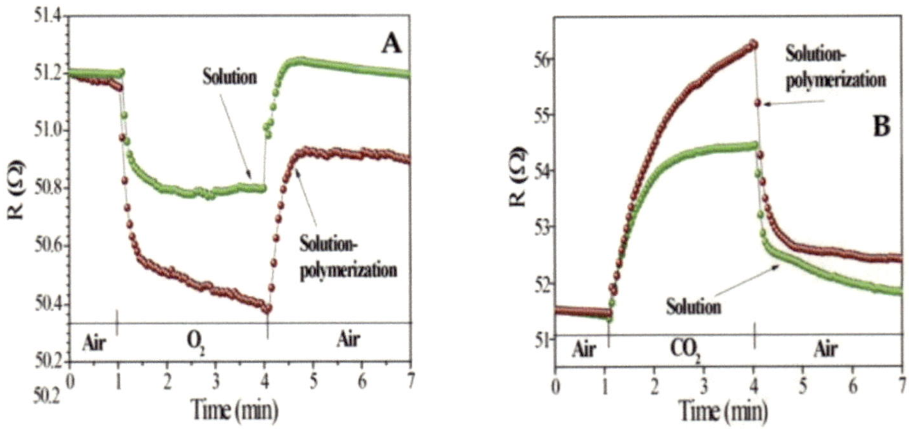

Fig. 14. Resistance *vs.* time graphs recorded in: (A) air/O_2, and (B) air/CO_2, for $Gd_{0.9}Sr_{0.1}CoO_3$ prepared by solution and solution-polymerization methods (380°C).

Fig. 15A shows the variation of resistance with time measured when air, CO_2 and O_2 were alternatively supplied. This experiment was made in order to test the gas selectivity of $Gd_{0.9}Sr_{0.1}CoO_3$ prepared by solution-polymerization. Fig. 15A shows a combination of the results displayed in Fig. 14; having a decrease of R in oxygen, and the opposite behavior in carbon dioxide. The variation of resistance registered in each gas was nearly the same to that observed in Fig. 14. The response pattern of Fig. 15 was highly reproducible, revealing also that the gas detection involves reversible processes.

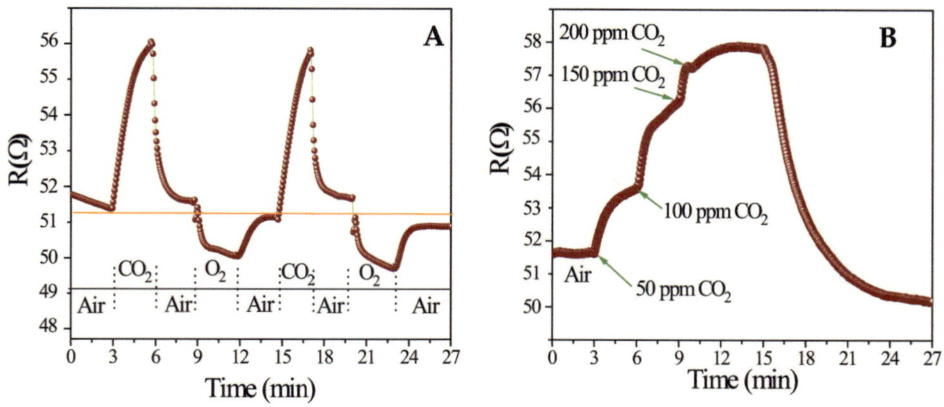

Fig. 15. Resistance *vs.* time plots acquired in: (A) air, CO_2 and O_2; and (B) different concentrations of CO_2 (380°C).

Fig. 15B displays the ability of $Gd_{0.9}Sr_{0.1}CoO_3$ to detect variations in the concentration of CO_2. In this experiment, dry air was supplied for about 10 min, until stable resistance measurements were obtained. Then, CO_2 with a concentration of 50 ppm was injected; this produced an increase of R of about 2 Ω. Later, the concentration of CO_2 was increased to 100 ppm, which increased R in ~2.5 Ω. This result agrees with that shown in Fig. 14B; however, when the concentration of CO_2 was increased to 150 and 200 ppm, smaller increments of R were registered. The latter suggests **that the concentration limit at which $Gd_{0.9}Sr_{0.1}CoO_3$ can satisfactorily detect CO_2 is 100 ppm**. **However,** for large concentrations, a saturated state was **observed. This behavior can be attributed to several factors, which include surface area, gas diffusivity and temperature, among others. On the other hand,** this kind of experiment was also performed to detect variations in the concentration of O_2; however, unreliable results were produced. Future improvements on the gas sensor device may result in a better performance in oxygen. Therefore, polarization curves were focused on the ability of $Gd_{0.9}Sr_{0.1}CoO_3$ to detect variations in the concentration of CO_2.

3.4.3 Polarization curves

Polarization curves (or I-V curves), were recorded to test the ability to detect changes in the concentration of CO_2 (static mode). These curves were obtained at 380°C, using CO_2 at concentrations of 50, 100, 150 and 200 ppm. Measurements in air were also performed to obtain a reference curve. Fig. 16 shows typical I -V curves recorded from -5 V to 5 V. It is worth to mention that beyond this voltage range, reliable measurements were not acquired. These curves show that the presence of CO_2 produced an abrupt decrease of current. This is in agreement with the results presented in the previous section, where CO_2 increased the resistance. By increasing the concentration of CO_2, smaller values of current were registered; at both cathodic and anodic voltages.

On the other hand, the nonlinear behavior of I-V curves of Fig. 16 is similar to that reported for varistor gas sensors. According to Lin et al., the increase of nonlinearity of I-V curves of varistor gas sensors is caused by the decrease of grain size (Lin et al., 1995). Due to the small

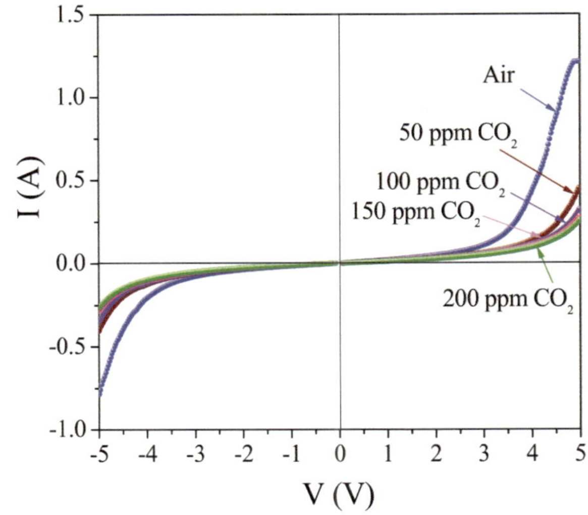

Fig. 16. Polarization curves obtained at different concentrations of CO_2, on $Gd_{0.9}Sr_{0.1}CoO_3$ films (380°C).

particle size of $Gd_{0.9}Sr_{0.1}CoO_3$ (solution-polymerization), the nonlinearity of I-V curves is similar to that reported by Lin and coworkers.

3.4.4 Dynamic gas sensing response in alternating current

Another useful approach to analyze the gas sensing properties of materials, is by means of the use of alternating current. One of the advantages of using alternating current is that some materials respond satisfactorily to gases, at high frequencies, but not at direct current. In the present experiments, the magnitude of the impedance (|Z|) was measured using several frequencies, from 100 Hz to 100 kHz. The amplitude of the signal was 1 V. From several results it was observed that the best graphs were recorded at 100 kHz.

Fig. 17 shows |Z| *vs.* time graphs obtained after the successive injection of air and 50 ppm O_2 (100 kHz, 380°C). The difference between the graphs 17A and 17B is the exposition time to the test gas. One of the reasons to perform these experiments was to investigate if the stabilization of |Z| can be obtained after a prolonged exposition to the test gas. Fig. 17B shows that the longer exposition to 50 ppm O_2 produced a continuous decrease of |Z|; which is in agreement with Fig. 14A. **The incomplete desorption of oxygen may be explained by the fact that $Gd_{0.9}Sr_{0.1}CoO_3$ possess oxygen vacancies in its crystal structure, which is commom in transition metal oxides with the perovskite-type structure. Is possible that part of the oxygen supplied in these tests occupies the oxygen vacancies of $Gd_{0.9}Sr_{0.1}CoO_3$. About this issue, it is well known that the oxygenation of $YBa_2Cu_3O_{7-\delta}$ produces the high T_c superconducting phase; which indicates the ability of perovskites to retain oxygen. The stabilization of |Z| with time was not observed in these experiments, but some improvements in the gas sensor device may enhace the detection of O_2. These improvements include the increase in the surface contact between the metal electrode and the oxide powder. Another relevant issue is the use of gold or platinum as electrodes.**

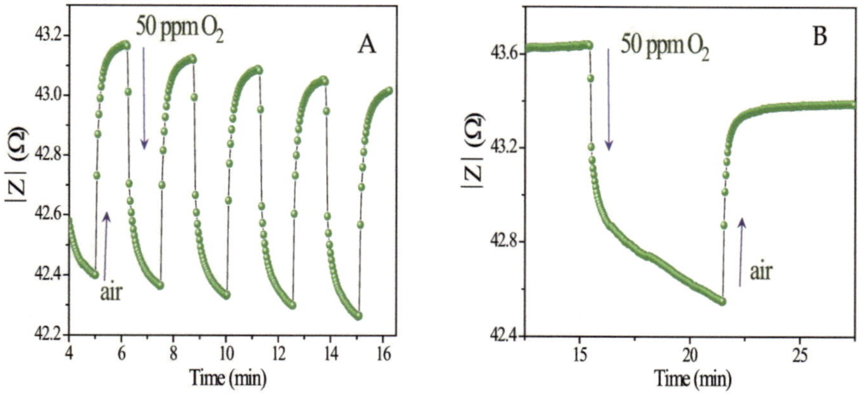

Fig. 17. |Z| *vs.* time plots recorded at different exposition times to 50 ppm O_2 (100 kHz, 380°C).

Fig. 18 shows typical results obtained when a film of $Gd_{0.9}Sr_{0.1}CoO_3$ was exposed to air and 50 ppm CO_2 (100 kHz, 380°C). Analogously to the previous results, the difference between the figures 18A and 18B is the exposition time to CO_2. Notable repeatibility in the detection of CO_2 can be observed in Fig. 18A. Full recovery of the original |Z| value can also be noticed. Fig. 18B revealed a lack of a saturated state, similar to that observed in oxygen. Even so, we can conclude that a better performance in CO_2 was obtained.

Fig. 19 shows |Z| *vs.* time graphs acquired at different concentrations of: (A) oxygen and (B) carbon dioxide. It is worth to mention that these experiments were also performed at direct current; however, the results were far to be satisfactory, and they were not included in this work. Fig. 19 reveals that the increase of the gas concentration in 50 ppm produced a proportional variation of |Z|. In oxygen, a continuous decrease of |Z| with time can be observed; which is more attenuated in carbon dioxide. In summary, based in the preceding results, $Gd_{0.9}Sr_{0.1}CoO_3$ can be considered as an alternativey CO_2 sensor material. As an oxygen gas sensor material, a rather limited performance was observed.

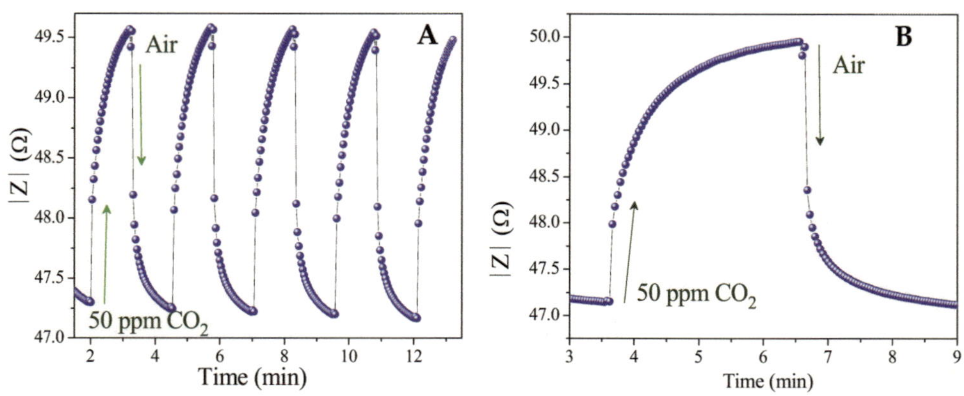

Fig. 18. |Z| *vs.* time graphs acquired using two different exposition times to 50 ppm CO_2 (100 kHz, 380°C).

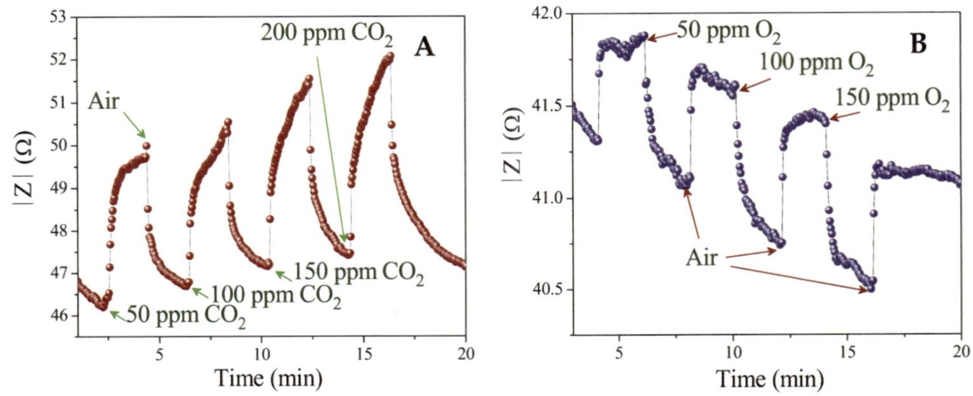

Fig. 19. $|Z|$ *vs.* time graphs acquired at different concentrations of (A) O_2 and (B) CO_2 (100 kHz, 380°C).

3.5 Gas sensing mechanisms

The explanation of the gas sensing mechanism in semiconductor oxides is frequently based in the change of resistance (or impedance) produced by adsorption processes. It has been reported in the literature that gas adsorption depends, among others, on the operation temperature, particle size, specific surface area and gas partial pressure.

Oxide semiconductors can be classified as p-type or n-type, according to the variation of conductivity caused by a reducing or oxidizing gas (Moseley, et al., 1991). In an oxidizing atmosphere, a p-type semiconductor material adsorbs oxygen molecules. These molecules capture electrons, increasing the number of charge carriers (holes). As a result, the electrical conductivity increases (Baraton, et al., 2003). The oxygen adsorption can occur by means of one of the following reactions:

$$O_2\,(gas) + e^- \rightarrow O_2^-(ads) \qquad (2)$$

$$O_2\,(gas) + 2e^- \rightarrow 2O^-\,(ads) \qquad (3)$$

$$O_2\,(gas) + 4e^- \rightarrow 2O^{2-}(ads) \qquad (4)$$

About the CO_2 gas sensing mechanism, the increase of $|Z|$ suggests a change in the dielectric constant (κ) of the film. CO_2 can not be adsorbed on the surface of $Gd_{0.9}Sr_{0.1}CoO_3$ because is a stable molecule. The formation of a thin layer of carbonate decreases κ, due to the dielectric constant of carbonates is frequently smaller than the oxides (Ishihara, et al., 1991). The decrease of $|Z|$ with κ can be explained in terms of the following equation:

$$|Z| = \sqrt{R^2 + \left(\frac{1}{\kappa\omega C_o}\right)^2} \qquad (5)$$

Where R is the electrical resistance, κ is the dielectric constant, ω is the angular frequency and C_o is the capacitance in air (Halliday, et al., 2001). The decrease of κ produces the increase of |Z|; then, when air is reintroduced, |Z| returns to its original value. Some authors have studied the formation of bicarbonate, bidentate or monodentate on MgO, CaO and ZrO_2 (Tsuji, et al., 2003); however, the identification of the specific type of carbonate formed on $Gd_{0.9}Sr_{0.1}CoO_3$ is currently investigated.

On the other hand, some authors have reported that the process by which charge carriers are transported through a material, depends on the particle size (Bochenkov, et al., 2005). First, the thickness of the charged layer (L_S) depends on the surface potential (V_S), according to the equation:

$$L_s = L_D\sqrt{\frac{eV_S^2}{kT}} \tag{6}$$

Where L_D is the Debye length, e is the electron charge, k is the Boltzmann´s constant and T is the temperature. L_D depends on other physical parameters through the following relation:

$$L_D = \sqrt{\frac{\kappa kT}{2\pi e^2 N_d}} \tag{7}$$

Where κ is the dielectric constant and N_d is the concentration of donor impurity. Typical values of L_S are in the range 1 to 100 nm. If D is the crystallite size, three possible mechanisms are usually proposed:

1. D >> $2L_S$, the conductance is limited by Schottky barriers
2. D = $2L_S$, the conductivity is controlled by the necks formed among particles
3. D < $2L_S$, in this case the entire volume of each crystallite participates in the charge transport

Fig. 20 shows a schematic illustration of the last two mechanisms. The dark green area represents the volume through which charge carriers move. The third mechanism may be a possible explanation of the improved gas sensing behavior in nanostructured $Gd_{0.9}Sr_{0.1}CoO_3$ (solution-polymerization).

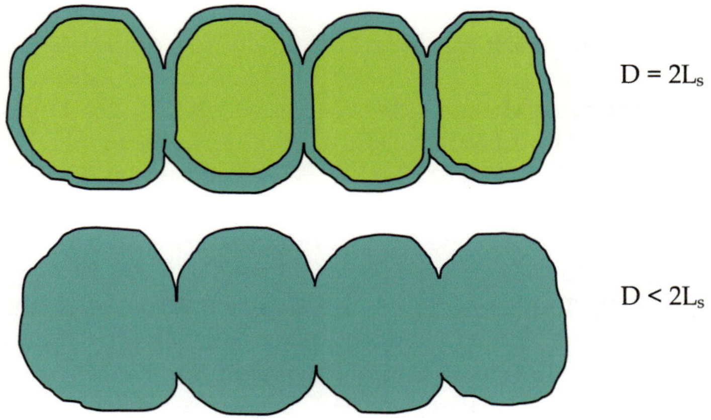

Fig. 20. Scheme of two mechanisms of conductance in semiconductor materials.

4. Conclusions

The solution-polymerization process is a simple and cost effective method for preparing nanostructured $Gd_{0.9}Sr_{0.1}CoO_3$. Due to PVA and sucrose are inexpensive materials, this method can be used to obtain other nanostructured compounds; which can be used in fields like gas sensors, gas separation membranes, electrodes for solid oxide fuel cells and catalysis, to mention a few. In this work, single phase $Gd_{0.9}Sr_{0.1}CoO_3$ prepared by solution-polymerization was obtained from 600°C. However, by the solution method, it was obtained at 800°C. Extensive porosity caused by the decomposition of PVA and sucrose, as well as lower calcination temperatures, produced a nanostructured network of $Gd_{0.9}Sr_{0.1}CoO_3$. The latter was observed by transmission electron microscopy.

Through the analysis of the gas sensing results it was possible to conclude that $Gd_{0.9}Sr_{0.1}CoO_3$ (solution-polymerization), detects oxygen and carbon dioxide at lower operation temperature. **The stability and repeatability observed in the detection of CO_2 was notably enhanced when alternating current was used. However, the stability in the detection of O_2 needs improvements in the gas sensor device. The response and recovery times as a function of operation temperature are important parameters and are under investigation.**

According to recent publications, the decrease of particle size, at the nanometer scale, causes that charge carriers are transported through the entire volume of the crystallites. This is relevant in solid state gas sensor materials because small resistance changes can be detected using standard instrumentation.

5. Acknowledgment

Financial support from the Coordinación General Académica of the Universidad de Guadalajara and COECYTJAL (Grant PS-2008-847) is greatly acknowledged. The authors are grateful to CONACYT for two doctorate scholarships (N.L.L.C. and E.R.L.M.).

6. References

Baraton, M.I. & Merhari L. (2003). Electrical Behavior of Semiconducting Nanopowders versus Environment, *Rev. Adv. Mater. Sci.* Vol. 4, pp. 15-24.

Bochenkov, V.E. & Sergeev G.B. (2005). Preparation and Chemiresistive Properties of Nanostructured Materials, *Adv. Colloid. Interface Sci.* Vol. 116, pp. 245-254.

Brosha, E.L.; Mukindan, R.; Brown, D.R.; Garzon, F.H.; Visser, J.H.; Zanini, M.; Zhou, Z. & Logothetis, E.M. (2000). CO/HC Sensors Based on Thin Films of $LaCoO_3$ and $La_{0.8}Sr_{0.2}CoO_{3-\delta}$ Metal Oxides, *Sens. Actuators B*, Vol. 69, pp. 171-182.

Das, R.N. (2001). Nanocrystalline Ceramics from Sucrose Process, *Mater. Lett.* Vol. 47, pp. 344-350.

Dutta, A.; Ishihara, T.; Nishiguchi, H. & Takita Y. (2004). Amperometric Solid-State Gas Sensor using $LaGaO_3$ Based Perovskite Oxide Electrolyte for Detecting Hydrocarbon in Exhaust Gas, *J. Electrochem. Soc.* Vol. 151, pp. H122-H127.

Halliday, D.; Resnick, R. & Walker, J. (2001). *Fundamentals of Physics*, Wiley, New York, USA.

Karl, T.R. & Trenberth, K.E. (2003). Modern Global Climate Change, *Science*, Vol. 302, pp. 1719-1723.

Kerr, R.A. (2007). Global Warming is Changing the World, *Science*, Vol. 316, pp. 188-190.

Kong, L.B. & Shen, Y.S. (1996). Gas-Sensing Property and Mechanism of $Ca_xLa_{1-x}FeO_3$ Ceramics, *Sens. Actuators B*, Vol. 30 pp. 217-221.

Kosacki, I. & Anderson, H.U. (1998). Nanostructured Oxide Thin Films for Gas Sensors, *Sens. Actuators B*, Vol. 48, pp. 263-269.

Gülgün, M.A.; Nguyen, M.H. & Kriven, W.M. (1999) Polymerized Organic–Inorganic Synthesis of Mixed Oxides, *J. Am. Ceram. Soc.* Vol. 82, pp. 556-560.

Ishihara, T.; Kometani, K.; Hashida, M. & Takita, Y. (1991) Application of Mixed Oxide Capacitor to the Selective Carbon Dioxide Sensor, *J. Electrochem. Soc.* Vol. 138, pp. 173-176.

Lee, S.J. & Kriven, W.M. (1998) Crystallization and Densification of Nano-Size Amorphous Cordierite Powder Prepared by a PVA Solution–Polymerization Route, *J. Am. Ceram. Soc.* Vol. 81, pp. 2605-2612.

Lin, F.C.; Takao, Y.; Shimizu, Y. & Egashira, M. (1995) Zinc Oxide Varistor Gas Sensors: Effect of Bi_2O_3 Content on the H_2-Sensing Properties, *J. Am. Ceram. Soc.* Vol. 78, pp. 2301-2306.

Mahata, P.; Aarti, T.; Madras, G. & Natarajan, S. (2007) Photocatalytic Degradation of Dyes and Organics with Nanosized $GdCoO_3$, *J. Phys. Chem. C*, Vol. 111, pp. 1665-1674.

Michel, C.R.; Martínez, A.H. & Morán, J.P. (2009). Carbon Dioxide Gas Sensing Behavior of Nanostructured $GdCoO_3$ Prepared by a Solution-Polymerization Method, *J. Alloy. Compd.* Vol. 484, pp. 605-611.

Moon, J.W.; Seo, W.S.; Okabe, H.; Okawa, T. & Koumoto, K. (2000). Ca-Doped $RCoO_3$ (R = Gd, Sm, Nd, Pr) as Thermoelectric Materials, *J. Mater. Chem.* Vol. 10, pp. 2007-2009.

Moseley P.T.; Stoneham, A.M. & Williams, D.E. (1991). *Techniques and Mechanisms in Gas Sensing*, Adam Hilger, Bristol, UK.

Parry, M.; Canziani, O. & Palutikof, J. (2008). Key IPCC Conclusions on Climate Change Impacts and Adaptations, *World Meteorological Organization Bulletin*, Vol. 57, pp. 78-85.

Post, M.L.; Tunney, J.J.; Yang, D.; Du, X. & Singleton D.L. (1999). Material Chemistry of Perovskite Compounds as Chemical Sensors, *Sens. Actuators B*, Vol. 59, pp. 190-194.

Rey-Cabezudo, C.; Sánchez-Andújar, M.; Mira, J. Fondado, A.; Rivas, J. & Señarís-Rodríguez, M.A. (2002). Magnetotransport in $Gd_{1-x}Sr_xCoO_3$ ($0 < x < 0.30$) Perovskites, *Chem. Mater.* Vol. 14, pp. 493-498.

Suo, H.; Wu, F.; Wang, Q.; Liu, G.; Qiu, F.; Xu, B. & Zhao M. (1997). Influence of Sr Content on the Ethanol Sensitivity of Nanocrystalline $La_{1-x}Sr_xFeO_3$, *J. Solid State Chem.* Vol. 130, pp. 152-153.

Takeda, Y.; Ueno, H.; Imanishi, N.; Yamamoto, O.; Sammes, N. & Phillipps, M.B. (1996). $Gd_{1-x}Sr_xCoO_3$ for the Electrode of Solid Oxide Fuel Cells, *Solid State Ionics*, Vol. 86-88, pp. 1187-1190.

Tsuji, H.; Okamura-Yoshida, A.; Shishido, T. & Hattori, H. (2003) Dynamic Behavior of Carbonate Species on Metal Oxide Surfaces: Oxygen Scrambling Between Adsorbed Carbon Dioxide and Oxide Surface, *Langmuir*, Vol. 19 pp. 8793-8800.

Wang, C.C.; Akbar, S.A. & Madou M.J. (1998). Ceramic Based Resistive Sensors, *J. Elecroceram.* Vol. 2:4, pp. 273-282.

Wienhöfer, H.D.; Bremes, H.G. & Nigge, U. (2004) Mixed Conduction and Electrode Properties of Doped Cobaltites and Chromites, *Solid State Ionics*, Vol. 175, pp. 93-98.
Yamazoe, N. (2005). Toward Innovations of Gas Sensor Technology, *Sens. Actuators B*, Vol. 108, pp. 2-14.

9

Survey of the Application Nanoscale Material in Chemical Sensors

Mahboubeh Masrournia and Zahra Ahmadabadi
Department of Chemistry, Faculty of Science, Mashhad Branch, Islamic Azad University, Iran

1. Introduction

Sensors and sensor arrays for the detection of chemical and biological substances have attracted much attention in recent years. The ultimate goal is to fabricate sensors that can determine the presence of a wide range of substances at relevant concentration levels with sufficient selectivity and sensitivity. Such research would ultimately produce technology that could be applicable in many segments including food processing, environment alremediation, agriculture, medical diagnostics and defense. The main requirements besides selectivity and sensitivity are fast response, low fabrication costs, robustness and portability. Hence intensive research activities around the world are focused on developing new sensing materials and technologies (Fam et al., 2011, Sinh et al., 2006). With the development of nanotechnology, there is a growing demand for advanced electronics based on functional nanomaterials. To date, various nanomaterials have been investigated from both fundamental and practical perspectives, and their unique chemical and physical characteristics have been continuously discovered. Important characteristics and the quality parameters of nanosensors can be improved over the case of classically modeled sensors merely reduced in size.

2. Nanostructure sensors

2.1 Carbon nanotube sensors

Since discovery, carbon nanotubes (CNTs) (in 1991) have been extensively explored for numerous applications. Of all the nanomaterials reported, carbon-based nanomaterials like fullerenes, graphene and carbon nanotubes (CNTs) show a huge potential in bringing sensor technology to the next level. Of these carbonaceous materials, CNT stand out as the most promising material for deployment in electronic sensing platforms due to its superior chemical and electronic properties. Furthermore, CNT possess a great potential for being employed both as a part of the transducer element as well as a functional receptor element in an electronic device. Despite the potential associated with CNT-based sensors, the efforts devoted to commercialization are still being limited by the challenges involved in CNT synthesis and device fabrication. Over the years, several attempts have been made to address these issues, for example, controlled synthesis of CNT. Although the scientific breakthroughs are numerous in this area, much of the CNT sensor research still remains at a proof-of-concept and/or prototype stage (Fam et al. 2011, Sinh et al., 2006).It could be observed from patents filed over the recent years on CNT array sensors (Kovtyukhova et al., 2003) that CNT platform has the potential to become mercialized in the near future as challenges in device fabrication and understanding of the functioning mechanisms are being overcome. This is further evidenced by a recent report on commercializing CNT based sensors introduced by Nano mix Inc., emphasizing on fabrication approaches for obtaining uniform devices for detection of biomolecules and toxic substances resulting from industrial emissions. There are generally three techniques used for producing CNTs, namely, carbon arc-discharge technique (Ebbesen et al.1992), laser-ablation technique (Thess et al.1996) and growth by either (CVD) or other methods (Jose-Yacaman et al.,1993, Li et al.,1996, Star et al., 2004).

2.1.1 Sensor fabrication

The synthesis of isolated carbon nano tubes with uniform outer diameters and ordered spacing over wafer-scale areas was investigated for fabrication of nano-electrode arrays on silicon wafers for field emission and sensor devices. Multi-walled carbon nano tubes (MWCNTs) were grown on TiN electrode layer with iron catalyst patterned by nano-imprint lithography (NIL), which allows the precise placement of individual CNTs on a substrate. The proposed techniques, including plasma-enhanced chemical vapor deposition (PECVD) and NIL, are simple, inexpensive, and reproducible methods for fabrication of nano-scale devices in large areas. CNT sensor device fabrication essentially integrates the CNT morphology obtained via the chirality, diameter and length separation using controlled synthesis techniques with a supporting substrate and electrode interconnections in a transistor (Fig. 1) or resistor configuration.

The ability to sense lies in the nature of the interaction between the sensitive material and the analyte; whether the analyte molecules bind specifically to the sensitive material and hence change its intrinsic properties, electronic or mechanical. Due to the difficulty in optimization of such a device, commercialization is still elusive one notable contribution to commercialization could be found in the works of Gruner and Star et al.

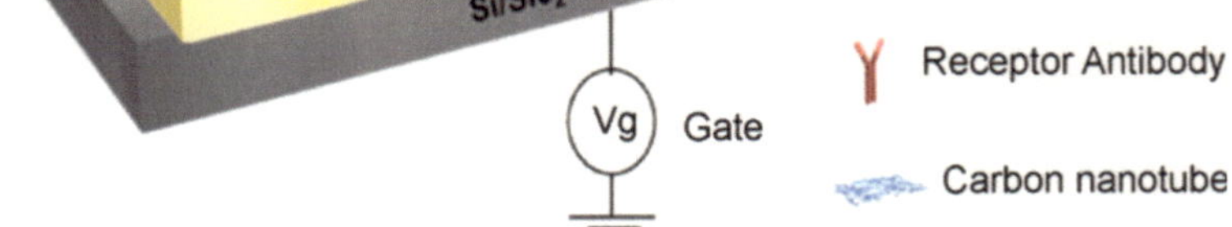

Fig. 1. Schematic of CNT transistor device (Star et al., 2004)

The catalyst patterns were defined by an array of circles with 200nm in diameter, and variable lengths of pitch. The nano-patterned master and Fe catalyst were observed with good pattern fidelity over a large area by atomic force microscope (AFM) and scanning electron microscopy (SEM) (Fig.2). Nano-electrodes of MWCNTs had diameters ranging from 50nm to 100nm and lengths of about 300 nm. Field emission tests showed the reducing ignition voltage as the geometry of nano tube arrays was controlled by catalyst patterning. These results showed a wafer-scale approach to the control of the size, pitch, and position of nano-electrodes of nanotubes for various applications including electron field-emission sources, electrochemical probes, functionalized sensor elements (Chang et al.,2011).

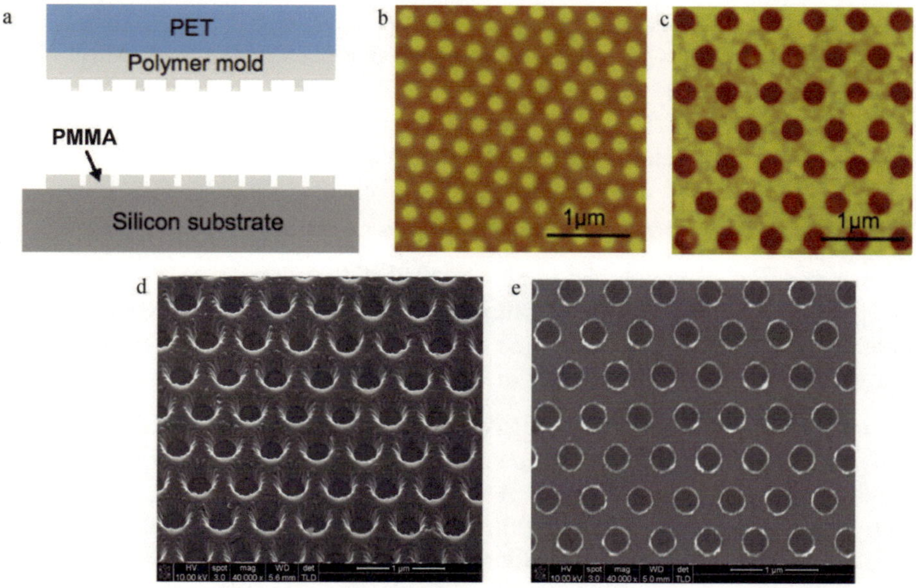

Fig. 2. (a) Schematic of UV curable resin mold. AFM images show the surface morphologies of polymer mold upper part (b) and bottom part (c). SEM images show (d) the deposition layer of Fe catalyst on a nano-patterned substrate and (e) nano-patterned Fe catalyst dots after lift-off(Chang et al., 2011).

Electrochemical (EC) sensing approaches have exploited the use of carbon nanotubes (CNTs) as electrode materials owing to their unique structures and properties to provide strong electrocatalytic activity with minimal surface fouling. Nanofabrication and device integration technologies have emerged along with significant advances in the synthesis, purification, conjugation and biofunctionalization of CNTs. Such combined efforts have contributed towards the rapid development of CNT-based sensors for a plethora of important analytes with improved detection sensitivity and selectivity. The use of CNTs opens an opportunity for the direct electron transfer between the enzyme and the active electrode area. Of particular interest are also excellent electrocatalytic activities of CNTs on the redox reaction of hydrogen peroxide and nicotinamide adenine dinucleotide, two major by-products of enzymatic reactions.

This excellent electrocatalysis holds a promising future for the simple design and implementation of on-site biosensors for oxidases and dehydrogenases with enhanced selectivity. To date, the use of an anti-interference layer or an artificial electron mediator is critically needed to circumvent unwanted endogenous electroactive species. Such interfering species are effectively suppressed by using CNT based electrodes since the oxidation of NADH, thiols, hydrogen peroxide, etc. by CNTs can be performed at low potentials (Chang et al., 2011).

2.1.2 Binding of CNTs with polymer

CNT reinforced polymeric matrix nanocomposites are prepared by polymerization of poly (methyl methacrylate) and polystyrene. This typical procedure includes the admixture of CNTs and the polymers, drying of the mixture and dissolving it with toluene. A uniform suspension of CNTs (up to 10 mg/mL) can be prepared using carboxymethylcellulose or kappa-carrageenan, which may wrap around CNTs. The strategy based on kappa-carrageenan-CNT network is used for entrapping GOx in a stable manner to form a biosensing scheme for detection of glucose (Kumar Vashist et al., 2011).

2.1.3 Polymer coated carbon nanotube-based gas sensors

For polymer coatings of nanotubes, similar to SWNT film formation, drop casting can be used. A polymer solution is placed onto a SWNT film and the solvent is allowed to evaporate. While simple, this technique also demonstrates similar reproducibility and uniformity problems to the drop casting of SWNT films.

Another method is the formation of a SWNT–polymer composite. Instead of coating the SWNTs with a polymer, the SWNTs are embedded in the polymer film, which helps with film uniformity and sensor selectivity. However, the deposition of the SWNT–polymer composite film is typically conducted with drop casting, and does not necessarily improve reproducibility. Chemical functionalization improves reproducibility of the polymer film formation, but chemically modified SWNTs can ultimately degrade sensor performance and significantly raise the recovery time of sensors. Finally, electropolymerization was selected as a simple and promising way for coating SWNT films with polymers. In this technique, the polymer film thickness can be controlled by the amount of charge applied during the process, while the uniformity and morphology of the polymer can be tuned by the monomer concentration, pH, ionic strength, and electrical parameters (Field et al., 2010).

2.1.3.1 Sensor fabrication

Following Fig. 3a, a 2.00mm diameter circle of SWNT-MCE membrane was removed from the vacuum filtered film using a hole punch (Harris, Uni-Core), as seen by the holes in Fig. 1b, and transferred to a sensor surface. The sensor substrate was made from a 100mm Silicon wafer, on which a 500nm thermal oxide layer was grown. On the wafer, 170Å chromium and 1000Å gold layers were deposited by DC magnetron sputtering (~10−2 Torr of argon background pressure) and patterned via standard lift-off photolithography. The exposed SiO2 was treated with 1% 3-aminopropyltriethoxysilane (APTES) (Gelest, Inc.) in acetone to increase the interaction between the SWNT film and sensor surface (Field et al., 2010).

Fig. 4c shows a SWNT-based sensor after nanotube transfer and deposition from the MCE membrane. In the center of the sensor, a ~2mm diameter dark circle can be seen of a 4.00 mg/L SWNT film (Field et al. 2010).

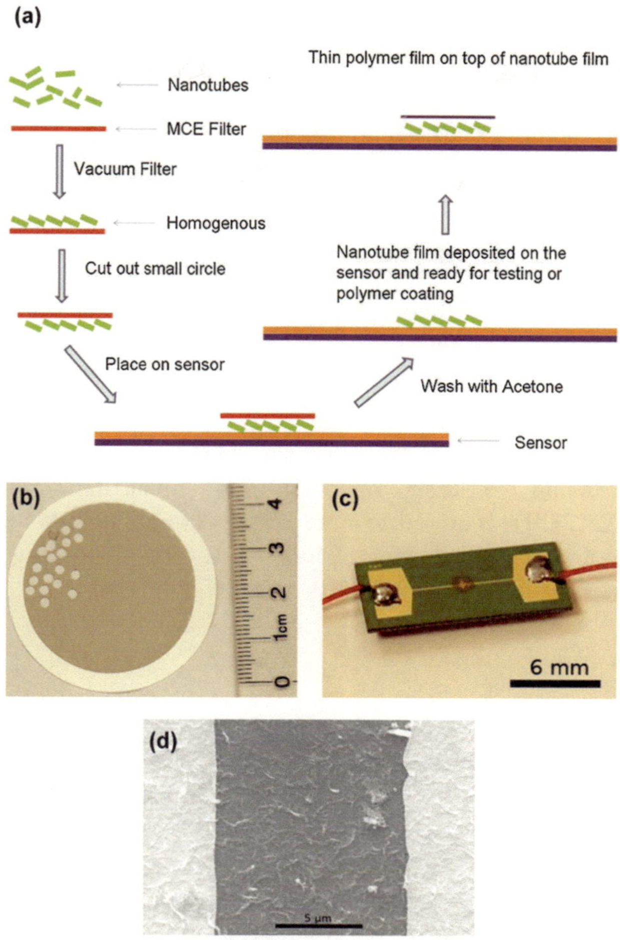

Fig. 3. (a) A flow chart of the fabrication process for the SWNT-based chemi-resistor, (b) a photograph of a vacuum-filtrated SWNT film on a 47-mm diameter MCE membrane,using 25mL of 4.00 mg/L SWNT solution (2-mm diameter circles punched out for subsequent transfer to sensors), (c) a photograph of a SWNT sensor on silicon substrate with 500nm thick thermal oxide and 100nm thick gold electrodes (a dark circle of 2-mm diameter representing a transferred SWNT film), (d) a SEM image of the SWNT film on the 10μm gap between the electrode(Field et al. ,2010).

The relatively high reproducibility of the sensor fabrication process allowed for the response testing to be conducted in a series of three batches. As an example, the response for a sensor coated with N-MP and exposed to various concentrations of DMMP can be found in Supplementary material. It is seen that different polymer coated SWNT sensors would help to produce diverse modulated sensitivity profiles against several gases. The results can be extended to establish a comprehensive histogram of modulated response to each analyte. The response from each polymer coating in an array creates a unique molecular signature that can be used to detect and identify unknown gas molecules. Efforts are underway to establish a library of responses and improve the polymer coatings to maximize the detection and identification power of an array of polymer coated SWNT sensors (Field et al., 2010).

2.1.4 Other strategies

Clay and ceramic materials are also useful for modification of CNT based electrodes. NiO/Na-montmorillonite (clay) binds to CNTs and is used to modify GCE using a Nafion solution (Zhao et al., 2011). A CNT ceramic composite is prepared by admixing CNTs and silica sol. Toluidine blue (Tb) adsorbs on CNTs to improve the dispersion of CNTs in water. The Tb-CNT suspension is then dropped on the electrode surface for subsequent modification with horseradish peroxidase. Ionic liquid (IL) modified CNTs are highly resistant towards biomolecular fouling and have high electron transfer kinetics. The IL/CNT modified electrode is prepared by adding CNTs to trihexyltetradecylphosphonium bis (trifluoromethylsulfonyl)-imide and dispersing in DMF with ultrasonication (Hsu et al.,2009). The IL-CNT suspension is dropped on the electrode surface for subsequent electrodeposition of metallic nanoparticles.

Ferrocene and its derivatives are widely used as artificial mediators of electron transfer in CNT-EC sensors owing to their relatively low molecular mass and reversible voltammograms. $NaCNBH_3$ is added to a suspension of aminated MWCNT and ferrocene carboxaldehyde (FcCHO) with stirring for 24 h. The resulting MWCNT-FcCHO nanocomposite is used to modify GCE (Xiao et al.2009).

2.1.5 CNT sensors applicaction

Biomolecular sensing (Amino acids)

CNT-EC sensors have been extensively reported for biomolecular sensing, especially for the detection of glucose and neurotransmitters. There are also many reports where researchers have successfully demonstrated the detection of proteins, DNA, biomarkers, cells, microorganisms and other biomolecules.

A MWCNT-Cu_2O paste electrode can detect 19 amino acids. Combined capillary electrophoresis with amperometric detection is used for the detection of various amino acids such as arginine, tryptophan, histidine, threonine, serine, and tyrosine, which had high oxidation current responses. A rapid glutamate biosensor based on the oxidation of nicotinamide adenine dinucleotide (NADH) is prepared by immobilizing glutamate dehydrogenase (GLDH) on the SWCNT thionine mediator (Fam et al., 2011).

Besides sensing applications, the use of CNTs as catalyst support to construct functionalized nanocomposite interface for oxidation of NADH suggests great potential applications for the design of biofuel cells and bioelectronics.

Carbon nanotubes humidity sensor

A multiwall carbon nanotubes (MWCNTs) humidity sensor based on alternating current (AC) testing is made, and its frequency characteristic is studied. The mechanism of the sensor response to RH under alternating current testing can be interpreted by the theory of capillary condensation and dielectric physics, and the electrical model of the sensor is explored by both experimental and theoretical analysis. The stability, repeatability and response time of the sensor are investigated in details. Experiment results show that the sensor under AC testing demonstrates higher sensitivity and better reversibility (Zhao et al., 2011).

Carbon nanotubes based gas sensor

Besides liquid phase applications, the electrical resistivities of SWCNTs change sensitively on exposure to NO_2, NH_3 and O_2 with very fast response times compared to metal-oxide and polymer based sensors. This feature might offer an opportunity for the development of gas phase sensors using CNTs. On this line of argument, various configurations employing CNTs as chemo-sensitive overlay have been impressively studied to fabricate innovative gas and vapor phase chemical sensors (Someya et al.2003, Penza et al. 2004, 2005, Santhanam et al., 2005). In particular, the peculiarity of SWCNTs to change their optical properties due to the adsorption of environmental pollutant molecules was for the first time demonstrated in 2004. In that case, nano-scale thin films of SWCNT bundles were transferred on the optical fiber termination by means of the Langmuir–Blodgett (LB) technique and used as sensitive coatings for the development of volatile organic compounds (VOCs) optical chemo-sensors. The feasibility of exploiting novel chemo-sensitive coatings based on cadmium arachidate (CdA)/singlewalled carbon nanotube (SWCNT) composites for the development of high performance optical chemo-sensors has been experimentally investigated and proved(Consales et al., 2009).

Recently, several groups reported that CNT could be used as a miniature chemical sensor (Trojanowicz, 2006). Kong et al. demonstrated that the conductance of semiconducting SWNT (S-SWNT) could be substantially increased or decreased upon exposure to gaseous molecules, such as NO_2 or NH_3 (Kong et al. 2000). The presence of gas molecules could be determined by monitoring the electrical resistivities of nanotubes. The individual S-SWNT sensors were very small and had good sensitivity as high as 103 at room temperature. The nanotube sensors have shown to several have advantages over presently available metal oxide and polymer sensors. The reversibility of the nanotube sensor could be achieved by slow recovery under ambient conditions or by heating at a high temperature. Collins's group reported that the isolated semiconducting SWNT could be converted into apparent metals through exposure to oxygen at the room temperature (Collins et al.,2000). It was demonstrated recently that a thermoelectric "nano-nose" could be built from tangled bundles of SWNTs (Adu et al., 2001). The response was specific and gases such as He, N_2, H_2 could be easily detected. Experiments showed the thermoelectric power and resistivity of a bundle of SWNTs varied with different gases, such as He, N_2, H_2, O_2 or NH_3. In the design of gas sensors, some unusual electronic, mechanical and thermal properties of CNTs are utilized and the unique feature characteristics of most reported sensing transduction modes related to conjugated polymers and carbon nanotubes have previously discussed (Dai et al., 2002). The sensors developed so far have used different mechanisms of interacting analytes with CNTs, as well as different modes of preparing CNTs in sensors. For constructing gas sensors, most often, changes in the electrical properties of CNTs as result of their interaction with analytes are used.

2.1.6 Technical challenges and future directions of CNT

Numerous reports on evaluation of toxicity of CNT and its biocompatibility are indicative of the potential of CNT for commercial device fabrication. Recent years have witnessed a growing number of CNT related patents emphasizing commercialization possibilities. A large number of research groups are actively pursuing CNT related research to study their interaction capabilities with potential analytes. The relatively less expensive CNT related devices and fabrication processes also have attracted interest from other field of science and technology. In summary, there is a huge potential for CNT in the field of bio and chemical sensors. Numerous reports on remarkable sensing properties of nanotubes would envisage feasible commercialization strategies. Although ultra-sensitive detection methodologies and detection capabilities at fM concentration levels have been reported, substantial research efforts are still needed to eliminate high matrix interference from real samples.

2.2 Other nanostructure sensors

As far as chemical sensing is concerned, it has been known, from more than five decades, that the electrical conductivity of metal oxides semiconductors varies with the composition of the surrounding gas atmosphere. The sensing properties of semiconductor metal oxides in form of thin or thick films other than SnO_2, like TiO_2, WO_3, ZnO, Fe_2O_3 and In_2O_3, have been studied as well as the benefits from the addition of noble metals - Pd, Pt, Au, Ag - in improving selectivity and stability.

2.2.1 ID ZnO nano-assemblies as chemical sensors

ID ZnO nano-assemblies were prepared on Al_2O_3 substrates by plasma enhanced-chemical vapor deposition (PE-CVD), and characterized in their morphology and chemical composition by field emission-scanning electron microscopy (FE-SEM), energy dispersive X-ray spectroscopy (EDXS) and X-ray photoelectron spectroscopy (XPS). For the first time, the sensing performances of PE-CVD ZnO nanosystems were tested in the detection of toxic/combustible gases (CO, H_2 and CH_4), revealing very good responses already at moderate working temperatures. In particular, carbon monoxide and hydrogen detection was possible already at 100 °C, whereas methane sensing required a minimum temperature of 200 °C. The performances of the present ZnO nanosystems, that make them attractive candidates for technological applications, are presented and discussed in terms of their unique and controllable morphological organization (Barreca et al., 2010).

Various ZnO nanostructures, such as nanobelts, nanorods, and nanowires, have been grown on presynthesized SnO_2 nanobelts via a simple thermal evaporation of Zn powders, without using any catalysts, producing various heterostructures (Liangyuan et al., 2010). The evaporation temperature is the critical experimental parameter for the formation of different morphologies of these nanostructures. Room temperature photoluminescence spectra of the heterostructures show that the relative intensity of ultraviolet emission to the green band can be tuned by controlling the morphologies and sizes of the secondary-grown 1D ZnO nanostructures, suggesting that the nano-heterostructures of these nanostructures grown on SnO_2 nanobelts may have potential applications in nano-optoelectronic devices (Krishnakumar et al., 2009).

High degree of crystallinity and atomic sharp terminations make nanowires very promising for the development of a new generation of gas sensors reducing instabilities, typical in polycrystalline systems, associated with grain coalescence and drift in electrical properties. These sensitive nanocrystals may be used as resistors, and in FET(field-effect transistors) based or optical based gas sensors.

2.2.2 Applications of metal oxide nanostructures

Metal oxide gas sensors Metal oxides semiconductors are normally high gap metal oxides in which the semiconducting behavior arises from deviation of stoichiometry. They should always be regarded as compensated semiconductor: cation vacancies are acceptors, yielding holes and negative charged vacancies, shallow states made up of oxygen vacancies acts as n-type donors, since the bonding electrons on the adjacent cation are easily removed and donated to the conduction band.

The termination of the periodic structure of a semiconductor at its free surface may form surface localized electronic states within the semiconductor band gap and/or a double layer of charge, known as a surface dipole. As for conduction, metal oxide gas sensors are generally operated in air in the temperature range between 500 and 800 K where conduction is electronic and oxygen vacancies are doubly ionized and fixed. In quasi 1D gas sensors the current flows parallel to the surface. When the nanowire is fully depleted, carriers thermally activated from surface states are responsible for conduction. Indeed when considering nanowires bundles, the conduction mechanism is dominated by the inherent intercrystalline boundaries at nanowires connections - like in polycrystalline samples - rather than by the intracrystalline characteristics; the intergranular contact provides most of the sample resistance (Lupan et al., 2009).

The metal semiconductor junction that forms at the interface between the layer and the contacts can play a role in gas detection, enhanced by the fact that the metal used for the contact acts also as a catalyst. The contact resistance is more important for single nanowires since it is in series to the semiconductor resistance that for bundles where it is connected to a large number of resistances.

The development of sensitive and reliable gas sensing devices for the detection of flammable/toxic chemical agents, such as CO, H_2 and CH_4, has gained an increasing attention for safety, energy and environmental applications. In particular, in spite of various works, continuous efforts are devoted to the development of low-cost, lightweight sensors for H_2 detection wherever hydrogen is produced, stored or used, including industrial plants, spacecrafts and fuel cells vehicles. Furthermore, the detection of leakages of flammable/explosive fuels, like methane, and toxic combustion products, such as CO, is of great importance to control domestic gas boilers, mining environments and combustion processes in automotive engines and industrial plants (Mukherjee et al. 2009, Bhattacharyya et al., 2008).

Hydrogen, when absorbed by a metal, occupies interstitial sites in the metal lattice causing this lattice to expand and thereby changing the physical properties of the metal. Micro-fabricated palladium coated cantilevers utilize the expansion of palladium following hydrogen absorption as the detection principle. These sensors consist of a palladium film coated on one side of a cantilever. However volume expansion on absorption of hydrogen is prohibited by the substrate on which the film is coated and the induced stresses are transduced into the mechanical bending or curvature of the cantilever. There is a limited amount of research published on cantilever hydrogen sensors and one problem associated with this type of sensor arises from interfacial sliding, delamination and even detachment of the palladium film as a result of expansion/contraction induced weakening of the palladium/ substrate adhesion. The initial adhesion strength between the palladium film and the underlying substrate depends on the method used to deposit the palladium film. It has been shown that palladium films fabricated by sputtering bombardment have superior adhesion characteristics and ultimately exhibit superior hydrogen detection characteristics as a result of efficient mechanical transduction (Chou et al. 2008). Further enhancement of the Pd adhesion by the fabrication of trenches in the surface of the cantilever beams led to improve hydrogen sensing down to 50 ppm and faster response.

Another novel mechanical hydrogen sensor is described and comprises a micromachined cantilever fabricated out of the cleaved edge of an optical fiber (Iannuzzi et al.,2007).

The sensing tungsten nanowire is synthesized by focused ion beam-chemical vapor deposition (FIB-CVD) using $W(CO)_6$ precursor gas on batch-fabricated microelectrodes. The nanowire measures 10μm in length and 150 nm in diameter, and is suspended 50μm above the wafer substrate connecting two microelectrodes to enhance the exposure of nanowire surface to target gas molecules for improved chemisorptions. The nanowire is then functionalized by palladium–platinum (Pd–Pt) sputtering process to be sensitive to the hydrogen molecules. The significant increase of nanowire resistance is observed whenever it is exposed to hydrogen molecules, and the resistance is recovered to the original value after hydrogen molecules are purged out. The experimentally characterized hydrogen sensor shows 11.52% increase of resistance to the exposure to 1% (10,000 ppm) hydrogen and 99% nitrogen gas mixture at room temperature and atmospheric pressure. The lowest hydrogen concentration measured is 0.001% (10 ppm) with 0.58% corresponding resistance increase. The schematic view and electric circuit configuration of the hydrogen sensor based on a suspended tungsten nanowire bridge on microelectrodes is shown in Fig. 4(a). The tungsten nanowire is synthesized 50μm above the substrate on top of batch-processed microelectrodes. The microelectrodes are fabricated on a heavily doped silicon-on-insulator (SOI) wafer with 50μm device layer and 1μm buried oxide using single photolithography step, deep reactive ion etching (DRIE) and the following wet etching in hydrofluoric acid. Scanning ion microscope (SIM) image in Fig4 (b) shows the nanowire bridge connecting two microelectrodes with bonded aluminum wire for signal measurement (Choi et al., 2009).

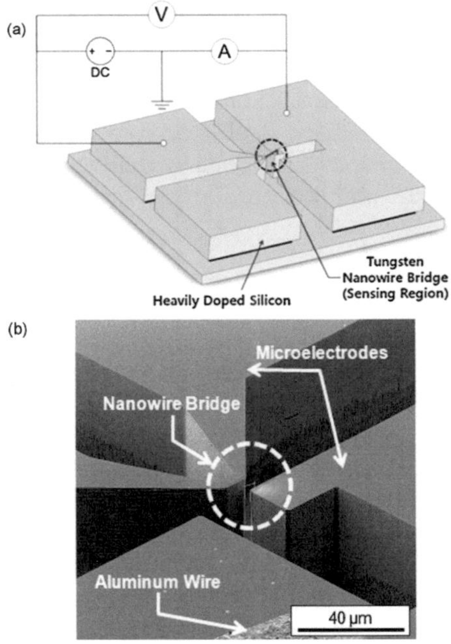

Fig. 4. (a) Schematic view of suspended single tungsten nanowire bridge hydrogen sensor. The nanowire is synthesized by FIB-CVD on microelectrodes and constant current is applied between the microelectrodes and through the nanowire to monitor the voltage drop when it is exposed to hydrogen gas. (b) SIM image of microelectrodes and suspended single tungsten nanowire bridge with bonded aluminum wire (Choi et al., 2009).

Focused ion beam-chemical vapor deposition (FIB-CVD) is an alternative to fabricate nanostructures on desired position with controllable dimensions. The most common application of FIB has been repairing and wiring the defected semiconducting circuits. Recently, efforts were undertaken to fabricate three-dimensional nanostructures using FIB.

2.3 Nano-grass polyimide-based humidity sensors

This section deals with a micro-capacitive-type relative humidity sensor with nano-grass polyimide as a dielectric sensing material. Humidity sensor achieves key performance indices such as quick response, high sensitivity, and stability enabled by the modification of polyimide into nano-grass morphology where the dimension of a typical individual pillar is 387nm×40 nm. A low hysteresis operation is also achieved by integrating a micro-heater in the sensing area. The nano-grass morphology is created with an oxygen plasma treatment on polyimide surface which increases surface to volume ratio by more than 280 times larger than that of a simple flat-film. This amplification of the surface to volume ratio leads to the rapid adsorption and desorption of water into the sensing material. Furthermore the oxygen plasma treatment introduces a carbonyl group that facilitates an enhanced affinity of the polyimide surface to water molecules. XPS analysis is used to confirm the emergence of carbonyl groups as a result of the treatment. The total response time of a nano-grass sensor is 11 s which is improved by about 2.5 times than a normal flat-film sensor. The sensitivity of the nano-grass sensor is 0.08 pF/%RH (relative humidity) which is improved by 8 times compared with the flat-film one. In stability test for 200 h, the signal of the nano-grass sensor is fluctuated ±1.0%RH (Lee et al., 2011).

2.3.1 Sensor structure

Humidity sensor consists of interdigitated gold electrodes and a polyimide layer coated on top as a moisture sensing material. Since a pure capacitive detection was used in our configuration, a leakage current path between the electrodes was prevented by a SiO_2 passivation layer between the electrode (Au) and the sensing material (polyimide). The size of the whole device was 6mm×3mm, and the typical dimension of sensing polyimide area was 2.3mm×1.0mm×2μm. The interdigitated electrodes consist of nine parallel fingers, 1.9mm long, 0.2μm thick, 20μm in width, and 50μm apart from each other (Fig.5,6). A heater pattern was also included which was 20μm in width, 0.2μm thick, and 12.8mm in length, running along the centre of the gap. The total resistance of the heater line was 78Ω (Lee et al. ,2011).

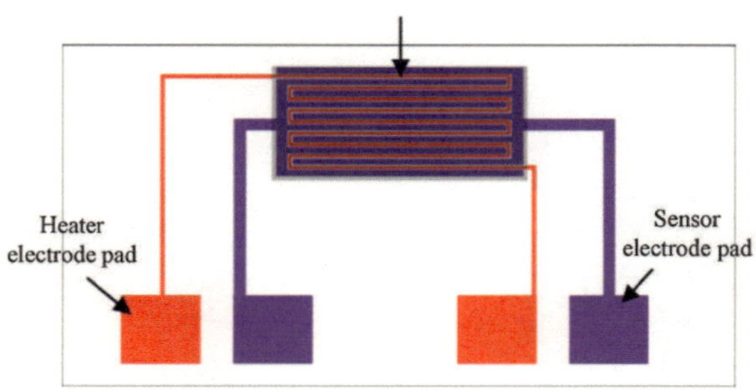

Fig. 5. Mask pattern and photograph of device(Lee et al. ,2011).

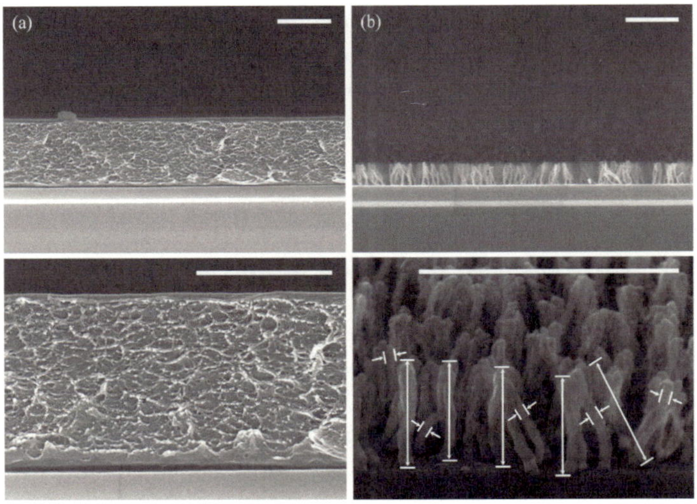

Fig. 6. Polyimide nano-grass produced by an oxygen plasma (scale bar; 1μm): (a) flat-film polyimide, (b) nano-grass polyimide (Lee et al. ,2011). .

2.3.2 Device fabrication

Fig. 7 shows the fabrication process of polyimide-based humidity sensor. The process began with the wet oxidation of 2000Å silicon oxide on a silicon wafer as an insulating layer (Fig.7a). Cr/Au (100/1000 Å) layers were e-gun deposited on the substrate, followed by a wet etching to pattern the sensing electrodes and the heater configuration (Fig. 6b). TEOS (Tetra-ethyl-orthosilicate) was deposited on top of the electrodes as a passivation layer with PECVD. Polyimide was spin-coated at 5000 rpm, for 60 s on the substrate as a sensing layer. The polyimide layer was patterned by using a wet etching with a photoresist as an etching mask. Since the polyimide could be directly etched with the diluted developing solution, the developing and etching was carried out simultaneously. After removing the photoresist, the polyimide layer originally shows flat-film morphology (Fig. 7e). A reactive ion etching was further used to modify the flat polyimide into the nano-grass (Fig. 7f). Fig.7 shows the fabrication process of polyimide-based humidity sensor. The process began with the wet oxidation of 2000Å silicon oxide on a silicon wafer as an insulating layer (Fig. 7a).

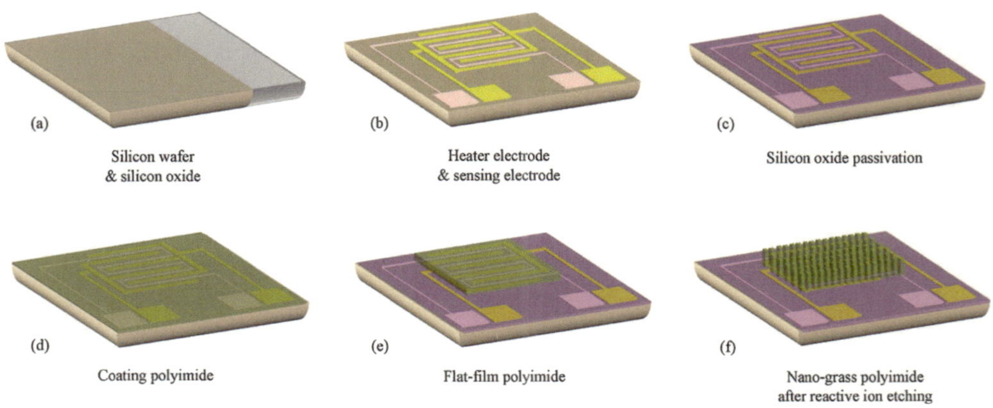

Fig. 7. Fabrication process of device (Lee et al., 2011).

A humidity sensor with high performance compared to a normal flat film type humidity sensor was developed with an increased surface area and water affinity on a nano-grass polyimide morphology enabled by a plasma etching. The nano-grass morphology has the estimated surface to volume ratio 280 times larger than that of the flat-film surface, resulting in the high sensitivity and efficient adsorption of moisture. XPS spectra indicated the increase of carbonyl groups on the plasma treated polyimide surface. The increase of carbonyl groups is responsible for further increasing the sensitivity to moisture. A heater was used to reduce the hysteresis of our nano-grass sensor. The nano-grass sensor demonstrated a clear improvement over the normal flat-film sensor in key specifications such as quick response and sensitivity, low hysteresis, and longterm stability. According to our analysis with the model equation which combined the Looyenga and the Dubinin's equations, the nano-grass sensor appears to have larger free energy of adsorption and maximum volume fraction of water than those of the flat-film sensor. This enhanced performance together with a possibility for an array fabrication may enable many new applications, e.g., a fine real-time tracking of the relative humidity distribution in a fuel cell stack (Lee et al., 2011).

2.4 Gas sensor device fabrication and testing

The schematic view of a hydrogen sensor based on Pd modified graphene is shown in Fig. 8a. CVD grown graphene with a typical size of 3mm×4mm was transferred onto a highly doped Si wafer covered by a 300 nm-thick SiO_2 film. The graphene sample on SiO_2/Si was then decorated with a thin Pd film (1nm thickness) using electron beam evaporation. To define the contacts, Ti/Au (5 nm/100 nm) films were also deposited by electron beam evaporation through a shadow mask. Subsequently, the sensor sample was assembled into a chip carrier by wire bonding, as shown in Fig. 8b were demonstrated on Pd-decorated CVD graphene films. The hydrogen sensors show high sensitivity, fast response and recovery, and are usable for multiple cycles. The mechanism of hydrogen detection can be attributed to carrier density in graphene induced by PdHx (Wu et al. ,2010).

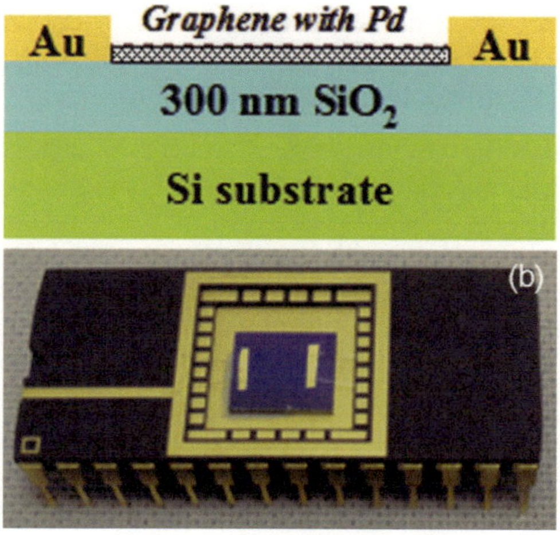

Fig. 8. (a) Schematic cross-sectional view of graphene deposited with Pd nanoparticles on SiO_2/Si substrate for H_2 sensor. (b) Photograph of a typical H_2 sensor device. The size of SiO_2/Si substrate is about 1 cm (Wu et al., 2010).

2.5 Diamond nano particles

Modified diamond nano particles (DNP) are investigated as sensitive coatings for use with Surface Acoustic Wave (SAW) transducers. DNP were surface treated (e.g. hydroxylation, hydrogenation) to enhance their affinity to specific compounds such as nitro-aromatic compounds, nerve-agent stimulant, or toxic gases. They were deposited onto SAW transducers by a layer-by layer deposition method to assess their performances for chemical vapor sensing. The DNP coated sensors were tested towards dinitrotoluene (DNT), dimethyl methyl phosphonate (DMMP), ammonia (NH_3) and relative humidity (RH). The sensors showed high sensitivity to the target compounds, good repeatability and low detection limit at sub-ppm levels, at room temperature. SAW sensors are promising for the detection of toxic chemicals due to their extremely high sensitivity. Nevertheless the sensitive coating deposited on such transducers is a limiting element. Indeed in most cases the sensitive coatings are based on polymers that are generally difficult to deposit homogeneously onto the transducer's surface .The use of these transducers requires a strict control of the coatings' properties in terms of thickness, uniformity, viscosity and film adherence in order to obtain reliable performances. However, the deposition methods that may potentially satisfy these conditions are limited to some polymer/surface systems with specific properties (Chevallier et al., 2011).

The DNP coated sensors were tested towards dinitrotoluene (DNT), dimethylmethylphosphonate (DMMP), ammonia (NH_3) and relative humidity (RH). The sensors showed high sensitivity to the target compounds, good repeatability and low detection limit at sub-ppm levels, at room temperature.

The diamond coatings' surface was either oxidised or reduced to see the effect on the response to ammonia gas, ethanol, DNT or DMMP vapours exposures. The sensors were generally very sensitive to the target chemicals and the response fully reversible. Oxidation of the surface promoted hydrogen-bond formation and therefore enhanced the response to most vapours under test. Even though the sensors were not very selective, demonstrated the suitability of DNP coatings as stable and reliable sensing interface. This opens up wide opportunities for immobilizing more selective and highly sensitive chemical/biochemical receptors onto SAW transducer surfaces via strong covalent binding of those receptors on diamond nanoparticles deposited homogeneously on the SAW sensors surfaces. Modified diamond nanoparticles (DNPs) as an alternative sensitive coating that could solve some of the issues encountered with other known sensitive coatings.

Indeed, DNPs may be found in nanometer sizes and can be deposited as single or multiple layers on a variety of sensor surfaces with uniform thicknesses and high surface area, which is in favour of high sensitivity. DNPs feature also other attractive properties for chemical detection: on the one hand they are mostly made of sp^3 carbon which is very stable in time; it is intrinsically inert hence it can be considered as a stable sensor platform onto which specific receptors are immobilized. On the other hand the carbon terminated surface of diamond offers wide perspectives from organic chemistry and biochemistry for strong attachment of specific receptors on the diamond surface via covalent bonding (Krueger, 2008). Those receptors will be chosen for their high affinity with the target species that need to be detected. The sensor response will then be either reversible or non-reversible depending on the type of interactions between the receptor molecule and the target analyte.

The DNP coatings exposed shortly to CH_4/H_2 plasma were observed using an optical microscope in order to verify the homogeneity of the layer across the entire surface of the transducer.

2.6 Nano-structured conducting polymers

Conducting polymer (CP) nano tubes and nano wires can be prepared by physical approaches like electro spinning, chemical routes like hard physical template-guided synthesis and soft chemical template synthesis (e.g., interfacial polymerization, template-free method, dilute polymerization, reverse emulsion polymerization, etc.), and a variety of lithography techniques(Skotheim et al. ,2007).

In order to fulfill the potential applications of CP nanostructures, it is necessary to precisely address their physical properties. Electrical and optical proper ties of non-individual 1D-CP nano structures, have been extensively studied, showing unusual physical and chemical behavior due to the nano size effects. There are some key questions to be clarified, for example, the effects of the nano contacts on the electrical measurements, the tuning and the control of the electrical properties of individual nano fibers and nano tubes, etc. These questions are particularly important to develop nano devices based on individual nano wires and nano tubes (Rong et al., 2009).

Since the electrical and optical properties of CP can be reversible changed by a doping/dedoping process, CP nano structures have been widely explored as chemical sensors, optical sensors and biosensors due to greater exposure area than the conventional bulk polymers. some studies demonstrate that the gas(HCl,NH_3, NO,H_2S, ethanol, liquefied petroleum gas (Jang et al., 2005), etc.) sensors based on poly aniline or PEDOT nano fibers usually how a higher sensitivity and shorter response time because of higher surface areas (Lu et al., 2009). For example, it was found that the response times of the polyaniline nano fiber sensors exposed to chloroform, toluene and tri ethylamine were about a factor of 2 faster ,with the current variations up to 4 times larger than those of the bulk poly aniline sensors.

Drug delivery devices have flourished during the last few decades and are extensively used in various kinds of treatments. Conducting polymer-based devices have been investigated to examine how they can serve as electrically controlled drug delivery devices inside the body. One major challenge is to develop a drug delivery system that allows strict control of the ON/OFF state. In addition, such a device must be able to deliver the drug of interest at doses that are required to obtain the therapeutic effect.

Nanomaterials of conjugated polymers are found to have superior performance relative to conventional materials due to their much larger exposed surface area (Rajesha et al., 2009).

The developments in nano-structured conducting polymers and polymer nanocomposites have large impact on biomedical research. Significant advances in the fabrications of nanobiosensors/sensors using nano-structured conducting polymers are being persistently made (Maity et al., 2006, Price et al. ,2008).

The study also demonstrates the role of nano-structured conducting polymers in the emerging field of nanosensors/biosensors. A detail analysis has been carried out on the latest research advancement made in the development of nano-structured conducting polymers and polymer nanocomposites based sensors/biosensors. As the surface nano-structure becomes more demanding and complex, more synthetic methods for the construction of nano-structured materials will be required. These methods will use new nano technological approaches to conducting polymers and their applications in biomedical research. Increasing interest in and practical use of nanotechnology, especially, in conducting polymers and polymer composites have lead the researchers to the rapid development of nano sensors/biosensors with improved process ability and functionality over previously developed sensors(Maity et al. 2006, Price et al. 2008, Ma et al. 2006).

2.7 Organic thin film transistor chemical sensors

Organic thin film transistor (OTFT) chemical sensors rely on the specific electronic structure of the organic semiconductor (OSC) film for determining sensor stability and response to analytes. The delocalized electronic structure is influenced not only by the OSC molecular structure, but also the solid state packing and film morphology. Phthalocyanine (H_2Pc) and tetrabenzoporphyrin (H_2TBP) have similar molecular structures but different film microstructures when H_2Pc is vacuum deposited and H_2TBP is solution deposited. The difference in electronic structures is evidenced by the different mobilities of H_2TBP and H_2Pc OTFTs. H_2Pc has a maximum mobility of 8.6×10^{-4} cm^2 V^{-1} s^{-1} when the substrate is held at 250 °C during deposition and a mobility of 4.8×10^{-5} cm^2 V^{-1} s^{-1} when the substrate is held at 25 °C during deposition. Solution deposited H_2TBP films have a mobility of 5.3×10^{-3} cm^2 V^{-1} s^{-1}, which is consistent with better long-range order and intermolecular coupling within the H_2TBP films compared to the H_2Pc films. Solution deposited H_2TBP also exhibits a textured film morphology with large grains and an RMS roughness 3–5 times larger than H_2Pc films with similar thicknesses. Despite these differences, OTFT sensors fabricated from H_2TBP and H_2Pc exhibit nearly identical analyte sensitivity and analyte response kinetics. The results suggest that while the interactions between molecules in the solid state determine conductivity, localized interactions between the analyte and the molecular binding site dominate analyte binding and determine sensor response (Royer et al., 2011).

Group of sensor materials were characterized by their good thermal stability, high optical absorbance and functionality. Porphyrins are structurally related to phthalocyanines, with the four -CH groups in the inner porphyrin ring replacing the four meso nitrogens of phthalocyanine. Several studies have investigated differential chemical sensing by changing the central metal atom in the Por/Pc core or by changing the peripheral substituents (Wang et al. 2011, Bohrer et al. 2009, 2007) Sensing.

By analyzing the transient recovery in OTFT chemical sensors, the analyte binding kinetics can be elucidated. The t_{60} values are normalized with respect to MeOH to illustrate the consistent relative recovery times for each sensor. The relative recovery time normalizes the t_{60} for each recovery with respect to the t_{60} for MeOH. This eliminates run-to-run variations in flow rate or temperature which could influence recovery rates. Both H_2Pc and H_2TBP sensors exhibit rapid recovery following doses of weak binding analytes such as MeOH and ACN, whereas for strong binding analytes such as TMP and DMMP there is a slower recovery. The H_2TBP sensor recovery immediately following a 15.8 ppm DMMP

H_2Pc sensors show significantly shorter t_{60} for isophorone than DMMP and TMP despite having a high sensitivity to isophorone; however, anomalous recovery characteristics for isophorone in H_2Pc chemiresistors have been reported previously. It has been suggested that physisorbing analytes can interact with MPc films by preferential binding or by weak van der Waals interactions with the conjugated π system of H_2TBP and H_2Pc(Gundlach et al. 2008,). Therefore, it is possible that the meso nitrogens present in H_2Pc, and absent in

H_2TBP, contribute to sensor response and recovery. However, the fast recovery and low sensitivity for each sensor to doses of MeOH and ACN suggest that the molecular structure of the extended π system does not significantly alter sensor response (Royer et al., 2009).

OTFT device properties are highly dependent on fabrication methods which influence film electronic structure by affecting grain size and intermolecular coupling (Bohrer et al., 2009). The different surface morphologies of 100 nm H_2Pc films with T_{sub} = 25–250 °C and a solution processed H_2TBP film are shown in atomic force microscopy (AFM) images presented in Fig. 9a–e. Line profiles of a 500 nm segment are shown as insets for each image. The height scale on the line profiles is held constant to demonstrate the larger surface roughness for H_2TBP films. OTFT sensors based on solution processed H_2TBP were found to have enhanced mobilities while yielding chemical sensing properties nearly identical to OTFT sensors based on vapor deposited H_2Pc. The mobilities of the films were strongly affected by differences in film microstructure, but this had little influence on chemical sensor behavior. This is consistent with analyte binding being chiefly a function of interactions with individual molecules of the sensor film. This study suggests the feasibility of preparing nonvolatile metal coordination complex sensor arrays with solution processed films. Consistent chemical sensor response can be obtained despite dramatic changes in field-effect mobility, which implies that relative chemical sensor response is a more robust property than field-effect mobility in OTFT sensors (Royer et al., 2011).

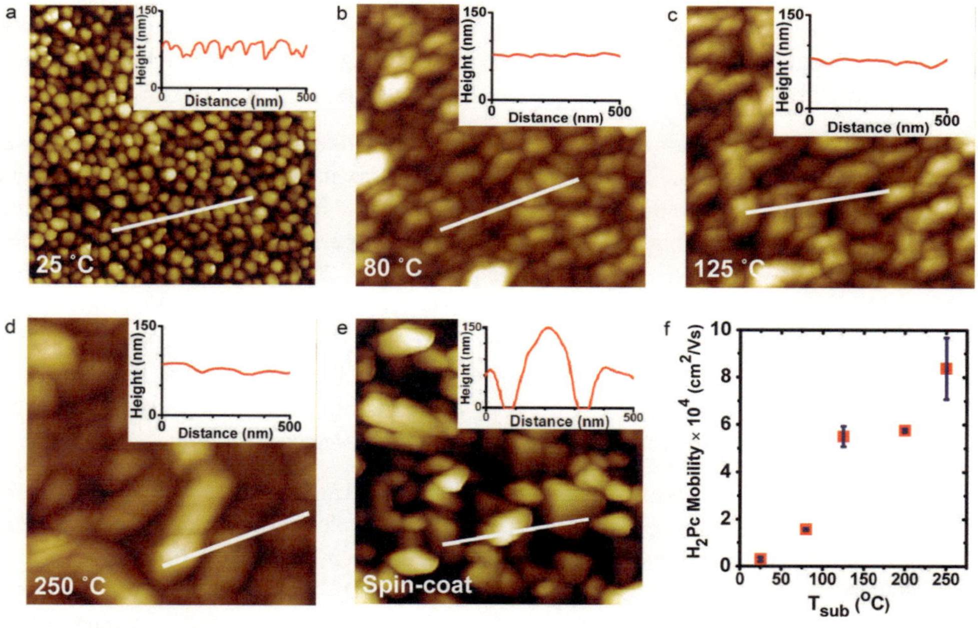

Fig. 9. H_2TBP and H_2Pc film microstructure characterization. Atomic force microscopy (AFM) images of a 1 μm × 1 μm region of 100 nm H_2Pc deposited on SiO_2 substrates held at (a) 25 °C, (b) 80 °C, (c) 125 °C, (d) 250 °C. (e) AFM image of 1 μm × 1 μm region of solution processed H_2TBP. The insets show line profiles of a 500 nm segment indicated by a white line in the image. The height scale on the line profiles is held constant to demonstrate the larger surface roughness for H_2TBP films. (f) Mobility of H_2Pc OTFTs with different substrate temperatures (T_{sub}). Some error bars are smaller than the markers (Royer et al., 2011).

2.8 Flexible chemical sensors

Flexible chemical sensors are not a new concept. Traditional chemical sensors based on conductive polymers over polymer substrates are generally flexible. However, with the advent of high-quality nanowires and nanotubes and the rapid development of flexible electronics which involves many breakthroughs in the synthesis and manipulation of these nanoscale materials, significant progress has been made in modern flexible chemical sensors.

In traditional conductive polymer chemical sensors, self-organization in many solution-processed, semiconducting conjugated polymers results in complex microstructures, in which ordered microcrystalline domains are embedded in an amorphous matrix. This has important consequences for the electrical properties of these materials, i.e. charge transport is usually limited by the most difficult hopping processes and is therefore dominated by the disordered matrix, resulting in low charge carrier mobilities ($<10^{-5}$ $cm^2/V.s$).

Mobility as high as 104 $cm^2/V.s$ in SWNT has been reported (Zhou et al. 2005). Owing to the superior quality of SWNTs, high sensitivity, low detection limit and fast response time have been observed in these nanowire- and nanotube-based chemical sensors. For highly crystalline metal oxide nanowires, a few sensing reactions occurred at room temperature instead of at high temperature. Ingenious nanowire and nanotube assembly and alignment open up many possibilities for high-performance multifunctional flexible chemical sensor development. Large integrated sensors with built-in intelligence are also on the horizon. The future of flexible chemical sensors is indeed very promising.

There are a few common features in these successful results with the use of ZnO, SnO_2 and In_2O_3 metal oxide nanowires. The use of nearly single-crystalline (high-quality) nanowires appears to be essential. A small diameter (~10 nm), with a high surface-to-volume ratio, is also important for reaching high sensitivity at room temperature. The use of metal oxides for gas sensing has been extensively reviewed (Shen et al., 2009, Huang et al., 2007, Graf et al., 2006). Although the majority of the work uses polycrystalline metal oxides and requires temperature higher than 300°C, promising results based on high-quality single-crystalline metal oxides have been achieved at ambient temperature (see above). It is foreseeable that high quality metal oxide nanowires, nanotubes and possibly nanoparticles may be attached to flexible substrates and developed into flexible chemical sensors.

A Si nanowire array, prepared separately, has been transferred onto a polyethylene terephthalate (PET) substrate with use of a poly(dimethylsiloxane) (PDMS) contact printing technique. The flexible Si nanowire FETsensor can detect NO_2 down to 20 ppb. Both Si and Ge nanowires have been transferred onto hard (Si/SiO_2) and soft (Kapton, polyimide) substrates on a wafer scale through contact printing with sliding and lubrication. The process prepares nanowire arrays with high density (~8 nanowires/μm) and 95% directional alignment. Therefore, it is likely that flexible chemical sensors based on crystalline metal oxide nanowires will be developed in the future.

A new class of nanostructured plasmonic chemical sensors has been developed in recent years (Tong et al., 2005, Jiang et al., 2008). Gold and silver nanoparticles are frequently studied because they exhibit a strong surface plasmon resonance (SPR) in visible wavelength. The aggregation of these nanoparticles will result in a significant color change that serves as the basis of sensitive chemical detection. For example, the surfaces of Au nanoparticles may be derivatized with oligonucleotide and in the presence of complementary target DNA, Au nanoparticles will aggregate and change their color from red to blue. A similar approach was applied to trinitrotoluene (TNT) detection where Au nanoparticles prepared from trisodium citrate were first treated with cysteamine ($SHCH_2CH_2NH_2$). Upon addition of TNT solution, the Au nanoparticles solution turned from red to violet blue owing to TNT-induced Au nanoparticle aggregation (Fig. 10). Although these chemical sensing techniques were demonstrated in solution, it is likely that solid-state thin-film methods will be developed in the future (Tong et al., 2005).

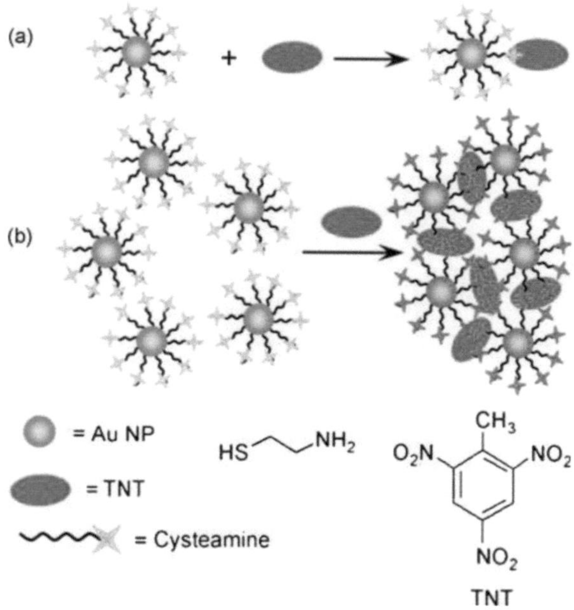

Fig. 10. Schematic drawing showing the working principle of nanoparticle-based chemical sensors. (Tong et al., 2005).

3. Conclusion

CNT have become one of the most extensively studied nanostructures because of their unique properties. CNT can enhance the electrochemical reactivity of important biomolecules and can promote the electron-transfer reactions of proteins. The remarkable sensitivity of CNT conductivity with the surface adsorbates permits the use of CNT as highly sensitive nanoscale sensors. These properties make CNT extremely attractive for a wide range of electrochemical sensors ranging from amperometric enzyme electrodes to DNA hybridization biosensors. Nevertheless, the major future challenges for the development of CNT-EC sensors include miniaturization, optimization and simplification of the procedure for fabricating CNT based electrodes with minimal non-specific binding, high sensitivity and rapid response followed by their extensive validation using "real world" samples. A high resistance to electrode fouling and selectivity are the two key pending issues for the application of CNT-based biosensors in clinical chemistry, food quality and control, waste water treatment and bioprocessing.

Diamond nano-wires can be a new approach towards next generation electrochemical gene sensor platforms. DLCs have also been applied to electrochemical measurements as electrode materials. DLC microelectrodes have been reported to commercial ELISA kits for detection of HIV antigen, HBV antigen and HCV antigen.

In the recent years, sensors research has become more materials oriented and the emphasis has been on advanced functional nano materials that serve as specific sensing layers. These novel materials include organic and polymer complexes and different oxides of Sn, Zn and Ba.

Synthesizing CP nanostructures with improved electrical conductivity and/or mobility will enhance the performance of the devices and many of the applications discussed. Several potential applications and corresponding challenges of these nanofibers and nanotubes in chemical, optical and bio-sensors, nano-diodes, field effect transistors, field emission and elec- trochromic displays, super-capacitors and energy storage, actuators, drug delivery, neural interfaces, and protein purification have been known.

4. References

Barreca D., Bekermann D., Comini E., Devi A., Fischerb R. A., Gasparotto A., Maccato C., Sberveglieric G., Tondellod E. (2010). *1D ZnO nano-assemblies by Plasma-CVD as chemical sensors for flammable and toxic gases*, Sensors and Actuators B, Vol.149. pp.1–7.

Bhattacharyya P., Basu P. K., Mondal V., Saha H.(2008). *A low power MEMS gas sensor based on nanocrystalline ZnO thin films for sensing methane,* Microelectronics Reliability, Vol. 48, No. 11-12, pp. 1772–1779.

Bohrer F.I, Colesniuc C.N, Park J., Schuller I.K, Kummel A.C, Trogler W.C.(2008). *Selective Detection of Vapor Phase Hydrogen Peroxide with Phthalocyanine Chemiresistors.* Journal of the American Chemical Society Vol.130,pp. 3712

Bohrer F.I., Sharoni A., Colesniuc C., Park J. Schuller I.K., Kummel A.C., Trogler W.C.(2007). *Gas Sensing Mechanism in Chemiresistive Cobalt and Metal-Free Phthalocyanine Thin Films,* Journal of the American Chemical Society ,Vol.129, pp. 5640–5646

Bohrer F.I., Colesniuc C.N., Park J., Ruidiaz M.E., Schuller I.K., Kummel A.C., Trogler W.C. (2009). *Comparative Gas Sensing in Cobalt, Nickel, Copper, Zinc, and Metal-Free Phthalocyanine Chemiresistors* .Journal of the American Chemical Society, Vol.131, pp.478–485.

Chang W.S, Kim J.W, Choi D.G, Han C.S. (2011). Fabrication of nano-electrode arrays of free-standing carbon nanotubes on nano-patterned substrate by imprint method, Applied Surface Science,Vol.257, pp.3063–3068.

Chevallier E., Scorsone E., Bergonzo P. (2011). *New sensitive coating based on modified diamond nanoparticles for chemical SAW sensors.* Sensors and Actuators B, Vol.154 ,pp. 238–244.

Choi J., Kim J., (2009)Highly sensitive hydrogen sensor based on suspended, functionalized single tungsten nanowire bridge, Sensors and Actuators B , Vol.136,pp. 92–98.

Chou Y.I., Chiang H.C., Wang C.C. (2008). *Study on Pd functionalization of microcantilever for hydrogen detection promotion,* Sensors and Actuators B, Vol.129, pp. 72–78.

Clement K. W. Adu., Sumanasekera G.U., Pradhan B.K., Romero H.E., Eklund P.C. (2001). *Carbon nanotubes: A thermoelectric nano-nose*, Chemical Physics Letters Vol.337 No(1–3), pp. 31-35.

Collins P.G., Bradley K., Ishigami M., Zettl A. (2000). Extreme Oxygen Sensitivity of Electronic Properties of Carbon Nanotubes Science , Vol.287 , 5459,pp. 1801-1804.

Consales M., Crescitelli A., Penza M., Aversa P., Delli Veneri P., Giordano M., Cusano A.(2009) *SWCNT nano-composite optical sensors for VOC and gas trace detection* ,Sensors and Actuators B, vol.138,pp. 351–361.

Dai L.M., Soundarrajan P., Kim T. (2002). *Sensors and sensor arrays based on conjugated polymers and carbon nanotubes,* Pure and Applied Chemistry, Vol .74 (9),pp. 1753-1772.

Ebbesen T.W, Ajayan P.M, (1992). *Large-scale synthesis of carbon nanotubes,* Nature Vol. 358,pp.220–222.

Fam D.W.H., Palaniappan Al., Tok A.I.Y., Liedberg B., Moochhala S.M.(2011) *A review on technological aspects influencing commercialization of carbon nanotube sensors* ,Sensors and Actuators B, Vol.157,pp. 1– 7.

Field C. R., Yeomb J., Salehi-Khojin A., Masel R. I. (2010). *Robust fabrication of selective and reversible polymer coated carbon nanotube-based gas sensors,* Sensors and Actuators B, Vol.148, pp. 315–322.

Graf M. , Gurlo A. , Barsan N ., Weimar U. , Hierlemann A. (2006). *Microfabricated gas sensor systems with nanocrystalline metal-oxide sensitive films, J Nanopart Res,* Vol.8,pp.823–39.

Gundlach D.J., RoyerJ.E., Park S.K., Subramanian S., Jurchescu O.D., Hamadani B.H., Moad A.J., Kline R.J., Teague L.C., Kirillov O., Richter C.A., Kush-merick J.G., Richter L.J., Parkin S.R., Jackson T.N., Anthony(2008) *Contact-induced crystallinity for high-performance soluble acene-based transistors and circuits.* Nature Materials. Vol. 7, pp. 216–221.

Hsu H.L., Jehng J.M., Liu Y.C (2009). *Synthesis and characterization of carbon nanotubes synthesized over NiO/Na-montmorillonite catalyst and application to a hydrogen peroxide sensor* ,Materials Chemistry and Physics, Vol. 113, pp.166-171.

Huang X-J., Choi Y.K. (2007) .*Chemical sensors based on nanostructured materials* .Sensors and Actuators B , Vol.122, pp. 659-671.

Hubert T., Boon-Brett L., Blackb G., Banach U. (2011). *Hydrogen sensors – A review* .Sensors and Actuators B Vol.157,pp.329– 352.

Iannuzzi D., Slaman M., Rector J.H., Schreuders H., Deladi S. Elwenspoek M.C. (2007). *A fiber-top cantilever for hydrogen detection,* Sensors and Actuators B Vol. 121, pp.706–708.

Jang J., Chang M., Yoon H. (2005) .*Chemical Sensors Based on Highly Conductive Poly(3,4-ethylenedioxythiophene) Nanorods* .Adv Mater . Vol. 17, pp. 1616–1620.

Jiang Y., Zhao H. , Zhu N. , Lin Y. , Yu P ., Mao L. (2008). *A simple assay for direct colorimetric visualization of trinitrotoluene at picomolar levels using gold nanoparticles,* Angew Chem Int Ed Engl Vol.47, pp.8601–8604.

Jose-Yacaman M., Miki-Yoshide M., Rendon L., Santiesteban J.G. (1993). *Catalytic growth of carbon microtubules with fullerene structure,* Appl. Phys. Lett.,Vol.62, pp. 202–204.

Kong J., Franklin N.R., Zhou C.W., Chapline M.G., Peng S., Cho K.J., Dai H.J. (2000). *Nanotube molecular wires as chemical sensors,* Science, Vol. 287, No. 453, pp. 622-625.

Kovtyukhova N. I., Mallouk T. E., Pan L., Dickey E. C. (2003). *Individual Single Walled Nanotubes and Hydrogels Made by Oxidative Exfoliation of Carbon Nanotube Ropes*, J. Am. Chem. Soc., Vol. 125,pp. 9761-9769.

Krishnakumar T., Jayaprakash R., Pinna N., Donato N., Bonavita A., Micali G., Neri G. (2009) .*CO gas sensing of ZnO nanostructures synthesized by an assisted microwave wet chemical route*. Sensors and Actuators B, Vol. 143,pp. 198–204.

Krueger A. (2008) *Diamond Nanoparticles: Jewels for Chemistry and Physics*. Advanced Materials, Vol. 20, pp. 2445–2449.

Kumar Vashist References and further reading may be available for this article. To view references and further reading you must purchase this article S., Zheng D., Al-Rubeaan K., Luong H.T., Sheu. J.(2011). *Technology behind commercial devices for blood glucose monitoring in diabetes management:* A review, Biotechnology Advances,Vol.29, pp.169–188.

Lee H., Lee S., Jung S., Lee J. (2011). *Nano-grass polyimide-based humidity sensors*, Sensors and Actuators B Vol.154,pp. 2–8.

Li References and further reading may be available for this article. To view references and further reading you must purchase this article.Z-F., Blum Frank D., Bertino Massimo F., Chang-Soo K.m, Pillalamarri S. K.. (2008).One-step fabrication of a polyaniline nanofiber vapor sensor, Sensors and Actuators B, Vol.134,pp.31–35

Liangyuan C., Zhiyong L., Shouli B. , Kewei Z., Dianqing L., Aifan C. Liu C. C. (2010). *Synthesis of 1-dimensional ZnO and its sensing property for CO*. Sensors and Actuators B, Vol. 143,pp. 620–628.

Li .W.Z, Xie .S.S, Qian .L.X, Chang .B.H, Zou .B.S, Zhou .W.Y, Zhao .R.A, Wang .G.(1996). *Large-Scale Synthesis of Aligned Carbon Nanotubes*, Science, Vol. 274, 1701.

Lupan O., Chow L., Chai G. (2009). *A single ZnO tetrapod-based sensor*, Sensors and Actuators B,Vol.141, pp.511-517.

Lu H. H., Lin C. Y., Hsiao T. C., Fang Y. Y., Ho K. C., Yang D. F., Lee C. K., Hsu S. M., Lin C. W. (2009). *Electrical Properties of Single and Multiple PEDOT Nanowires for Sensing Nitric Oxide Gas*, Anal. Chim Acta,Vol. 640, pp. 68-74.

Maity A., Biswas M. (2006). *Recent progress in conducting polymer, mixed polymer-inorganic hybrid nanocomposites*. J Ind Eng Chem, Vol.12, pp. 311–351.

Ma X., Zhang X., Li Y., Li G., Wang M., Chen H., Mi Y. (2006). *Preparation of Nano-Structured Polyaniline Composite Film via, Carbon Nanotubes Seeding Approach and its Gas-Response Studies*, Macromol. Mater. Eng. Vol. 291, pp. 75–82.

Mukherjee N., Ahmed S.F., Chattopadhyay K. K.(2009) .Role of solute and solvent on the deposition of ZnO thin films,*Electrochimica Acta*, Vol. 54(16), pp. 4015-4024. References and further reading may be available for this article. To view references and further reading you must purchase this article.

Penza M., Antolini F., Vittori-Antisari M. (2004). *Carbon nanotubes as SAW chemical sensors materials*, Sensors and Actuators B, vol. 100, pp.47–59.

Penza M., Cassano G., Aversa P., Cusano A., Cutolo A., Giordano M., Nicolais L. (2005).*Carbon nanotubes acoustic and optical sensors for volatile organic compound detection*, Nanotechnology, Vol. 16, pp. 2536–2547.

Price T.D., Rahman A.R.A., Bhansali.S, (2008). *Design rule for optimization of microelectrodes used in electric cell-substrate impedance sensing (ECIS), Biosensors and Bioelectronics* , Sensors and Actuators B, Vol.132, pp. 623–630.

Rajesha, Ahujab T., Kumar D. (2009). *Recent progress in the development of nano-structured conducting polymers/nanocomposites for sensor applications,* Sensors and Actuators B, Vol. *136, pp. 275–286.*

Rong J., Oberbeck F. , Wang X ., Li X., Oxsher J., Niu Z. , Wang Q. (2009). *Tobacco mosaic virus templated synthesis of one dimensional inorganic–polymer hybrid fibres,* J Mater Chem, Vol. 19 pp. 2841–2845.

Royer J. E., Lee S., Chen C., Ahn B., Trogler W. C., Kanicki J.D., Kummel. A.C. (2011). Analyte selective response in solution-deposited tetrabenzoporphyrin thin-film field-effect transistor, Sensors and Actuators B (In press).

Santhanam K.S.V., Sangoi R., Fuller L. (2005). *A chemical sensor for chloromethanes using a composite of multiwalled nanotubes with poly(3-methylthiophene),* Sensors and Actuators B, Vol.106, pp. 766–771.

Shen G., Chen P-C., Ryu K., Zhou C. (2009). *Synthesis, Optical, and Mesomorphic Properties of Self-Assembled Organogels Featuring Phenyl Ethynyl Framework With Elaborated Long-Chain Pyridine-2,6-Dicarboxamides (an invited contribution to a special issue of the Gels and Fibrillar Networks: Molecular and Polymer Gels and Materials with Self-Assembled Fibrillar Networks.* J Mater Chem. Vol.19,pp. 828–39.

Sinh N., Ma J., Yeow J.T.W., Nanosci J. (2006). *Carbon Nanotube Based Sensors,* Journal of Nanoscience and Nanotechnology. Vol 6, pp.573–590.

Skotheim T., Reynolds J.(2007) J. *Handbook of conducting polymers.* 3rd ed. Boca Raton: CRC Press.

Someya T., Small J., Kim P., Nuckolls C., Yardley J.T. (2003). Alcohol vapor sensors basedon single-walled carbon nanotube field effect transistors, Nano Lett., Vol. 3,pp.877–881.

Star A., Bradley K., Gabriel J.-C. P, Gruner G. (2004). *Carbon nanotube nanoelectronic devices for chemical detection in liquid hydrocarbons,* Vol.49 (2), pp.887-888.

Thess A., Lee R., Nikolaev P., Dai P., Petit H. J, Robert .J, Xu .C. H, Lee .Y. H, Kim .S. G, Rinzler.A. G, Colbert D. T., Scuseria G. E., Tomanek D., Fischer J. E. , Smalley R. E. (1996) . *Crystalline Ropes of Metallic Carbon Nanotubes.*Science Vol. 273, pp.483-487.

Tong L., Lou J., Gattass R., He S., Chen X., Liu L. , Mazur E . (2005). *Assembly of Silica Nanowires on Silica Aerogels for Microphotonic Devices.* Nano Lett, Vol.5, pp. 259–262.

Trojanowicz M. (2006) *Analytical applications of carbon nanotubes: a review* Trac-Trends in Analytical Chemistry vol.25 (5),pp. 480-489.

Wang B., Zuo X., Wu Y.Q., Chen Z.M., He C.Y., Duan W.B. (2011). *Comparative NH_3-sensing in palladium, nickle and cobalt tetra-(tert-butyl)-5,10,15,20-tetraazaporphyrin spin-coating films,*Sensors and Actuators B: Chemical Vol.152, pp.191–195.

Wu W., LiuZ., Jaureguic L. A., Yu Q., Pillai R., Cao H., Bao J., ChenY. P., Sh.-S Pei. (2010) *.Wafer-scale synthesis of graphene by chemical vapor deposition and its application in hydrogen sensing,* Sensors and Actuators B,Vol.150 , pp.296–300.

Xiao F., Zhao F., Mei D., Mo Z., Zeng B. (2009) *.Nonenzymatic glucose sensor based on ultrasonic-electrodeposition of bimetallic PtM (M = Ru, Pd and Au) nanoparticles on carbon nanotubes–ionic liquid composite film* , Biosens Bioelectron, Vol.24, pp.3481–3486.

Xu T., Zach M. P. , Xiao Z. L., Rosenmann D, Welp U, Kwok W. K., Crabtree G. W. (2005) *.Self-assembled monolayer-enhanced hydrogen sensing with ultrathin palladium films.* Appl Phys Lett, Vol.86,pp. 203104.

Zhao Z., Liu X., Chen W., Li T. (2011). *Carbon nanotubes humidity sensor based on high testing frequencies,* Sensors and Actuators A Vol. 168, pp.10–13.

Zhou X., Park Y., Huang S., Liu J., McEuen P.L. (2005). Band structure, phonon scattering, and the performance limit of single-walled carbon nanotube transistors, Phys.Rev Lett Vol. 95, pp.146805.

Part 3

Electrical Chemical Sensor

10

Polymer Thin Film Chemical Sensors

Renat Salikhov[1] and Aleksey Lachinov[2]
[1]*Bashkir State Pedagogical University,*
[2]*Institute of Molecule and Crystal Physics, URC RAS,*
Russia

1. Introduction

Polymer and other organic thin films and multilayers are important for a wide range of applications, including electronics, optoelectronics and sensors. Most modern chemical sensors use semiconducting and conducting polymers because they offer cheap technology, mechanical and size advantages (Gerard et al., 2002; Hangarter et al., 2010; McQuade et al., 2000). They can form selective layers in which the interaction between the analyte substance and the polymer leads to change of a such physical parameter as conductivity. Also they can be used in devices that form circuit elements such as transistors (Dimitrakopoulos & Malenfant, 2002; Qiu et al., 2009). The widespread literature dealing with various applications of these polymers can be divided into two groups: polymers in electronic devices on the one hand (Angelopoulos, 2001) and polymers in chemical sensors based on various transduction mechanisms on the other (Bailey & Persaud, 2001). The sensor applications take advantage of the physical changes that take place in the polymers when they are exposed to various chemicals. This property has its origin in the molecular and macroscopic structure of polymers. In this paper we shall discuss only sensors based on changing of electronic properties of polymers resulting from their interaction with different chemical agents.

Semiconducting and conducting polymers can be used as the selective layer in sensors or as the transducer itself. For example, change of polymer conductance on exposure to a gas is the sensing mechanism in a chemiresistor. They are relatively easily and inexpensively fabricated. The interaction between organic semiconductor and electrically neutral gas is used as the transduction principle in field-effect transistor (FET) sensors during two last decades (Josowicz & Janata, 1986).

There are many different formats of chemically sensitive FETs both for gas and liquid applications (Gaponik et al., 1997). In FETs the current magnitude flowing through the polymer is modulated by the interaction with the analyte. The response may depend on both the conductivity and the work function of the polymer. It is difficult to separate influence of the various forms of modulation. For example, in (Polk et al., 2002) localized energy states can affect the value of the work function but do not affect the conductivity of the polymer.

FETs, fabricated with single-walled carbon nanotubes, have been found (Collins et al., 2000) to be sensitive to various gases - for example, oxygen, nitrogen dioxide, ammonia, etc. FETs of this type can operate as gas sensors (Qi et al., 2003) on account of their high sensitivity, their fast response time, and their compatibility with dense-array fabrications.

Chemically sensitive capacitors (Sundgren et al., 1992) are closely related to FETs. This is not surprising because the FET is a capacitor often fabricated from the same materials and with the same dimensions. Chemically sensitive capacitors, on the other hand, are excited by an applied a.c. voltage, the voltage-dependent capacitance is measured by an impedance bridge and the analytical information is extracted from the shift of the capacitance–voltage curve. Capacitors are obviously easier to fabricate, but in FETs and Schottky diodes, particular attention must be paid to the quality of the contacts between the polymer and the metal.

Most used polymers are p-type semiconductors having rather large values of work function, so the choice of suitable metals to form Schottky contacts is limited. Sensors based on Schottky diodes are relatively simple to fabricate, but the choice of metals is critical and the supporting electronics and data interpretation are more complicated (Potje-Kamloth, 2001)

Semiconducting polymers are useful as building materials in electronic thin film sensors. Because of their porosity they are easily penetrated by gases that can change their electronic properties. Such interactions are advantageous in sensing for gases. Polymer thin films are highly attractive for fabrication of multi-sensing arrays, which makes sensors based on them very suitable for commercial applications. These arrays involve diverse interactive coatings on multiple sensors with the resulting multivariate data analyzed by pattern-recognition techniques. Array detectors are sometimes referred to as electronic noses because the nose uses multiple receptors whose signals are processed by neuronal pattern-recognition processes (Grate, 2000).

In conclusion, polymers are useful materials for gases sensing for several reasons, including their capability to accumulate atoms and molecules on sensor surfaces by reversible sorption and their selectivity due to their chemical structure, which can be easily varied through synthesis. Polymers can yield sensors with rapid, reversible, and reproducible responses. In addition, sets of different polymers can be collected for use in sensor arrays, providing the variable selectivity across the array surface. A sensing polymer must also have a number of desirable physical properties. For example, solubility in organic solvents may be necessary, or a certain viscosity for using coating processes.

2. Polyarylenephthalides (diphenylphthalide-based polymers)

Over the last decades, a new scientific area has been developed, the electronics of thin films of electrically active polymers. The possibility of creating organic films, including polymer ones, with fantastic physical properties exists. With decreasing the material thickness below a certain value, various factors, previously insignificant, start affecting the appearance of new properties (including electronic ones), the surface effect increases significantly. This manifests itself especially strongly in the case of metal - insulator contacts. In what follows, it was not distinguished the concepts of an insulator and a polymer, because it will be considered polymers having a wide energy gap E_g (E_g > 2 eV). The main paradox of the phenomena described in this review consists in the fact that the efficient charge transfer (electric conductivity) occurs in materials all of whose properties indicate that they are insulators.

In the early 1980s, a new polymer-poly(diphenylenephthalide) (Salazkin et al., 2008)-was synthesized at the Institute of Organic Chemistry (Ufa Research Center, Russian Academy of Sciences), which opened the way to synthesizing a number of new compounds. Its structural formula and optimized three-dimensional geometry are shown in Fig. 1, 9.

As was established later using the method of quantum-chemical calculations (Johansson et al., 1994), a remarkable feature of these polymers is that with a change in the length of one of the bonds in such a molecule to a certain magnitude, a second energetically stable state can arise that is characterized by the existence of electron levels in the region of previously forbidden energies. To perform quantum-chemical calculations of the polymer structure the semi-empirical Austin Model 1 (AM1) method (Dewar et al., 1985) was used to optimize the chemical structure of organic molecules and the pseudopotential method of the valence effective Hamiltonian (VEH) (Bredas et al., 1981) was used with the data obtained in terms of the AM1 method as the initial parameters. The authors of (Johansson et al., 1994) studied properties of an isolated chain, although it is obvious that taking the intermolecular interactions into account may be important for the understanding of the relevant phenomena. In all cases, the geometry was optimized using monomer or dimer units, whereas the electron structure was determined for periodic boundary conditions with a monomer unit. Calculations in terms of the VEH method yield the electron energy band structure.

Figure 1 shows the optimized geometry for three different cases. In a neutral system (Fig. 1 a), the central carbon atom in the backbone chain is quaternary (sp^3 hybridization), and the C_6H_4 group (which must enter the main chain of the polymer) is in the benzenoid state. Benzenoids are monocyclic aromatic compounds with delocalized p electrons, because p bonds in the cyclic group are conjugated. Moreover, the distance from both phenyl groups (in the main chain) to the central quaternary carbon atom turned out to be 1.50 Å. The distance between the central atom and the nearest oxygen atom in the side group (C - O spacing) is equal to 1.47 Å . The torsion angle between the planes of phenyl groups inside the monomer is about 65°, which indicates a weak conjugation degree along the polymer chain.

The calculation started from the initial geometry in which the central carbon atom was in the sp^2 hybrid state, the bonds surrounding the carbon atom lay in the same plane, and the oxygen atom was not bound to the central carbon atom. The procedure of optimizing the molecule geometry was stopped at a local minimum with a slightly increased formation heat compared with that of the neutral ground state. The phenyl groups retained the benzenoid type. The structure around the central carbon atom was contracted, which led to a decrease in the length of the bonds. The distance from the carbon atoms of the phenyl groups to the central carbon atom was now about 1.44 Å, whereas in the geometry of the neutral ground state, it was equal to 1.50 Å. The C - O spacing in this case was 3.10 Å. A separate preliminary calculation of the reaction coordinates for isolated polymer chains showed that the magnitude of the activation barrier is not more than the actual difference in the heats of formation of these two optimum configurations. The local energy minimum is very shallow; therefore, the shape with an open cycle in the neutral molecule is very unstable. However, intermolecular interactions, including those caused by the dipole moment, reduce the attraction between the positive carbon atom and negative carboxyl group and lead to a stabilization of the chemical structure. This apparently indicates that the shape with an open bond can be formed as a result of the action of external forces in the solid state, when the surroundings of the polymer chains are capable of stabilizing the higher energy state.

The addition of an electron and of a compensating ion to any of the three geometric configurations under consideration shows that the structure with an open cycle becomes stable. The energy of stabilization of the monomer is about 113 kcal mol^{-1}. The bond lengths indicate that the phenyl group continues to be of the benzenoid type as before. The largest change is observed in the distances from the phenyl groups to the central carbon atom, which decrease from 1.50 to 1.43 Å. The bond lengths along the polymer chain become almost equal to each other. It can be seen from Fig. 2 c that the C - O bond is broken (the C - O spacing is equal to 3.65 Å). This suggests that the central carbon atom has the sp^2 rather than sp^3 hybridization, as in the neutral state. We note for comparison that the opening of the C - O bond is a well-known phenomenon in the case of phenolphthalein, which has a chemical structure resembling the structure of the monomer of the polymer under consideration.

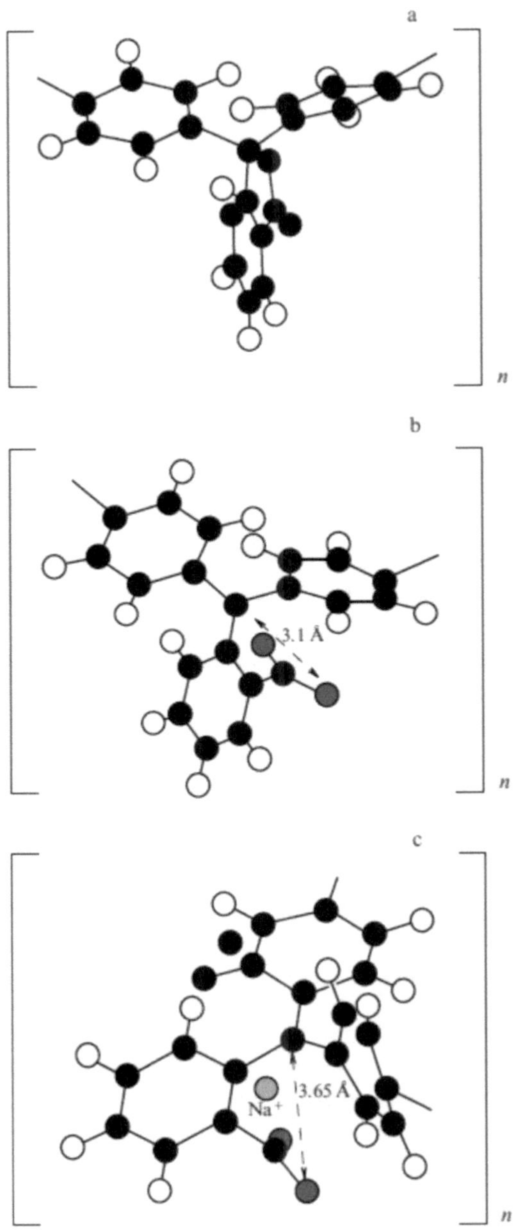

Fig. 1. Optimized geometry for three different cases: (a) a neutral molecule in the ground state; (b) the molecule in the state with a greater total energy, which corresponds to an open lactone cycle and (c) alloying by adding one electron and a counter-ion Na^+ to the molecule

The investigation of geometric optimization with a monomer used as the unit cell shows that two different geometric structures with a relatively small difference in the total energy of the ground states can be realized in polyarylenephthalides. Thus, numerical investigations have shown that the individual molecules of poly(diphenylenephthalide) (PDP) have two energetically stable geometric configurations. The first corresponds to the neutral ground state, which is characterized by the widths of the energy gap between the top occupied and bottom vacant orbitals equal to 4.2 eV. The second configuration is determined by a modified set of lengths of intramolecular bonds and a nonzero spectrum of electron states in the region of forbidden energies. However, the stability of this state is insufficient to ensure its actual existence. But its stability increases (the local minimum becomes relatively deep) if the molecule captures an electron and becomes a negative-charge ion. In this case, the calculations predict the appearance of a new deep electron level in the energy gap.

However, the results of calculations based on the experimental data on measurements of the energy characteristics of the polymer by X-ray photoelectron spectroscopy (XPS) (Zykov et al., 1992) were insufficient to correctly understand the properties of this material.

In the language of the energy band model, this process must lead to the appearance of deep electron levels located, according to the calculations in (Johansson et al., 1994), almost at the Fermi level. It follows from these calculations and experiments that by changing the rate of injection of charge carriers into the polymer film, one can substantially control the electric conductivity of wide-band nonconjugated polymers. The available experimental work (Lachinov et al., 1990; Kornilov & Lachinov, 1997) generally confirmed this conclusion.

Optical properties of poly(diphenylenephthalide) films near the electronic switching threshold have been studied by modulation spectroscopy (Lachinov et al., 2010). The results obtained have been interpreted using quantum chemical calculations of the electron energy characteristics of the polymer molecule. It has been established that the injection of charge carriers into the polymer film and interaction of this excess charge with the macromolecule can lead to the formation of deep trap states near the center of the band gap.

The electroabsorption modulation spectra were obtained for the PDP films at different amplitudes of the modulating voltage. The spectra consist of two bands. At a voltage of 0.5 V across the sample, the maximum of the first band is located in the range 490–515 nm and the maximum of the second band is located in the range 580–610 nm. An increase in the voltage changes the shape of the bands and positions of their maxima. At a voltage of 0.8 V, the intensity of the first band increases and the position of the maximum is shifted toward the long-wave spectral range. In this case, the intensity of the second band slightly decreases and the position of its maximum is shifted toward the long-wave range.

It was assumed that the modulation of absorption is associated with the presence of electronic states in the mobility gap of the polymer (Fig. 2). The same states can participate in the transfer of the charge injected into the polymer from electrodes. According to (Lachinov et al., 2006), the conductivity in a thin polymer film is caused by the appearance of a narrow charge transfer band located near the center of the band gap of the polymer. In this case, the modulation of optical absorption results from the change in the population of deep electronic states by injected charge carriers. The occupation of the states leads to a decrease in the absorption, and their depletion results in an increase in the absorption. It is worth noting that the position of the specific features in the first region of the modulation spectrum coincides well with the position of the emission bands in the photoluminescence and electroluminescence spectra of PDP.

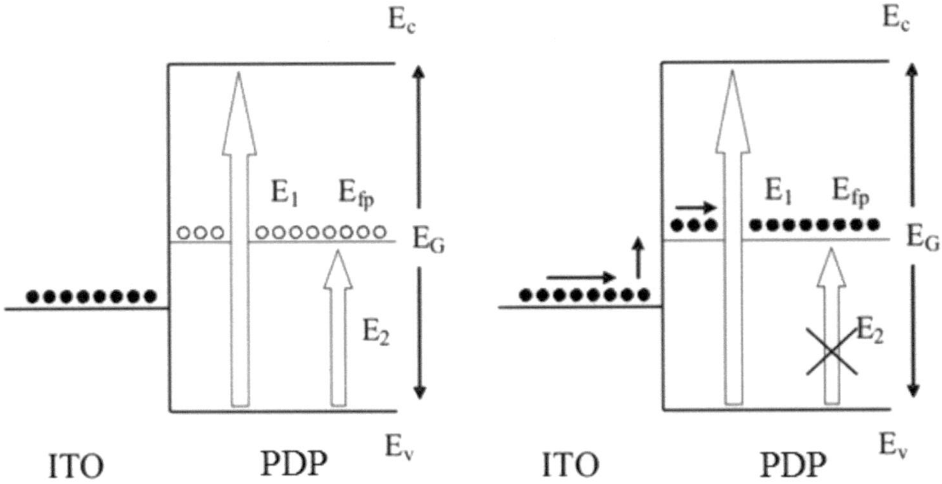

Fig. 2. Band diagrams of the ITO/PDP interface for (a) vacant and (b) occupied electronic states in the polymer film: E_{fp} is the Fermi level of the polymer, E_C is the conduction band, E_V is the valence band, E_1 is the transition from the valence band to the conduction band, and E_2 is the transition from the valence band to the narrow charge transfer band.

Thus, it was established using modulation spectroscopy that the charge injection into the polymer film near the electronic switching threshold leads to the formation of the additional structure in the absorption spectrum in the transparency range of the polymer. The spectral position of the new structure correlates well with the position of maxima in the photoluminescence and electroluminescence emission spectra. The analysis of experimental data with due regard for our quantum chemical modeling of the interaction between the excess electron and the polymer molecule within one monomer unit demonstrated that this interaction can result in the appearance of new deep electronic states. The energy position of these states agrees satisfactorily (with allowance made for a relatively large error in the calculation) with the results of measurements of the electroabsorption.

Complex investigations permitted revealing many new physical phenomena in organic polymeric objects with a nontraditional conductivity. To emphasize the unusual character of such materials, it is convenient to call them electroactive, in contrast to well-known electroconducting materials, which are mainly conjugated polymers. This means that most electroconducting polymers have double bonds in the molecule structure, whose reactivity (and, consequently, energy) depends on the number of such bonds (i.e., these bonds affect one another), whereas in polymers based on diphenylphthalide, the active double bonds of the different parts of the molecule are independent of each other.

The transition from a dielectric state to a high conductive state (HCS), induced by abnormal low external influences is observed in thin electroactive polymer (polyheteroarylines) films. The feature of this phenomenon is that the temperature dependence of the HCS conductivity is of the metal type and the conductivity can be extremely high $> 10^5$ $(\Omega{\cdot}sm)^{-1}$. Also polymer films in the HCS possess a high anisotropy of conductivity. In the HCS the polymer film reveals a domain pattern: insulating matrix surrounds conductive domains. The domain diameter ranges from 25 to 1000 nm (depending on the polymer film thickness). The main effects of electron switching were considered in sufficient detail in (Lachinov et al., 2006).

The unusual properties of the thin polyheteroarylines films open broad prospects for their application in variety of industrial and technological branches. Mechanical sensors can be based on the pressure dependence of the polymer film conductivity. In switches, keyboards and digitizers the high sensitivity of the polymer film to uniaxial pressure can be used. The effect of switching from the metallic to the insulating state when the current exceeds a critical value can be applied for protecting electric circuits from current overload. Since a working element of such a device is a single conducting domain of ≈ 1 μm diameter, it can be formed as part of an integrated circuit. The polymer compound can be utilized as glue with anisotropic conductivity for surface assembly instead of soldering in the electronics industry. The effect of bistable switching in an electric field can be used for producing devices similar to ovonic memory switches. Temperature sensors can be based on the effect of thermostimulating switching. Pristine polymer film in the HCS can be used as a substitute for precious metals to protect contacts in various switches.

3. Charge transport in poly(diphenylenephthalide) thin films

Currently there is great interest in the use of polymers for the fabrication of various electronic devices (Halik et al., 2004; Salleo et al., 2004) due to the significant processing, mechanical, and size advantages. As compared with inorganic materials, polymers exhibit a greater variety of properties related to charge transport mechanisms offering more ways to improving performance. Most devices use silicon/polymer/metal structure (Musa & Eccleston, 1999), which also suits the goals of our study. Keeping in mind possible applications we chose Si/PDP/Cu "sandwich" type structure to carry out current-voltage (*I-V*) characteristic measurements in a range of temperatures to identify the predominant current transfer mechanisms and further elucidate the electronic structure of the polymer under investigation (Salikhov et al., 2007a, 2007b).

Organic-inorganic multilayer structures of the silicon/polymer/metal type have been fabricated. The current-voltage characteristics of these structures versus temperature are measured. The obtained results are treated within different models of charge transport such as Richardson-Schottky thermionic emission, Fowler-Nordheim tunneling, and Pool-Frenkel and hopping mechanisms. The conclusions about the mechanisms of charge transport at different electric fields and temperatures have been made. The Richardson-Schottky model is in good quantitative agreement with the experimental data in the case of the metal-polymer interface with the energy barriers of 0.26 - 0.10 eV. The charge transport in the polymer is explained by the hopping mechanism near the Fermi level with the activation energies being of 0.01- 0.04 eV.

I-V characteristics of the three-layer Cu/PDP/ *p*–Si structure have been measured over a range of temperatures. The data obtained were analyzed by fitting theoretically predicted dependencies of conductance on the electric field to the experimental ones. Relevance of the corresponding model was then assessed by comparison of the fitting parameters with the measured activation energies and the other polymer properties. It was concluded that the dominant charge injection mechanism, at least at higher temperatures, is the thermionic emission. Also, it was shown that the tunneling through the contact-polymer interface gives no significant contribution to the charge injection into the film. It was also concluded that the charge transport through the film thickness may be satisfactorily explained by the hopping of carriers between localized states in a narrow energy band near the Fermi level.

In a series of works, Bassler and his colleagues (Bassler, 1993; Richert et al., 1989) developed a detailed model of charge carrier transfer. The organic matrix with embedded transfer centers is simulated by a cubic lattice with a lattice constant *a* in which each site is occupied by a transfer center. The position of the energy level *E* involved in the charge transfer for each center is an independent random quantity due to the stochastic and uncorrelated influence of the environment. The corresponding density of states is described by a Gaussian distribution,

$$\rho(E) = (2\pi\sigma^2)^{-1/2} \exp(-E^2 / 2\sigma^2). \tag{1}$$

This choice of the density of states $\rho(E)$ is additionally confirmed by the Gaussian shape of the absorption and fluorescence bands experimentally observed for many polymers. The specific calculations within the described model were performed using only Monte Carlo numerical simulation. It was revealed that the field dependence of the mobility is an S-shaped curve. The central portion of this curve is approximately described by a linear dependence of the quantity $\ln\mu$ on $F^{1/2}$. The results of these calculations made it possible to propose the model of hopping transport over centers with a Gaussian distribution of energy levels. Analysis of the data obtained in our work can affirm the aforementioned hopping model of charge transfer in the temperature range preceding the transition to the high conductivity state (Salikhov et al., 2007c, 2007d).

Considerable attention has been paid recently to analysis of the properties of the contact between two organic materials. This is primarily due to the need for matching the energy parameters of individual contacting layers in multilayer systems of the type of organic electroluminescent diodes. Transport properties of a multilayer system containing such an interface are considered as a rule in the direction perpendicular to the interfacial plane, which can be explained by the type of charge transport in an external electric field. However, it should be noted, that there have been problems associated with the inability to obtain sufficient mobility of charge carriers necessary for practical applications.

4. Conducting polymer/polymer interface

The properties of a near-surface region in films are usually different from those in the bulk. This difference is related to several factors, including the excess free Gibbs energy of the surface and a decrease in the entropy. The mutual influence of contacting layers takes place at their interface (boundary) and involves both the structure of a transition layer and its electron properties. Changes in the electron properties can be traced by measuring the transport characteristics of the resulting two-dimensional (2D) structures. However, the study of the electron transport properties of boundary layers is frequently complicated by a relatively high intrinsic conductivity of the initial polymers, which masks the conductivity of the boundary layer. Therefore, for the experimental investigation of such systems, it is expedient to select polymers with minimum intrinsic conductivity (e.g., dielectrics), so as to minimize the current component passing through the volume of polymers and study a change in the conductivity along the contacting layers.

Recently it was reported about the interface between two insulators $LaAlO_3$ and $SrTiO_3$ with anomalous electronic properties, in particular, the charge carrier mobility μ = 10^4 $cm^2V^{-1}s^{-1}$, was investigated (Ohtomo & Hwang, 2004). Later on (Reyren et al., 2007), superconductivity and giant magnetoresistance were revealed in such structures. These phenomena were explained in the framework of the hypothesis that a two-dimensional electron gas is formed in the transient layer between two insulators. The mechanism of the formation of this electron gas is not clearly understood, but active studies made it possible to reveal many interesting properties of this interface including superconductivity at 200 nK, magnetoresistance, ferromagnetic ordering of the electrons, field control, and many other properties inherent in a two-dimensional electron gas. In 2008, similar results were obtained in (Alves et al., 2008), where the transient layer was formed at the interface between two organic insulators -tetrathiafulvalene (TTF) and 7,7,8,8, tetracyanoquinodimethane (TCNQ). The mobility in these structures was about a few $cm^2V^{-1}s^{-1}$, and the metal conductivity was revealed. At the same time anomalously high conductivity at the interface between two identical polymer layers of poly(difenylenephthalide) has been demonstrated (Salikhov et al. 2008).

This paper presents the results of further investigation of the charge transport along the boundary of two organic layers of PDP (Gadiev et al., 2010; Salikhov et al., 2009). It was expected that the bending of energy levels in this region must lead to increased local charge concentration and, hence, modify the transport properties of the 2D structure under consideration. In particular, the electric conductivity of the interface layer must be several orders of magnitude larger than the bulk one. In this connection, it seems important to study the electronic properties of the interface between two dielectric polymer films. The objective of this work was to form the polymer–polymer interface and to investigate its electrophysical properties including the sensory ones.

We used poly(diphenilenephthalide)s of the polyheteroarylene polymer class as the objects of investigation. These polymers are highly soluble in organic solvents and possess high film-forming properties. In our experiments, polymer films were obtained by spin coating of the polymer solution in cyclohexanon of various concentrations. The film thickness varied from 30 nm to 1 μm. Metallic copper electrodes were prepared by vacuum thermal-diffusion sputtering. The current–voltage (*I*–*V*) characteristics of the samples were measured using the standard technique. Temperature measurements were taken in a range of 90–300 K using a cryostat based on a Dewar vessel. The system under investigation was placed in the Dewar vessel, and the temperature was varied by changing the position of the sample relative to the nitrogen surface. Temperature was monitored with the help of a thermocouple.

The choice of the polymer material was dictated by the following considerations. Analysis of electron transport properties of the boundary layers proper is complicated, since conducting polymer materials possess a high intrinsic conductivity, which may shunt the conduction of the boundary layer. In this connection, measuring conditions must be ensured such that the conductivity along the interface being studied is higher than the bulk conductivity of the material. Consequently, for our experiments we chose polymers with a low intrinsic conductivity. The optical band gap of the polymers used is 3.9–4.2 eV and the electron work function is approximately 4 eV. Obviously, such polymers are poor conductors in the initial state. For this reason, any insignificant improvement of their transport properties can easily be monitored.

The experimental samples were the multilayer structures presented schematically in Fig. 3a. The sample was prepared by consecutive operations as follows. The polymer film was deposited on the surface of the purified polished glass substrate by spin coating from the PDP solution. The solvent residuals were removed from the film by successive drying first in air at room temperature for 30 min and then at 150°C for 40 min. Then, the metal electrodes were deposited on the polymer film surface by vacuum thermal diffusion sputtering through the shadow mask. The final stage was the deposition of the second polymer layer adhering the entire sequence of technological operations described above. Each layer was about 300 nm thick. The thickness of polymer films and the state of the surface were monitored by a SMM-2000T atomic force microscope. The electrodes were made of copper. The distance between the electrodes was varied from 30 to 60 μm.

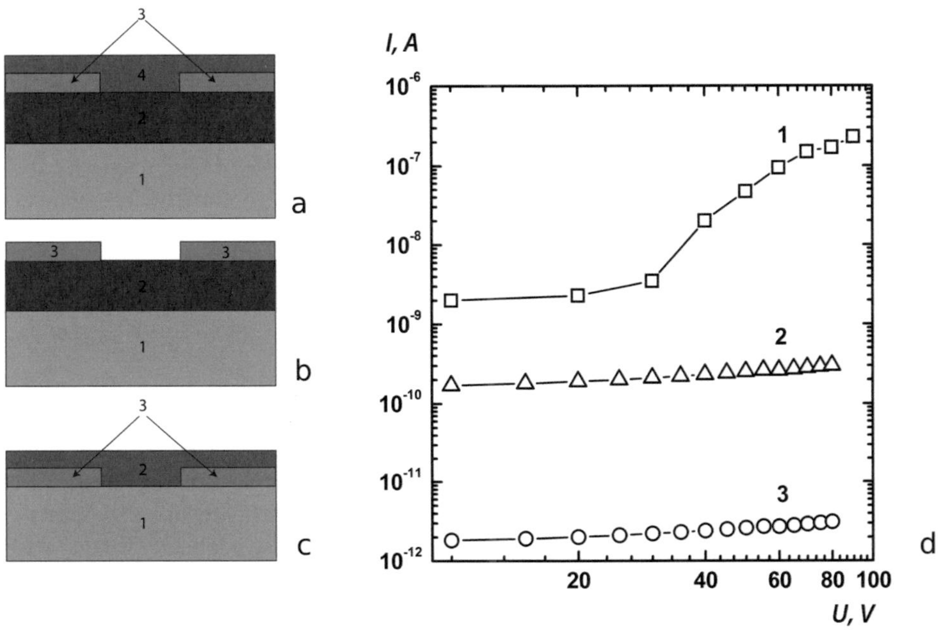

Fig. 3. Structures of the experimental samples with (a) two polymer layers, (b) one polymer layer and the electrodes on the polymer surface, (c) one polymer layer and the electrodes under the polymer film, where 1- substrate, 2- polymer layer, 3- metal electrodes, 4- polymer layer, and (d) current-voltage characteristics for samples: (1) - a, (2) - b, and (3) - c

In the preparation of the sample, the resistivity was controlled by the two-probe method. As soon as the electrodes were deposited on the first film surface (see Fig. 3b), the resistivity was $\rho \sim 10^7$ Ω·cm (Fig. 3d, curve 2) and decreased down to $\rho \sim 10^3$ Ω·cm when the second layer was deposited (see Fig. 3d, curve *1*). To refine the possible contribution of the charge carrier transport along the polymer–substrate interface, the samples with the electrode on the glass substrate surface were additionally prepared (see Fig. 3c). In this case, the resistivity was $\rho \sim 10^8$ Ω·cm (see Fig. 3d, curve *3*). An analysis of the current-voltage characteristic obtained in the measurement of the charge carrier transport along the polymer–polymer interface (see Fig. 3d, curve *1*) demonstrated that it is well approximated by the function of the form $I = kU^n$. Such dependences can be interpreted in the injection model. This conclusion does not contradict the results of the earlier studies of the electron transport properties of the metal–poly(diphenylphthalide)–metal system [10].

Two regions with different exponents n are clearly seen on the current–voltage characteristic: n =1 at low voltages, and the current–voltage characteristic at higher voltages is superlinear with $n \sim 2$. The voltage at which the transition from the linear to the superlinear region occurs corresponds to the situation when the concentration of the thermally generated equilibrium free charges becomes comparable to that of the injected charges (Bunakov et al., 2003). This balance of concentrations allows us to estimate the electron mobility μ,

$$\mu = JL^3 / \theta\varepsilon\varepsilon_0 U^2. \quad (2)$$

The current density J was estimated in terms of the current and the surface area through which the current was flowing, and which, in turn, was calculated in terms of the electrode width and the transport layer thickness; θ is defined as the ratio of the maximum and minimum currents of the limiting trap filling region. According to the AFM recording (see Fig. 4), the transport layer thickness was set equal to 10 nm and U is the voltage at the point of the transition from the linear to the square law. The charge carrier mobility calculated by formula (2) was $3.76 \cdot 10^{-2}$ cm^{2}V^{-1}s^{-1}. It is very high for such materials and is almost five orders of magnitude higher than the bulk mobility $\mu \sim 10^{-6}$ cm^{2}V^{-1}s^{-1} (Lachinov et al., 2004) measured by the time-of-light method. The value obtained is close to the maximum charge carrier mobilities in organic materials, in particular, in organometallic compounds such as phthalocyanines ($\mu \sim 0.4 \div 1$ cm^{2}V^{-1}s^{-1}).

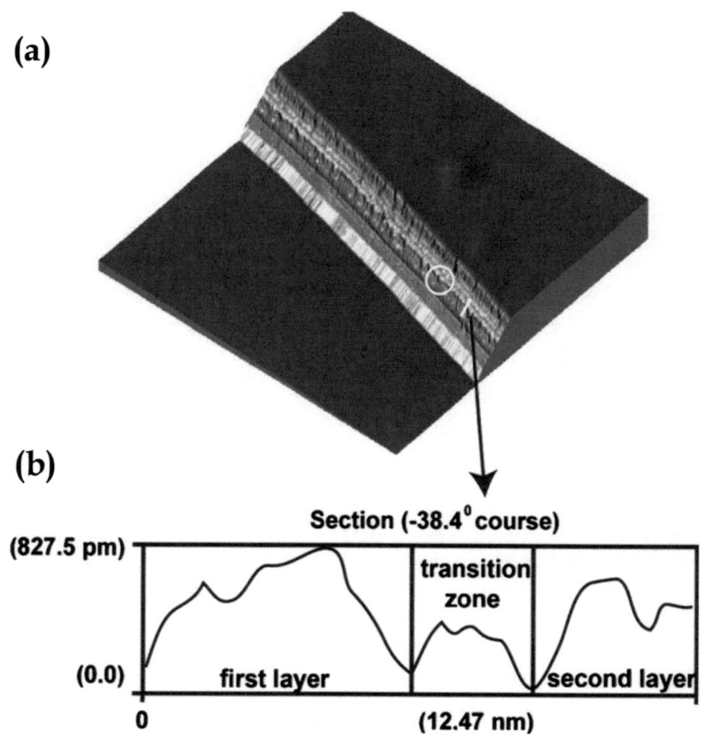

Fig. 4. The interfaces between two polymer layers (atomic-force image). a) 3D image of the two-layer polymer film, (polymer/polymer interface marked by a white circle). b)

Transition zone profile between two polymer films (the same area at the 3D image marked by a white line).

It is well known that the bulk conductivity of the polymers studied here is of the hole type. At the same time, the formation of the highly conducting state is associated with electron conduction. The question arises about the type of charge carriers ensuring an elevated conduction of the boundary transport layer. The answer to this question could be obtained from analysis of transport characteristics of transistor structures. Thus, we consider the results of measurements of the *I–V* curves for samples prepared on silicon plates with an arrangement of electrodes similar to the configuration of a field transistor (see Fig. 5a). Samples with polymer films on silicon plates with an oxide layer were prepared using an analogous technique. The substrates for the transistor structures were *n*++-Si plates with an ohmic electrode on one side of the plate and a SiO_2 layer (100 nm in thickness) formed by thermal oxidation on its other side.

We measured the dependence of the current between the drain and the source (lateral electrodes) on the potential supplied to the gate (lower ohmic electrode on the silicon plate) (Fig. 5b). The measured dependence is referred to as the transfer characteristic, which shows that the current increases by about five orders of magnitude as compared to the structure of a one-layer polymer film (in this case, the currents are small and amount to less than 10 fA). The shape of the resultant curve indicates that it is electrons which are carriers responsible for charge transport. Indeed, an increase in current is observed for positive values of potential across the shutter.

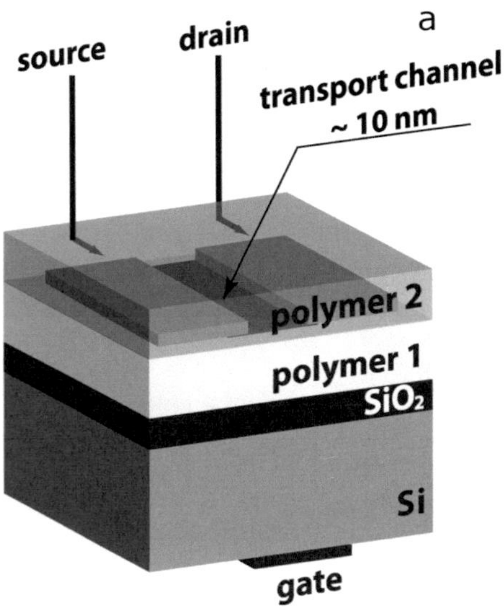

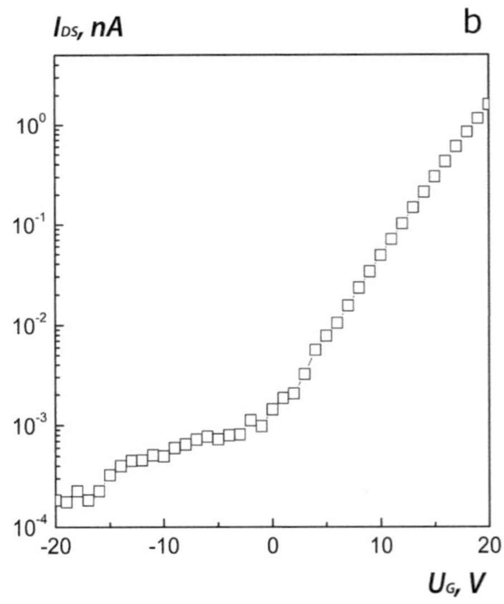

Fig. 5. Organic field-effect transistor based on two polymer films interface as a transport layer (a) and its transfer characteristic (b).

Furthermore, it was also found that it is possible to control the conductivity along the interface by the external electric field using the field-effect transistor circuit. The current–voltage characteristics obtained in this experimental configuration allowed for additional estimates of the charge carrier mobility (Horowitz, 1998). The calculation was performed by the formula:

$$I_d = \frac{Z}{L}\mu C_i (V_g - V_t) V_d, \tag{3}$$

where I_d is the drain–source current; Z and L are the width and length of the conducting channel, respectively; μ is the mobility; C_i is the dielectric layer capacitance; and V_g, V_d, and V_t are the gate, source, and threshold voltages, respectively. The mobility calculated by formula (3) was $\mu = 4 \cdot 10^{-2}$ cm^2 V^{-1} s^{-1}, and this agrees well with the earlier obtained value.

The temperature dependences of the conductivity measured for two different experimental configurations. Using the two-probe method we obtain the conductivity increases with the temperature. This behavior of the current-voltage characteristic can be attributed to the charge carrier injection from the metal electrodes into the polymer film limited by Schottky barrier. This conclusion agrees with the above analysis of the current–voltage characteristics (Salikhov et al., 2008). This means that the two-electrode method does not allow one to estimate the true electric conductivity along the interface between two polymer films. In view of this circumstance, the four-probe measurement of the conductivity was held. As the temperature increases, the electric conductivity decreases, possibly indicating its metal type. We note that a similar difference in the behavior of the conductivity temperature curves measured by the two- and four-probe methods was observed in the investigation of the properties of the quasi-two-dimensional electron gas formed along the $LaAlO_3$-$SrTiO_3$ and TTF–TCNQ interfaces (Kirtley & Mannhart, 2008).

5. Sensory properties

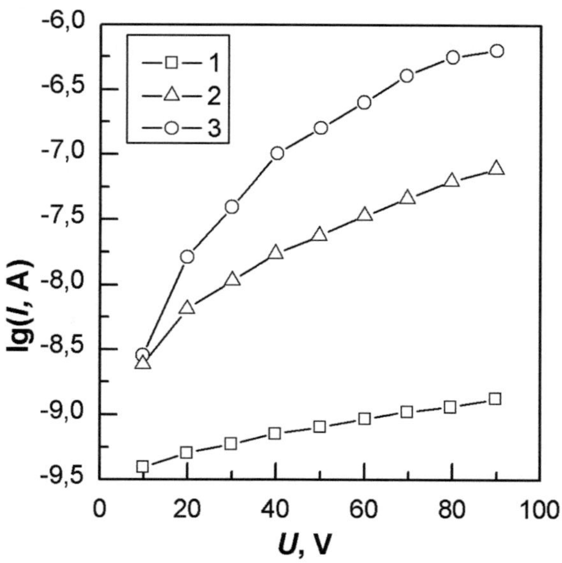

Fig. 6. Current–voltage characteristics of structures with aluminum electrodes, measured under different conditions: *1*) with air relative humidity of 20%; 2) in ethanol vapor; *3*) with relative humidity of 80%

The following are the sensory properties of samples with interface between two dielectric polymer films (Salikhov et al, 2009). In order to explain what part of a specimen is sensitive to a change in humidity (interface layer at the boundary of two polymer films or bulk material), the current-voltage (*I-V*) characteristics were measured for specimens with a single film and lower arrangement of paired electrodes. In this case, humidity does not affect conductivity. As the results of measurements have shown, the greatest change in conductivity occurs with arrangement of electrodes between polymer layers.

I-V characteristics are presented in Fig. 6 for a specimen with aluminum electrodes measured under different conditions. Curve *1* corresponds to measurement in an open atmosphere with 20% relative humidity. With an increase in relative humidity to 80% (curve *3*), conductivity increases by three orders of magnitude. Curve *2* was obtained in ethanol C_2H_5OH vapor, whose molecule also contains a hydroxyl group. In this case, conductivity differs by two orders of magnitude from the original.

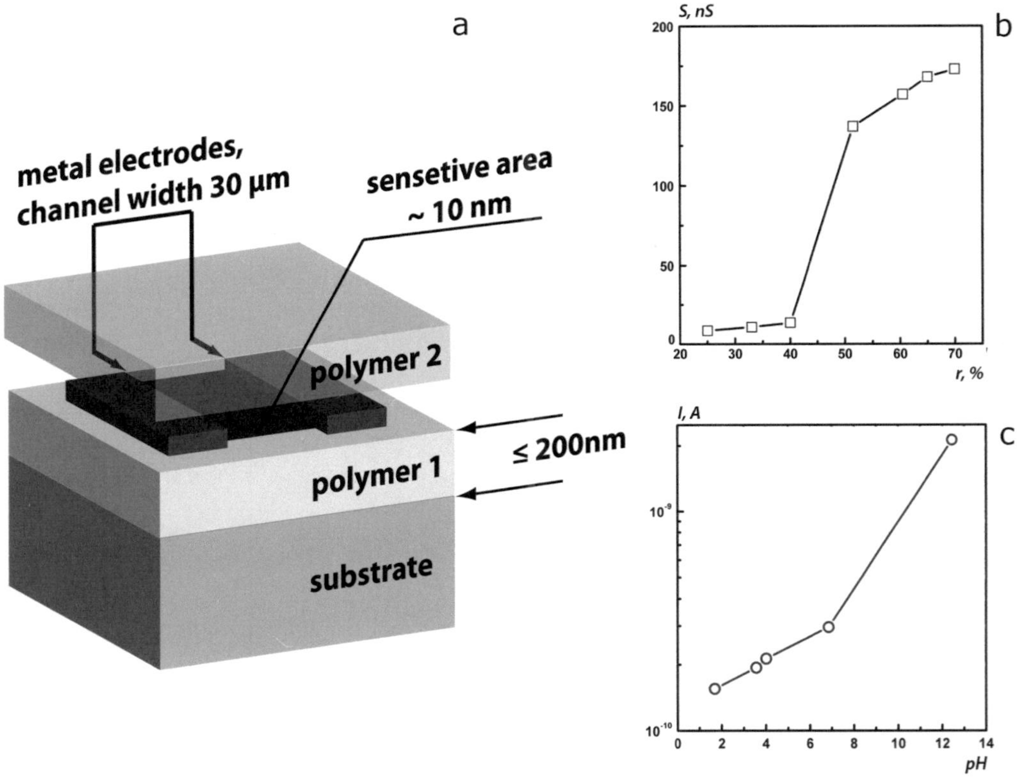

Fig. 7. Sensor on the base of two polymer films interface (a) and its characteristics: dependence of conductivity on air relative humidity *r* (b), dependence of current with a fixed voltage on pH index (c)

The next stage of the work (Salikhov et al., 2011) was creation of an acting model of a sensor for air relative humidity (see Fig. 7a). The current-voltage characteristics of the structures obtained were measured with different amounts of atmospheric humidity. A dependence of conductivity on relative humidity *r* was constructed on the basis of measured curves, and it is presented in Fig. 7b. As follows from Fig. 7b, an increase in humidity affects the increase in specimen conductivity. It is possible to separate three sections on the curve in different ranges of humidity values. With relatively low humidity (25–40%), there was little increase in conductivity. The next range (40–50%) corresponds to a sharp increase in conductivity. With a further increase in humidity (50–70%), there is an insignificant increase in conductivity with subsequent saturation. Rapid response sensors for relative humidity were studied. It follows from experiments that the response time is not more than 5 sec (see Fig. 8). Most likely, this time corresponds to diffusion levelling with rapid opening of a cover beneath which there is a sensor with increased humidity of 80%, with 20% humidity for the surrounding air. Use as dopants of substances whose molecules do not contain a hydroxyl group has demonstrated that in this case there is no change in conductivity. Thereby the selectivity has been demonstrated for the interface between polymer films to substances with hydroxyl groups. By using this unique feature of a transport layer, the experimental specimens obtained may be used as multiphase sensors for determining the hydrogen index (pH-meter), and this has been done by experiment. Compositions with a known pH value were added successively to a specimen. Then the current voltage characteristics were measured. The results obtained are shown in Fig. 7c in the form of a dependence of the current with the same fixed voltage on pH value. It follows from Fig. 3c that there is an increase in conductivity with a reduction in acidity. This confirms the assumption made about the elective sensitivity towards the hydroxyl group. The use of new polymer materials opens extensive possibilities for creating chemical nano-sized sensors for different substances. The selectivity is achieved by adjusting the upper polymer layer.

To explain the high conductivity over the interface between organic materials the various physical models were proposed. In (Nikitenko et al., 2010) the conductivity is due to the rather high surface density of geminate pairs formed at the interface. Conditions are established wherein the transitions of a significant portion of charge carriers between molecules do not require thermal activation or tunneling.

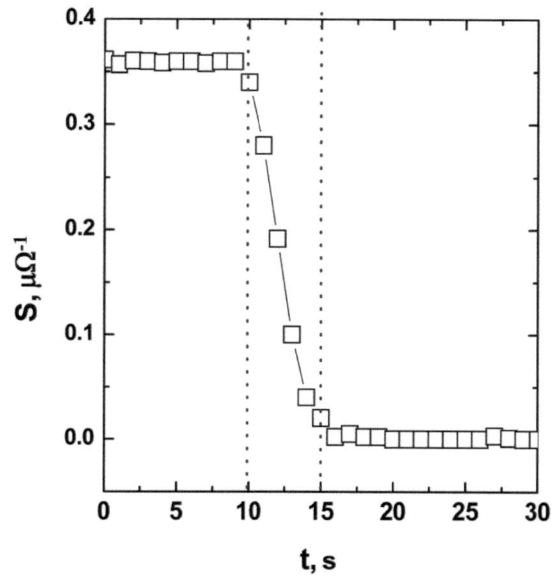

Fig. 8. Change in conductivity with time with a sharp reduction in humidity

We can assume that the structure of the interface between two polymer layers differs from the equilibrium structure in the bulk of the sample. The experiments described in (Schefefold et al., 1996) show that polymers with side functional groups must orient themselves toward the interface between these groups. The latter may lead to selective ordering of these groups at the surface and to an increase in their local concentration due to close stacking. In the polymers studied here, the phthalide group that can be appropriately oriented relative to the film surface plays the role of the side group (see Fig. 9). In principle, this conclusion does not contradict the assumption of the possibility of formation of localized levels of the transient layer with the participation of side phthalide groups of polymer molecules, which are electron traps according to (Johansson et al., 1994)

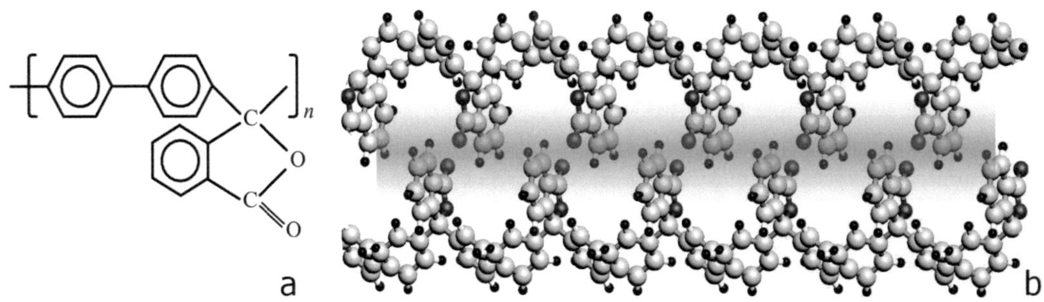

Fig. 9. Structural formula of poly(diphenylenephthalide)'s monomer (a) and three-dimensional model image of two polymer films interface (b)

In the framework of one of the models previously proposed in (Nakagawa, et al., 2006; Pauli & Willmott, 2008; Popović et al., 2008), it was assumed that to form the two- dimensional electron gas at the interface between two insulators, dipole groups should be present on at least one of the surfaces in contact. From this point of view, it seems possible to use materials having functional groups with a high dipole moment. Polymers can be taken as such materials. It is well known that the side functional groups of polymers are predominantly oriented along the normal to the polymer–air interface (Fischer et al., 1998). A similar orientation should occur when one polymer film is formed on the surface of the other. As a result, a nanolayer can appear in the near interface region, which is formed by the side groups of polymer molecules.

6. Conclusion

The sharp increase (by three to five orders of magnitude) in the conductivity of the samples after the deposition of the second polymer layer can be explained as follows. A thin transition layer with a structure differing from the bulk structure is apparently formed at the interface between the two polymers. We can assume that the interface between polymer films facilitates preferred antiparallel orientation of phthalide side fragments with an appreciable dipole moment (Lachinov et al., 2000). Such an orientation increases the local "2D" concentration of electron traps (phthalide side groups), which does not contradict the results obtained in (Schefefold et al., 1996). The trapping of the excess charge injected from the electrodes to phthalide groups of poly(diphenylenephthalide) leads to the emergence of deep localized states in the polymer band gap (Duke & Fabish, 1976; Lachinov et al., 2000). These states ensure a higher conductivity level and modifies the shape of *I–V* curves upon the passage of current. Thus, our studies lead to the conclusion on the possibility of formation of a transport layer at the interface of two polymer films, which possesses an elevated conductivity.

To summarize, we have shown that the interface between two dielectric polymer films has an anomalously high conductivity and charge carrier mobility. The investigation data such as the presence of a pronounced interface between the polymer films, the metal type of the conductivity, and the presence of the field effect suggest the presence of the quasi-two-dimensional electron gas at the interface between the polymer insulators. However, this statement requires additional investigations, first of all, at low temperatures. It is interesting that the values of the charge carrier parameters at room temperature were relatively lower than the parameters of the interfaces prepared based on inorganic materials (Ohtomo & Hwang, 2004; Reyren et al., 2007). Probably, this is attributed to the high defect content of the polymer–polymer interface due to the imperfect technology used in this interface preparation. Although the technology should be improved, the advantages of the polymer organic interfaces in the manufacture of the electronic materials with unique electronic properties are apparent. These materials are very promising for the production of new electronic devices, including chemical sensors for environmental monitoring, alarm systems and automation.

7. Acknowledgment

This work was supported by the Russian Foundation for Basic Research (projects no. 09 03 00616a and 11 02 01445a).

8. References

Alves H.; Molinari A.S.; Xie H. & Morpurgo A.F. (2008). Metallic conduction at organic charge-transfer interfaces. *Nature Mater.* Vol. 7, pp. 574-580 (July 2008), ISSN 1476-1122

Angelopoulos, M. (2001). Conducting polymers in microelectronics. *IBM J. Res.Dev.*, Vol. 45, No. 1, (January 2001), pp. 57–75, ISSN 0018-8646

Bailey, R. A. & Persaud, K. C. (2000). Sensing volatile chemicals using conducting polymer arrays, In: *Polymer Sensors and Actuators,* Osada,Y. & De Rossi, D. E., (Eds), pp. 149–181, Springer, Berlin.

Bassler H. (1993). Charge Transport in Disordered Organic Photoconductors a Monte Carlo Simulation Study. *Phys. Status Solidi* B. Vol. 175, Is. 1, pp. 15-56 (January 1993), ISSN 1521-3951

Brédas J. L.; Chance R. R.; Silbey R.; Nicolas G. & Durand Ph. (1981). A nonempirical effective Hamiltonian technique for polymers: Application to polyacetylene and polydiacetylene. *J. Chem. Phys.* Vol. 75, Is.1, p. 255 (13 pages) (July 1981), ISSN 0021-9606

Bunakov, A. A.; Lachinov, A.N. & Salikhov, R. B. (2003). Current–voltage characteristics of poly(diphenylenephthalide) thin films. *Techn. Phys.* Vol. 48, No. 5, pp. 626-630 (May 2009), ISSN 1063-7842

Collins P. G.; Bradley K.; Ishigami M. & Zettl A. (2000). Extreme Oxygen Sensitivity of Electronic Properties of Carbon Nanotubes. *Science* Vol. 287, No. 5459, pp. 1801-1804 (March 2000), ISSN 0036-8075

Dewar M.J.S.; Zoebisch E.G.; Healy E.F. & Stewart J.J.P. (1985). Development and use of quantum mechanical molecular models. 76. AM1: a new general purpose quantum mechanical molecular model. J. Am. Chem. Soc. Vol. 107, Is. 13, pp. 3902-3909 (June 1985), ISSN 0002-7863

Dimitrakopoulos, C.D. & Malenfant, P. R. L. (2002). Organic thin film transistors for large area electronics. *Adv. Mater.*, Vol. 14, Is. 2, pp. 99–117 (January 2002), ISSN 0935-9648

Duke, C. B., & Fabish, T. J. (1976). Charge-Induced Relaxation in Polymers. Phys. Rev. Lett. Vol. 37, Is. 16, pp. 1075-1078 (October 1976), ISSN 0031-9007

Fischer, D. A. ; Mitchell, G. E. ; Yeh, A. T. & Gland J.L. (1998). Functional group orientation in surface and bulk polystyrene studied by ultra soft X-ray absorption spectroscopy. *Appl. Surface Sci.*, Vol. 133, Is. 1-2, pp. 58-64 (May 1998), ISSN 0169-4332

Gadiev R.M.; Lachinov, A. N.; Kornilov V.M.; Salikhov, R. B.; Rakhmeev, R. G. & Yusupov A.R.(2010). Anomalously high conductivity along the interface of two dielectric polymers. *JETP Lett.* Vol. 90, No. 11, pp. 726–730 (February 2010), ISSN 0021-3640

Gaponik, N. P.; Shchukin, D. G.; Kulak, A. I. & Sviridov, D. V. (1997). A polyaniline based microelectrochemical transistor with an electrocatalytic gate. *Mendeleev Commun.* Vol. 7, Is. 2, pp. 70–71, ISSN 0959-9436

Gerard, M.; Chaubey, A. & Malhotra, B. D. (2002). Application of conducting polymers to biosensors. *Biosens. Bioelectron.*, Vol. 17, Is. 5, (May 2002), pp. 345–359, ISSN 0956-5663

Grate, J. W. (2000). Acoustic wave microsensor arrays for vapor sensing. *Chem. Rev.* Vol. 100, Is. 7, pp. 2627–2648 (July 2000), ISSN 0009-2665

Halik M.; Klauk H.; Zschieschang U.; Schmid G.; Dehm C.; Schütz M.; Maisch S.; Effenberger F.;, Brunnbauer M. & Stellacci F. (2004). Low-voltage organic transistors with an amorphous molecular gate dielectric. Nature. Vol. 431, pp. 963-966 (October 2004), ISSN 0028-0836

Hangarter, C.M.; Bangar, M.; Mulchandani, A. & Myung N.V. (2010). Conducting polymer nanowires for chemiresistive and FET-based bio/chemical sensors. *J. Mater. Chem.*, Vol. 20, Is. 16, pp. 3131-3140 (February 2010), ISSN 0959-9428

Horowitz G. (1998). Organic fielf-effect transistors. *Adv. Mater.* Vol. 10, Is. 5, pp. 365-377 (March 1998), ISSN 0935-9648

Johansson N.; Lachinov A.; Stafström S.; Kugler T. & Salaneck W.R. (1994). A theoretical study of the chemical structure of the non-conjugated polymer poly(3,3'-phthalidylidene-4,4'-biphenylene). *Synth. Met.* Vol. 67, Is. 1-3, pp. 319-322 (November 1994), ISSN 0379-6779

Josowicz, M. & Janata, J. (1986). Suspended gate field-effect transistor modified with polypyrrole as alcohol sensor. *Anal. Chem.* Vol. 58, Is. 3, (March 1986), pp. 514–517, ISSN 0003-2700

Kirtley, J. R. & Mannhart, J. (2008). Organic electronics: When TTF met TCNQ. Nature Mater. Vol. 7, pp. 520-521 (July 2008), ISSN 1476-1122

Kornilov V. M. & Lachinov A. N. (1997). Electrical conductivity in the metal-polymer-metal system: The role of boundary conditions. JETP Lett. Vol. 52, No. 4, pp. 833-841 (April 1997), ISSN 0021-3640

Lachinov A. N.; Zherebov A. Yu. & Kornilov V. M. (1990). Anomalous electron instability of polymers due to uniaxial pressure. JETP *Lett.* Vol. 52, No. 2, pp. 103-106 (July 1990), ISSN 0021-3640

Lachinov, A. N.; Zagurenko, T. G.; Kornilov, V. M.; Fokin A.I.; Aleksandrov I.V. & Valiev R.Z. (2000). Charge transfer in a metal-polymer-nanocrystalline metal system. Phys. Solid State. Vol. 42, No.10, pp. 1935-1941 (October 2000), ISSN 1063-7834

Lachinov, A. N. ; Salikhov, R. B. ; Bunakov A.A. & Tameev, A. R. (2004). Charge carriers generation in thin polymer films by weak external influences. *Non-Linear Optics, Quantum Optics.* Vol. 32, No. 1-3, pp. 13-20, ISSN 1543-0537

Lachinov A. N. ; Kornilov V. M.; Zagurenko T.G. & Zherebov A. Yu. (2006). On the high conductivity of nonconjugated polymers. *JETP.* Vol. 102, No. 4, pp. 640-645 (April 2006), ISSN 1063-7761

Lachinov A. N.; Zhdanov E.R.; Rakhmeev R.G.; Salikhov R.B. & Antipin V.A. (2010). Modulation of optical absorption in poly(diphenylenephthalide) films near the switching threshold. *Phys. Solid State.* Vol. 52, No. 1, pp. 195–200 (January 2010), ISSN 1063-7834

McQuade, D. T.; Pullen, A. E. & Swager, T. M. (2000). Conjugated polymer-based chemical sensors. *Chem. Rev.*, Vol. 100, No. 7, (July 2000), pp. 2537–2574, ISSN 0009-2665

Musa I. & Eccleston W. (1999). Electrical properties of polymer/Si heterojunctions. *Thin Solid Films.* Vol. 343–344, pp. 469-475 (April 1999), ISSN 0040-6090

Nakagawa, N.; Hwang, H. Y. & Muller, D. A. (2006). Why some interfaces cannot be sharp. *Nature Mat.*, Vol. 5, No. 3, pp. 204-209 (March 2006) ISSN 1476-1122

Nikitenko, V. R.; Tameev, A. R. & Vannikov, A. V. (2010). The mechanism of metal conductivity over the interface between organic insulators. Semiconductors. Vol. 44, No. 2, pp. 211-217 (February 2010), ISSN 1063-7826

Ohtomo, A., & Hwang, H. Y. (2004). A high-mobility electron gas at the LaAlO3/SrTiO3 heterointerface. Nature. Vol. 427, pp. 423-426 (January 2004), ISSN 0028-0836

Pauli, S. A. & Willmott, P. R. (2008). Conducting interfaces between polar and non-polar insulating perovskites. J. Phys.: Condens. Matter., Vol. 20, No. 26, 264012 (July 2008), ISSN 0953-8984

Polk, B. J.; Potje-Kamloth, K.; Josowicz, M. & Janata, J. (2002). Role of protonic and charge transfer doping in solid-state polyaniline. *J. Phys. Chem.* Vol. 106 (44), pp. 11457–11462 (November 2002), ISSN 1520-6106

Popović Z. S.; Satpathy S. & Martin R. M. (2008). Origin of the two-dimensional electron gas carrier density at the LaAlO3 onSrTiO3 interface. Phys. Rev. Lett. Vol. 101, Is. 25, p. 256801 (4 pages) (December 2008), ISSN 0031-9007

Potje-Kamloth, K. (2001). Conducting polymer-based Schottky barrier and heterojunction diodes and their sensor application, In: *Handbook of Surfaces and Interfaces of Materials*, Nalwa, H. S., (Ed), Vol. 5, pp. 445–494, Academic, San Diego

Qi, P; Vermesh O.; Grecu M.; Javey A.; Wang Q.; Dai H.; Peng S. & Cho K. J. (2003). Toward large arrays of multiplex functionalized carbon nanotube sensors for highly sensitive and selective molecular detection. *Nano Lett.* Vol.3, Is. 3, pp. 347-351 (March 2003), ISSN 1530-6984

Qiu L.; Lee W. H.; Wang X.; Kim J. S.; Lim J.A.; Kwak D.; Lee S. & Cho K. (2009). Organic thin-film transistors based on polythiophene nanowires embedded in insulating polymer. *Adv. Mater.*, Vol. 21, Is. 13, pp. 1349-1353 (April 2009), ISSN 0935-9648

Reyren N.; Thiel S.; Caviglia A. D.; Kourkoutis L. F.; Hammerl G.; Richter C.; Schneider C. W.; Kopp T.; Rüetschi A.-S.; Jaccard D.; Gabay M.; Muller D. A.; Triscone J.-M. & Mannhart J. (2007). Superconducting Interfaces Between Insulating Oxides. *Science*. Vol. 317, pp. 1196-1199 (August 2007), ISSN 0036-8075

Richert R.; Pautmeier L. & Bassler H. (1989). Diffusion and drift of charge carriers in a random potential: Deviation from Einstein's law. Phys. Rev. Lett. Vol. 63, Is. 5, pp. 547-550 (July 1989), ISSN 0031-9007

Salazkin S. N.; Shaposhnikova V. V.; Machulenko L. N., Gileva N. G., Kraikin V. A. & Lachinov A. N. (2008). Synthesis of polyarylenephthalides prospective as smart polymers. *Polymer Sci. A.* Vol. 50, No. 3, pp. 243-259 (March 2008), ISSN 0965-545X

Salleo A.; Chen T. W.; Völkel A. R.; Wu Y.; Liu P.; Ong B. S. & Street R. A. (2004). Intrinsic hole mobility and trapping in a regioregular poly(thiophene). *Phys. Rev. B.* Vol. 70, p. 115311 (10 pages) (September 2004), ISSN 1098-0121

Salikhov R.B.; Lachinov A.N. & Rakhmeyev R.G. (2007a). Electrical properties of heterostructure Si/poly(diphenylenephthalide)/Cu. *J. Appl. Phys.* Vol. 101, Is. 5, p. 053706 (5 pages) (March 2007), ISSN 0021-8979

Salikhov R.B.; Lachinov A.N. & Rakhmeyev R.G. (2007b). Mechanisms of conductivity in metal-polymer-Si thin film structures. *Mol. Cryst. Liq. Cryst.* Vol. 467, pp. 85-92, ISSN 1542-1406

Salikhov R.B.; Lachinov A.N. & Bunakov A.A. (2007c). Charge transfer in thin polymer films of polyarylenephthalides. *Phys. Solid State.* Vol. 49, No. 1, pp. 185-189 (January 2007), ISSN 1063-7834

Salikhov, R. B.; Lachinov, A. N. & Rakhmeev, R. G. (2007d) Conduction mechanisms in silicon-polymer-metal heterostructures. *Semiconductors.* Vol. 41, No. 10, pp. 1165-1169 (October 2007), ISSN 1063-7826

Salikhov, R. B.; Lachinov, A. N. & Rakhmeev, R. G. (2008). Transport layer at the boundary of two polymer films. *Techn. Phys. Lett.*, Vol. 34, No. 6, pp. 495-497, ISSN 1063-7850

Salikhov, R. B.; Lachinov, A. N.; Kornilov V.M. & Rakhmeev, R. G. (2009a). Properties of the transport layer formed at the interface between two polymer films. *Techn. Phys.* Vol. 54, No. 4, pp. 575-579 (April 2009), ISSN 1063-7842

Salikhov R.B.; Lachinov A.N.; Rakhmeyev R.G. ; Gadiev R.M.; Yusupov A.R. & Salazkin S.N. (2009b). Chemical sensors based on nano-polymer films. Measurement Techniques. Vol. 52, No. 4, pp. 427-431 (April 2009), ISSN 0543-1972

Salikhov R.B.; Lachinov A.N.; Rakhmeyev R.G. & Gadiev R.M. (2011). High Conductivity of the Interface Between Two Dielectric Polymer Films. Mol. Cryst. Liq. Cryst. Vol. 535, Is. 1, pp. 74-81, ISSN 1542-1406

Schefefold, F.; Budkowski, A.; Steiner, U.; Eiser E.; Klein J. & Fetters L.J. (1996). Surface phase behavior in binary polymer mixtures. II. Surface enrichment from polyolefin blends. J. Chem. Phys., Vol. 104, Is. 21, p. 8795 (12 pages) (June 1996), ISSN 0021-9606

Sundgren, H.; Lundstroem, I.& Vollmer, H. (1992). Chemical sensor arrays and abductive networks. *Sens.Actuat. B* Vol. 9, Is. 2, pp. 127–131 (August 1992), ISSN 0925-4005

Zykov B.G.; Baydin V.N.; Bayburina Z.Sh.; Timoshenko M.M.; Lachinov A.N. & Zolotuchin M.G. (1992). Valence electronic structure of phthalide-based polymers. *J. Electron. Spectrosc. Rel. Phenom.* Vol. 61, Is. 1, pp. 123-129 (December 1992), ISSN 0368-2048

11

Photo-Assisted Organic Pollutants Sensing by a Wide Gap pn Heterojunction

Yoshinobu Nakamura[1], Yusuke Morita[2], Yui Ishikura[2], Hidenori Takagi[1] and Satoru Fujitsu[3]
[1]The University of Tokyo,
[2]Shonai Institute of Technology,
[3]Tokyo Institute of Technology,
Japan

1. Introduction

Since the industrial revolution, living environment is little by little changed to the worse due to the air and soil pollution. The etiological compounds are chemical bi-product from industries, whose accumulation in atmosphere have been negative legacy of overreached materials civilization for a long time. The advanced emission control technology in the advanced nations gradually decreases the total amount of harmful chemical substances spread to atmosphere, however, more than two hundred of harmful compounds have still been contaminating our living environment. Among these compounds, the emission of toluene, benzene, xylene and formaldehyde are in higher level. We have a risk of the health hazard by continuous breathing in these compounds.

Recently, not only outdoors, but also indoor air pollutions by these compounds are regarded as social problems. Man suffers from the strange disease named "sick building syndrome", associated with continuous breathing in benzene, toluene, formaldehyde, or other harmful volatile organic compounds (VOCs), which are evaporated from adhesive, paint, or synthetic building materials, etc. From the viewpoint of maintenance of safe environment, there is currently great interest in the development of VOC sensors for the applications involving environmental monitoring. Up to now, gas chromatography (GC) or gas chromatography mass-spectroscopy (GC-MS) systems have been applied for this purpose, however, such an analytical method had disadvantage of higher running cost and complicated operation including gas sampling by engineers or surveyors. In addition, their slow response is not appropriate for real-time environmental monitoring.

For the simplest and low-cost solid-state device for VOC detection, a semiconductor gas sensor and its peripheral technology have been paid much attention. The working mechanisms of a semiconductor gas sensor are described later, but simply speaking, the modulation in electric resistance accompanied by the catalytic oxidation reaction of testing gases over a sensor body. For the operation of a traditional semiconductor gas sensor, the activation energy for the chemical reaction is provided by the lattice vibration activated under the elevated temperature ranging 150-600 °C. As a heat source for the gas sensor operation, on-chip micro-heaters are often adopted, however, additional heating system causes the complexity in the sensor system and it should be omitted if possible. If a semiconductor gas sensor has sufficient sensitivity for VOCs at room temperature, both sensing element and its operation system can be embedded in one IC tip. Such a system will become a powerful tool for *in situ* environmental monitoring. The most desirable sensor system for environmental monitoring is a VOC sensor especially for toluene and benzene. Thermodynamically stable aromatic VOCs hardly interact with the surface of the solid-state sensor devices at moderate temperature region and even if an oxide semiconductor could work as an aromatic VOC sensor, severe operating conditions, higher operation temperature (typically > 450°C), will usually be required.

How to activate the thermodynamically stable molecules is a key for designing a solid-state VOC sensor with lower operation temperature. For this purpose, the authors will propose new concepts for gas sensing under a mild condition, whose working mechanism is based on the assistance of "external" energy, electric potential energy and photon energy. A semiconductor pn junction devise is applied for the reaction field of a gas sensor system, because the active carrier (hole in VB: Valence Band) can efficiently injected into VOC molecule over the junction interface. Detail in our strategy is shown in the next chapter. Both electron and photon can supply the several eV order of energy, corresponding to the thermal energy of over 10000K. A pn junction devise is appropriate for the use of such colossal energy supplied by photons or electron holes. Our results will provide a "chemical" application of wide gap oxide semiconductor heterojunction devises, which had been paid much attention for their application to UV-LDE or UV detectors.

2. VOC sensing using an oxide semiconductor -How to oxidize of thermodynamic stable molecules

The working principle of a semiconductor gas sensor is based in the catalytic oxidation of testing gas molecules over a sensor body. A simple model for the gas sensor operation of n-type semiconductor, for example, SnO_2, ZnO, etc. is summarized in Figure 1. In air, the surface of an oxide semiconductor tends to be charged negatively due to the surface excess anions, *i.e.*, adsorbed oxygen ions coordinated to surface cations (Figure 1(A)). For the compensation of this excess negative charge (δ^- in Figure 1(B)) originated from adsorbates, the space charge layer with opposite charge (δ^+) is induced near surface, then, in the case of a *n*-type semiconductor, the surface conductivity becomes smaller than that of the semiconductor bulk. Guessing from this situation, the conductance of a sensor element changes as a function of the total amount of the effective charge by the adsorbates, which would be changed by the catalytic reaction with gas molecules in atmosphere.

When combustible gas molecules are introduced in atmosphere, the surface band bending of an oxide semiconductor will become gentle depending on the nature of the surface adsorbed species as is illustrated in the lower part of Figure 1(B). For example, (C_mH_n) with an oxide semiconductor surface and sometimes molecularly adsorbed species ($C_mH_{n\ ads}$) donates electrons to the semiconductor bulk. This process is regarded as partial oxidation of C_mH_n.

$$C_mH_{n\ ads} \rightarrow (C_mH_n)_{ads}^+ + e^- \quad \ldots \qquad (1)$$

where the subscription "ads" means the adsorbed chemical species over the surface. Positively charged adsorbates $(C_mH_n)^+$ distill the original surface negative charge and the total electric conductivity of an *n*-type oxide semiconductor is increased as is illustrated in Figure 1(C). In the case of a *p*-type oxide semiconductor such as NiO, the upward band bending is also expected due to the surface negative charge. In this case, the surface conductivity is enhanced because of the surface hole accumulation layer.

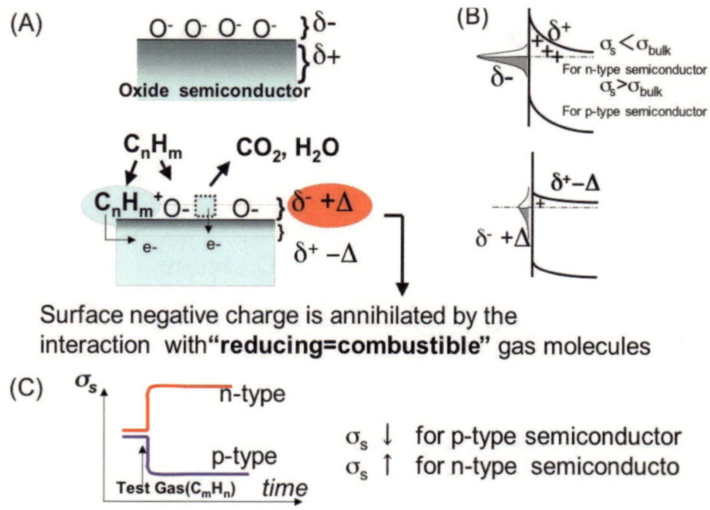

Fig. 1. Working principle of a traditional semiconductor gas sensor made by an oxide semiconductor.

Supposing Langmuir–Hinshelwood mechanisms, $(C_mH_n)^+$ is successively oxidized to CO_2 and H_2O using the oxidation power of surface oxygen ions (O_{ads}^-).

$$(C_mH_n)_{ads}^+ + (2m + n/2)O_{ads}^- \rightarrow mCO_2 + (n/2)H_2O + (2m+n/2-1)e^- \quad \ldots \qquad (2)$$

Totally, (2m + n/2) electrons are donated to the semiconductor surface during the full oxidation of a C_mH_n molecule. After the removal of the surface adsorbed oxygen (O_{ads}^-), oxygen vacancies (□ in Figure 1(A)) are produced near surface but presently re-occupied by the oxygen molecules in atmosphere. When the rate of oxygen re-adsorption process,

$$\square + 1/2O_2 + e- \rightarrow O^-_{ads} \qquad (3)$$

is sufficiently slow rather than that of oxidation process (2), the surface negative charge decreases to $\delta^- + \Delta$, where δ^- is the original surface negative charge by the adsorption of oxygen and Δ is the positive charge produced during the oxidation reaction, that is the summation of effective positive charge of adsorbate ($(C_mH_n)_{ads}^+$) and oxygen vacancies induced by the reaction of C_mH_n. Most of the traditional semiconductor gas sensor materials are *n*-type semiconductors the response is positive, *i.e.*, the conductivity increases by the introduction of combustible gases. As is mentioned before, if we use a *p*-type semiconductor as a gas sensor element, the effect of space charge on the surface conductivity is completely opposite and its surface conductivity is enhanced by oxygen adsorption. When C_mH_n gas is introduced, the surface conductivity σ_s decreases by the effect of positive charge induced by partial or full oxidation of C_mH_n and the response in electric conductivity is negative as is shown in Figure 1(C).

The necessary condition for gas sensor operation by an oxide semiconductor is the procedure of the gas-surface reaction and both partial oxidation (1) and full oxidation (2) are available. If the surface of an oxide semiconductor has sufficient oxidation power against an aromatic VOC, for example benzene, it works as a benzene sensor. In the other words, oxidation power of a semiconductor surface determines whether the sensor is sensitive to benzene or not. According to the Hückel's rule, aromatic rings are thermodynamically stable and they are difficult to be fully oxidized to CO_2 in a mild condition. Tin oxide (n-SnO_2) based gas sensors are usually insensitive for benzene or toluene due to its lower oxidation power for a catalyst at the operating temperature of 200-300°C. For the detection of aromatic VOCs, severe temperature condition and the specific surface morphology having ultra-fine (sub-μm - nm order) grains is required in the sensor body.

3. The strategy for designing an aromatic VOC sensor -The "electro" activation and "photo" activation of VOC molecules for the procedure of the oxidation reaction-

As is shown in the previous section, the most important subject for designing an aromatic VOC sensor is how to activate or ionize the target compound adsorbed over the sensor body. High temperature operation is the most easygoing way for adding the aromatic VOC sensitivity to an oxide semiconductor gas sensor. However, in this method, a high-power consumption heat source must be set in the sensor system. Therefore, this type of sensor system is not appropriate for a micro-sensor devise manufactured by the circuit integration technology.

Here, we propose two novel methods for obtaining the active sensor surface for the reaction with aromatic VOCs. One is "Electro activation" of an adsorbate of target molecules and the basic concept is summarized in Figure 2. In this process, a semiconductor pn junction interface is used as a reaction field for target gases, i.e., aromatic VOCs. The electric field at the junction interface would modify the carrier concentration of the near interface region and assist the "active" carrier injection/extraction from/to VOC molecules adsorbed over the junction region. This process is speculated by the electric theory of catalysis proposed by Wolkenstein. A semiconductor gas sensor whose principle is based on his theory had been proposed by the authors and first applied to selective CO gas sensing by a p-CuO/n-ZnO heterojunction.

The other method is "photo activation" of an adsorbate by UV light incidence. This process is "direct" activation of testing gas adsorbates. It is well known that the π–π^* transition of aromatic ring occurs by the UV-C light incidence and an electron occupied in π^* orbital is so active as to be extracted by an oxidant such as holes (h^+) in VB of a p-type oxide semiconductor. Photo-driven oxidation process for aromatic VOCs over p-NiO surface under UV light incidence is summarized in Figure 3. The vertical axis represents the electron potential vs. vacuum level, which is proportional to the instability of holes, i.e., the oxidation power of holes. The excited states of π electrons of benzene (π^* state) is located 4.13eV with respect to the vacuum level and its electric state is near the VB edge of a p-NiO. This situation enables this system to the electron transfer from the adsorbate to VB of p-NiO.

The important result of our study is that "electro- " and "photo-" activation never works for the VOC sensing operation by itself. The synergy effect of "electro- " and "photo-" activation is indispensable for the sensor. Detail is shown in the latter chapters.

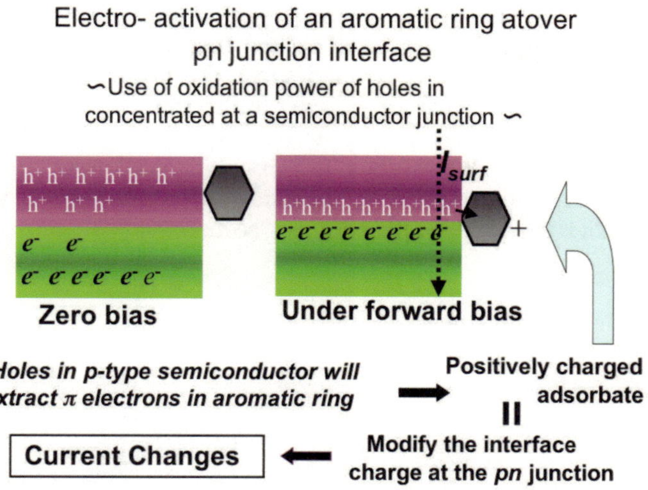

Fig. 2. The concept of "electro-" activation of combustible (reducing) gas molecules over a pn junction diode and its application to a chemical sensor.

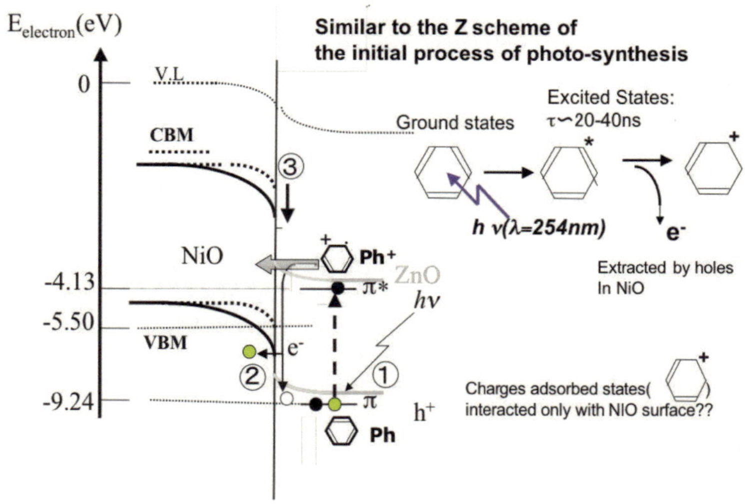

Fig. 3. Photo-driven oxidation process for aromatic VOCs over p-NiO surface under UV light incidence. Because of the large potential gap between VB edge of NiO and ground state of aromatic VOC (**Ph**), Ph is not oxidized under dark, however oxidation of exited state (=electron transfer from Ph* to VB of NiO) is possible.

4. "Electro-" activation of aromatic VOCs using a wide gap pn junction diode

The basic concept of the electro-activation of aromatic ring: Ph by the assist of active hole injection at a pn junction interface is summarized in Figure 4(A). Applying the bias voltage causes the displacement of Fermi level at the junction interface and when the bias is applied to the forward direction, the quasi-Fermi level of holes in *p*-type semiconductor will be pulled down relative to the original level. As a result, holes in *p*-type semiconductor are concentrated at the junction region as is illustrated in Figure 2. Supposing the electronic theory of catalysis, total amount of charged adsorbates:Ph^+ and related their catalytic reaction rate varies as a function of Fermi level. Charged adsorbed species over the junction region are regarded as the surface or interface states (Ph^+/Ph) and the carrier transport process through the junction will be modified as a function of their density and energy distributions. For the oxidation of thermodynamically stable molecule such as aromatic VOC, use of wide gap *p*-type semiconductor is appropriate, because wide gap oxide semiconductor has deep VB rather than narrow gap one. Detail is explained in the following sections.

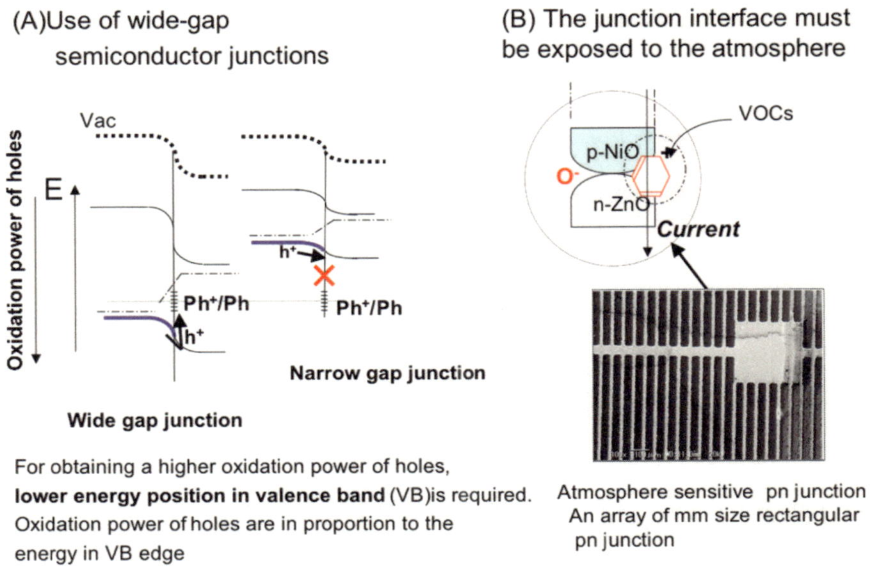

Fig. 4. How to activate an aromatic VOC (Ph) through the "electro-"activation process at a wide gap pn junction diode. The potential gap between VB of *p*-type semiconductor and the redox potential of Ph becomes smaller by using a wide gap oxide semiconductors for a sensor device. For adding an atmosphere sensitive properties to a pn junction diode, the junction interface must be exposed to atmosphere by making a micro sensor array shown in Figure (B).

This phenomenon will be applied to gas sensors or other chemical sensors. Up to now, some pn heterojunctions are able to be applied for humidity monitoring, CO gas sensing, Cl_2 gas sensing, NO_x sensing, *etc.* When the nano-spaces between *p*-type and *n*-type semiconductor are filled with organic solvent molecules, the electrical current passing through the contact interface dramatically changes and characteristic hysteresis loop is observed in its current-voltage (I-V) relation. By the mathematical analysis of the obtained I-V curves, molecular recognition of the organic solvents is possible.

In the next section, we will design an atmosphere sensitive pn heterojunction diode using the continuous film preparation and patterning by photolithography as is shown in Figure 4(B).

5. Preparation of a *p*-NiO/*n*-ZnO heterojunction for the chemical sensor use

In this section, we will explain how to process a pn junction diode to a chemical sensor element. According to our concept, the hetero-interface between a wide-gap *p*-type and *n*-type oxide semiconductor is appropriate for the reaction field for the oxidation of aromatic ring, because the contribution of "electro-" activation is expected in this field as is shown in Figure 4(A). Holes in a wide gap *p*-type oxide semiconductor with larger escape depth of VB electron is expected to have high oxidation power, because the stability of holes would be anti-proportional to the escape depth of VB electrons. Instability of holes indicates its higher oxidation power. The energy levels of electrons in CB of semiconductors can be closely correlated to the reduction activity of a semiconductor photo-catalyst and the similar theory can be valid for the electro-catalyst for oxidation using a *p*-type oxide semiconductor. We should select a wide gap *p*-type semiconductor with "deeper" VB.

Here, we select a wide gap *p*-type semiconductor, nickel oxide (NiO) as a pn heterojuchtion diode for VOC sensor application and prepare a diode structure having non-epitaxial junction interface by the continuous film preparation by sputtering technique. The junction interface prepared by this method is composed of the assembly of numerous number of point junctions and there is "nano-space" around the point junctions. The carriers (electrons and/or holes) are transferred through these point junctions and their transport properties will be strongly modified in the "nano-space". If combustible gas molecules are introduced, they are transferred through "nano-space" around the point junctions. Then, combustible gases adsorb near the junction interface and their catalytic oxidation proceeds in the vicinity of the point junctions. The atmosphere sensitive pn junction diode for the use of aromatic VOC sensing is illustrated in Figure 4(B). For the efficient gas transport to pn contact junction, we design the micro sensor alley, whose dimension is several μm in length and several μm in width.

The transport properties and related gas sensing properties of an oxide semicondcutor pn junction diode depend on the crystal axis relationship between p-type and n-type semiconductor. It is appeared that the sensor properties of a pn junction based on ZnO strongly depend on the interface index of ZnO. Due to the structural anisotoropy, crystal planes of ZnO are roughly classified to Zn-terminated polar surface ((0001) plane, here, it is named Zn-face), O-terminated polar surface ((000-1) plane, named O-face) and non-polar surface. Zn-face and O-face has opposite surface charge. In the ZnO based pn junction diode, the polarity of surface charge will strongly modify its basic junction properties.

Zinc oxide (ZnO) substrate was prepared by vapor transport technique using ZnO sintered compact as a ZnO vapor source. Use of ZnO single crystal substrates for the device preparation will provide a sophisticated interface, however, such the junction device would not have gas sensing properties because the junction is too sophisticated and it prevents the introduction of "nano-space" for transferring testing gas molecules to the active sites.

Zinc oxide (ZnO) crystals grown by the method proposed by Noritake were polycrystalline but highly oriented to the c-axis direction, whose orientation angle was limited within 0.5 degree. The photograph of an as grown crystal is shown in Figure 5 and top and bottom surface of the crystal was Zn-terminated polar surface (Zn-face) and O-terminated polar surface (O-face), respectively. Nickel oxide film was deposited on this substrate by rf magnetron sputtering technique with the thickness of 100nm using a 2at% lithium doped NiO sintered compact as a sputtering target. Nickel oxide (NiO) films by this process were colored gray and its carrier concentration was evaluated to be 5×10^{17} cm^{-3} by Hall effect measurement. The conductivity of the p-NiO film is small compared with a ZnO substrate, whose carrier concentration was 2×10^{19} cm^{-3} and the potential drop at the junction region mainly occurred at the *p*-NiO side.

After the Ohmic electrode (Ni and Au film) was deposited over the NiO film, the devise was processed into atmosphere sensitive one. Inter-digital electrode pattern was drawn by photolithography technique and chemical etching was conducted for exposing the junction interface to the atmosphere. The fabrication process of *p*-NiO/*n*-ZnO sensor element was summarized in Figure 5. After chemical etching process in 0.1M HCl aqueous solution, the transport properties of a *p*-NiO/*n*-ZnO heterojunction became sensitive to the atmosphere. The top view of the inter-digital electrode was shown in Figure 4(B) and 5.

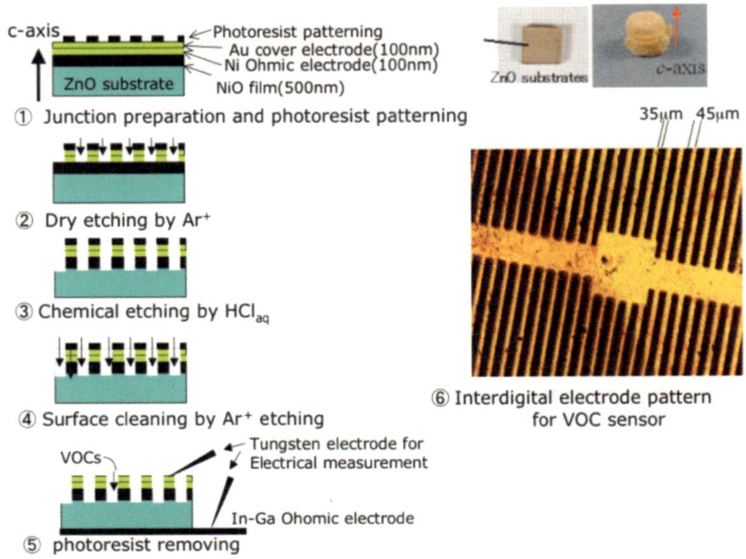

Fig. 5. Schematic view of the preparation of a *p*-NiO/*n*-ZnO heterojunction VOC sensor element. Pn junction is prevared by depositing a NiO film on a highly oriented ZnO polycrystalline substrate. After depositing an Ohmic electrode on the junction, sensor array is prepared by photolithography patterning and chemical etching. We can prepare a *p*-NiO/*n*-ZnO heterojunction by more symple process, that is, continuous depsositing of NiO and ZnO film. Similar etching process is possible for preparing a sensor element shown in figure 6.

6. VOC sensing characteristics of a p-NiO/n-ZnO heterojunction in the dark -Effect of "Electro-" activation of adsorbate of VOCs-

The *I-V* characteristics of an original, i.e., before processing to sensor array, p-NiO/n-ZnO(Zn-face) heterojunction is curve 1 in Figure 6. The I-V curve looks like "good" rectifying character apparently, however, the diode factor n is calculated to n=6.2, that means large number of interface states exist and that would contribute to the recombination current across the junction. The I-V curve is modified again by processing into a sensor element and its rectifying character is completely spoiled after the chemical etching process (Curve 2 in Figure 6). The rectifying character is restored by low temperature (200°C) annealing in oxygen and simultaneously, large hysteresis is appeared in the I-V response. This characteristic I-V is the result of the extrinsic interface states, which is introduced by low temperature annealing in oxygen, therefore, carrier transport through the pn junction interface is governed by the nature of this extrinsic interface states.

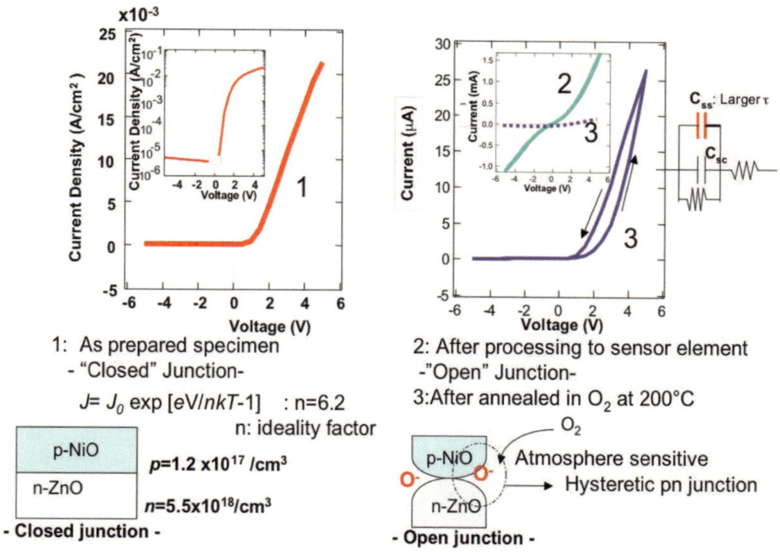

Fig. 6. The basic junction properties of a p-NiO/n-ZnO(Zn-face) heterojunction prepared by continuous NiO and ZnO film deposition on a glass substrate. The rectifying character of the junction is spoiled by processing it into a micro-sensor array(curve 2). The rectifying characteristics are restored by annealing the device in oxygen atmosphere as is shown in curve 3.

Is the "electro-" activation effective for an aromatic VOC sensing at a p-NiO/n-ZnO heterojunction with such high density of extrinsic interface states? That is the main subject for the discussion in this section. We checked the effect of applied bias on various VOC sensing properties of a p-NiO/n-ZnO heterojunction sensor element. In the case of the detection of simple inorganic combustible gas, such as carbon monoxide (CO), the electro-activation plays an important role for the sensor operation. In our hypothesis, positively charged CO adsorbate (CO^+) over the p-CuO surface promotes the carrier transport through p-CuO-n-ZnO junction interface, therefore, the junction works as a CO gas sensor. Supposing the electronic theory of catalysis, the surface concentration of CO^+ adsorbates over p-type semiconductor changes as a function of surface Fermi level and in the case of a pn junction type gas sensor, it is enhanced by applying the forward bias. This hypothesis is well supported by our experimental results.

The question is whether such the reaction scheme is valid for the detection of a thermodynamically stable molecules such as aromatic VOCs. The p-NiO/n-ZnO(Zn-face) junction resistance is not in response to aromatic VOCs, while very small response is obtained by acetone and ethanol vapor introduction(Figure 7). The current response is rather complex as is shown in Figure 7A. Just after the acetone vapor introduction, the forward current (V=+5.0V) rapidly increases (point B in Figure 7(A)), then, gradually decreases to current revel lower than the original value (point C). Similar sensing characteristic are obtained by the introduction of ethanol vapor. A key to understanding this phenomena is "unipolar" sensing characteristics, that is a p-NiO/n-ZnO(Zn-face) heterojunction sensor element works as acetone sensor only when the junction is forward biased, while it dose not work under reverse bias. As is shown in Figure 7B, only the forward current changes A→B→C. The forward current shows about 7% decrease under 4300ppm of acetone (curve C in Figure 7(B)).

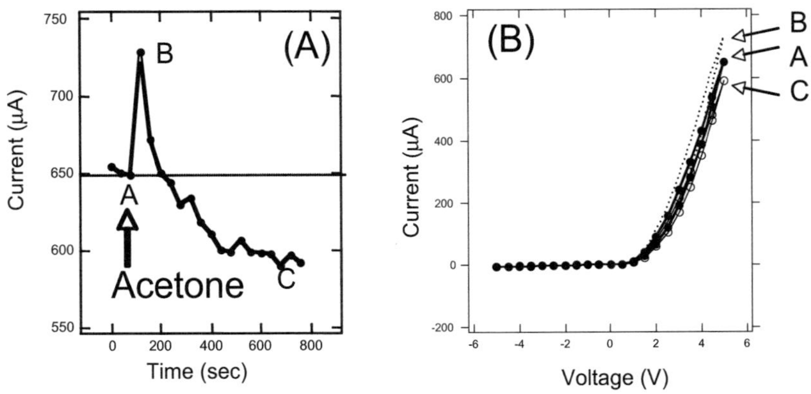

Fig. 7. The effect of acetone vapor introduction on the transport properties of a *p*-NiO/*n*-ZnO(Zn-face) heterojunction sensor element under dark. The current –time behavior at V=+4.5V forward bias is shown in figure (A). At the point (A), acetone vapor arrives at around the junction interface and after 600sec (point C) the acetone concentration reaches 4300ppm. The I-V curve B is that of point B in figure (A).

In our acient study, we had evaluated the effect of organic solvent introduction on the I-V characteristics of a *p*-CuO/*n*-ZnO heterojunction. When the acetone or ethanol is introduced to the junction region, forward current dramatically increases, at that time, characteristic hysteresis is grown in the reverse biased region. Initial forward current enhancement of *p*-NiO/*n*-ZnO in Figure 7(A) is supposed to be due to the anodic oxidation (hole injection from *p*-NiO) of ethanol and acetone over *p*-NiO surface as is predicted in ethanol detection by a *p*-CuO/*n*-ZnO heterojunction. As we discussed, thermodynamically stable molecules, aromatic VOCs are difficult to be detected by using *p*-NiO/*n*-ZnO heterojunction at room temperature. These results mean that adsorbed aromatic VOC molecules are not ionized or activated by the chemical interaction with positively (or negatively) biased *p*-NiO surface at room temperature. Anodic oxidation of VOCs at a pn junction interface had thought to be profitable process for acetone or ethanol sensing, however, holes in VB of p-NiO did not work an oxidant for aromatic VOCs in the ground states.

It had been known that under higher applied voltage, titanium dioxide (TiO_2) films behaves like *p*-type semiconductor because of its space charge limited current (SCLC) and turns to be sensitive for benzene at room temperature. Its working mechanism is speculated to be the creation of surface donor states due to the interaction between carriers (holes) and benzene adsorbates. In the case of the *p*-NiO/*n*-ZnO heterojunction sensor element, though high concentration of holes are injected into the junction region at least under forward bias, electrical activation of benzene or toluene is invalid under dark. This process is valid only for orgatic compound which are easy to be oxidized.

7. Photo-assisted aromatic VOC sensing of a *p*-NiO/*n*-ZnO sensor element under UV light irradiation

As is discussed in the previous section, the oxidation power of holes in the valence band (VB) of *p*-type semiconductor is a key to designing a pn junction type VOC sensor. The oxidation power of holes is suggested to be proportional to the energy of VB edge vs.

vacuum level. From the empirical rule, the VB position of an oxide semiconductors is at most -7eV vs. vacuum level and that of *p*-type semiconductor shows 1-2eV higher position, therefore, even in a wide gap *p*-type semiconductor, holes in VB do not have sufficient oxidation power for extracting electrons from aromatic VOC adsotbates. Relative relation between the conduction/valence band of *n*-ZnO, *p*-NiO, and ground/excited states of benzene is illustrated in Figure 8. The ground state of toluene lies -8.8eV below the vacuum level and that of benzene is -9.24eV and those are far larger than VB edge of *p*-NiO (-5.5eV). Benzene, toluene or other VOCs having six-membered aromatic ring is known to molecularly adsorb over a transition metal oxide (TMO) surface, whose aromatic ring is parallel to the surface as is illustrated in the upper part of Figure 8. At lower temperature, the activation energy is not sufficiently supplied due to the large potential gap (>3eV) between π ground states and VB of *p*-NiO, therefore, ionization of aromatic ring followed by the spontaneous electron transfer between adsorbate and TMO surface does not occur. Charge neutral chemical species hardly affect the carrier transport through the pn junction interface. This is the limit of the ability of "electro-activation" of adsorbates and we must consider novel concept for activating an adsorbate, that is "photo-activation".

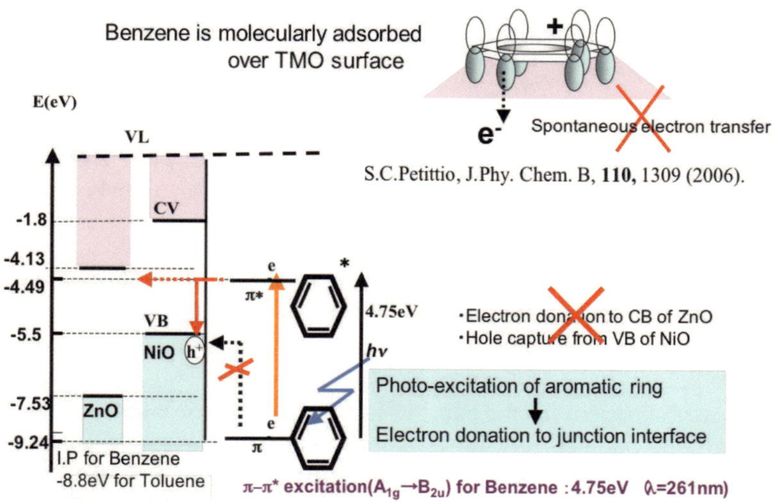

Fig. 8. Possible working principle of aromatic VOC sensing in a *p*-NiO/*n*-ZnO heterojunction. π* electron in the excited states of benzene can interact with holes in VB of *p*-NiO. From the experimental data, applying the forward bias is necessary for the electron transfer from π* state to the pn junction interface (red arrow in the figure). Spontaneous electron transfer dose not occur without the assist of applied bias.

Under UV light irradiation, π electrons in aromatic ring is excited to π* level, which is located 1eV above the VB of *p*-NiO. The π* electron is allowed to interact with holes in VB of *p*-NiO and the electrons in aromatic ring is suggested to be actively transferred following the scheme in Figure 8. Such a photo-driven electron transport between semiconductor pn junction and gas adsorbate is analogous to Z scheme (the light reaction in photosynthesis in plants), in which one molecule of the pigment chlorophyll absorbs one photon and loses one electron having strong reduction power. In the case of the photo-driven electron transport discussed here, the process is "get one hole" having strong "oxidation" power using the photon energy.

The analogy between the photo driven electron transport at the pn junction and Z-scheme is illustrated in Figure 9. After the electron transfer, positively ionized adsorbed species, for example, $C_6H_6^+$(benzene adsorbate) will work as interface states and their positive charge will modify the interface barrier height. Therefore, the junction device works as an aromatic VOC sensor. Of course, similar scenario is expected in toluene adsorption.

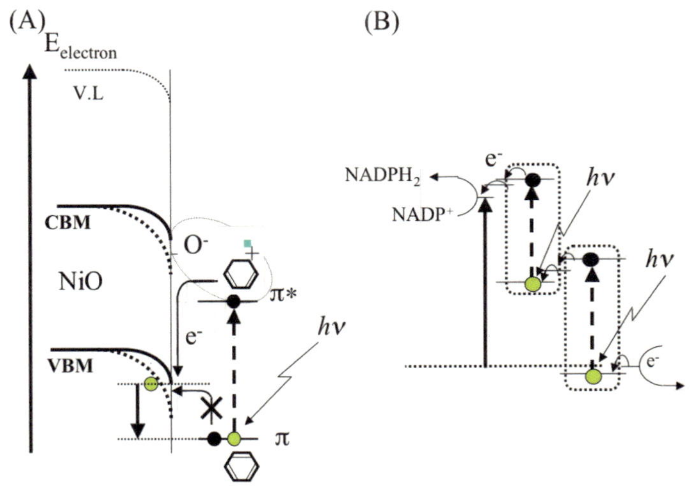

Fig. 9. The analogy between (A) VOC oxidation with the assist of photon energy and (B) light reaction (synthesis of reduction regent: $NADPH_2$) in photosynthesis , which is named Z-scheme.

The effect of toluene introduction on the I-V characteristics of a *p*-NiO/*n*-ZnO(Zn-face) and *p*-NiO/*n*-ZnO(O-face) heterojunction sensor elements are summarized in Figure 10. The UV light source used in this experiment is a low pressure mercury lamp whose characteristic emission-lines are known to be λ(wavelength)=254, 295,302,313 and 365nm in the UV region. The photon energy is sufficient for the band-to-band excitation for both NiO and ZnO. Under the UV irradiation, the photocurrent flows both forward and reverse directions and the I-V curve is changed from 1(under dark) to 2 (under UV irradiation) as is shown in Figure 10.

Under UV irradiation, the toluene sensing properties are added to a *p*-NiO/*n*-ZnO sensor element and only the forward current decreases as a function of toluene concentration in atmosphere. The I-V curve is changed from 2 to 3 after the 3325ppm of toluene vapor introduction. There is characteristic difference in the basic junction properties between *p*-NiO/*n*-ZnO(Zn-face) and *p*-NiO/*n*-ZnO(O-face) junction. The junction properties of a *p*-NiO/*n*-ZnO(Zn-face) sensor element keep good rectifying character even after processing into a sensor element, while that of a *p*-NiO/*n*-ZnO(O-face) sensor element shows a little leaky character and its onset potential in forward current is far smaller than that of a *p*-NiO/*n*-ZnO(Zn-face) element. Though *p*-NiO/*n*-ZnO heterojunctions show different transport and photo response properties depending on the crystal plane of the ZnO substrate, there is no characteristic differences in the toluene sensing properties between *p*-NiO/*n*-ZnO(Zn-face) and *p*-NiO/*n*-ZnO(O-face) sensor element. This result suggests that the nature of the ZnO crystal plane does not largely contribute to the toluene sensing mechanisms and the ZnO layer works only for the electron -injector/hole-collector at the junction region.

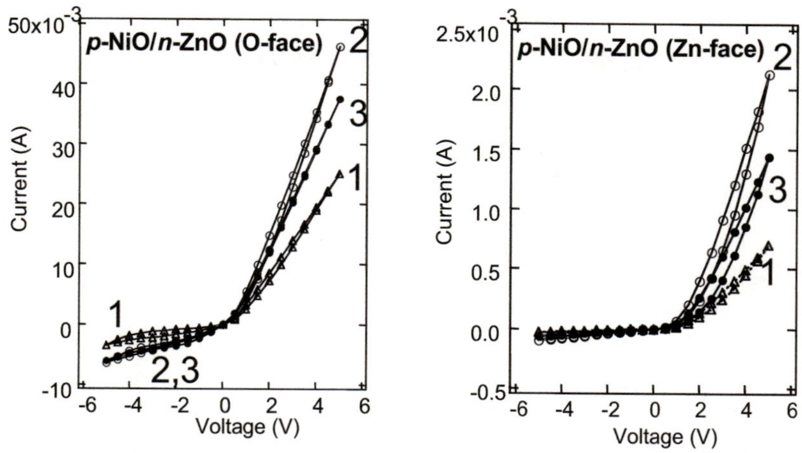

Fig. 10. The effect of toluene introduction on the I-V characteristics of a *p*-NiO/*n*-ZnO(Zn-face) and *p*-NiO/*n*-ZnO(O-face) heterojunction sensor elements. Curve 1: The I-V curve under dark, Curve 2: The I-V curve under UV light irradiation and Curve 3: The I-V curve in toluene 3325ppm atmosphere under UV light irradiation. Toluene sensing properties are added in both junctions under UV irradiation, though there are characteristic differences in the basic junction properties.

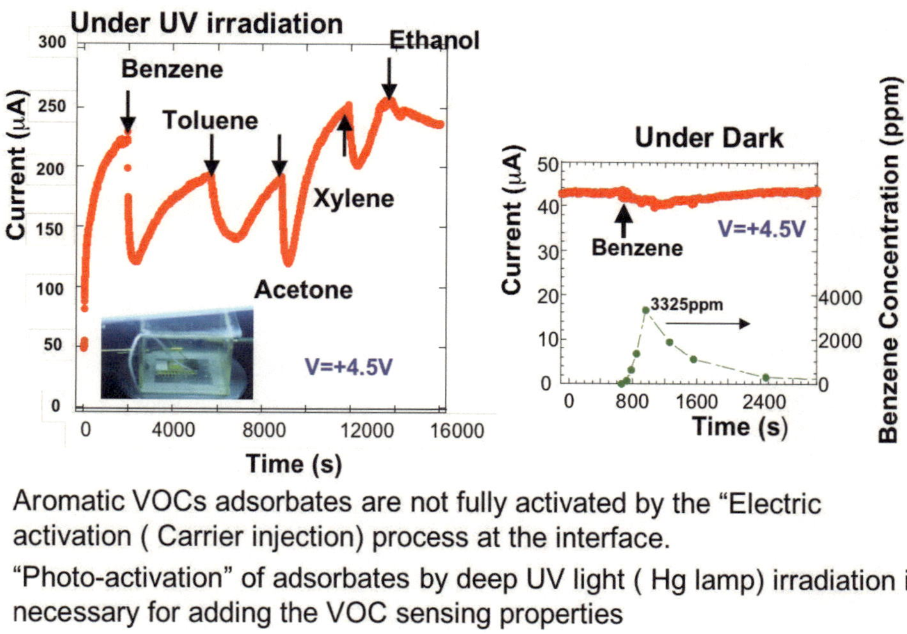

Fig. 11. The effect of various VOC introduction on the photo current through the *p*-NiO/*n*-ZnO(Zn-face) junction interface under full spectra of Hg lamp. The *p*-NiO/*n*-ZnO(Zn-face) junction is sensitive for aromatic VOCs (Benzene, toluene, xylene… and acetone) under UV irradiation at room temperature. The sensitivity for ethanol vapor is never enhanced under UV light irradiation, while that for acetone vapor is extremely enhanced under UV irradiation.

Independent of the crystal plane of ZnO, the electrical current suppression by toluene introduction is clearly observed only when the *p*-NiO/*n*-ZnO junction is biased to the forward direction. Such the unipolar gas sensing characteristics are feature of molecular sensing by a pn junction device, whose working principle is based on the "electro-" activation. Under forward bias, holes will be injected from *p*-NiO bulk into the junction interface and they will extract the π* electrons in the excited states of toluene.

This process would not occur without the assist of carrier injection by applied bias. A *p*-NiO thin film in itself dose not show toluene sensing characteristics even under UV irradiation. Both "electro-" and "photo-"activation is indispensable for the detection of VOCs at the pn junction interface. The important subject for the design of an aromatic VOC sensor is the synergy effect of two independent activation processes, "electro-" and "photo-" activation. Similar effects are obtained by benzene, xylene and acetone introduction, because all these compounds have absorption spectra in UV region, which locate near the line spectra of mercury lamp. On the other hand, methanol and ethanol sensitivity are not enhanced under UV light irradiation. These results are summarized in Figure 11. The result under reverse bias is also shown in the right panel of this figure and you can see the reverse current did not respond to toluene introduction even under UV irradiation.

The *p*-NiO/*n*-ZnO heterojunction shows rapid response for VOC in atmosphere and the current change, ΔI ($=I_0-I(t)$, where I_0 is the current before toluene introduction and I(t) is the current at t second after VOC introduction) directly related to the toluene concentration in atmosphere. The time dependence of the current change ΔI (= the magnitude of the current decrease by the introduction of toluene) under forward bias is shown in Figure12. The rise and decay of the toluene concentration in the ambient after the toluene injection is overplotted. The current decreased rapidly following the toluene concentration and ΔI was appeared to be proportional to the square root of the toluene concentration, whereas no remarkable current modulation was observed under reverse bias. Rapid response is the great advantage for the use of environmental monitoring.

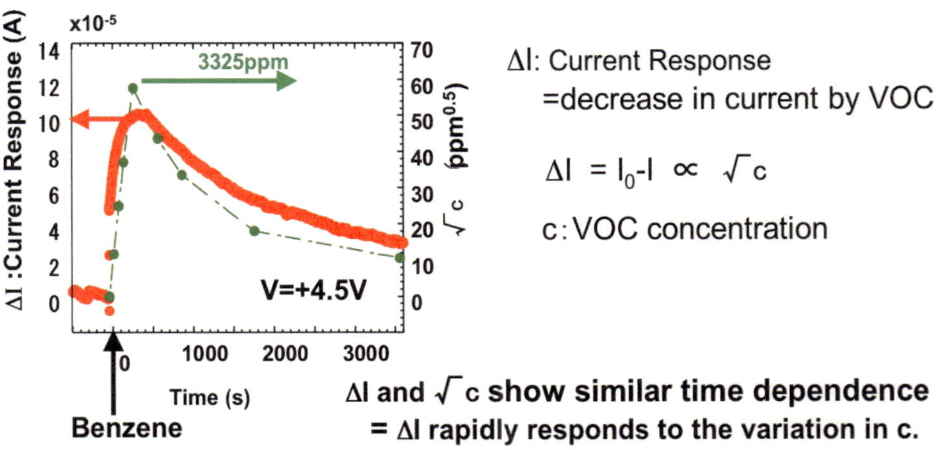

Fig. 12. The time dependence of the current decease by toluene introduction (ΔI) and toluene concentration in atmosphere, c. ΔI is proportional to the square root of c ($\sqrt{c}$).

Here, we discuss the working mechanisms of toluene sensing by a *p*-NiO/*n*-ZnO heterojunction under UV irradiation in detail. The next experiment gives us important information on the VOC sensing mechanism. That is, a simple photo-catalytic reaction will be invalid for promotion of toluene or benzene sensing characteristics in a *p*-NiO/*n*-ZnO sensor element. Figure 13 shows the effect of the spectrum of UV light source on the toluene sensing characteristics of *p*-NiO/*n*-ZnO(Zn-face) sensor element. When the UV light source with full spectrum is switched on, band-to-band excitation of NiO and ZnO occurs and the photocurrent flows through the junction. The measurement in Figure 13 is conducted under forward bias (V=+4.5V) and in this condition, the current rapidly decreases by the introduction of aromatic VOC (benzene) vapor in atmosphere. However its sensing characteristics are completely spoiled by cutting off λ=254nm UV light using PYREX glass filter. Even if λ=254nm UV light is missing, the photocurrent still flows because the band to band excitation of NiO and ZnO is still possible by the adsorption of λ= 295,302,313 and 365nm line of the UV source.

It had been reported that La doped ZnO film having nm sized grains shows toluene sensing characteristics under UV (λ=356nm) irradiation by UV-LED. Photo-catalytic reaction will promote its gas sensing characteristics, however, the effect is limited at elevated temperature (>150°C). Taking a look at the result in Figure 13, photo-generated carriers at the NiO/ZnO interface dose not directly promote the activation and reaction of aromatic VOCs. That is, λ=254nm deep UV light incidence is essential for aromatic VOC sensing operation and it might cause the direct activation of aromatic ring of toluene or benzene. In wet atmosphere, the photo-catalytic activity of TiO_2 are known to be promoted by deep UV light incidence, however, in that case, UV light (λ=254nm) is used for the production of OH radicals, and that would not valid in our cases, because all our measurement is conducted in dry conditions.

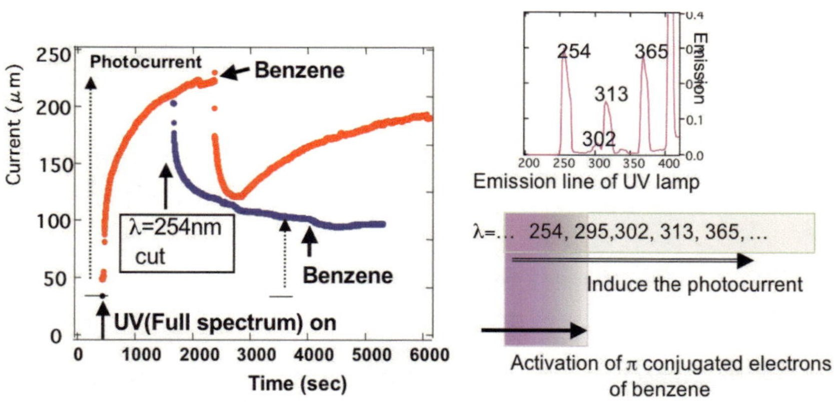

- Photo current (=carrier injection)do not assist the VOC sensing independently.
- λ=254nm deep UV light
- incidence is indispensable for benzene sensing of a p-NiO/n-ZnO heterojunction

Fig. 13. Effect of λ=254nm irradiation of mercury UV lamp on the aromatic VOC (benzene) sensing characteristics of *p*-NiO/*n*-ZnO(Zn-face) heterojunction sensor element. The sensitivity for benzene is disappeared when the λ=254nm emission line is cut off.

The first excited state of π conjugated electron, π* state ($^1B_{2u}$) of aromatic ring for toluene is 4.69eV higher than π ground state ($^1A_{1g}$), corresponding to the photon energy of λ=266.8nm. The 254nm line of mercury lamp can excite the ground state, π electrons to π* state, as a result, aromatic ring will become unstable and activated. The π* ($^1B_{2u}$) state of aromatic ring is located 4.13eV with respect to vacuum level (V.L.) for toluene and 4.34-4.53eV for benzene. The lifetime of the excited state of toluene is evaluated to be 26~45 ns and these are possible values for the electron transfer from π* excited state to the surface of NiO (or ZnO), if the orbital overlap between the adsorbate and NiO (or ZnO) surface is sufficiently large. Judging from the energy of CB edge of ZnO, electron donation from π* of toluene to CB of ZnO would be theoretically possible, however it is against the experimental results, because the positively ionized adsorbate would decrease the surface barrier height of *n*-ZnO.

Our speculation is supported by the following experimental result. Figure 14 shows the absorption spectra of toluene, benzene, acetone and ethanol. The line spectra of UV light source are over-plotted in this figure. From the result of Figure 11, the current rapidly decreases by toluene, benzene or acetone introduction, while no enhanced sensitivity for ethanol is observed under UV irradiation. VOCs which are detectable by a *p*-NiO/*n*-ZnO sensor element have absorption line located around λ =254nm emission line, while ethanol have no absorption lines coincidence with Hg emission spectrum.

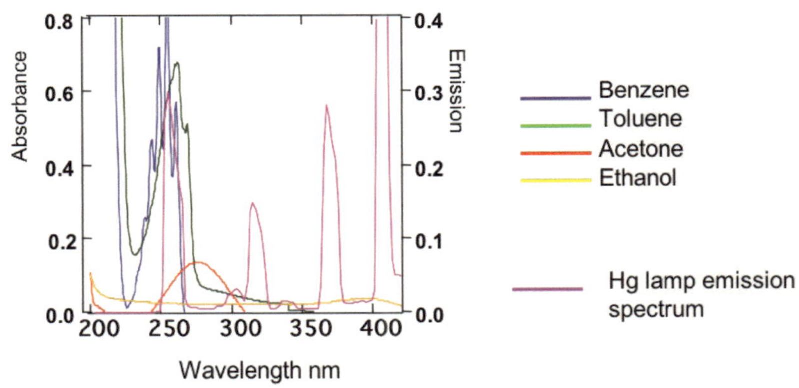

- π→π* excited states of aromatic ring were induced by λ=254nm UV light incidence

Benzene, Toluene, Acetone → Detectable by p-NiO/n-ZnO heterrojunction
Ethanol → Extremely low sensitivity

VOCs having an absorption around λ=254nm line can be detectable using a p-NiO/n-ZnO heterojunction under UV irradiation

Fig. 14. The absorption spectra of toluene, benzene, acetone and ethanol in the UV region. The line spectrum of UV light source is over-plotted in this figure. Ethanol has no absorption in this energy region and the enhanced ethanol sensitivity of *p*-NiO/*n*-ZnO sensor element is not observed even under UV irradiation (shown in Figure 11).

The toluene sensing mechanism is speculated as follows. Under λ=254nm UV light irradiation, adsorbed toluene at the pn junction interface is activated and π conjugated electrons are excited to π^*. Due to the long lifetime, π^* electrons of aromatic ring can react with holes artificially injected and concentrated in *p*-NiO layers by the forward bias applied to the pn junction interface. Active hole injection by applied bias is indispensable for the sensor operation and this speculation is supported by the fact that NiO film itself shows no response for toluene vapor even under full spectrum of UV irradiation. Positive extra-charge introduced by positively ionized toluene adsorbates would decrease or neutralize the interface negative excess charge by adsorbed oxygen or something, then the barrier height of the near-interface of *p*-NiO increases and the current passing through the junction interface decreases as a function of the aromatic VOC concentration.

These processes will be the primary step in full oxidation of aromatic ring over the *p*-NiO/*n*-ZnO junction interface. When one electron is extracted from aromatic ring, extra stability of aromatic compound, which is explained by Hückel's rule is diminished and then successive oxidation reaction would proceed accompanied by the consumption of the adsorbed oxygen over the NiO/ZnO interface. However, further discussion will be needed for determining whether the UV assisted full oxidation of aromatic VOCs proceeds over forward biased *p*-NiO-*n*-ZnO junction interface.

8. Conclusions and perspective

The authors succeeded to detect aromatic VOCs by using *p*-NiO/*n*-ZnO heterojunction sensor element under UV irradiation. The operation temperature is very low and it works even at room temperature. The core mechanism for aromatic VOC detection by a wide gap pn junction is the synergy effect of "electro-" and "photo-" activation of VOC adsorbates, that is, and active carrier injection followed by the photo-driven oxidation (positive ionization) of aromatic ring, which is triggered by remarkable UV irradiation of λ=254nm by a mercury lamp.

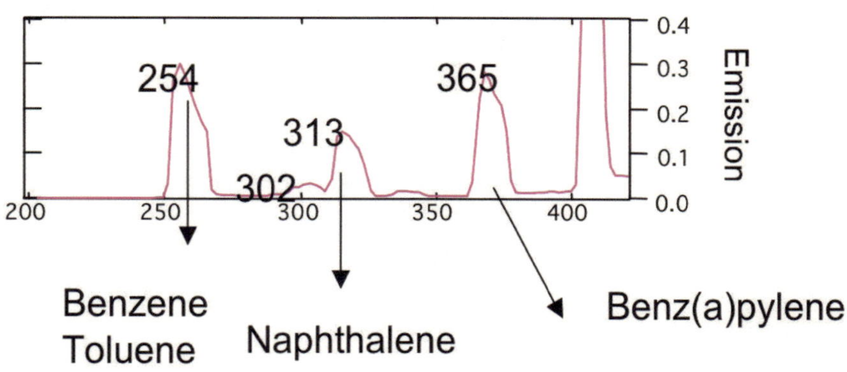

Fig. 15. The relation between the emission spectrum of mercury lamp and the π–π^*absorption line of various aromatic VOCs.

Another novel function, molecular recognition will be expected and molecular recognition of π conjugated electrons systems would be possible by tuning the wavelength of excitation light source. The π–π*absorption line of various aromatic VOCs are over-plotted in Figure 15. For example, molecular recognition of benz(a)pylene, a deadly poisonous molecule, would be possible by tuning the wavelength of excitation light to λ=365nm.

9. References

C.A. Redlich, J.Sparer, M.R. Cullen, (1997), Sick-building syndrome, *The Lancet*, Vol.349, Issue.9057, pp.1013–16, ISSN 0140-6736

U. Schlinka, M. Rehwagena, M. Dammb, M. Richtera, M. Bortec, and O.Herbarth, (2004), Seasonal cycle of indoor-VOCs: comparison of apartments and cities, *Atmospheric Environment*, Vol. 38, pp.1181–90, ISSN 1352-2310

N.Yamazoe, G.Sakai, and K.Shimanoe, (2003), Oxide semiconductor gas sensors, *Catalysis Surveys from Asia*, Vol.7 Issue 1, pp.63-75, ISSN 1571-1013.

C.J.Martinez, B.Hockey, C. B. Montgomery, and S. Semancik, (2005), Porous Tin Oxide Nanostructured Microspheres for Sensor Applications, *Langmuir*, Vol.21, Issue 17, pp.7937-44, ISSN 0743-7463

H.Yanagida, (1988), Intelligent Materials—A New Frontier, *Ang. Chem.*, Vol.100, Issue 10, pp.1143-46, ISSN 0044-8249

H.Ohta, M.Hirano, K.Nakahara, H.Maruta, T.Tanabe, M.Kamiya, T.Kamiya, and H.Hosono, (2003), Fabrication and photoresponse of a *pn*-heterojunction diode composed of transparent oxide semiconductors, *p*-NiO and *n*-ZnO,*Appl. Phys. Lett.*, Vol.83, Issue 5, pp.1029-31 (2003).ISSN 0003-6951

H.Ohta, M.Orita, M.Hirano and H.Hosono, (2000), Current injection emission from a transparent p –n junction composed of p-SrCu2O2 Õn-ZnO,Appl.Phys.Lett. Vol.77, Issue4, pp.425-27, ISSN 0003-6951

W. H. Weinberg,(1991), Kinetics of Surface Reactions, in Dynamics of Gas- Surface Interactions, edited by C. T. Rettner and M. N. R. Ashfold, Royal Society of Chemistry, London, pp. 171, ISBN 0-85186-853-3.

S.Kikuchi, (1997),A History of the Structural Theory of Benzene - The Aromatic Sextet Rule and Huckel's Rule, *J. Chem. Education,* Vol.74, pp.194-201 ISSN 0021-9584

T.H.Wolkenstein, (1960), "Electronic Theory of Catalysis", *Advances in Catalysis*, Vol.12, pp.189, Academic Press, NY, ISBN 0-12007-813-9

Y.Nakamura, H.Yoshioka, M.Miyayama, H.Yanagida, T.Tsturutani, O.Okada, and Y.Nakamura, (1990), Selective CO Gas Sensing Mechanism with CuO/ZnO Heterocontact, *J.Electrochem.Soc.*, Vol.137,Issue 3, pp.940-43, ISSN 0013-4651

T.Hashimoto, H.Nakano, and K.Hirao, (1996), Theoretical study of the valence π→π□ excited states of polyacenes: Benzene and naphthalene, *J.Chem.Phys.*, Vol,104, Issue 16, 6244-58, ISSN 0021-9606

Y.Ishida, A.Fujimori, H.Ohta, M.Hirano and H.Hosono,(2006), Potential profiling of the nanometer-scale charge-depletion layer in *n*-ZnO/*p*-NiO junction using photoemission spectroscopy, *Appl. Phys.Lett.*, 89, paper No.153502, ISSN 0003-6951

Y.Nakamura, Y.Ariga, A.Kishimoto, M.Miyayama, O.Okada, and H.Yanagida, (1999), Electric Control of the Chemical Reaction in Ceramic-Ceramic Interface-Modeling of the Reaction Scheme, *J. Electroceramics*, Vol.4, Issue S-1, pp.105-112, ISSN 1385-3449

R.A.Marra, Y.Nakamura, S.Fujitsu and H.Yanagida,(1986), Humidity Sensitive Electrical Conduction of ZnO-$Ni_{1-x}Li_xO$ Heterocontacts,*J. Am. Ceram. Soc.*, Vol.69, Issue 7, pp.C-143 -47, ISSN 0002-7820

Y.Nakamura, A.Ando, T.Tsurutani, O. Okada, M.Miyayama, K.Koumoto and H. Yanagida,(1986), Gas Sensitivity of CuO/ZnO Heterocontact, *Chem. Lett.*, Vol.15, Issue 3, pp.413-416, ISSN 0366-7022

M.Miyayama, H.Yatabe, Y.Nakamura and H.Yanagida, (1987), Chlorine Gas Sensing Using ZnO/SiC Hetero-contact, *J. Ceram Soc., Jpn*,Vol.95, Issue 11, pp.1145-47, ISSN 0914-5400.

Y.Nakamura, S.Ishikawa, R.Koyama, and H.Yanagida, (1994), NO_x Sensing Characteristics of SiC/ZnO Heterocontact, *J. Ceram. Soc., Jpn, International Edition*, Vol.102, pp.272-78, ISSN 0914-5400.

K.Hirota, M.Miyayama, Y.Nakamura and H.Yanagida, (1994), Quantitative Analysis of Ethanol in Water Solution by a Heterocontact-type Sensor of CuO/ZnO, *Chem. Lett.*, Vol.23, Issue 10, pp.1793-96, ISSN 0366-7022

Y.Nakamura, M.Ikejiri, M.Miyayama,K.Koumoto, and H.Yanagida, (1985), The Current-Voltage Characteristics of a CuO/ZnO Heterojunction, *Nippon Kagakukai-shi*, Vol.1985, Issue 6, pp.1154-59, ISSN 0369-4577

H.Hosono and T.Kamiya,(2003), Research on Ceramic Materials based on the Analysis of Electronic States, *Ceramics, Jpn.*, Vol.38, Issue 10, pp.825-31, ISSN 0009-031X

M.Fujii, T.Kawai and S.Kawai,(1984), Photocatalytic activity and the energy levels of electrons in a semiconductor particle under irradiation, *Chem. Phys. Lett.*, Vol.106, Issue 6, pp.517-22, ISSN 0009-2614

Y.Nakamura, H.Aoki, A.Kishimoto, and H. Yanagida, (2000), The Effect of Ambient Gases on the Carrier Transport through the $YBa_2Cu_3O_{7-d}$ / ZnO Heterocontact Interface, Mass and Charge Transport in Inorganic Materials, Part B, Edited by P.Vincenzini and V.Buscaglia, pp.1463-70, Techna Srl., Faenza, ISBN 88-86538-39-1

Y.Nakamura, Y.Yoshida, Y.Honaga and S.Fujitsu, (2005), Design of a Gas Sensitive Transparent Heterojunction -the system $SrCu_2O_2$-ZnO-, *J. Euro. Ceram. Soc.*, Vol.25, Issue 12, pp.2167–70, ISSN 0955-2219

F. Noritake, N. Yamamoto, Y. Horiguchi, S. Fujitsu, K. Koumoto and H.Yanagida, (1991), Vapor-Phase Growth of Transparent Zinc Oxide Ceramics with c-axis Orientation, *J. Am. Ceram. Soc.* Vol.74 Issue 1, 232–33 ISSN 0002-7820

Y.Hou, X.Wang, L.Wu, Z.Ding, and X.Fu, (2006), Efficient Decomposition of Benzne Over a b-Ga_2O_3 Photocatalyst under Ambient Conditions, *Environ. Sci. Tech.*, Vol.40, Issue 18, pp.5799-5803, ISSN 0013-936X

M.J.Dewitt, D.W.Peters, and R.J.Levis, (1997), Photoionization/dissociation of alkyl substituted benzene molecules using intense near-infrared radiation *Chem. Phys.*, Vol.218,Issue 1-2, pp.211-23, ISSN 0301-0104

S.C.Petittio, E.M. Marsh, and M.A. Langell, (2006), Adsorption of Bromobenzene on Periodically Stepped and Nonstepped NiO(100), *J. Phy. Chem. B*, Vol.110, Issue 3, pp.1309-18, ISSN 6106/1089-5647

Part 4

Artificial Chemical Sensor

12

Inspiration from Nature: Insights from Crustacean Chemical Sensors Can Lead to Successful Design of Artificial Chemical Sensors

Kristina S. Mead
Denison University,
USA

1. Introduction

The chemical sensors possessed by crustaceans are aesthetically beautiful and are deployed in many ways, depending on the ecological task being carried out by the animal, the nature of the chemical signal, and the dispersal characteristics of the environment. As many studies have indicated, they are engineering marvels, able to detect very low concentration of odorant molecules under difficult conditions, and also able to respond to evolutionary pressures, so that over time, populations that migrate to different habitats have evolved sensor characteristics that match the features of their new surroundings.

While it may not be possible at this point to create "smart" materials that can evolve, we can learn from patterns of sensor structure observed in nature. This work typically involves comparative functional anatomy and biomechanics. We can also learn from patterns of sensor deployment. These studies examine the role of sensor movement relative to the animal, or the role of animal movement relative to the environment. These approaches often involve comparative behavioral observations of the animals followed by the development of algorithms that are then tested using biomimetic robots. The ability of the robots to perform basic olfactory tasks- such as finding an odor source- can indicate the validity of the programmed search strategies determined from animal behavior.

While interesting in their own right, these discoveries can also have profound medical, technical, and military implications. For example, researchers are investigating the use of animal-based search strategies to map chemical plumes, to pinpoint dangerous domestic gas leaks, to monitor hazardous water decontamination, to fine-tune the efforts of homeland security in detecting chemical traces, etc.

In this chapter, I plan to discuss approaches and key findings that detail the synergistic approach taken by biologists and engineers in tandem to analyze crustacean chemical sensors. I will start by describing how crustaceans use chemical sensors in a variety of impressive, ecologically critical tasks such as finding food, mates, habitat, avoiding predators, and even recognizing specific individual conspecifics. I will then detail the findings from mathematical models synthesizing these observations, as well as results from programmed robots and autonomous underwater vehicles. Lastly, I will examine the societal benefits and outcomes of this work. I will conclude with a section on promising future directions.

2. Engineers and biologists work together to study biological chemical sensors

Engineers and biologists can share their expertise about structures in order to learn how animals have evolved strategies to solve ecological problems. This knowledge can be translated to the development of technology benefiting humans or human society.

2.1 High performing biological chemical sensors

Often, these projects start when biologists identify species of organisms that perform a particular physiological or ecological task at a high level. For example, male sphinx moths, after encountering a very small number of pheromone molecules, can identify and follow the odor signal released by a female moth for 1-2 miles (Ache & Young, 2005). This ability has led to significant research of insect and crustacean chemical sensors, and the strategies used by these animals to track the odors of interest to their source.

2.2 Fruitful team efforts

Because of the complexity of the science, these experiments often require teams of people including organismal biologists, biomechanicians, neurophysiologists, chemists, mathematical modelers, and engineers.

2.3 Many techniques

In addition to requiring a team of researchers, this work gains strength from a variety of experimental approaches. In order to properly replicate the natural environment, behavioral observations supplementing those made in the field are often made with living animals tracking odors in a carefully calibrated flume. Once the behavioral studies have been completed, the team may rely on scanning electron micrography to understand the fine structure of the chemical sensors. Because successful olfaction requires that the odor molecule come in contact with the chemical sensor, it is imperative to understand the interaction of the sensor with the surrounding fluid (air or water). Thus, a common approach is to make large, dynamically scaled models of the sensor. The resulting information can then be used in the creation of mathematical models.

3. Crustaceans interpret their world through chemical cues

As humans, we are used to our perception of the world being dominated by our senses of sight and hearing. While crustaceans use those senses as well, their dominant senses are those relying on chemical sensors: smell and taste. This is especially true of crustaceans living in low light or noisy conditions.

3.1 Smell vs. taste

For terrestrial animals, smell is typically used to describe the modality of detecting volatile compounds in low concentrations, while taste refers to the chemical detection that occurs during the physical manipulation of highly concentrated material. This distinction between smell and taste is less clear in aquatic species (Ache & Young, 2005), since neither sense deals with volatile compounds in aqueous environments, and because the same compounds, or mixtures of compounds can be of interest in both senses (Caprio, 1977).

However, there may be some relevant distinctions at the chemical sensor level. All crustacean chemosensory organs for both smell and taste are chemosensory sensilla, which are small, hair-like structures containing chemosensory cells and supporting cells. These sensilla are divided into two types: bimodal chemosensory and mechanosensory sensilla located on mouthparts, walking legs, and other parts of the body, and unimodal chemosensory sensilla located on the antennules (Hallberg & Skog, 2011). The bimodal sensilla are thought to function when the chemical source is in direct contact with the animal, while the unimodal sensilla are thought to detect molecules traveling through the water. The following treatment will focus on the unimodal sensilla.

3.2 Roles of chemical cues in animal ecology

Many marine crustaceans rely on water-borne chemical cues in a variety of ecologically critical activities such as finding food, mates, suitable habitat, detecting predators, and communicating with conspecifics (e.g. Atema & Voigt 1995, Caldwell 1979, 1985, Weissburg 2000, Weissburg & Zimmer-Faust 1993, 1994, Zimmer-Faust, 1989). Most crustaceans, including lobsters, crabs, crayfish, prawns, leptostracans, anaspidans, mysids, amphipods, tanaids, isopods, ostracods, phyllopods, cumaceans and stomatopods detect odors from distant sources by using unimodal chemosensory sensillae called aesthetascs (Hallberg & Skog, 2011). The following descriptions describe these activities in more detail.

3.2.1 Chemical cues help crustaceans obtain food and habitat

Environmental odors help organisms find food and suitable habitat. For example, blue crabs (Weissburg, 2000), crayfish (Moore & Grills 1999) and mantis shrimp (Mead et al., 2003) use chemical plumes to find food. At smaller spatial scales, copepods do the same (Koehl & Stricker, 1981). Spiny lobsters use chemical cues to decide whether or not to use a particular shelter (Horner et al., 2006), and anomurans are attracted by the odors emitted by large predatory snails consuming prey snails. In this case, although a food odor is the cue, the goal is not food but a potential exchange of shells negotiated at a trade "market" made up of all the attracted crabs. Even one more available shell may cause a cascade of shell exchanges (Gherardi & Tricarico, 2011). Similarly, despite the danger associated with such a signal, odors emitted by dead or damaged conspecifics are also attractive, with the goal to induce acquisition of better shells (Gherardi & Tricarico, 2011).

3.2.2 Chemical cues help crustaceans find mates

Crustaceans also use chemical substances dissolved in water to find and recognize mates. For example, male copepods can accurately and rapidly follow odor trails left by females, over hundreds of body lengths (Yen et al., 1998, Yen & Laskey, 2011). Shore crabs and blue crabs find mates and form pairs in the presence of sex pheromones (Hardege & Terschak, 2011, Kamio & Derby, 2011). Female reproductive pre-molt decapod shrimp release water soluble substances that elicit mate searching behavior in males. In some species of caridean shrimp, large dominant males release a substance that is attractive to females (Bauer, 2011). In low density systems, the pheromone release is premolt, but in high density systems it appears to be postmolt, possibly limiting harassment (tumults) in dense populations (Bauer, 2011). Lobsters also release and detect chemical cues to find mates. Males release compounds in their urine that help female lobsters locate a potential mate. Once the females are nearby, they release chemicals that drive subsequent mating behaviors (Aggio & Derby 2011).

3.2.3 Chemical cues mediate social interactions in crustaceans

Once the potential partners are in proximity, pheromones mediate sexual attraction and mating readiness, but can also signal social status, aggregation, and alarm (Ache & Young, 2005). Decapod shrimp (Bauer, 2011) and anomuran crabs and lobsters (Gherardi & Tricarico, 2011) can recognize members of the opposite sex via contact chemoreception, perhaps (at least in decapod shrimp) using cuticular hydrocarbons and glycoproteins (Bauer, 2011). Chemical cues may also indicate advanced reproductive state, i.e. spawning readiness in females and sperm availability in males (Gherardi & Tricarico, 2011). Contact cues may also mediate initial pair formation in monogamous shrimp species that establish permanent bonds (Bauer, 2011).

In contrast, mantis shrimp rely on chemical cues that act over a distance to identify sex and relative social position (Mead & Caldwell, 2011). For example, mantis shrimp enter empty burrow cavities spiked with the odor of a mantis shrimp that lost a dominance battle with the test mantis shrimp much more rapidly than they enter an empty burrow cavity containing the odor of a victorious mantis shrimp (Caldwell, 1985). Dominant or subordinate status is also mediated by odors that act over a distance in crayfish (Breithaupt, 2011). During dyadic contests, crayfish emit more urine containing chemical signals as the aggression escalates, and the eventual winner emits more than the eventual loser (Breithaupt, 2011). Since urine contains the metabolites of current and recent biochemical processes, it is quite feasible that these cues contain important physiological information about the signaler. Biogenic amines such as serotonin and octopamine, implicated in aggression in crustaceans, are candidate molecules (Breithaupt, 2011).

3.2.4 Chemical cues are important in sibling and individual recognition

Several crustaceans employ chemical cues in sibling or individual recognition. Decapod shrimp can recognize previous pair partners, also via contact chemoreception, (Bauer, 2011). They appear to spend less time on courtship behaviors when reintroduced to previous mating partners than when encountering individuals of the opposite sex which they had never met (Johnson, 1977). A similar effect is also seen in the mantis shrimp *Pseudosquilla ciliata* (Mead & Caldwell, 2011). In other species of mantis shrimp, males are less aggressive with previous mating partners than they are with females with which they have not been paired (Caldwell, 1992). As mentioned above, mantis shrimp also recognize previous winners and losers of cavity contests (Caldwell, 1985). Lobsters (Aggio & Derby, 2011, Karavanich & Atema 1998a) and crayfish (Breithaupt, 2011) have this ability as well. Hermit crabs may have an even more refined ability to remember individuals, able to identify a conspecific as familiar outside the winner/loser dichotomy (Gherardi & Tricarico, 2011). This recognition appears to be associated with properties of the shell.

3.2.5 Chemical cues assist in avoiding predators

Chemical cues are not necessarily attractive; in the case of predator cues or of damaged conspecifics, they are usually repulsive. For instance, blue crabs avoid traps baited with injured conspecifics (Hay, 2011). Crustaceans can exhibit a variety of responses to chemical cues (crushed conspecifics, or predator odor), such as decreasing their forward movement, reducing movement in any direction and assuming protective positions, moving away from the source of odor, increasing defensive displays, and increasing preference for protective habitats (Hazlett, 2011). When crayfish are exposed to a combination of food odors and predator odors, they reduce their foraging, increase their use of refuges, and avoid the alarm odor by spending more time outside of the odor plume (Tomba et al., 2001). Of course, this results in a decrease in foraging success.

4. Crustacean chemosensor structure

All chemosensory sensilla contain ciliated bipolar sensory cells. The bimodal sensory cells, are thought to function in contact chemosensation. In contrast, most of the cues described above that act over a distance are detected by unimodal chemical sensors called aesthetascs. The description below focuses on aesthetascs. In order for distance chemoreception mediated by aesthetascs to work, the odor molecules must first be transported to the surface of the aesthetasc, the cuticle. The morphology of the aesthetascs, their arrangement on the antennule, and the movement of the antennule relative to ambient water motion affect the flow of water bearing odorants around the sensillae. This in turn affects the transport of odorants to the aesthetasc surface and hence to the chemoreceptors within the sensillae. Thus, the structure and deployment of the antennules affect the ability of the animal to identify the contents of the plume and locate its source

4.1 Contact chemoreception is usually mediated by bimodal sensilla

Bimodal sensilla contain both chemosensory and mechanosensory cells. These structures typically have a thick, dense cuticle with an apical pore, and are arranged all over the body but especially densely on the mouthparts and walking legs (Hallberg & Skog 2011). They can be large and robust, as in the lobster hedgehog sensilla, or small and slender, as in some crab bimodal sensilla. Often the former are found on the walking legs while the latter are found on mouthparts, antennae, and antennules (Hallberg & Skog 2011).

4.2 Chemoreception of odor from distant sources is usually mediated by aesthetascs and antennules

The most common unimodal sensillum found in crustaceans is the aesthetasc. Aesthetascs are stiff, cuticular hair-like structures organized in arrays found exclusively on the antennules (Hallberg et al., 1992). Aesthetascs are typically long and slender, varying from 5-30 um in diameter and 10-1400 um long, depending on the species and the age, sex, and condition of the individual, and the flow habitat. For example, aesthetascs in low flow environments tend to be longer and more spaciously arrayed, to facilitate odorant access (Ziemba et al., 2003, Mead, 2008). In contrast to bimodal sensilla, aesthetascs have a porous cuticle that allows low molecular weight molecules to pass through, thus acting like a molecular sieve (Gleeson et al., 1993). Male-specific sensilla, another type of unimodal olfactory sensilla, are found in only a few groups, including mysids (Hallberg & Skog, 2011).

4.3 Two signaling pathways

The dichotomy of the two types of chemoreceptors is reflected in the organization of the CNS pathways. Evidence from spiny lobsters and other crustaceans suggests that many crustaceans possess two pathways that contribute to chemically mediated responses (Mellon, 2007, Schmidt & Ache, 1996, Schmidt *et al.*, 1992). In the olfactory pathway, olfactory neurons within the aesthetascs project into a synaptic region of the midbrain called the olfactory lobe which is organized in dense regions called glomeruli (Schmidt & Ache, 1992). The non-olfactory chemoreceptive pathway involving contact chemoreception via bimodal sensilla projects onto a variety of synaptic areas distributed throughout the brain and ventral nerve cord (Schmidt & Mellon 2011).

4.4 Aesthetasc cellular structure and receptors

Within the cuticle, each aesthetasc typically has between a few to a hundred bipolar sensory cells, which may have highly branched outer dendritic surfaces. This multiplicity of neuronal processes is thought to increase the surface area for stimulus capture. (Ache & Young, 2005). The outer dendritic surfaces are sprinkled with odor receptors, which are G protein coupled receptors (Buck & Axel, 1991). It is thought that each sensory bipolar neuron expresses one receptor type, and projects to one glomerulus.

4.5 Variable number of sensors in the array

The number of aesthetascs per antennules is variable. Burrowing decapods such as *Upogebia* spp. and *Callianassa* spp. have 15-22 aesthetascs per antennules (Hallberg & Skog, 2011). Decapods such as *Panulirus argus* have between 1,200 and 2,400 aesthetascs, depending on the size of the individual (Hallberg & Skog, 2011). It is not yet known to what extent this difference in sensor number is reflected in olfactory use or ability to use olfaction in the various tasks outlined in section 3.

4.5.1 Role of number of sensors

Especially when considering how knowledge of animal sensors can be applied to artificial sensors, it can be interesting to speculate on the performance advantage conferred by multiple sensors (Derby & Steullet, 2001). This question can be considered biologically, chemically, and physically. In terms of biology, crustaceans rely on olfaction for a number of critically important ecological tasks, sometimes involving exquisitely complicated behavior. It stands to reason that there would be a many-sensored apparatus capable of resolving nuanced information to support these tasks. Also, given the developmental or molt stages in which aesthetascs are nonfunctional and also the inevitable damage likely to sensor arrays belonging to crustaceans that engage in aggressive encounters, it would be practical to have some redundancy. Chemically, more sensors might mean that more odor molecules, or more mixtures of compounds, are distinguishable. Similarly, many sensors allows for the possibility of range fractionation. If different populations of sensors sensitive to the same molecule are responsive at different concentrations, more sensitive measurements and responses might be possible. In terms of physics, range fractionation could also occur in terms of temporal dynamics. As a result, different populations of sensors responsive to the same molecule might respond and thus adapt over different time courses, resulting in subsets of receptors best able to detect slow, medium, or rapid changes in chemical concentrations. This ability could also be important in some models of odor plume navigation. As we will see later, the arrangement of sensor arrays is important too. Some of the methods of following odor plumes to their source require at least two sensor arrays, for spatial comparison of odor concentration. Lastly, larger arrays, with more sensors, have an increased probability of encountering chemical signal. This increase in size could be especially important in flow conditions that give rise to sparse odor filaments within the odor plume envelope, as described below.

4.5.2 Role of shape of sensor

Crimaldi et al. (2002) used physical and mathematical models to analyze the effect of sensor size and shape, among other characteristics. When they used similar-sized sensors that were either square-shaped or long, skinny rectangles, they found that the linear arrangement was much more efficient at capturing odor signals from the environment when the average odor concentration was low, but that the square arrangement had better performance when the average odor concentration was high.

4.5.3 Sensor response time

The odor-containing filaments within the odor plume are often very narrow, and thus of short duration. Therefore, chemical sensors with a slow response time (e.g. slower than 100 Hz, for a small sensor) are apt to register lower concentrations than the actual peak concentration encountered because the peak is temporally "smeared" over the entire response time of the sensor. Thus, the likelihood of detecting large concentrations decreases dramatically as the temporal response of the sensor becomes slow (Crimaldi et al., 2002). Interestingly, large sensors are less sensitive to this temporal smearing, with a cut-off of about 10 Hz rather than the 100 Hz for the small sensors. Smearing also occurs spatially. Because odor filaments are highly localized, large sensors operating at a low frequency (e.g. slower than 1 Hz) are apt to register lower concentrations than the actual peak signal encountered because the signal is averaged over the full extent of the sensor and effectively diluted.

5. Odor plume structure

The dispersal of chemical signals is controlled by the physics governing flow. At low flow velocities and at small length scales, flow is dominated by viscous processes. In this laminar regime, experienced by animals such as copepods, streamlines do not cross, and the odor plume is relatively coherent (Weissburg, 2011). Animals attempting to follow such an odor plume would experience sharp gradients in chemical concentration, especially in the cross-stream direction. In contrast, many animals, especially those in coastal and/or wavy environments, experience more turbulent flow regimes. These environments are characterized by unpredictable velocity fluctuations, leading to fluctuations in the chemical concentrations of odor plumes downstream from the odor source. In these circumstances, experienced by many larger benthic crustaceans including crabs, lobsters, crayfish, and mantis shrimp, turbulent odor plumes are composed of filaments of odor-laden water interspersed with odor-free fluid (e.g. Crimaldi et al., 2002, Crimaldi & Koseff, 2001, Koehl, 2006, Moore & Crimaldi 2004, Moore et al., 1991, Moore et al., 1994, Webster et al., 2001,

Weissburg, 2000). The structure of an odor plume (and thus how the plume is encountered by navigating animals) is affected by several flow and substrate characteristics including the mean velocity, the turbulent intensity, the presence of waves and the gradient of flow speed above the substratum (the current boundary layer) (Crimaldi & Koseff 2001). As flow becomes more turbulent, the odor filaments typically thin, elongate, and have lower odorant concentration. This is probably because the more numerous eddies stretch the odor-containing fluid into longer elements that increase the surface area across which diffusion can act. Often the odor filaments are more numerous (Finelli, 2000), and more often above a concentration threshold (Crimaldi & Koseff 2001). Different substrates, such as sand, gravel or cobble, and the type of vegetation growing on the substrate, give rise to different boundary layer features and different chemical signal structures due to changes in the turbulent intensity (Finelli, 2000, Moore et al., 2000).

5.1 Effects of plume structure on plume tracking

These differences in hydrodynamics affect plume tracking success. Blue crabs are more successful at finding odor sources when the flow speed is above 1 cm/s, and walk straighter and more directly upstream as flow velocity increases (Weissburg & Zimmer-Faust, 1993, Finelli et al., 2000). Stomatopods are more successful at finding odor sources in wave-affected flow than in unidirectional flow (Mead et al., 2003). Crabs find odor sources more efficiently on a cobble substrate rather than on sand (Moore & Grills, 1999).

5.2 Features of odor plume structure of interest to navigating crustaceans

It is not clear precisely what aspects of environmental odor signals are important to the animal, but progress has been made in terms of understanding the responses of olfactory neurons. Different olfactory cells probably respond to different aspects of odor filament structure, but at least some olfactory neurons appear to be sensitive to peak odor concentration and odor pulse duration (Gomez & Atema, 1996a, 1996b; Gomez et al., 1999). In addition, models and experiments suggest that sensory cells can distinguish between different rates of increasing odor concentration (onset slope) (Kaissling, 1998a, 1998b , Rospars et al., 2000, Zettler & Atema, 1999).

6. How crustaceans deploy their sensors

Many crustaceans create some kind of movement or current to facilitate olfaction. In some cases, the movement is by the signaler, as in the case of chemicals emitted in urine streams directed forward by strong gill currents in crayfish (Breithaupt, 2011). Since urinary distribution of chemical cues, especially sex pheromones, and gill currents, are common in crabs and lobsters (Aggio & Derby, 2011), it is probable that these odor-laden streams or jets are common among the decapod crustaceans. These jets can impart significant energy to the surrounding flow (Weissburg 2011). Pleopod fanning, used by crayfish (Breithaupt, 2001), male blue crabs (Weissburg, 2011) and mantis shrimp (Mead & Caldwell, 2011), and maxilliped pulsing, used by mantis shrimp (Mead & Caldwell, 2011) are other movements used to increase the transport of scent away from the animal.

Animal movements can facilitate chemoreception as well. Pleopod fanning and maxilliped pulsing can be used by mantis shrimp to create currents that draw water past chemosensory surfaces (Mead & Caldwell, 2011). Crayfish use a fan organ for the same task (Breithaupt, 2001).

Other movements or currents are generated only by the receiver. For example, many crustaceans move their antennules- and thus their aesthetascs- through the surrounding medium to facilitate odor reception. One common movement found in crayfish, lobsters, crabs, mantis shrimp, and other crustaceans is an olfactory flick (Koehl, 2011). Other receptive sensor movements include antennulation in mantis shrimp (Mead & Caldwell, 2011). These currents or movements of the sensory appendages allow chemical cues to better penetrate the sensory arrays. The sensor movements implicated in signal reception and the ways in which the motions benefit olfaction are further described below.

6.1 Pleopod fanning and other fan organs

Pleopod fanning or use of crayfish fan organs can create large, local currents greater than the ambient flow. When crayfish fan organs are restrained so that no current is produced, they are unable to localize odor sources (Breithaupt, 2001). Recent field observations of *Squilla empusa* (Mead, *unpublished*) indicate that mantis shrimp often lurk at the entrance of their burrows with their eyes and antennules near the substrate surface. In this orientation, pleopod fanning draws fluid into the burrow past the antennules. Mantis shrimp appear to engage in pleopod fanning when a variety of olfactory stimuli- food odors, predator odors, odors from conspecifics- are presented. Since these mantis shrimp currents are also essential for keeping oxygenated water moving through the burrow, it has not been possible to test the effect of pleopod fanning currents on olfactory response by restraining the pleopods, because this would quickly be lethal. However, since the benthic boundary layer near the mantis shrimp burrows is typically on the order of 2-3 cm, it is likely that mantis shrimp lurking in burrows would be less likely to detect, let alone respond to, olfactory stimuli in the absence of pleopod fanning.

6.2 Maxilliped pulsing

Maxilliped pulsing creates more local and more transient changes in water flow and thus in chemical signal distribution. Maxilliped pulsing is often seen in burrow occupants as an intruder is approaching the entrance. The movement consists of a series of simultaneous anterior extensions of multiple pairs of maxillipeds toward the intruder. The appendages move in circular strokes, with cycle rate, extension distance, and the openness of the maxilliped dactyls all increased the closer the male came to the entrance. Although dye studies indicate that most of the resultant fluid motion is away from the occupant, small return currents bring water from the side and across the female's antennules (Caldwell 1992). Maxilliped pulses, therefore, may function to both send and to receive signals.

6.3 Antennulation

Antennulation consists of the forward extension of the paired antennules coupled with short, quick oscillations in the vertical plane (Caldwell, 1992). Antennulation typically occurs as a male is approaching a cavity inhabited by a female, possibly to garner information about the inhabitant's reproductive status and to determine, if female, if the occupant is a previous mate. Antennulation appears to be a subcategory of olfactory flicking used by mantis shrimp in many investigatory situations, such as in response to a food odor or in tracking an odor to its source.

6.4 Olfactory flicking

Olfactory flicking describes an antennule odor sampling movement observed in many crustaceans. In all crustaceans studied to date, including shrimp, mantis shrimp, crabs, crayfish, and lobsters (Koehl, 2011), each flick has two parts: a rapid, outward lateral movement followed by a slower return medial movement. The functional importance of flicking, and indeed of all the other sensor movements described in this section, is based on the idea of a boundary layer.

6.5 A layer of slow-moving fluid coats the antennules

An antennule immersed in a fluid can be roughly thought of as having layers of fluid coating every surface. Because of the "no-slip" condition, the layer of fluid closest to the antennule's surface does not move with respect to that surface. Thus, a velocity gradient forms in the fluid between the antennule and the mainstream flow. Typically, the distance from the surface to the point where the velocity is 99% of the mainstream velocity is called the boundary layer (e.g., Vogel, 1994).

6.6 Olfactory sampling currents and movements "thin" the boundary layer, speed up the delivery of chemical cues, allow discrete sampling, and increase the probability of signal encounter

When the appendage moves rapidly through the fluid (in a flick or an antennulation), or when the ambient flow is fast relative to the antennule (because of environmental flows or animal-created currents such as pleopod fanning), the boundary layer of slow-moving fluid coating the antennules gets thinner. The thickness of the boundary layer is very important for olfaction because it can act as a barrier to odorant access to the aesthetascs. While odorants can be very rapidly transported through the environment in currents (at speeds on the order of centimeters to meters per second), the time required to cross the boundary layer relies on the much slower process of molecular diffusion.

Thus, flicking can speed up the delivery of chemical cues to the surface of the aesthetasc. Advection-diffusion models using data from dynamically scaled physical models of mantis shrimp antennules indicate that odor molecules in filaments moving into the sensor array arrive at the surfaces of the aesthetascs within milliseconds (Stacey et al., 2003). In contrast, during the slower return stroke of the flick and during the stationary pause between flicks, the just-sampled fluid remains in a layer coating the sensors, and new, odor-containing fluid flows around rather than into the array of chemosensory hairs (Mead et al., 1999; Mead & Koehl, 2000). Thus, an antennule of a mantis shrimp takes a discrete sample in time and space of its odor environment only during the flick outstroke (Mead & Koehl, 2000, Stacey, et al., 2003). This pattern of discrete sampling appears to be widespread among crustaceans (Koehl, 2011). Other ventilatory mechanisms in other phyla have similar functions, suggesting that the ability to take discrete samples of the chemical environment is integral to the success of odor discrimination (Ache & Young, 2005). Lastly, physical and mathematical modeling of lobster flicks indicate that flicking increases the likelihood that the antennule encounters peak concentrations of signal (Crimaldi et al., 2002).

6.7 The frequency of sensor movements

When a crustacean samples water with a flick outstroke, the movement of the antennule changes the rate at which chemical signals are encountered. If the flicking is against the mainstream flow, (i.e. the animal is angled upstream) the velocity of the antennules relative to the ambient flow will be increased, so that the speed of signal encounter will increase. In addition, since flicking into the direction of flow also increases the area of the plume that is sampled, flicking antennules encounter more odor relative to stationary antennules (Crimaldi et al., 2002).

7. Crustaceans are able to localize odor sources

Once crustaceans have detected chemical cues from distant sources, identified them, and determined that they are of interest, they often have to navigate the odor plume to its source to find food, habitat, or a mate. The difficulty of this task depends both on the flow environment, which sculpts the odor signal, and on the sensory and locomotory abilities of the animal. Organisms tracking odors must sample the odor signal, usually by moving their chemosensitive organs through the odor-containing medium, or by creating currents that bring the odor to the chemosensors. Since odor plumes consist of fine filaments of concentrated odor molecules interspersed with the surrounding fluid, and since the filaments can be twisted along all axes and the entire plume can meander, the organism has to continuously sample the plume to get updated information about possible locations of the source. This can lead to more complicated tracking strategies than those found in organisms orienting to light or sound. Various taxes and odor localization schemes have been proposed to describe these odor plume search strategies.

7.1 Navigating using presence/absence of odor and flow direction

One of the simplest navigational strategies proposed is odor-gated rheotaxis. In this strategy, an organism moves upstream as long as it detects odor above a threshold concentration or frequency. This strategy requires only the ability to distinguish between absence and presence, and the ability to determine the direction of flow. Behavioral data sets in support of this idea include the fact that some crustaceans do not respond to odor cues in the absence of flow (Weissburg & Zimmer-Faust, 1993).

7.2 Navigating using time-averaging and temporal comparisons

An organism searching odor sources using klinotaxis moves upstream as long as sequential measurements show a stable or increasing concentration of odor. For this to work, each measurement has to reflect a time-averaged signal, since instantaneous measurements of odor concentration fluctuate over orders of magnitude. Most crustaceans do not appear to sample long enough or fast enough in one location to be able to reliably estimate average odor concentrations. This strategy has been hypothesized to be most effective in large, slowly moving predators such as starfish, whelks, and catfish.

7.3 Navigating using spatial comparisons

Many crustaceans appear to be using a variation of strategies in which the organism compares inputs to more than one chemosensory structure, for example left and right antennules. Removing the chemosensory structures on one or both sides limits plume-tracking success in crayfish (Kraus-Epley & Moore, 2002). In the version of the spatial comparison strategy used by lobsters, the animal appears to turn slightly to the sensor that intercepted the highest chemical concentration. The lobster thus remains in the "center" of the odor plume, where there is a greater likelihood of encountering concentrated odor filaments (Weissburg, 2011). Other navigational strategies relying on spatial comparisons seem to cause animals to move upstream when the left and right chemosensory inputs are different. This mechanism could result in animals tracking the "edges" of plumes, as blue crabs (Weissburg, 2011) and mantis shrimp (Mead et al., 2003) appear to do, because the antennules closest to the plume "center" would have a greater likelihood of encountering high concentration than the antennules sampling in a more lateral position. Note that plume "center" and plume "edge" are statistical concepts only evident after long time-averaging.

7.4 Other navigational strategies used by very small crustaceans

While probably all organisms can detect some classes of chemicals, organisms that are very small may have difficulty detecting the differences in concentration across space that could assist in orienting to an odor source. It is possible that some very small crustaceans, like bacteria, adopt a kinesis strategy in which they move in larger randomly oriented steps when subject to low concentration, and smaller steps when exposed to high concentration. Eventually, this 'random walk' positions them in areas of higher concentration. Other microorganisms alter their rate of rotation, with similar effect (Vogel, 1994).

8. What can we learn from animal sensors to apply to artificial sensors?

Table 1 sums up some of the factors affecting performance in crustacean chemical sensors discussed in sections 4-7. The summary is broken into 1) structural factors affecting the isolated sensor, flow factors affecting chemosensor performance, and animal factors affecting sensor performance.

8.1 Design principles

Studies of crustacean chemosensor design suggest that artificial chemical sensors should be composed of multiple small elements, distributed spatially across the platform. The sensors should be a sensitive as possible, and sample at a high frequency, in order to detect the maximum number of encountered odor filaments. The ability to make simultaneous spatial comparisons between at least two sensors facilitates plume tracking (Webster et al., 2001).

8.2 Role of movement

Artificial sensors will probably be the most effective if they can move through the environment or have the fluid medium flow past them. A repetitive, asymmetrical movement analogous to the ubiquitous crustacean flick will 1) hasten the arrival of the chemicals at the sensor array, 2) facilitate the creation of discrete samples, and 3) increase the size of the sample space.

Factor	Effect	Rationale
Factors affecting sensor performance		
Number of sensors	More sensors better	Redundancy, can detect more chemicals, accuracy, range fractionation, if can compare between arrays can probably navigate to source more effectively
Size of sensor array	Bigger array better	Can intercept more signals; many small sensors better than one big one, to avoid spatial smearing
Ability to resolve discrete samples	Essential	Can potentially make temporal comparisons, necessary for some navigational schemes
Sample frequency	Higher frequency	Higher frequency leads to less temporal smearing, so more odor filaments encountered
Factors affecting flow and/or odor plume structure		
Ambient current	Can be essential	Current aids in establishing directionality
Current or sensor movement produced by animal	Aids in signal reception	Can improve chemical cue reception by favoring advective transport over slow molecular diffusion, but will affect signal downstream
Turbulent mixing	Can increase or decrease navigation success	Turbulent mixing increases plume volume, increasing the likelihood that animals will encounter signal, but decreasing signal strength via mixing
Substrate roughness	Can increase or decrease navigation success	Turbulence increases as substrate roughness increases; this can positively or negatively affect a sensor's ability to detect signal
Factors affecting odor plume navigation		
Animal size	Large animals have better spatial sampling	Bigger animals can use spatially distributed sensors
Animal mobility	More search strategies available to slow animals	Slow animals may be able to employ time-averaging
Animal antennules flicking	Facilitates plume tracking	Increases encounter of peak concentration

Table 1. Factors affecting crustacean chemical sensor performance

8.3 Applications in medical, technical and military realms

There are many potential applications of knowledge about crustacean chemosensors to the medical, technical, and military fields. Recently, several groups have developed autonomous, chemical sensing robots that can track odors using strategies derived (in part) from crustaceans (Grasso, 2001, Ishida et al., 2001). Large scale devices could work in open environments to find and destroy unexploded mines or repair leaks of toxic substances. Small scale devices could potentially detect and identify odors of life-threatening chemicals in enclosed hospital environments, or even within the human body. Other applications include quality control in industrial settings. Most of these developments have been based on developing sensors appropriate for the task at hand, and combining them with the tactic strategies employed by living organisms.

8.3.1 Human health applications

Pathfinding using chemical cues is an important process with many applications in normal growth and development. For instance, the location of certain chemoattractants in the extracellular matrix (ECM) can be critical for the appropriate laying down of new neurons (neurogenesis) and blood vessels (angiogenesis). The subversion of these processes is implicated in some forms of cancer metastasis, and in many other diseases and syndromes, including atherosclerosis, arthritis, infertility, and tooth decay. If we better understand sensors and odor trail navigation, we might be more equipped to predict and treat malfunctions of chemical sensor operation in the body. Furthermore, the body emits characteristic odors via the skin, blood, urine, and breath. These odors reflect genetics, physiological state, diet, metabolic condition, age, the immune response, and the effects of stress by virtue of VOCs (volatile organic compounds). Typically, there is no one particular marker of state or disease, but changes in the relative abundances of key VOC components. For example, VOC changes due to schizophrenia can be detected in skin odor, alterations in blood VOCs may indicate liver cancer, breath VOCs may predict diabetes, breast cancer, and cystic fibrosis, and urine VOCs may help diagnose urinary tract infections (Oh, et al., 2011). It has been possible to study at least some of these VOCs using gas chromatography, mass spectrophometry and other methods, but these techniques have been time-consuming and expensive. Electronic noses are small, portable, cheaper to use once developed, and give results more quickly.

8.3.1.1 Structure of electronic and bioelectrical noses for biomedical applications

Currently, there are research and commercially available electronic noses based on a variety of nanotechnologies including metal oxide semiconductor sensors, conducting polymer-based sensors, piezoelectric sensors, and optical sensors, described beautifully in Oh, et al. (2011). A new generation of bioelectrical sensors is also being developed. These bioelectrical noses consist of harvested or cultured cells or odorant receptors grown or otherwise attached to microchips that respond with an electrical output as part of the odor signal transduction process. Both the electronic noses and bioelectrical noses represent areas of active research. Two of the most promising health applications using electrical noses will be described below.

8.3.1.2 The detection of bacterial populations

Dutta et al. have used an electronic nose consisting of 32 black carbon composite conducting polymer sensors to distinguish between populations of bacteria (2002). When VOCs are introduced to the system, any of the 32 individual sensors that bind to any of the VOCs swell and change their conductivity. These changes create a combinatorial profile in the 32 sensors that can be analyzed via principle component analysis to identify specific species and types of bacteria. One of the most interesting (of many!) applications of this sensor has been to detect and distinguish MRSA (methicillin resistant *Staphylococcus aureus*) and MSSA (methicillin susceptible *S. aureus*; Dutta et al., 2005). In this report, swabs from infected areas of patients were placed in 15 ml vials for 5 minutes, to allow the aromatic compounds to volatilize. The resulting sa,ple was then "sniffed" by the electronic nose. Using innovative data clustering techniques, Dutta et al. (2005) were able to identify bacterial classes with very high accuracy. If placed in hospital ventilation systems, this electronic nose could theoretically detect and prevent contamination of other patients or equipment.

8.3.1.3 The use of electronic noses to detect lung cancer

Peng et al. (2009) have developed an electronic nose based on cross-reactive chemiresisters in which the organic molecules that respond to particular odorants are adsorbed onto gold nanoparticles. The gold/organic film lies between adjacent electrodes; the resistance between the electrodes changes in response to odorant binding, which probably causes the film to swell. The sensors were able to distinguish between breath samples from healthy and cancerous individuals with > 86% accuracy and >90% repeatability.

8.3.2 Industrial and military use

Artificial noses are also used to monitor VOCs in industrial settings. Recent uses include monitoring the quality of tea (Dutta et al., 2003), and checking pork products for the presence of boar taint (Wäcker et al., 2010). The electronic noses used by Dutta and others are typically used in a stationary manner, with the sampled air drawn into the apparatus much like a vertebrate sniff. Other artificial chemical sensors are being developed for robotic platforms, so that they could not only identify chemical compounds, but search and find them, much as insects and crustaceans follow odor plumes in the environment.

8.3.2.1 Autonomous electronic nose searching platforms

In addition to the high sensitivity, selectivity, and rapidity of response that are optimal in all electronic noses, autonomous sensors expected to move freely in an environment must also be small, consume power efficiently, and be robust enough to withstand the elements and any other hazards. Common approaches for these sensors include tin oxide sensors and quartz microbalance sensors (Russell, 2001). Tin oxide sensors work on the principle that when the heated sensor encounters reducing gases, resistance decreases. Quartz crystal microbalances function by detecting changes in crystal resonant frequency as the chemical of interest is adsorbed onto the quartz. Robotic platforms equipped with at least two of these sensors and a inward drawing fan to facilitate odor interception have been programmed with variations on the plume searching and plume tracking algorithms used by crustaceans described in section 7.3. (Kowadlo and Russell, 2008). These robots were 70%-100% successful, depending on the nature of the flow field, the odor source, the particular robot platform, and the signal processing algorithm.

Encouraged by the success of trained dogs at finding explosives, researchers are excited about the possibility to developing odor-searching robots that could find traces of TNT and other explosives. These robots could potentially find and signal the presence of unexploded ordinance, old abandoned mines, and other hazards (Russell, 2001). Other potential uses include implants or roving sensors that could warn of the presence of natural gas, radioactive material, or could monitor water quality, especially during the clean-up of hazardous materials. In addition to the unexploded ordinance application mentioned above, field-deployed sensors could also help to detect the presence of biological toxins and other agents of war. These latter applications would have the advantage of limiting human exposure to potentially lethal situations. In an interesting, iterative relationship, one of the most potent applications of the artificial sensors coupled with robotic, motile platforms is to test the search algorithms predicted from animal behavior. This approach has been used effectively by Grasso (2001), Ishida (2001), and others.

9. Future directions

While much progress has been made, much work remains to be done, in terms of both basic science and its applications.

9.1 Basic science

From a basic research perspective, the chemicals used in olfactory signaling are still unknown, in all but a few cases. Furthermore, it is suspected that in most cases, the cues consist of mixtures of compounds. The ways in which nuanced physiological condition could be communicated via small alterations in ratios of mixtures will no doubt be a fruitful area of future study. Another important future topic for both basic and applied scientists should be the role of the multiple sensory modalities used simultaneously by many crustaceans and available for sensor designers. For instance, mantis shrimp probably use vision as well as their sense of smell, at least when light conditions are favorable, and copepods and crabs rely on hydromechanical cues as well as olfactory signals.

Other basic questions involve the role of deception using chemical signals, the energetic costs to chemical signaling, and the social costs of chemical signaling. These questions are connected. For instance, when is it against an organism's interest for information about its physiological state to be broadcast into the environment?

9.2 Applied aspects

While the chemical sensors deployed in a industrial or hospital setting may face predictable environments, one of the biggest challenges to using chemical sensors in the field is the unpredictability of the flow environment. Many features of the odor plume- peak odor concentrations, filament width, filament sharpness, etc. vary both with environmental factors and with distance from the source. Different odor plume tracking strategies used by crustaceans may work best under different flow conditions, and some crustaceans even seem to vary their strategy based on odor plume structure (Grasso, 2001). Thus, it will be important to develop autonomous devices that can flexibly and intelligently adopt the best navigational strategy given the task, the habitat and the available odor signal, even more so than as described in Kowadlo and Russell (2008). Lastly, the ability to deploy different sensors given the exigencies of the environment would be an asset. Here, too, the use of multiple sensory modalities should be a future topic (Grasso, 2001, Grasso & Atema, 2002).

10. Conclusions

In conclusion, crustacean chemical sensors are remarkably effective structures that are essential in helping these animals complete a variety of critical ecological tasks. By studying crustacean chemical sensors, we can both contribute to the understanding of important biological questions and provide inspiration to engineers trying to create artificial sensors that can carry out important and dangerous tasks for humans.

11. Acknowledgements

I am grateful to InTech for the invitation to submit this chapter, to Denison University for funding during part of the writing period of this chapter, to the Marine Biological Laboratory for access to their wonderful library holdings, and to Breithaupt and Thiel's book *Chemical communication in crustaceans*, a great resource.

12. References

Ache, B. W. & Young, J. R. (2005). Olfaction: diverse species, conserved principles. *Neuron* Vol. 48: 417-430.

Aggio, J. & Derby C. D. (2011) Chemical communication in lobsters. Chapter 12 in *Chemical communication in crustaceans.* Eds T. Breithaupt and M Thiel. Pp. 239-256, Springer ISBN 978-0-387-77100-7. 565 pp.

Atema, J. & Voigt, R. (1995) Behavior and sensory biology. In *Biology of the Lobster Homarus americanus* (ed JR Factor), pp. 313- 348. San Diego: Academic Press.

Bauer, R. T. (2011). Chemical communication in decapods shrimp. Chapter 9 in *Chemical communication in crustaceans.* Eds T. Breithaupt and M. Thiel. Pp. 277-296, Springer ISBN 978-0-387-77100-7. 565 pp.

Breithaupt, T. (2001). Fan Organs of Crayfish Enhance Chemical Information Flow. *Biol. Bull.* 200: 150-154.

Breithaupt, T (2011). Chemical communication in crayfish. Chapter 13 in *Chemical communication in crustaceans.* Eds T. Breithaupt and M Thiel. Pp. 257-276, Springer ISBN 978-0-387-77100-7. 565 pp.

Buck, L., & Axel, R. (1991). A novel multigene family may encode odorant receptors: a molecular basis for odor recognition. *Cell* 65:175–187.

Caldwell, R. L. (1979). Cavity occupation and defensive behaviour in the mantis shrimp Gonodactylus festae: evidence for chemically mediated individual recognition. *Anim. Behav.* 27: 194-201.

Caldwell, R. L. (1985). A test of individual recognition in the mantis shrimp *Gonodactylus festae. Anim. Behav.* 33: 101-106.

Caldwell, R. L. (1992). Recognition, signaling, and reduced aggression between former mates in a stomatopod. *Anim. Behav.* 44: 11-19.

Caprio, J. (1977). Electrophysiological distinctions between the taste and smell of amino acids in catfish. Nature 266, 850–851.

Crimaldi, J. P. & Koseff, J. R. (2001) High-resolution measurements of the spatial and temporal scalar structure of a turbulent plume. *Exp. in Fluids* 31, 90-102.

Crimaldi, J. P., Koehl, M. A. R., Koseff, J. R. (2002) Effects of the resolution and kinematics of olfactory appendages on the interception of chemical signals in a turbulent odor plume. *Env. Fluid Mech.* 2, 35-63.

Derby, C.D., & Steullet, P. (2001). Why do animals have so many receptors? The role of multiple chemosensors in animal perception. *Biol. Bull.* 200: 211-215

Dutta, R., Hines, E. L., Gardner, J. W., Boilot, P. (2002). Bacteria classification using Cyranose 320 electronic nose. *Biomed. Eng. Online* 1:1–7.

Dutta, R., Kashwan, K. R., Bhuyan, M., Hines, E. L. & Gardner, J. W. (2003) Electronic nose based tea quality standardization. *Neural Networks* 16: 847-853

Dutta, R. (2005) Identification of Staphylococcus aureus infections in hospital environment: electronic nose based approach. *Sensors and Actuators B* 109: 355–362.

Finelli, C. M. (2000). Velocity and concentration distributions in turbulent odor plumes in the presence of vegetation mimics: A flume study *Mar. Ecol. Prog. Ser.* 207: 297-309.

Gherardi F. & Tricarico, E (2011). Chemical ecology and social behavior of anomura. Chapter 15 in *Chemical communication in crustaceans.* Eds T. Breithaupt and M Thiel. Pp. 297-312, Springer ISBN 978-0-387-77100-7. 565 pp.

Gleeson, R.A., Carr, W.E.S., & Trapido-Rosenthal, H.G. (1993). Morphological characteristics facilitating stimulus access and removal in the olfactory organ of the spiny lobster, *Panulirus argus*: insight from the design. *Chem. Senses* 18: 67–75.

Gomez, G. & Atema, J. (1996a). Temporal resolution in olfaction: stimulus integration time of lobster chemoreceptor cells. *J. Exp. Biol.* 199: 1771-1779.

Gomez, G. & Atema, J. (1996b). Temporal resolution in olfaction II: Time course of recovery from adaptation in lobster chemoreceptor cells. *J. Neurobiol.* 76: 1340-1343

Gomez, G., Voigt, R., & Atema, J. (1999). Temporal resolution in olfaction III: flicker fusion and concentration-dependent synchronization with stimulus pulse trains of antennular chemoreceptor cells in the American lobster. *J. Comp. Physiol. A* 185: 427-436.

Grasso, F. (2001). Invertebrate-Inspired Sensory-Motor Systems and Autonomous, Olfactory-Guided Exploration *Biol. Bull.* 200: 160–168.

Grasso, F. & Atema J (2002) Integration of flow and chemical sensing for guidance of autonomous marine robots in turbulent flows. *Journal of Environmental Fluid Mechanics* 2: 95–114.

Hallberg, E. and Skog, M. (2011). Chemosensory sensilla in crustaceans. Chapter 6 in *Chemical communication in crustaceans.* Eds T. Breithaupt and M Thiel. Pp. 103-121, Springer ISBN 978-0-387-77100-7. 565 pp.

Hardege, J. D. & Terschak, J. A. (2011). Identification of crustacean sex pheromones. Chapter 19 in *Chemical communication in crustaceans.* Eds T. Breithaupt and M Thiel. Pp. 373-392, Springer ISBN 978-0-387-77100-7. 565 pp.

Hay, M. E. (2011). Crustaceans as powerful models in aquatic chemical ecology. Chapter 3 in *Chemical communication in crustaceans.* Eds T. Breithaupt and M Thiel. Pp. 41-62, Springer ISBN 978-0-387-77100-7. 565 pp.

Hazlett, B. A. (2011). Chemical cues and reducing the risk of predation. Chapter 18 in *Chemical communication in crustaceans.* Eds T. Breithaupt and M Thiel. Pp. 355-370, Springer ISBN 978-0-387-77100-7. 565 pp.

Horner, A. J., Nickles, S. P., Weissburg, M. J., Derby, C. D. (2006). Source and specificity of chemical cues mediating shelter preference of Caribbean spiny lobsters (*Panulirus argus*). *Biol. Bull.* 211: 128-139.

Ishida, H., Nakamoto, T., Moriizumi, T., Kikas, T., & Janata, J. (2001). Plume-Tracking Robots: A New Application of Chemical Sensors. *Biol. Bull.* 200: 222-226.

Johnson, T.R. Jr. (1977). Individual recognition in the banded shrimp *Stenopus hispidus* (Olivier). *Pac. Sci* 23:40-50.

Kaissling, K. E. (1998a) Flux detectors versus concentration detectors: Two types of chemoreceptors. *Chem. Senses* 23: 99–111.

Kaissling, K. E. (1998b). Pheromone deactivation catalyzed by receptor molecules: a quantitative kinetic model. *Chem. Senses* 23: 383-395.

Kamio, M. & Derby, C. D. (2011). Approaches to molecular identification of sex pheromones in blue crabs. Chapter 20 in *Chemical communication in crustaceans.* Eds T. Breithaupt and M. Thiel. Pp. 393-412, Springer ISBN 978-0-387-77100-7. 565 pp.

Karavanich, C. & Atema, J. 1998. Individual recognition and memory in lobster dominance. *Anim Behav.* 56: 15553-1560.

Koehl, M. A. R. (2006) The fluid mechanics of arthropod sniffing in turbulent odor plumes. *Chem. Senses* 31, 93-105.

Koehl, M.A.R. (2011). Hydrodynamics of sniffing by crustaceans. Chapter 11 in *Chemical communication in crustaceans.* Eds T. Breithaupt and M Thiel. Pp. 85-102, Springer ISBN 978-0-387-77100-7. 565 pp.

Koehl, M. A. R. & Stricker, J. R. (1981). Copepod feeding currents: food capture at low Reynolds number. *Limnol. Oceanog.* 26: 1062-1073.

Kowadlo, G., & Russell, R. A. (2008). Robot Odor Localization: A Taxonomy and Survey. *The International Journal of Robotics Research* 27: 869–894.

Kraus-Epley, K. E. & Moore, P. A. (2002) Bilateral and unilateral lesions alter orientation abilities of the crayfish, Orconectes rusticus. *Chem. Senses* 27: 49–55.

Mead, K. S. (2008). Do antennule and aesthetasc structure in the crayfish *Orconectes virilis* correlate with the flow environment? *Int. and Comp. Biol.* 48:823-833.

Mead, K., Koehl, M. A. R., & O'Donnell, M. J. 1999. Mantis shrimp sniffing: the scaling of chemosensory sensilla and flicking behavior with body size. *J. Exp. Mar. Biol. Ecol.* 241: 235-261.

Mead, K. & Koehl, M. A. R. 2000. Particle image velocimetry measurements of fluid flow through a model array of mantis shrimp chemosensory sensilla. *J. Exp. Biol.* 203: 3795-3808.

Mead, K. S., Wiley, M. B., Koehl, M. A. R., & Koseff J. R. (2003). Fine-scale patterns of odor encounter by the antennules of mantis shrimp tracking turbulent plumes in wave-affected and unidirectional flow. *J. Exp. Biol.* 206, 181-193.

Mead, K. S. & Caldwell, R. L. (2011). Mantis Shrimp: Olfactory apparatus and chemosensory behavior. Chapter 11 in *Chemical communication in crustaceans.* Eds T. Breithaupt and M Thiel. Pp. 219-238, Springer ISBN 978-0-387-77100-7. 565 pp.

Moore, P. A., Scholz, N., & Atema, J. (1991). Chemical orientation of lobsters, *Homarus americanus,* in turbulent odor plumes. *J. Chem. Ecol.* 17: 1293-1307.

Moore, P. A., Weissburg, M. J., Parrish, J. M., Zimmer-Faust, R. K. & Gerhardt, G. A. (1994) The spatial distribution of odors in simulated benthic boundary layer flows. *J Chem Ecol* 20: 255-279.

Moore, P. A., & Grills, J. L. (1999) Chemical orientation to food by the crayfish *Orconectes rusticus*: influence of hydrodynamics. *Animal Behaviour*, 58, 953–963

Moore, P. & Crimaldi, J. (2004) Odor landscapes and animal behavior: tracking odor plumes in different physical worlds. *J. Mar. Syst.*, 49, 55-64.

Oh, E. H., Song, H. S., Park, T. H., (2011). Recent advances in electronic and bioelectronic noses and their biomedical applications. *Enzyme and Microbial Technology* 48 (2011) 427–437.

Peng, G., Tisch, U., Adams, O., Hakim, M., Shehada, N., Broza, Y., Billan, S., Abdah-Bornyak, R., Kuten, A., Haick, H. (2009). Diagnosing lung cancer in exhaled breath using gold nanoparticles. *Nat. Nanotech.* 4:669–73.

Rospars, J.-P., Křivan, V. & Lánsky, P. (2000). Perireceptor and receptor events in olfaction. Comparison of concentration and flux detectors: a modeling study. *Chem. Senses* 25, 293-311.

Russell, R. A. (2001). Survey of Robotic Applications for Odor-Sensing Technology. *The International Journal of Robotics Research* 20: 144-162.

Schmidt, M., & Ache, B. W. (1992). Antennular projections to the midbrain of the spiny lobster. II. Sensory innervation of the olfactory lobe. *J. Comp. Neurol.* 319: 291–303.

Schmidt, M., & Ache, B. W. (1996). Processing of antennular input in the brain of the spiny lobster, *Panulirus argus.* I. Non-olfactory chemosensory and mechanosensory pathways of the lateral and median neuropil. *J. Comp. Physiol. A* 178: 579–604.

Schmidt, M., Van Eckeris, L. & Ache, B. W. (1992). Antennular projections to the midbrain of the spiny lobster. I. Sensory innervations of the lateral and medial antennular neuropils. *J. Comp. Neurol.* 318: 277–290.

Schmidt, M. & Mellon Jr, D. (2011). Neuronal processing of chemical information in crustaceans. Chapter 9 in *Chemical communication in crustaceans.* Eds T. Breithaupt and M Thiel. Pp. 123-147, Springer ISBN 978-0-387-77100-7. 565 pp.

Stacey, M. T., Mead, K. S. & Koehl, M. A. R. (2003). Molecular capture by olfactory antennules: mantis shrimp. *J. Math. Biol.* 44: 1-30.

Tomba, A.M., Keller, T.A., & Moore, P.A. (2001). Foraging in complex odor landscapes: Chemical orientation strategies during stimulation by conflicting chemical cues. *Journal of the North American Benthological Society* 20:211-222.

Vogel, S. (1994) *Life in Moving Fluids: The Physical Biology of Flow* (Second Edition) Princeton University Press 484 pp. ISBN: 9780691026169

Webster, D. R., Rahman, S., & Dasi, L. P. (2001). On the usefulness of bilateral comparison to tracking turbulent chemical odor plumes *Limnol. Oceanog.* 46, 1048-1053.

Weissburg, M. J. (2000) The fluid dynamical context of chemosensory behavior. *Biol. Bull.* 198: 188–202.

Weissburg, M. J. (2011) Waterborne chemical communication: stimulus dispersal dynamics and orientation strategies in crustaceans. Chapter 4 in *Chemical communication in crustaceans.* Eds T. Breithaupt and M Thiel. Pp. 63-83, Springer ISBN 978-0-387-77100-7. 565 pp.

Weissburg, M. J. & Zimmer-Faust, R. K. (1993) Life and death in moving fluids: Hydrodynamic effects on chemosensory-mediated predation. *Ecology* 74: 1428-1443.

Weissburg M. J., & Zimmer-Faust R. K. (1994) Odor plumes and how blue crabs use them in finding prey. *J Exp Biol* 197, 349-375.

Wäckers F., Olson, D., Rains, G., Lundby, F., & Haugen, J.-E. (2010) *Journal of Food Science* 76: Issue 1, S41-S47.

Yen, J., Weissburg, M. J., & Doall, M. H. (1998). The fluid physics of signal perception by a mate-tracking copepod. *Philos Trans R Soc B* 353: 787-804.

Yen, J. & Lasley, R. (2011). Chemical communication between copepods. Chapter 9 in *Chemical communication in crustaceans.* Eds T. Breithaupt and M Thiel. Pp. 177-197, Springer ISBN 978-0-387-77100-7. 565 pp.

Zettler, E. & Atema, J. (1999). Chemoreceptor cells as concentration slope detectors: preliminary evidence from the lobster nose. *Biol. Bull.* 197, 252-253.

Ziemba, R. E., Simpson, A., Hopper, R., & Cooper, R. L. (2003). A comparison of antennule structure in a surface- and a cavedwelling crayfish, genus Orconectes (Decapoda, Astacidae). *Crustaceana* 76:859–69.

Zimmer-Faust, R. K. (1989). The relationship between chemoreception and foraging behavior in crustaceans. *Limnol. Oceanogr.* 34, 1364-1374.

Part 5

Sensor Technology

13

Physical Vapour Deposition Techniques for Producing Advanced Organic Chemical Sensors

Michele Tonezzer[1,2] and Gianluigi Maggioni[3]
[1]*Università di Trento, Dip. Ingegneria dei Materiali e delle Tecnologie Industriali,*
[2]*Laboratori Nazionali di Legnaro - Istituto Nazionale di Fisica Nucleare,*
[3]*Università di Padova, Dip. Fisica at LNL - Istituto Nazionale di Fisica Nucleare,*
Italy

1. Introduction

In the organic chemical sensors field, the main focus to date has been on the molecular design of the receptor as a function of the analyte to be detected. Nevertheless chemical sensing requires an integrated approach, where both the molecular and the material properties of the sensing layer must be finely tuned to achieve the desired properties. Despite its great influence on the ultimate performances of the sensors, the material side has been largely neglected.

In this respect, chemical sensing of gases and vapours performed via thin solid films represents a particular challenge, as the desired recognition events need to operate at the gas-solid interface. Taking into account that the analyte recognition is mediated by the layer properties of the coated receptors, precise control and accurate characterization of these properties (e.g., thickness, permeability, and morphology) are critical for developing advanced sensing materials.

In this chapter, a description of several different real cases is carried out by the authors to highlight the significant advantages introduced into the sensing field by a bottom-up approach in which molecules purposely developed for detecting a specific analyte are deposited in well-controlled and designed architectures. The production methods, the properties and the sensing capabilities of novel advanced organic sensing materials grown by Physical Vapour Deposition (PVD) techniques are described and compared to the conventional ones. In particular the results concerning organic sensors developed with two main deposition techniques will be detailed and discussed, namely:

- Glow-Discharge-induced Sublimation (GDS), a novel patented plasma-based deposition technique recently developed at the Materials & Detectors Laboratory of INFN National Laboratories of Legnaro (Padua), which has demonstrated to produce extremely competitive, thin sensing organic films;
- High Vacuum Evaporation (VE), a more standard technique rarely used in the production of organic films despite the very interesting results obtained in producing organic and metallo-organic sensing elements.

Specific cases related to the improvement of the sensing capabilities of different conventional organic sensors by using these two techniques are displayed. In particular the chapter describes the production, the physical and chemical characterization and the sensing capabilities (by different transducing methods) of novel sensing materials based on three classes of macromolecular compounds: phthalocyanines, porphyrins and cavitands. Following the specific cases for each molecular class are reported:

- Phtahlocyanines. The physical and chemical features of copper (CuPc) and zinc (ZnPc) phthalocyanines thin solid films produced by GDS technique are reported. Their sensing capabilities towards several analytes (NO, NO_2 and ethanol) are analysed by means of electrical and optical transducing methods.
- Porphyrins. Different porphyrin compounds have been deposited by VE and GDS and their physical properties and sensing capabilities are reported and discussed. As regards VE technique, three porphyrin compounds have been deposited: free- (H_2TPP), cobalt- (CoTPP), and iron chloride- (Fe(TPP)Cl) 5,10,15,20 *meso*-tetraphenyl porphyrins. The chemical and morphological properties of VE samples have been analyzed by FT-IR and SEM images and their optical sensing capabilities towards different alcohol vapours (methanol, ethanol and iso-propanol) have been measured. As regards GDS technique, CoTPP and Fe(TPP)Cl samples have been produced and physically characterized. Moreover their optical responses towards ethanol vapours are discussed on the basis of their physical properties.
- Cavitands. VE deposition technique has been used for growing a novel supramolecular mass sensing film based on an ultimate insoluble receptor specifically designed for detecting short chain alcohols: the tetraphosphonate Tiiii [H, CH3, Ph] cavitand. Non-specific tetrathiophosphonate TSiiii [H, CH3, Ph] cavitand layers have been also deposited for comparison. The properties of deposited films were investigated by FT-IR analysis and their sensing capabilities were investigated by exposing Tiiii- and TSiiii-coated quartz crystal microbalances (QCMs) to ethanol in very low concentrations. Elovich kinetics were also used to analyze the sorption process occurring onto the different samples.

In order to underline the role of the molecular architectures on the final sensing capabilities of the samples, the sensing responses are explained on the basis of their physical properties. Moreover the authors, in order to point out the role of the deposition techniques on the final sensing capabilities of the new developed samples, compare in several cases their sesing performances with those of analogous samples produced by conventional methods.

2. Experimental

2.1 Film deposition

The organic chemical sensors were produced by Glow-Discharge-induced Sublimation (GDS), High Vacuum Evaporation (VE) and Spin Coating (SPIN).

2.1.1 Glow-Discharge-induced Sublimation (GDS)

The experimental setup used for the deposition of GDS films consists of a vacuum chamber evacuated by a turbomolecular pump to a base pressure of 10^{-4} Pa. The glow discharge is sustained by a 1-in. cylindrical magnetron sputtering source connected to a radio frequency power generator (600 W, 13.56 MHz) through a matching box. The organic powder is put on the surface of an aluminum target, placed on the sputtering source. The glow discharge feed gas is argon or helium (99.9999%), whose pressure inside the chamber is measured through a capacitance gauge. Typical values of rf power, target dc self-bias and working pressure are

10 to 40W, −20 to −500V and 5.00±0.05 Pa (Ar) and 20.0±0.1 Pa (He), respectively. A water cooled quartz crystal thickness sensor, placed close to the sample holder 7 cm above the source, has been used for measuring in real time the deposition rate of the organic layer.

2.1.2 High Vacuum Evaporation (VE)

The apparatus used for the evaporation of the organic films consisted of a stainless steel vacuum chamber equipped with a turbomolecular pump capable of base pressure of 10^{-4} Pa (Maggioni et al., 1997). The sample holder is a copper circular plate with a central aperture where a quartz crystal thickness sensor is located. The heater units are three copper crucibles: each of them is equipped with a thermocouple and a heating resistance so that temperatures up to 500°C can be reached and maintained within ±1°C. During the deposition the substrates were kept at room temperature.

2.1.3 Spin coating (SPIN)

Thin porphyrin films were produced by spin coating technique. Porphyrin solutions (0.1 wt.% of porphyrin powder in chloroform) were spun at 800 rpm for 30 s.

2.2 Physical and chemical characterization

The physical and chemical features of the samples were analyzed by Scanning Electron Microscopy (SEM), Atomic Force Microscopy (AFM), Infrared analysis (FT-IR), and UV-visible analysis (UV-Vis).

2.2.1 Scanning Electron Microscopy (SEM)

The surface morphology of the samples was investigated with a Philips XL-30 scanning electron microscope (SEM) at a working pressure of 10^{-3} Pa. SEM was also used for measuring the thickness of GDS samples through cross-section pictures.

2.2.2 Atomic Force Microscopy (AFM)

The surface morphology of the organic samples was investigated in air by a non-contact-mode atomic force microscope (AFM) model C-21 (Danish Micro Engineering), mounting a DualScope Probe Scanner 95-50. AFM measurements were also utilized to investigate the thickness of the samples by scratching a small area of the samples and acquiring an image of the borderline between the exposed substrate and the pristine film. The observations were performed with a Si_3N_4 pyramidal tip with a curvature radius lower than 50 nm.

2.2.3 Infrared analysis (FT-IR)

FT-IR spectra of the samples were recorded in the 4000-400 cm^{-1} range using a Jasco FT-IR 660 Plus spectrometer with a resolution of 16 cm^{-1}. The sample cell and the interferometer were evacuated to remove the adsorption peaks of water and atmospheric gases.

2.2.4 UV-Visible spectroscopy (UV-Vis)

A Jasco V-570 dual-beam spectrophotometer was used to perform the UV-visible absorption measurements in the 200-800 nm range, with a resolution of 2 nm.

2.3 Sensing properties

The sensing capabilities of the different sensing thin films were analysed by Electrical, Optical and Piezo-electrical (QCM) transducing methods.

2.3.1 Electrical sensors

The electrical responses were measured on samples grown onto two interdigitated combs of 1301 gold electrodes (650 + 651) deposited on 30mm×30mm×1mm silica slides (Maggioni et al., 2008). The experimental setup used to evaluate the gas-sensing capabilities of the films is equipped with two mass flow controllers (MFCs): the former allows to control the flow rate of analytes and the latter, which controls the flow rate of pure nitrogen, is used to suitably dilute the gas mixture to the desired composition. After heating the samples up to 150 °C under a nitrogen flow, the electrical current at fixed voltage (1 V) was measured every 6 s after setting the flow of both controllers at the desired values. For all the measurements reported the total flow was 1000 sccm. The gas mixtures in the cylinders used for the measurements were N_2 +NO_2 (0.98±0.05) ppm and N_2 +NO (98±3) ppm.

2.3.2 Optical sensors

The analysis chamber for the optical measurements of the CuPc samples to ethanol was the same used for the electrical measurements (Maggioni et al., 2008). The changes of reflectance upon ethanol vapour exposure was measured in the wavelength range 250-800 nm.

FT-IR response of the CuPc films was measured using a N_2+NO_2 500 ppm mixture. After the evacuation of the sample cell, the spectra were collected before gas admittance, after 10 min of gas exposure (10^5 Pa total pressure) and then after 15 h of cell evacuation.

The optical sensing performances of the porphyrin samples were measured in a purpose-built testing apparatus in which the sample chamber is placed inside a Jasco V-570 Spectrophotometer. The sample chamber is connected to two MFCs: the former (AERA FC-7800CD) allows to control the flow rate of the analyte mixture and the latter (AERA FC-7700CD) controls the flow rate of pure nitrogen. This apparatus allows to record both the whole 200-800 nm spectra and dynamic behaviours at fixed absorption wavelengths in different vapour concentration atmospheres.

2.3.3 Mass sensors (QCM)

Sensing measurements were performed using AT-cut quartzes with a fundamental frequency of 10 MHz and a crystal diameter of 8 mm. QCM sensors are mass transducers where the frequency of oscillation, for small increases of mass, linearly changes according to the Sauerbrey equation

$$\Delta f = k_q \Delta m$$

where the Δf is the frequency variation and Δm *the* increase in mass. The quartz constant is experimentally estimated to be kq=-0.46 Hz ng^{-1}. This value provides a nominal mass resolution of 1.6 ng Hz^{-1}, considering a minimum reliable frequency measurement of 1 Hz. To control the amount of the deposited films, QCM frequency was monitored during the deposition process: a total frequency variation of Δf=-20 ± 0.5 kHz was obtained for all samples produced.

The measurement system (Gaslab 20.1; IFAK, Magdeburg) is equipped with a flow chamber, containing four coated quartz crystals, a reference quartz crystal and a thermocouple. The chamber was thermostatted at 20 ± 0.1 °C. The QCM chamber is connected with two mass-flow controllers (Brooks 5850S) for controlling the flow rate of analyte mixture and pure nitrogen, respectively. During the analyses QCM frequency was measured versus time every 1 s. The ethanol vapour and NO gas were supplied by SAPIO Srl in gas cylinders with a certified concentration of 504 ppm and 150 ppm, respectively.

3. Phthalocyanines

Despite the prolonged studies and applications of metal phthalocyanines (MPcs) during the last forty years, the scientific research on these organic semiconductors is still vivid owing to their unique properties, such as thermal stability, chemical inertness, and biocompatibility. MPcs are thus interesting for several applications, including chemical sensing, photoconducting agents, photovoltaic cell elements, nonlinear optics, and electrocatalysis. In the gas sensing field, MPcs are mainly used as electrical gas sensors, because of the conductivity changes induced by the adsorption of oxidizing or reducing gases such as NO_x, halogens, and ammonia (Snow & Barger, 1989). More recently, MPcs have also been successfully tested as opto-sensing materials for the detection of volatile organic compounds (VOCs) (Spadavecchia et al., 2006).

The classical deposition methods of thin organic films such as spin coating, dip coating and the sol-gel method cannot be easily applied to the production of MPc coatings owing to the low solubility of these compounds in organic solvents. The most widely used deposition method for MPcs is then become the high vacuum evaporation (VE), which allows to avoid the drawbacks of the derivatization of MPcs aimed at improving their solubility. Plasma-based deposition methods have been also reported to succeed in depositing MPc films for gas sensing applications: plasma polymerized CuPc was deposited on piezoelectric quartz crystals (Kurosawa et al., 1990) and plasma was also used for the activation of CuPc molecules during the evaporation process (Choi et al., 1996).

In this section the results obtained in the production of phthalocyanine-based sensing films by means of the novel Glow-Discharge-induced Sublimation (GDS) technique are reported. In particular the authors report the experimental data concerning the physical characterization and the sensing capabilities of two MPcs: copper phthalocyanine (CuPc) and zinc phthalocyanine (ZnPc). Figure 1 displays the structures of the two compounds.

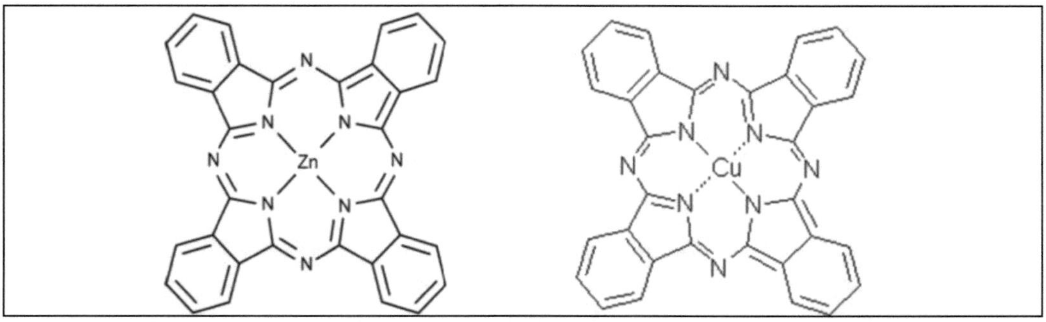

Fig. 1. Structure of: (left) copper phthalocyanine (CuPc); (right) zinc phthalocyanine (ZnPc).

3.1 Copper phthalocyanine (CuPc) deposited by GDS technique

3.1.1 Physical and chemical properties

GDS technique allowed to produce CuPc films which feature much rougher morphologies than VE ones and very high porosity (Maggioni et al., 2005a).

Fig. 2. SEM micrographs of VE (left) and GDS (right) CuPc samples.

In fact, as shown in Figure 2, while the VE films are relatively flat and compact, the GDS films are characterized by a rough structure with big grains (2 ìm and more) and a high porosity inside and between the grains. Nitrogen physisorption measurements confirm the high porosity of the GDS films (Maggioni et al., 2005a). Specific Surface Area (SSA) values, calculated for the GDS films, are higher than those for the VE samples: in fact, the VE films feature 31 m^2/g against 155 m^2/g of the GDS films. The pore size distribution curve of the VE films features a mono-modal and sharp distribution centered at around 4 nm. By contrast, GDS films are characterized by a bimodal distribution with micropores ($\approx$ 2 nm) and mesopores (> 30 nm). These structural characteristics of GDS films bring about their higher sensing capabilities with respect to the VE ones, owing to the increased interactive area between sensing receptors and analyte molecules.

3.1.2 Sensing properties

The sensing capabilities of the CuPc samples have been tested by several transducing methods. It is well known that the adsorption of reducing or oxidizing gases such as NO_x changes the electrical conductivity of CuPc, but both response time and recovery time are very long (in the range of hours). This problem can be solved by monitoring the change of the electrical current during the first tens of seconds after the gas exposure: this allows to shorten significantly the response times. Figure 3 reports that the slope of the curve current *versus* time is related to the gas concentration showing a linear relationship for low concentrations and a small divergence from linearity at higher concentrations (see Figure 3, right).

The detection limit of CuPc GDS sensor is around few tens of ppb for NO_2 and few ppm for NO (Maggioni et al., 2008). A strong change of the optical properties of MPc films can occur when they are exposed to a specific gas/vapour, because the electronic transitions within their π-aromatic system and the π -π interactions between contiguous MPc molecules can be highly affected by the interaction with the gas molecules. MPcs have been shown to optically respond to alcohol vapours (Spadavecchia et al., 2006; Maggioni et al., 2007): Figure 4 shows the changes of optical reflectance of a CuPc sample to ethanol vapours at a wavelength in the Q band (λ=580nm). The responses are characterized by a fast signal increase: t_{50} response time (defined as the time taken for the signal intensity to reach 50% of its final saturated value) is 7 s; moreover, when the ethanol vapour stream is switched off, the original signal yield is almost completely recovered after few tens of seconds (t_{50}= 12 s). Taking into account that the measurements are performed at room temperature, this evidence becomes particularly interesting.

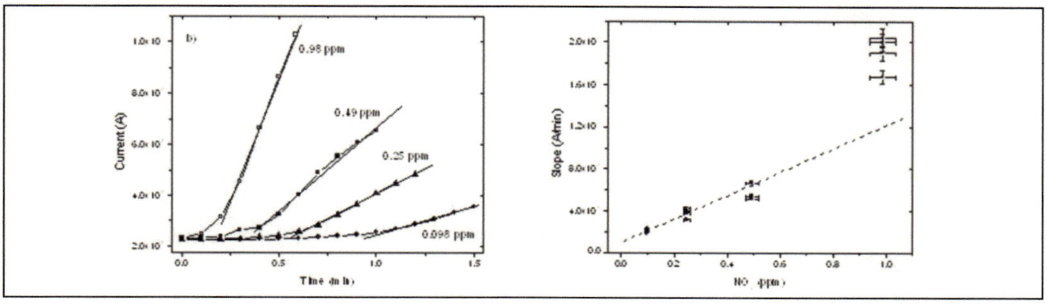

Fig. 3. (left) Electrical responses of a GDS CuPc film to NO_2-containing mixtures in the sub-ppm range. Fitting lines are reported for each curve. (right) Slopes of the electrical responses of a GDS CuPc film as a function of NO_2 concentration.

The slope of the calibration curve of this sensor (Figure 4, right) decreases at concentrations higher than 7000 ppm, suggesting a progressive saturation of the sites available for the analyte molecules. As the optical signal for a concentration of 1500 ppm is about 20 and the corresponding noise is about ±1, the detection limit for the present experimental configuration is around 200 ppm. This limit can be easily improved by changing some experimental parameters such as the active film surface area. In order to investigate the adsorption processes involved in the response of the CuPc sample to ethanol, the Langmuir isotherm model has been adopted. The linear fit of the Langmuir adsorption (Figure 4, right, red line) was done after neglecting the three data points at the higher concentrations.

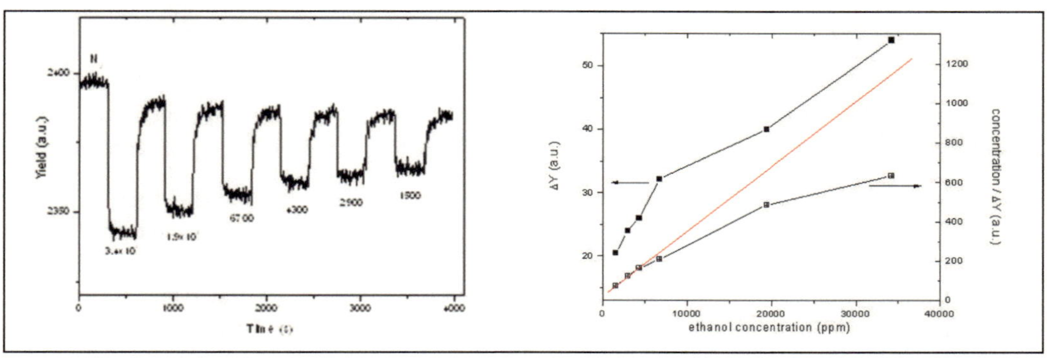

Fig. 4. (left) Optical response of CuPc sample towards different ethanol vapour concentrations at λ=580 nm. (right) Calibration curve and Langmuir plot of CuPc sample towards ethanol at λ=580 nm.

The linear correlation is very good at low concentrations while a strong deviation is observed at increasing concentration. The good linear correlation obtained at low concentrations shows that the Langmuir adsorption model, even with its limited assumptions, provides a basic understanding of the ethanol/CuPc interaction process.

The better performances of the GDS CuPc films as compared to the VE ones are pointed out by their response to NO when used as quartz crystal microbalance (QCM) sensors Figure 5 shows the dynamic response of two CuPc-coated QCMs to 40 ppm of NO: the response is measured as a change of QCM resonance frequency upon adsorption of the analyte molecules.

The VE-coated QCM exhibits responses characterized by low intensity (Figure 5, left) and its response remains the same for all the tested concentrations in the range from 10 to 150 ppm (Figure 5, right). On the contrary, the response of GDS-coated QCM is well pronounced even down to 10 ppm of NO and it is related to the NO concentration: the relation is quite linear in the low concentration range, while above 60 ppm the response reaches a constant value, which corresponds to the saturation of the available NO recognition sites.

The NO detection limit in the present configuration is around few ppm, taking into account that the noise is around 1 Hz. Moreover, GDS samples are characterized by short response (t_{50} = 6 s) and recovery times (t_{50} = 9 s).

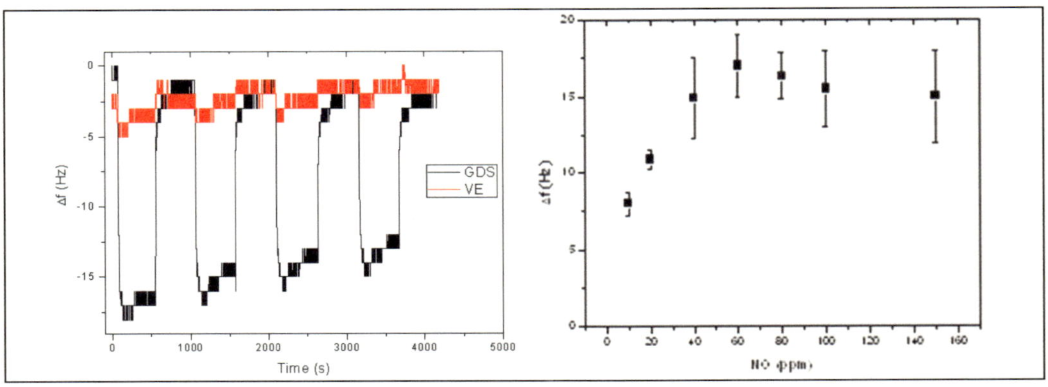

Fig. 5. (left) QCM response of GDS (black) and VE (red) CuPc samples in the presence of 40 ppm of NO. (right) Calibration curve of GDS sample in the NO range from 10 to 150 ppm

A transduction mechanism which can be also used for gas sensing involves the effects of analyte molecules on the IR features of the CuPc samples (Maggioni et al., 2005b). Figure 6 shows the FT-IR spectra of a GDS CuPc sample exposed to a N_2+NO_2 500 ppm atmosphere.

As can be seen, the NO_2 adsorption gives rise to clear spectral changes and these changes are almost completely reversible. Considering the out-of-plane vibrations, which are the most sensitive to the chemical environment, the γ(C-H) peak at 726 cm^{-1} decreases in intensity and shifts to 730 cm^{-1}. Moreover the relative intensity of the peak at 780 cm^{-1} slightly increases with respect to that of the peak at 773 cm^{-1}. These changes indicate that the NO_2 molecule strongly interacts with the benzene rings of the CuPc molecule, in agreement with the results of Sadaoka et al. (Sadaoka et al., 1982), which found that the sites for NO_2 adsorption were the ligand □ electron systems for CuPc and H2Pc and the central metal atoms for FePc and CoPc, respectively. Considering the in-plane-vibrations, the NO_2 effect is minor and is mainly a broadening of the peaks in the range from 1500 to 1000 cm^{-1}. The broadening is due to the disorder caused by the NO_2 doping in the film structure. As to the interaction between NO_2 and the central Cu atom, the Cu-O-NO vibration was reported at 1460 cm^{-1} (Sadaoka et al., 1982). In the spectrum of the NO_2-doped film a broad peak between 1440 and 1510 cm^{-1} with a maximum at 1457 cm^{-1} is found, while in the other spectra only one peak at 1466 cm^{-1} appears. Since the peak at 1457 cm^{-1} is distinct enough from that at 1466 cm^{-1}, it can be ascribed to a Cu-O-NO vibration thus suggesting a partial interaction between Cu and NO_2.

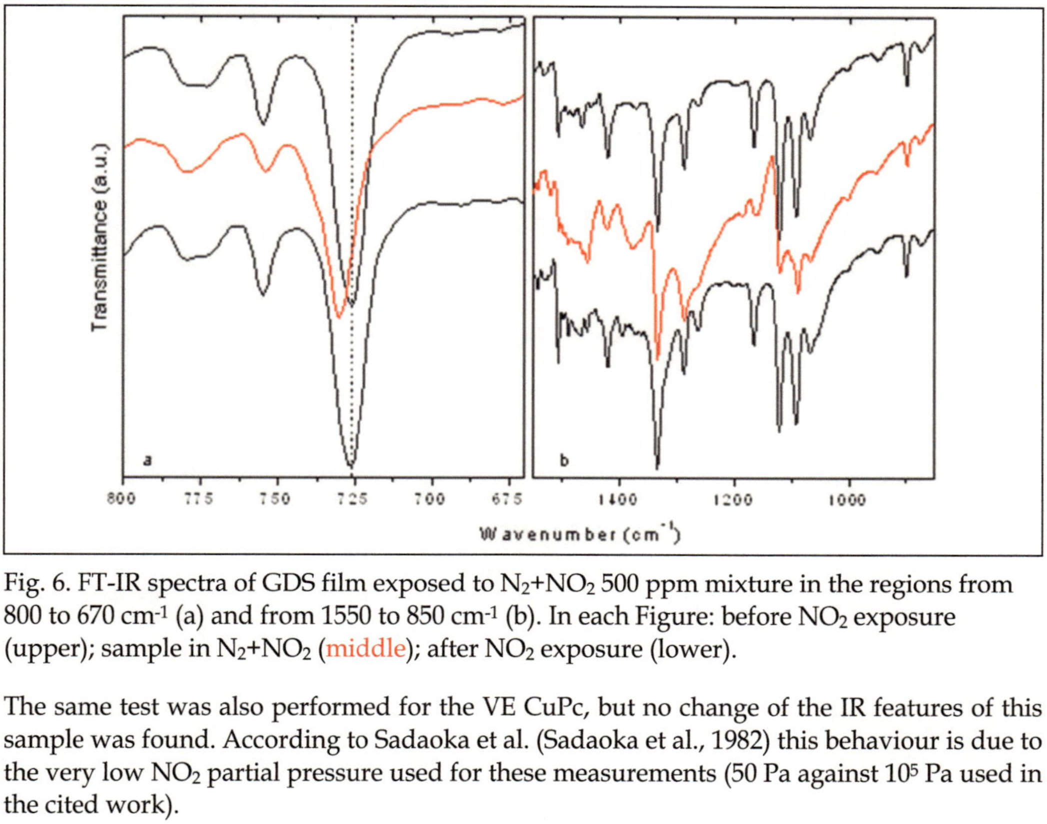

Fig. 6. FT-IR spectra of GDS film exposed to N_2+NO_2 500 ppm mixture in the regions from 800 to 670 cm^{-1} (a) and from 1550 to 850 cm^{-1} (b). In each Figure: before NO_2 exposure (upper); sample in N_2+NO_2 (middle); after NO_2 exposure (lower).

The same test was also performed for the VE CuPc, but no change of the IR features of this sample was found. According to Sadaoka et al. (Sadaoka et al., 1982) this behaviour is due to the very low NO_2 partial pressure used for these measurements (50 Pa against 10^5 Pa used in the cited work).

3.2 Zinc phthalocyanine (ZnPc) deposited by GDS technique

3.2.1 Physical and chemical properties

AFM images of the surface of a ZnPc film deposited on silicon substrate are shown in Figure 7. In spite of the high smoothness of the substrate, the morphology of the ZnPc film is characterized by the presence of grains with lateral sizes of hundreds of nanometers.

Like for CuPc films, a comparison with VE ZnPc films (Iwatsu et al., 1980) points out that the surface of GDS ZnPc films is much more porous. The roughness profile of the film surface and the mean height are 77±3 nm and 510±70 nm, respectively (Maggioni et al., 2007). Nitrogen physisorption measurements highlighted microporous solids in which the size of the micropores, as derived from the slit-like Horvath–Kawazoe model (Horvath &

Kawazoe, 1983), ranges between 1.1 and 1.4 nm. Specific surface area (SSA) of the sample is 58 m^2/g: this value is much higher than that of VE ZnPc sample, which is less than 1 m^2/g.

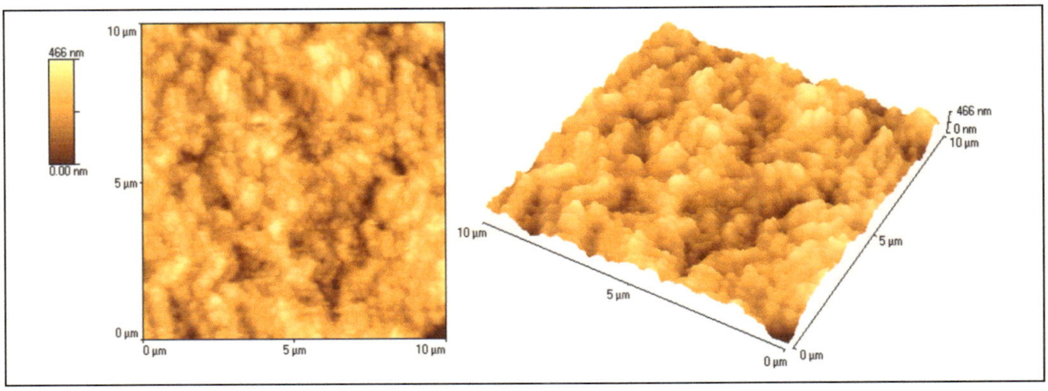

Fig. 7. AFM images of the ZnPc film: plane view (left) and 3D view (right) of the surface.

3.2.2 Sensing properties

Like for CuPc, the optical properties of ZnPc samples can be highly affected by the interaction with gas/vapour molecules. The exposure to ethanol, methanol and isopropanol (vapours changes the optical absorption in the spectral region 500–800 nm, where Q_I band (peaked at 680 nm) and Q_{II} band (peaked at 620 nm) of ZnPc are placed (Maggioni et al., 2007). Responses of GDS-deposited ZnPc films exposed to different ethanol concentrations are reported in Figure 8. The measurements were performed by recording the integral area (I) of the absorption spectrum in the selected wavelength region versus time.

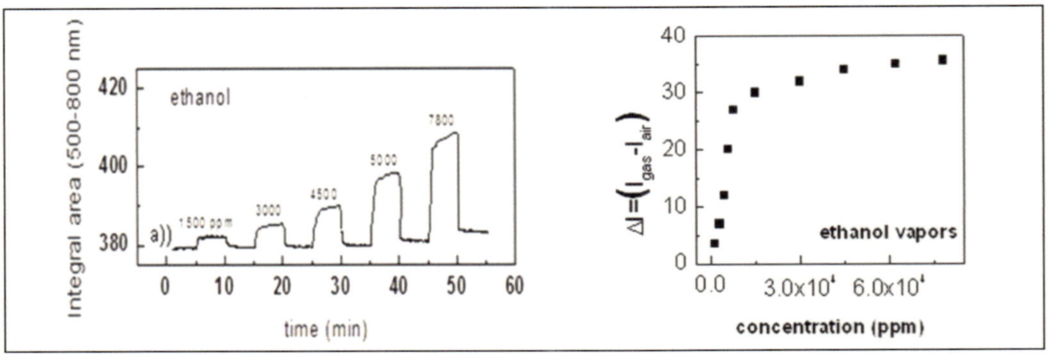

Fig. 8. (left) Optical response of GDS ZnPc sample in the presence of different ethanol vapour concentrations in the spectral region 500-800 nm. (right) Calibration curve of GDS ZnPc sample in the ethanol range from $1.5x10^3$ to $8x10^4$ ppm.

The response is fast and reversible: at an ethanol concentration of 5000 ppm the response time t_{90} (defined as the time taken for the signal intensity to reach 90% of its final saturated value) is around 60 s. The calibration curve obtained from the dynamic curves is reported in Figure 8., right. This curve represents the response of the sensors (intended as the difference, ΔI, between the intensity in presence of the ethanol vapour, I_{gas}, and in presence of dry air, I_{air}, i.e. $\Delta I = I_{gas} - I_{air}$) in the presence of different vapour concentrations. The response of the sensor is linear in the low concentration range, then it reaches a saturation value at higher concentrations. It has been shown that in the case of methanol and isopropanol the response keeps linear in the whole concentration range, although the concentrations are higher than those of ethanol (Maggioni et al., 2007). These differences in responses are due to the different affinity between the ZnPc macromolecular ring and the vapour molecules.

4. Porphyrins

Porphyrins, the "pigments of life", are perhaps the most important and widespread class of natural pigments. Nature provides a variety of tetrapyrrolic pigments with a similar ligand core, but differing in the metal centre and the side groups attached to the porphyrin rings. For instance, in heme the metal centre is an iron ion: in the case of chlorophyll it is magnesium and in vitamin B12 it is cobalt.

Tetraphenyl porphyrins have assumed a peculiar role in different fields of disciplines, ranging from molecular semiconductors to non-linear optics: in particular, their physical and chemical properties make these compounds good sensing materials for detecting different gaseous species, from NO_x and HCl to VOCs-(Brunik et al., 1999).

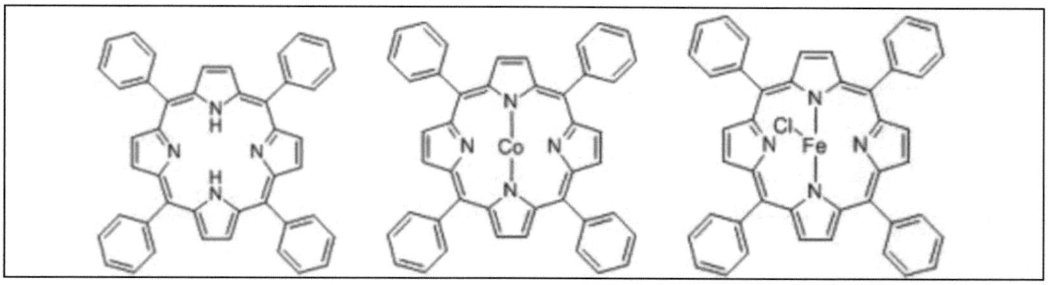

Fig. 9. Structure of: (left) free (H_2TPP); (centre) cobalt, and (right) iron chloride (FeTPPCl) 5,10,15,20 tetraphenyl porphyrins.

In fact, porphyrins are stable compounds and their physical and chemical properties can be finely tuned by simple modifications of their basic molecular structure. Moreover porphyrins present optical absorption and fluorescence bands in the visible region related to the electronic transitions of the aromatic systems. The interactions of analytes with porphyrin thin films affect both the optically active transitions of the single molecule and the π–π interactions between macrocycles, giving rise to detectable changes of the optical absorption spectra. This property allowed to develop several opto-chemical sensing procedures (Rakow & Suslick, 2000).

In this section the cases of three 5,10,15,20 tetraphenyl porphyrins will be treated: free (H_2TPP), cobalt (CoTPP), and iron chloride (FeTPPCl). Figure 9 displays the structures of these compounds.

4.1 Porphyrins deposited by VE technique

In order to be exploited as sensing materials, porphyrin compounds usually need to be deposited as solid films onto an appropriate substrate: a large number of chemical techniques (solvent casting, spin coating, Langmuir-Blodgett, electropolymerization, self assembled monolayers) have been studied for this purpose. Less attention has been paid to the use of physical deposition techniques in the production of sensing materials, in spite of their ability to significantly improve the recognition and sensing capabilities of the molecular precursors by growing highly competitive organic architectures.

The VE technique introduces several advantages in the production of porphyrin-based sensing thin films such as greater reproducibility, higher uniformity and stricter control of the film thickness in comparison with standard chemical techniques. Moreover vacuum evaporation, featuring the peculiarity to produce thin solid films without using any extraneous compound, allows to deposit samples characterized by a much higher purity than the common chemical techniques which need to use solvents. This aspect is particularly important in the gas sensing field because the unpredictable effects of the retained solvent on the final response of the sensing device are manifold, going from the occupation of adsorption sites to the interference in the analyte/material interaction.

In this section the chemical properties of VE thin films of H_2TPP, CoTPP, and (Fe(TPP)Cl) are described. Moreover their optical sensing capabilities upon exposure to vapours of three different alcohols, namely methanol (MeOH), ethanol (EtOH) and isopropanol (2- propanol) are discussed and compared to the sensing capabilities of conventional spin coated (SPIN) films (Bernini et al., 2008).

4.1.1 Chemical properties

Infrared (FT-IR) spectrum of VE H_2TPP film is reported in Figure 10 together with those of SPIN film and H_2TPP starting powder.

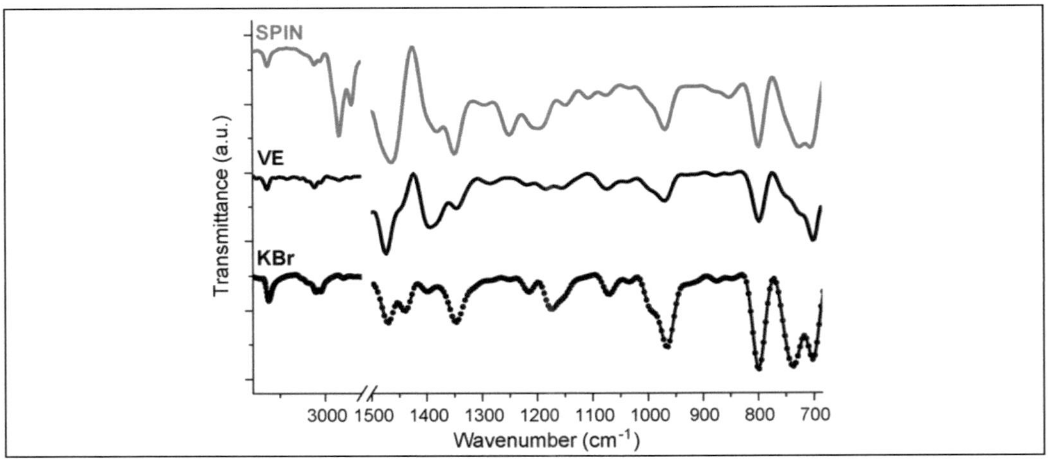

Fig. 10. FT-IR spectra of powder (lower line), VE film (middle line) and SPIN film (upper line) of H_2TPP.

The main results of this analysis are related to the molecular integrity and purity of the samples. In particular: i) VE samples are composed by integer porphyrin molecules, as demonstrated by the fact that FT-IR spectrum of the VE film shows all the characteristic peaks of this compound (Tonezzer et al., 2007b); ii) VE samples are characterized by high purity, as highlighted by the absence in the VE spectrum of any extraneous peak. On the contrary, the FT-IR spectrum of the SPIN sample shows two additional intense peaks at 2920 and 2854 cm^{-1}

due to the presence of retained chloroform, as confirmed by their disappearance after a combined purification treatment of 10 h at 100 °C in vacuum (P=10^{-3} Pa).

4.1.2 Sensing properties

The optical sensing capabilities of the produced porphyrin films have been tested by the following procedure: i) an optical absorption spectrum of the samples before analyte exposure was acquired in the 350–750 nm range under a nitrogen atmosphere (A_o); ii) the samples underwent a thorough conditioning procedure with exposures to saturated vapours of the various analytes (i.e. MeOH, EtOH and 2-propanol); iii) another absorption spectrum was acquired (A_t) when a steady absorbance signal was obtained.

This procedure allows to calculate the percentage absorbance variation patterns $\Delta A\% = ((A_t - A_o)/A_o \times 100)$ as a function of wavelength. This way it is possible to find, for each (sensing material-analyte) couple, the maximum optical absorbance change and the related wavelength. Figure 11 shows the case of VE and SPIN samples of H_2TPP exposed to ethanol.

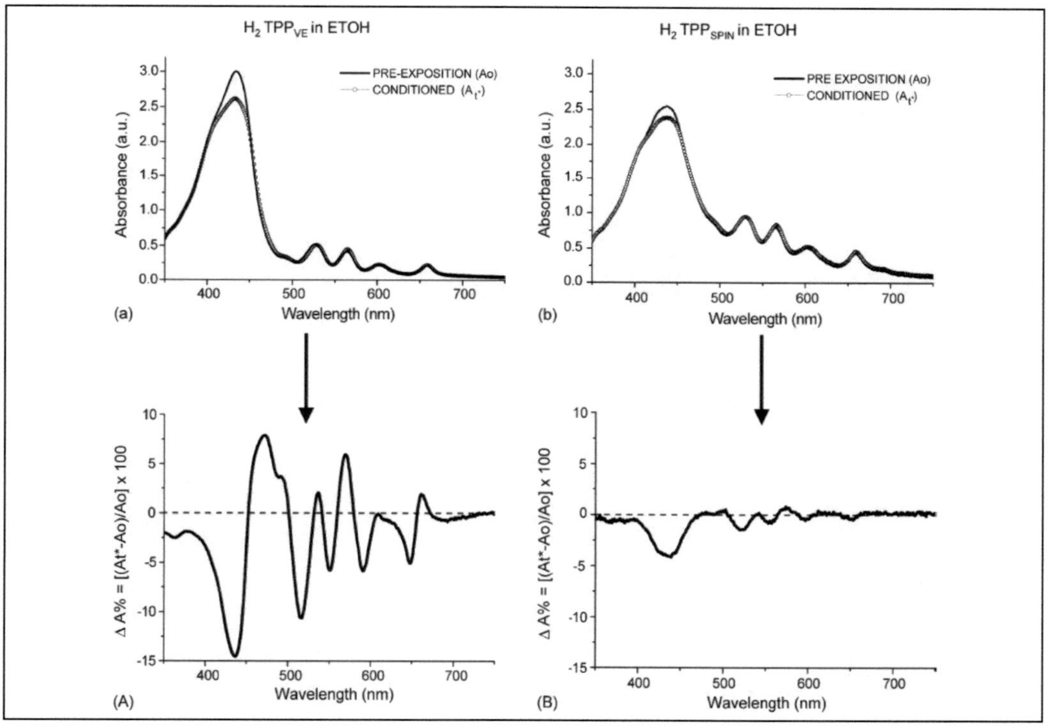

Fig. 11. Absorbance spectra before exposure (solid line) and after conditioning procedure in EtOH (dashed line) (a and b) and corresponding $\Delta A\%$ patterns (A and B) of VE and SPIN samples of H_2TPP, respectively.

Figure 12 reports the intensities of the maximum optical absorbance change $(\Delta A\%)_{MAX}$ of different porphyrin films grown by SPIN and VE technique after the conditioning procedure. VE samples showed higher optical absorbance changes than the SPIN ones with all the tested organic vapours, pointing out greater interactions with the analyte molecules: this different behaviour can be ascribed to the production method. In fact, while vacuum evaporation yields thin impurity-free films, SPIN technique produces samples containing traces of chloroform which hinder the adsorption process of the alcohol vapour molecules.

Furthermore dynamic responses of the samples have been investigated by recording the absorbance changes of the samples at the wavelengths corresponding to the maximum absorbance variations upon exposure to several cycles in different vapour atmospheres and concentrations. Figure 12 reports the evolution of the Soret band intensity of the Fe(TPP)Cl_{VE} and Fe(TPP)Cl_{SPIN} as a function of time upon EtOH exposure (3000 ppm). Both response and recovery phases are conducted at room temperature without any thermal treatment.

The samples responses are characterized by an increase of the absorbance after the EtOH vapour stream is switched on, followed by a slow increase until a saturation value is reached. At 700 s, the EtOH vapour stream is switched off and a dry nitrogen flush is activated: the absorbance rapidly decreases and the original band intensity is recovered. The sensing response of VE sample is characterized by a higher intensity (about three times) and faster behavior than the SPIN one. In particular, VE sample shows t_{50} much shorter (16 s) than SPIN ones (25 s). The faster responses of VE samples can be also ascribed to their higher purity: in fact, this feature is expected to improve their reactivity towards analyte molecules.

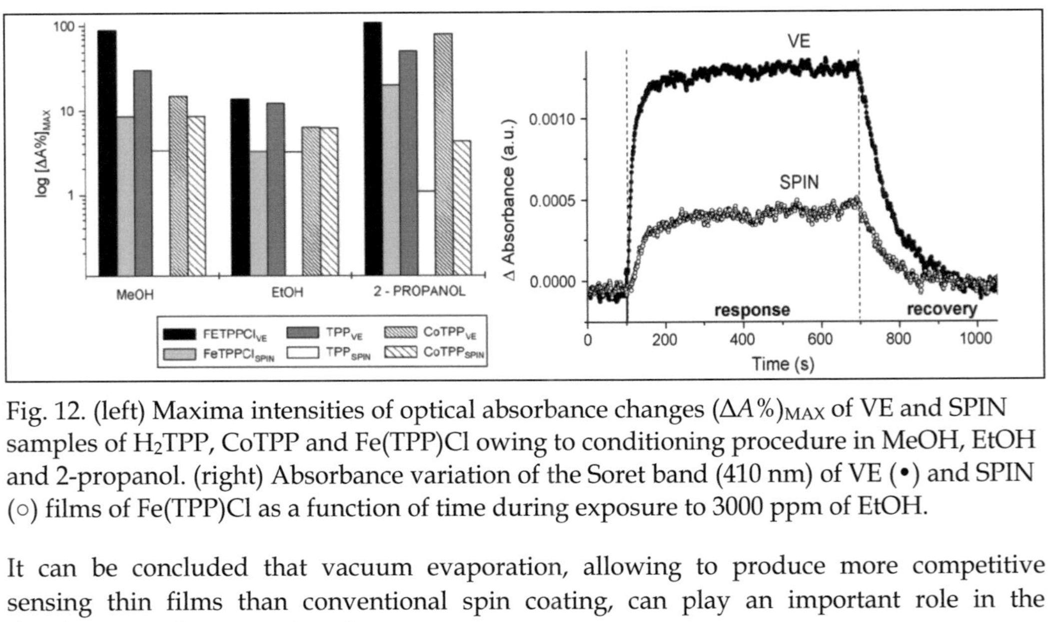

Fig. 12. (left) Maxima intensities of optical absorbance changes $(\Delta A\%)_{MAX}$ of VE and SPIN samples of H_2TPP, CoTPP and Fe(TPP)Cl owing to conditioning procedure in MeOH, EtOH and 2-propanol. (right) Absorbance variation of the Soret band (410 nm) of VE (•) and SPIN (○) films of Fe(TPP)Cl as a function of time during exposure to 3000 ppm of EtOH.

It can be concluded that vacuum evaporation, allowing to produce more competitive sensing thin films than conventional spin coating, can play an important role in the development of new sensing elements.

4.2 Porphyrins deposited by GDS technique

Like for phthalocyanines, Glow-Discharge-induced Sublimation (GDS) represents a new promising technique in the production of porphyrin-based sensing materials, allowing to obtain thin films with both high purity and very large surface area to volume ratio.

In this section the chemical properties of GDS films of CoTPP and Fe(TPP)Cl are reported. Moreover their optical sensing capabilities upon exposure to ethanol vapours are discussed and compared to the sensing capabilities of VE and SPIN films.

4.2.1 Physical and chemical properties

Figure 13 reports the visible absorption spectra of GDS and SPIN films of Fe(TPP)Cl, compared with the absorption spectrum of Fe(TPP)Cl compound dissolved in chloroform ($CHCl_3$) (10^{-5} M). The presence of the characteristic Fe(TPP)Cl optical B and Q bands into the GDS and SPIN films clearly indicates the presence of integer molecules into both the solid films: this experimental evidence points out the absence of molecular iron decomplexation effects into the produced films and demonstrates the integrity of Fe(TPP)Cl molecules within GDS films.

As optical properties of porphyrin compounds can be affected by changes in the aggregation state, UV-Vis spectra give also information about the film structures: in particular the main differences between GDS and SPIN spectra (Q bands of the samples show different red-shifting and both the B and Q bands show different broadening) clearly indicate different molecular arrangements within the two differently deposited Fe(TPP)Cl films (Tonezzer et al., 2009).

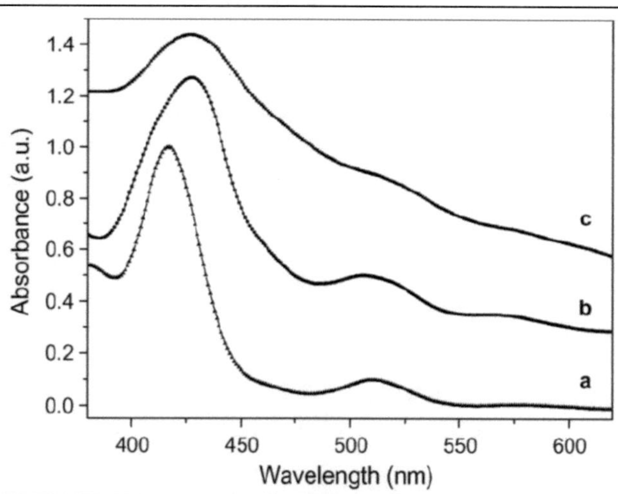

Fig. 13. UV-Visible (UV-Vis) absorption spectra of Fe(TPP)Cl in 10^{-5} M $CHCl_3$ solution (a) and Fe(TPP)Cl films: SPIN (b) and GDS (c).

In order to analyze the structures of differently deposited porphyrin films, several analyses were performed (Tonezzer, 2007c): among them SEM images give an important information about the samples morphology. Figure 14 shows SEM images of VE and GDS CoTPP films.

SEM pictures point out the two very different morphologies, which depend on the different deposition techniques. In particular VE sample shows surfaces characterized by flat morphology while GDS samples show a rough surface composed of micro-size domains.

Fig. 14. SEM images of VE (left) and GDS (right) films of CoTPP (magnification 2200x).

This unconventional rough morphologies are of paramount importance in sensing field: in fact they increase the surface interaction area between film and analyte molecules improving significantly the sensing capabilities of the final material.

4.2.2 Sensing properties

The optical sensing capabilities of the CoTPP and Fe(TPP)Cl thin films produced by GDS technique upon exposure to ethanol vapours were tested by using the same procedure already adopted for VE samples (see section 4.1.2).

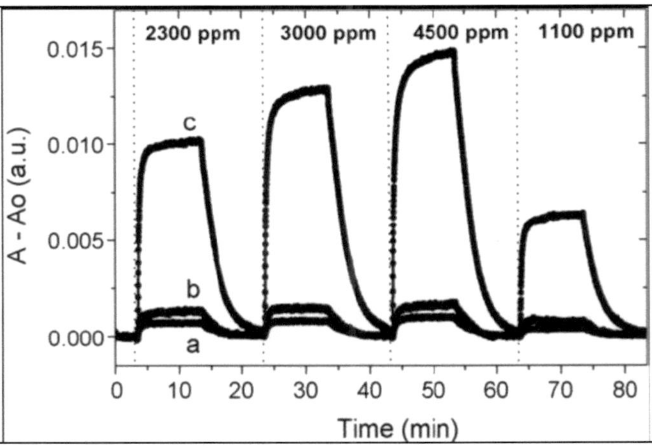

Fig. 15. Concentration dependence of SPIN (a), VE (b) and GDS (c) CoTPP films upon exposures to EtOH atmospheres over the concentration range 1100–4500 ppm.

Figure 15 shows the responses of different-CoTPP samples (GDS, VE and SPIN) upon exposure to EtOH in the range from 1100 to 4500 ppm. The responses increase with increasing the analyte concentration for all the samples. The response of the GDS sample is really fast, shows complete recovery and is much more intense than those of both SPIN (approximately 15 times) and VE (more than 10 times) samples at all the tested concentrations. This much higher response intensity can be attributed to its high surface morphology, as evidenced by the SEM images reported in Figure 14 (Tonezzer, 2007a).

Figure 16 shows the sensing responses of Fe(TPP)Cl samples towards EtOH: in particular Figure 16, left depicts the dynamic response of GDS- and SPIN-deposited Fe(TPP)Cl thin films upon exposure to 2300 ppm of EtOH and Figure 16, right reports the calibration curves of the samples. Fe(TPP)Cl responses are reported as the signal-to-noise ratio (SNR). SNR, expressed as the ratio between the signal intensity (I) and the signal noise (N), represents in fact a powerful dimensionless parameter which, indicating how the signal is higher than the noise by which it is affected, releases the sensing responses from the transducing method used to examine the sample sensing capabilities.

Fig. 16, left highlights the higher and faster response of GDS sample with respect to the SPIN one. In particular GDS film shows a SNR value of 33 which is 15 times higher than that of SPIN one (SNR = 2.2) and t_{90} value shorter (43 s) than SPIN one (63 s).

Fig. 16, right shows the plots of the response magnitude versus EtOH concentration for GDS and SPIN samples: GDS samples are characterized by optical responses much more intense than SPIN ones at all the tested concentrations. Moreover, the more leaning slope of the calibration curve of GDS sample in comparison with that of SPIN one highlights the higher sensitivity of the GDS sample than the SPIN one.

The optical behaviours of all the CoTPP and Fe(TPP)Cl samples presented in this section are characterized by a complete and fast recovery: at this regard it is worth to note that both CoTPP and Fe(TPP)Cl optical responses have been measured at room temperature. In fact, many authors have reported the requirement to heat porphyrin samples during the recovery phase, in order to obtain complete recovery in reasonable times (Pedrosa et al., 2002). Nevertheless if, on one hand, heating procedure gives the advantages to improve the recovery degree and speed, on the other hand it induces thermal stresses into the samples which can result in film degradation.

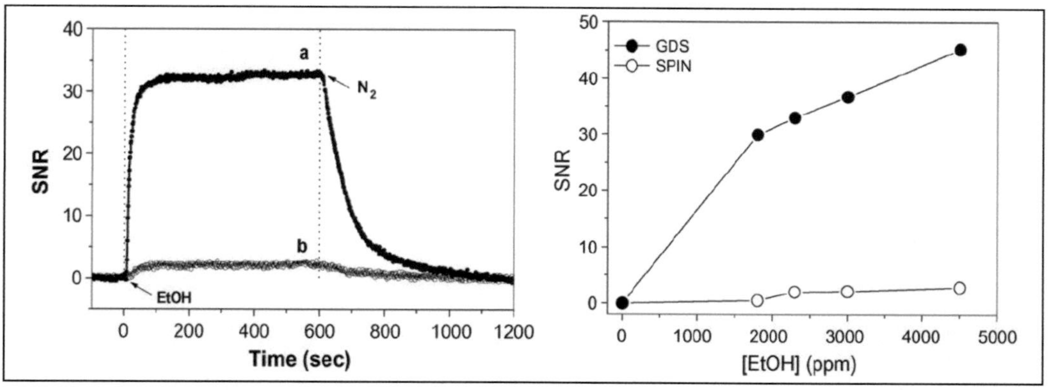

Fig. 16. (left) Single exposure/recovery cycle for GDS (a) and SPIN (b) films of Fe(TPP)Cl exposed to 2300ppm EtOH (T = 20°C). (right) Calibration curves of SPIN and GDS films of Fe(TPP)Cl upon exposures to EtOH atmospheres over the range 1800–4500 ppm.

These results-point out the fundamental role of the deposition technique used for the growth of sensing materials characterized by suited molecular structures and consequently by very competitive sensing capabilities.

5. Cavitands

Molecular recognition of gases is an emerging area of chemistry (Rudkevich, 2004). Cavitands, together with cyclodextrins and calixarenes, are the most studied receptors for gas/vapor sensing because of their outstanding host-guest properties, which are tunable for recognizing different classes of analytes (Cram & Cram, 1994). Among them phosphonate cavitands represent a third-generation emerging class of synthetic receptors whose molecular recognition properties toward alcohols and water were thoroughly demonstrated at the molecular scale by several analytical techniques such as ESI-MS and X-ray crystallography (Melegari et. al., 2008).

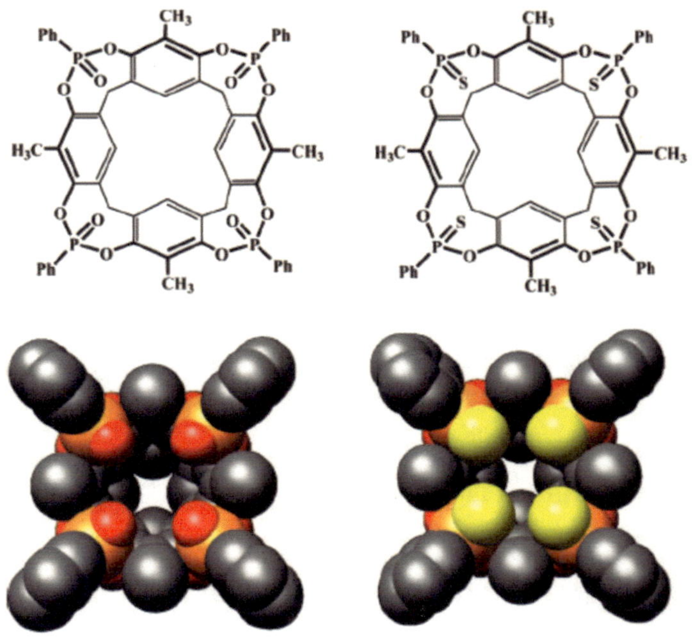

Tiiii[H, CH_3, Ph] TSiiii[H, CH_3, Ph]

Fig. 17. Structures (above) and 3D CPK models (below) of tetraphosphonate Tiiii[H,CH_3,Ph] and tetrathiophosphonate TSiiii[H,CH_3,Ph] cavitands.

The compounds described in this section are two novel phosphonate cavitands devoted to the detection of short chain alcohols synthetized by Dalcanale and co-workers (Biavardi et al., 2009): tetraphosphonate (Tiiii) and tetrathiophosphonate (TSiiii). Figure 17 shows the structures of the two compounds: they present an open, conformationally rigid cavity, delimited by four inward oriented P=O/P=S bridging groups at the upper rim. The structure of these cavitands is born out by previous studies on the sensing properties of tetraphosphonate cavitands toward short chain alcohols. Substitution of the four P=O groups with the P=S moieties completely prevents complexation by eliminating H-bonding interactions between the cavitand and the analyte. The structural similarity of the two cavitands allows for a valid comparison of the influence of molecular recognition on sensing performance. The structural variations of these novel compounds with respect to previously described ones (Melegari et al., 2008) are: (i) the four alkyl chains at the lower rim were removed, to minimize nonspecific interactions; (ii) four methyl groups were introduced in the apical positions to deepen the cavity and increase the strength of CH-π interactions.

5.1 Cavitands deposited by VE technique

In order to be exploited as sensing materials, cavitands are usually deposited as thin solid films by wet deposition techniques; in particular, spin-coating method is commonly employed (Feresenbet et al., 2004). Unfortunately wet deposition methods suffer from several drawbacks such as inhomogeneous surface morphology and uncontrollable thickness. Moreover, insoluble compounds cannot be deposited by such solution techniques. To date, the most common method of depositing insoluble compounds by solution techniques consists of decorating them with peripheral alkyl chains to improve their solubility: nevertheless this molecular derivatization introduces dispersion interactions that dilute the specific analyte response, thus significantly decreasing sensor selectivity. As alternatives, Langmuir-Blodgett and Langmuir-Schafer deposition methods provide ordered and reproducible monolayers: however, when a bulk response is required by the sensing layer (like in QCM sensors), deposition of thicker permeable coatings is mandatory.

In this section we report the employment of the VE technique for the deposition of Tiiii[H, CH_3, Ph] cavitand which represents the ultimate receptor for supramolecular mass sensing of short chain alcohols. VE technique is also used for producing TSiiii[H, CH_3, Ph]-sensing coatings for comparison. In the case of the Tiiii and TSiiii cavitands, the absence of alkyl chains at the lower rim implies a different film deposition method with respect to the standard ones (spray- and -spin coating) owing to the high insolubility of the two cavitands.

VE technique overcomes solubility problems, allowing for the direct formation of films from solid materials. The lack of residual solvent during the deposition process assures the formation of high purity films: this represents a basic requirement in the gas sensing field, because of the unpredictable effects of the retained solvent on the final response of the sensor, including occupation of adsorption sites and interference in analyte/material interactions. Moreover, VE technique guarantees good reproducibility, high uniformity, and homogeneity and provides accurate control over both the growth rate and the final thickness of the samples.

5.1.1 Physical and chemical properties

In order to investigate the purity of the VE cavitand films and to exclude the presence of impurities derived from cavitand decomposition, infrared spectra of Tiiii and TSiiii films have been collected and compared with the spectra of the respective powders (Figure 18).

Tiiii and TSiiii samples are characterized by similar infrared spectra due to the close resemblance of their molecular structures showing some differences in certain positions. Both Tiiii and TSiiii samples show all the characteristic peaks of the respective powders, indicating the absence of damaged molecules in both the VE samples, within the detection limits of the employed technique. Moreover, the lack of any additional peaks in the FT-IR spectra excludes the presence of extraneous compounds, demonstrating the high purity of the samples (Tonezzer et al., 2008).

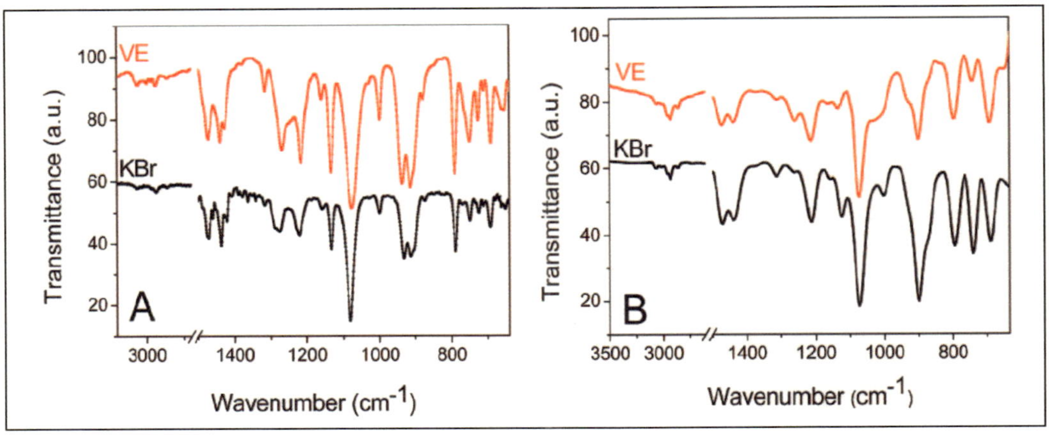

Fig. 18. FT-IR spectra of vacuum evaporated (VE) films of Tiiii (A) and TSiiii (B) cavitands. The spectra of the corresponding starting powders pressed in KBr pellet (KBr) are also reported as reference.

The integrity of the cavitand molecules into the sublimated films as well as the purity of the samples demonstrate the viability of the VE technique for the deposition of thermally stable molecular receptors. AFM measurements of the surface of the Tiiii and TSiiii films point out their high uniformity and homogeneity. Moreover, the similar thickness of the two films (350 nm for Tiiii and 355 nm for TSiiii), which were deposited with the same process parameters, highlights the high reproducibility of the VE deposition process (Tonezzer, 2008).

5.1.2 Sensing properties

Sensing measurements were performed by exposing cavitand-coated QCMs to different EtOH concentrations (ranging from 5 to 200 ppm) and monitoring the shift of the QCM fundamental resonance frequency induced by the mass change as a function of time. Figure 19, left shows the responses of Tiiii- and TSiiii- coated QCM sensors exposed to 25 ppm of EtOH. The most significant result is the high difference in response intensity between Tiiii (ΔF = -60 Hz) and TSiiii (ΔF = -7.5 Hz) mainly due to the following two key factors: (i) the preorganized cavity which provides a free volume available for the analyte, pivotal for effective H-bonding; (ii) the presence of synergistic CH-π interactions with the π-basic cavity and the energetically equivalent H-bonding options between the P=O groups at the upper rim and the analyte (specific interactions). The resulting mode of interaction between ethanol and the cavity shows how the ethanol chain fits into the cavity with its methyl residue (CH-π interactions), while the OH moiety undergoes H-bonding interactions with the P=O groups. The effective contribution of the H-bonds is absent in the TSiiii cavitand, because the P=S group is much less polarized than the P=O one and, consequently, it is ineffective as H-bond acceptor. Therefore the observed TSiiii responses are due to nonspecific dispersion interactions between the organic layer and EtOH and the large difference in the sensor responses can be attributed to the specific binding of EtOH by the Tiiii cavitand (Tonezzer et al., 2008).

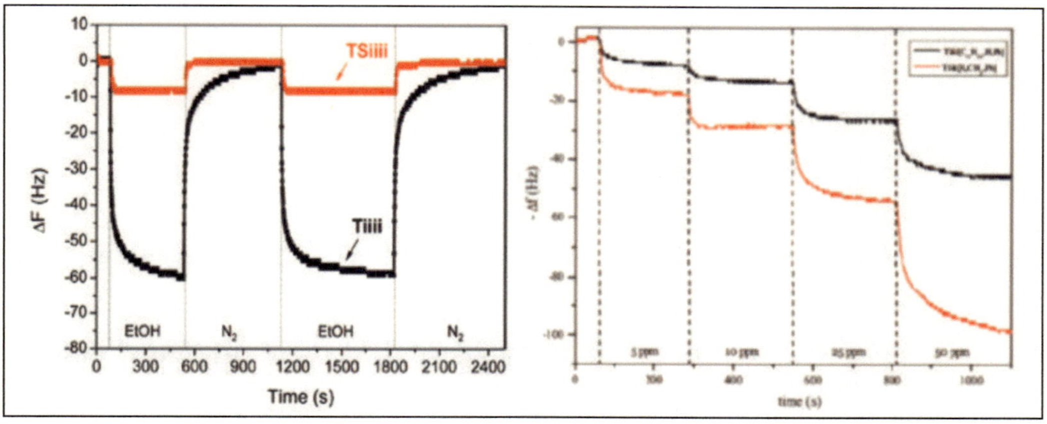

Fig. 19. (left) Change of the resonance frequency of VE Tiiii and TSiiii samples as a function of time during exposure to 25 ppm EtOH (T = 20 °C). (right) Change of the resonance frequency of SPIN Tiiii[$C_{11}H_{23}$, H, Ph] and VE Tiiii [H, CH_3, Ph] samples at increasing amount of EtOH.

The superior sensing performances of VE Tiiii[H, CH_3, Ph] sensor have been demonstrated by comparing its responses to those of the spin-coated long chain analogue Tiiii[$C_{11}H_{23}$, H, Ph], the best performer to date (Melegari, 2008). Figure 19, right shows that the combination of Tiiii[H,CH_3, Ph] receptor with VE deposition doubles the EtOH sensor sensitivity in the 5-50 ppm range in which nonspecific extracavity adsorption becomes significant. This result is remarkable, as it was obtained without altering the receptor site responsible for the alcohol complexation at the gas-solid interface.

To investigate the interaction between EtOH and the VE films, the recovery phase of the samples was analyzed by Elovich kinetics. According to this model, the surface uncovering Θ during the recovery phase as a function of time is given by the following formula:

$$\Theta(t) = \frac{1}{\beta} ln(t) + K$$

where β and K are constants. This model is based on the assumption that the desorption probability of an analyte molecule during surface purging decreases exponentially as a function of the number of analyte molecules already desorbed. Thus, by assuming that the change of resonance frequency (ΔF) is related only to the interaction between EtOH and cavitand coatings, the value of Θ should be proportional to ΔF. Hence, by plotting ΔF as a function of ln(t), a linear relationship should be obtained.

Figure 20 reports the plots of ΔF vs ln(t) for the recovery phases of the VE Tiiii and TSiiii layers, exposed to 5 and 100 ppm EtOH, respectively.

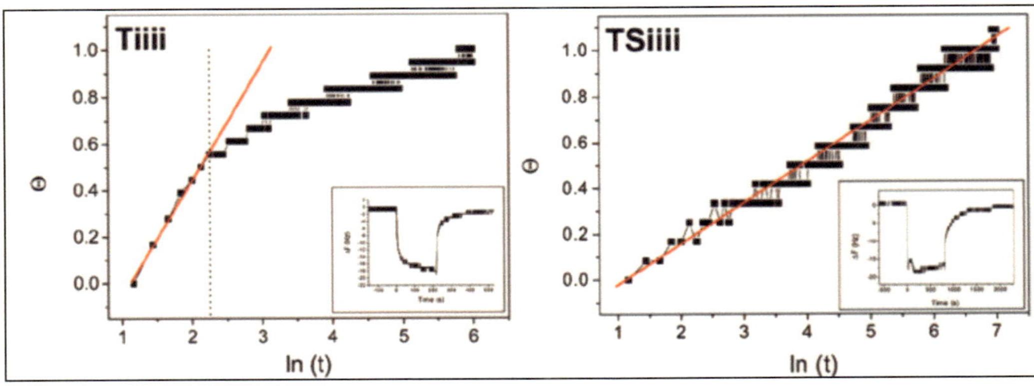

Fig. 20. Elovich recovery kinetics for vacuum-evaporated Tiiii and TSiiii films after exposing to 5 and 100 ppm EtOH, respectively.

As can be observed, the trends of Tiiii- and TSiiii-coated QCMs are characterized by different behaviors. The Tiiii behavior is approximately linear over the first 10 s, deviating from linearity at higher times. This indicates a change in activation energy dependent on surface coverage, supporting a two-step interaction process: (i) a weak interaction between analyte molecules and nonspecific sites and (ii) a stronger specific interaction due to cavity inclusion. In particular, the linear region represents the fast process when EtOH molecules are removed from the nonspecific sites of the Tiiii coating, while the nonlinear region represents slow EtOH release from the specific sites of the layer. By contrast, the TSiiii-coated QCM shows a completely linear behavior, indicating an interaction process dominated totally by nonspecific dispersion interactions.

These findings clearly indicate that TSiiii films are characterized mainly by nonspecific extracavity adsorption, while Tiiii layers feature intracavity complexation of EtOH.

6. Conclusions

In this chapter the authors exemplified the fundamental role played by the film deposition technique in the attainment of both specific physical properties and final detecting capabilities of organic sensing thin films. In this respect two physical vapour deposition techniques used in the production of sensing materials have been discussed: High Vacuum Evaporation (VE) and Glow-Discharge-induced Sublimation (GDS),the latter being a novel patented plasma-based deposition technique.

The results obtained for three classes of different organic and metallo-organic compounds, i.e. phthalocyanines, porphyrins, and cavitands, are here summarized.

- Phthalocyanines. Two phthalocyanine compounds, namely copper (CuPc) and zinc (ZnPc) phthalocyanines, have been deposited by GDS technique and thoroughly characterized. Morphological measurements, performed by SEM and AFM techniques, point out that both ZnPc and CuPc thin films have a very high surface roughness, much higher than that found for similar films deposited by conventional methods. Moreover both GDS samples feature peculiar microporous molecular architectures, as evidenced by nitrogen physisorption measurements.

The sensing capabilities of CuPc samples have been investigated towards different analytes and by different transducing methods. As electrical sensors CuPc samples exhibit detection limits down to 0.1 ppm for NO_2 (the standard attention level set by the European Union) and 10 ppm for NO and response times shorter than 30 s for 0.98 ppm NO_2 and 98 ppm NO. CuPc films have been also tested as optical sensors towards ethanol vapours within 1500 - 34000 ppm concentration range highlighting high-speed responses (t_{50} = 7 s and t_{90} = 12 s) and fast recovery (t_{50} =12 s and t_{10} =38 s). It has been shown that CuPc-coated QCMs can be used for NO detection down to few ppm. ZnPc films, tested as optical sensors towards different alcohol vapours (methanol, ethanol and isopropanol) showed stable and reproducible signals.

- Porphyrins. Porphyrin thin films were grown both by high vacuum evaporation (VE) and Glow-Discharge-induced Sublimation (GDS). In particular, three different porphyrin-based films were produced by VE technique: free (H_2TPP), cobalt (CoTPP), and iron chloride (Fe(TPP)Cl) 5,10,15,20 *meso*-tetraphenyl porphyrins. Their chemical properties and optical sensing capabilities towards different alcohol vapours (methanol, ethanol and isopropanol) have been analyzed and compared to those of conventional spin coated (SPIN) ones. FT-IR spectra pointed out that VE technique allows to produce samples characterized by a much higher purity than the SPIN ones and the study of the optical responses highlighted that their higher purity provide them with much larger sensitivity and much faster response times than the SPIN sensing elements.
  CoTPP and Fe(TPP)Cl films have been also produced by GDS technique: their chemical and physical features have been analyzed by FT-IR, UV-Vis and SEM analyses and their optical sensing capabilities towards ethanol have been measured. The chemical and physical characterization highlighted that GDS porphyrin films have a high purity, analogously to the VE samples, and feature an extremely large surface area to volume ratio, like the GDS phthalocyanine films. Optical sensing measurements confirmed that GDS technique allows to produce very competitive sensing films. In fact, both GDS CoTPP and Fe(TPP)Cl films feature much more intense response (up to 10 times) and much larger sensitivity with respect to conventional SPIN samples, high speed responses (t_{50} and t_{90} respectively of 10 s and 1 min approximately) and fast and complete recovery.
- Cavitands. VE deposition has been used for the first time in order to produce sensing thin films based on Tiiii[H, CH_3, Ph] cavitand, an ultimate receptor for supramolecular sensing. The characterization of the Tiiii films highlighted that VE technique is particularly suitable for the deposition of supramolecular compounds: in fact, VE allows to grow thin films starting from compounds characterized by low solubility (a quite common feature of highly specific recognition molecules). As regard the sensing capabilities, Tiiii films have been tested towards very low concentration of ethanol vapours (5-50 ppm) demonstrating that VE technique provides a significant improvement of the performances of cavitand receptor as sensitive material in mass sensors. Moreover Elovich kinetics, used to elucidate the sorption processes occurring onto the layers, highlighted the high specificity of the Tiiii sensors, which reduces the incidence of aspecific adsorption.

Chemical sensing requires an integrated approach, where both the molecular and the materials properties of the sensing layer must be finely tuned to achieve the desired properties. In this respect, thin film chemical sensors represents a particular challenge, taking into account that the analyte recognition is mediated by the layer properties of the coated receptors.

In this chapter, the authors describes different real cases in which significant advantages are introduced into the sensing field by a bottom-up approach in which recognition molecules are deposited in well-controlled and designed architectures by novel and accurate deposition techniques. The integrated approach used by the authors, where both the molecular and the materials properties of the sensing layers have been finely tuned, demonstrated to produce organic and metallo-organic sensors with improved performances over the existing ones.

This approach can be easily extended to many different classes of organic receptors, opening the way for the rational design of sensor materials tailored for the analytes to be detected.

7. Acknowledgment

The authors thank the research group of Prof. Dalcanale of Industrial Chemistry at the Department of Organic and Industrial Chemistry of Parma University (Italy) for providing them Tiiii and TSiiiii cavitands. M.T. thanks personally Prof. Dalcanale for the helpful discussions on supramolecular receptors. The research leading to these results has received funding by Fondazione Cassa di Risparmio di Trento e Rovereto (CARITRO) within GREEN project and from the European Community's Seventh Framework Programme FP7/2007-2013 under grant Agreement *Marie Curie 7th Framework Program - PCOFUND-GA-2008-226070, acronomy "Progetto Trentino"* within PHOTOFUTURE project.

8. References

Arai, T. & Kragic, D. (1999). Variability of Wind and Wind Power, In: *Wind Power,* S.M. Muyeen, (Ed.), 289-321, Scyio, ISBN 978-953-7619-81-7, Vukovar, Croatia

Bernini, R., Tonezzer, M., Mottola, F., Zeni, L., Quaranta, A., Maggioni, G., Carturan, S. & Della Mea, G. (2007). Volatile organic compounds detection using porphyrin-basedmetal-cladding leaky waveguides, *Sens. Actuators B*, Vol. 127, pp. 231-236

Biavardi, E., Favazza, M., Motta, A., Fragalà, I.L., Massera, C., Prodi, L., Montalti, M., Melegari, M., Condorelli, G.G. & Dalcanale, E. (2009). Molecular recognition on a cavitand-functionalized silicon surface, *JACS*, Vol. 131 (21), pp. 7447-7455.

Brunik, J.A.J., Di Natale, C., Campo dall'Orto, V., Macagnano, A., Angelaccio, A., Sgarlata, A., Hurst, J., Rezzano, I., Mascini, M. & D'Amico, A. (1999). Porphyrin thin films coated quartz microalances prepared by electropolymerization techniques, *Thin Solid Films,* Vol. 354, pp. 245–250.

Cram, D. J. & Cram, J. M. (1994). *Container Molecules and Their Guests,* The Royal Society of Chemistry, Cambridge, U.K.

Feresenbet, E.B., Dalcanale, E., Dulcey, C. & Shenoy, D.K. (2004) Optical sensing of the selective interaction of aromatic vapors with cavitands *Sensors and Actuators B,* Vol. 97 pp. 211–220.

Horvath, G. & Kawazoe, K. (1983). Method for the calculation of effective pore size distribution in molecular sieve carbon, *J. Chem. Eng. Jpn.,* Vol. 16, pp. 470–475.

Iwatsu, F., Kobayashi, T. & Uyeda, N. (1980). Solvent effects on crystal growth and transformation of zinc phthalocyanine, *J. Phys. Chem.*, Vol. 84, pp. 3223–3230.

Li, B.; Xu, Y. & Choi, J. (1996). Applying Machine Learning Techniques, *Proceedings of ASME 2010 4th International Conference on Energy Sustainability*, pp. 14-17, ISBN 842-6508-23-3, Phoenix, Arizona, USA, May 17-22, 2010

Maggioni, G., Carturan, S., Boscarino, D., Della Mea, G. & Pieri U. (1997). Polyimide and platinum containing polyimide thin films obtained by vapour deposition polymerization; effects of thermal treatments, *Mater. Lett.*, Vol. 32, pp. 147-150.

Maggioni, G., Quaranta, A., Carturan, S., Patelli, A., Tonezzer, M., Ceccato, R. & Della Mea, G. (2005). Deposition of Copper Phthalocyanine Films by Glow-Discharge-induced Sublimation, *Chem. Mater.*, Vol. 17, No. 7, pp. 1895-1904

Maggioni, G., Quaranta, A., Carturan, S., Patelli, A., Tonezzer, M., Ceccato, R. & Della Mea, G. (2005). Deposition of Copper Phthalocyanine Films by Glow-Discharge-induced Sublimation for Gas Sensing Applications, *Surf. Coat. Technol.*, Vol. 200, pp. 476-480

Maggioni, G., Manera, M. G., Spadavecchia, J., Tonezzer, M., Carturan, S., Quaranta, A., de Julián Fernández, C., Rella, R., Siciliano, P., Della Mea, G., Vasanelli, L. & Mazzoldi, P. (2007). Optical response of plasma-deposited zinc phthalocyanine films to volatile organic compounds. *Sens. Actuators B,* Vol. 127, pp. 150-156.

Maggioni, G., Carturan, S., Tonezzer, M., Quaranta, A. & Della Mea, G. (2008). Plasma-deposited copper phthalocyanine: A single gas-sensing material with multiple responses, *Sens. Actuators B*, Vol. 131, pp. 496-503

Melegari, M., Suman, M., Pirondini, L., Moiani, D., Massera, C., Ugozzoli, F., Kalenius, E., Vainiotalo, P., Mulatier, J.C., Dutasta, J.-P. & Dalcanale, E. (2008). Supramolecular sensing with phosphonate cavitands, *Chem. Eur. J.*, Vol 14 (19), pp. 5772-5779

Pedrosa, J.M. , Dooling, C.M., Richardson, T.H., Hyde, R.K., Hunter, C.A., Martin, M.T. & Camacho, L. (2002). The optical gas-sensing properties of an asymmetrically substituted porphyrin, *J. Mater. Chem.*, Vol. 12, pp. 2659–2664.

Rakow, N.A. & Suslick, K.S. (2000). A colorimetric sensor array for odour visualization *Nature*, Vol. 406 (6797), pp. 710-713

Rudkevich, D.M., (2004). Emerging Supramolecular Chemistry of Gases, *Angewandte Chemie - International Edition*, Vol 43 (5), pp. 558-571.

Sadaoka, Y., Sakai, Y., Yamazoe, N. & Shiyama, T. (1982). Effect of adsorbed oxidative gases on electrical properties of evaporated films of phthalocyanine. *Denki Kagaku*, Vol. 50, No. 6, pp. 457-462.

Siegwart, R. (2001). Indirect Manipulation of a Sphere on a Flat Disk Using Force Information. *International Journal of Advanced Robotic Systems,* Vol.6, No.4, (December 2009), pp. 12-16, ISSN 1729-8806

Snow, A.W. & Barger, W.R. (1989) *Phthalocyanines: properties and applications* in C.C. Leznoff, A.B.P. Lever (Ed.), Vol. 1, pp. 341-392, VCH Publishers, Inc., New York, U.S.A.

Spadavecchia, J., Ciccarella, G., Valli, L. & Rella, R. (2006). A novel multisensing optical approach based on a single phthalocyanine thin films to monitoring volatile organic compounds, *Sens. Actuators B*, Vol. 113, pp. 516–525

Tonezzer, M., Maggioni, G., Quaranta, A., Carturan, S. & Della Mea, G. (2007). Optical sensing properties of CoTPP thin films deposited by glow-discharge-induced sublimation, *Sens. Actuators B,* Vol. 122, pp. 613-619

Tonezzer, M., Quaranta, A., Maggioni, G., Carturan, S. & Della Mea, G. (2007). Optical sensing responses of tetraphenyl porphyrins toward alcohol vapours: A comparison between vacuum evaporated and spin-coated thin films, *Sens. Actuators B*, Vol. 122, pp. 620-626

Tonezzer, M., (2007). *Production, characterization and testing of solid organic and metallo-organic films for optical gas sensors*, University of Trento, Trento, Italy

Tonezzer, M., Melegari, M., Maggioni, Milan, R., Della Mea, G. & Dalcanale, E. (2008). Vacuum-Evaporated Cavitand Sensors: Dissecting Specific from Nonspecific Interactions in Ethanol Detection, *Chem. Mater.*, Vol. 20, pp. 6535–6542

Tonezzer, M., Maggioni, G., Quaranta, A., Carturan, S. & Della Mea, G. (2009). Growth, characterization and sensing capabilities of 5,10,15,20-meso-tetraphenyl iron (III) porphyrin chloride films obtained by means of a novel plasma-based deposition technique, *Sens. Actuators B*, Vol. 136, pp. 290-296

14

Drift Correction Methods for Gas Chemical Sensors in Artificial Olfaction Systems: Techniques and Challenges

S. Di Carlo[1] and M. Falasconi[2]
[1]*Department of Control and Computer Engineering*
Politecnico di Torino, Torino
[2]*SENSOR CNR-IDASC, Brescia, University of Brescia, Dept. of Chemistry and Physics for Engineering and Materials, Brescia*
Italy

1. Introduction

The human sense of smell is a valuable tool in many areas of industry such as perfumery, food and drink production, clinical diagnosis, health and safety, environmental monitoring and process control (Gobbi et al., 2010; Vezzoli et al., 2008). Artificial olfaction mimics human olfaction by using arrays of gas chemical sensors combined with pattern recognition (PaRC) systems (Pearce et al., 2003). When a volatile compound comes into contact with the surface of the array, a set of physical changes modifies the properties of the material from which each sensor is composed. This perturbation can be measured, digitalized and used as a feature for the specific compound. A preliminary training phase collecting samples from known volatile compounds is used to train a selected PaRC algorithm in order to map each concentration of gas to the responses from the sensor array. The trained model is then used for identification during later measurements. The classification rate of the PaRC system determines the final performance of the electronic olfaction system.

Gas sensor arrays represent a potentially low-cost and fast alternative to conventional analytical instruments such as gas chromatographs. However, successful applications of gas sensor arrays are still largely limited to specialized laboratories (Pardo & Sberveglieri, 2004). Lack of stability over time and the high cost of recalibration are factors which still limit the widespread adoption of artificial olfaction systems in real industrial setups (Padilla et al., 2010).

The sensor drift consists of small and non-deterministic temporal variations of the sensor response when it is exposed to the same analytes under identical conditions (Holmberg et al., 1997). The main result is that the sensor's selectivity and sensitivity decrease. The gas sensor drift changes the way samples distribute in the data space, thus limiting the ability to operate over long periods. PaRC models become useless after a period of time, in some cases weeks or a few months. After that time the artificial olfaction system must be completely re-calibrated to ensure valid predictions (Aliwell et al., 2001).

It is still impossible to fabricate chemical sensors without drift. In fact, drift phenomena afflict almost all kinds of sensors (Chen & Chan, 2008; Owens & Wong, 2009; Polster et al., 2009).

Sensor drift must be therefore detected and compensated to achieve reliable measurements from a sensor array.
Algorithms to mitigate the negative effect of gas sensor drift are not new in the field; the first attempt to tackle this problem dates back to the early 90s (Pearce et al., 2003). Nevertheless, the study of sensor drift is still a challenging task for the chemical sensor community (Padilla et al., 2010; Pearce et al., 2003). Solutions proposed in the literature can be grouped into four main categories: (i) periodic calibration, (ii) attuning methods, (iii) filtering of drift components and (iv) adaptive models. In this chapter the authors introduce the main challenges faced when developing drift correction techniques and will propose a deep overview of state-of-the-art methodologies that have been proposed in the scientific literature trying to underlying pros and cons of these techniques and focusing on challenges still open and waiting for solutions.

2. Gas chemical sensors and the drift phenomenon

A chemical sensor is a device that transforms chemical information, ranging from the concentration of a specific component to the total composition analysis, into an analytically useful signal (International Union of Pure and Applied Chemistry, 1991). In the gas chemical sensing, the input signal is the concentration of one or more gaseous species while the output signal depends on the transduction mechanism that is usually a variation of some physical properties of the sensing element such as: conductivity (or other electrical magnitudes), oscillation frequency (mass), temperature, electrochemical potentials, surface work function or optical properties (Janata, 2009).
Artificial olfaction systems, usually referred to as Electronic Noses (ENs) (Pearce et al., 2003), are machines designed for detecting and discriminating among complex odors using an array of broadly-tuned (non-specific) gas chemical sensors typically belonging to the above mentioned categories. An odor stimulus generates a characteristic fingerprint from the sensor array. Patterns from known samples can be used to construct a database (training set) and train a pattern recognition system so that unknown odor samples (test set) can be subsequently identified.
Attempts to measure odors with electronic instruments were made in the early 60s, but the "modern era" of artificial olfaction began in 1982 with the work of Persaud and Dodd (Persaud & Dodd, 1982), who used a small array of gas-sensitive metal-oxide devices to classify odors. The expression "electronic nose" appeared for the first time in 1987 (Shurmer et al., 1987), with its current definition given in the same year by Gardner (Gardner, 1987). Commercial instruments became available in the early 90s – first in Europe immediately followed by the U.S. – with pioneering machines developed by Alpha Mos[1] and Aromascan. ENs take their inspiration from the working mechanism of biological olfaction. However, current technologies based on chemical sensors are still far away from the capability of biological systems mainly because of their (still) poor selectivity and sensitivity with respect to biological receptors, and more importantly their lack of stability.
In recent years, classical chemical sensor technologies were complemented by new emerging technologies (Röck et al., 2008). In particular, machine olfaction has benefited from developments in several fields ranging from optical technologies developed by the telecommunications industry to the improvements in analytical chemistry such as: gas

[1] http://www.alpha-mos.com/

chromatography, mass spectroscopy and ion mobility spectrometry. This trend has also narrowed the gap between traditional ENs – used as a black box – and classical analytical techniques which aim to quantify individual volatile components.

Although the constant improvements in micro fabrication techniques and the rapid development of new nano fabrication techniques have allowed the production of functional micro and nanoscale chemical sensing devices with finer sensitivity (Comini & Sberveglieri, 2010) and selectivity (Haupt & Mosbach, 2000), signal repeatability over time still remains the real challenge in the chemical sensor field for all types of sensors. In order to have a reliable instrument it is of great importance that individual sensor signals are stable and reproducible. In practice, it is not worthy spending several weeks for training an EN system in a particular application if, as a consequence of changes in the sensor response, the system can only be used for a few days before recalibration is required.

The problem of chemical sensor stability over time is known as "sensor drift". It consists of (more or less) small and non-deterministic temporal variations of the sensor response when it is exposed to the same analytes under identical conditions (Holmberg et al., 1997). This is generally attributed to sensors aging (Sharma et al., 2001) or thermo-mechanical degradation (Mielle, 1996), but it can also be influenced by a variety of sources including environmental factors (Di Natale et al., 2002b; Ionescu et al., 2000). The main result is that sensors selectivity and sensitivity slowly decrease with time. The physical causes of this phenomenon are technology dependent and are strictly correlated to the sensing and transduction mechanisms.

2.1 Drift phenomenon in metal oxide semiconductor gas sensors

Among the different technologies used to fabricate chemical sensors, which in general lack enough information about the physical causes of the sensor drift, the physical meaning of the drift phenomenon in conductometric metal oxide sensors (MOX) has been deeply investigated in the past. In fact, signal drift is known to be a severe problem for these devices which have widespread commercial diffusion. A typical example is the sintered tin dioxide Taguchi Gas Sensor (TGS), a n-type semiconductor solid state device marketed by Figaro Engineering Inc.[2] since 1968 and widely applied for detection of oxidizing and reducing gases.

The sensing properties of TGS are based on the electronic (n-type) conductivity of tin dioxide (SnO_2) (Goepel & Schierbaum, 1995). The device consists of small SnO_2 grains which are in contact with each other. The sensing effect is due to an electronic depletion layer at the surface of the grains. The depletion layer is generated when oxygen is adsorbed, thus trapping electrons from the oxide. This induces an increased resistance at the grain surfaces. When the current passes from one grain to another it has to cross these depletion layers (Schottky barriers) which thus determine the sensor's resistance. The sensing effect, i.e., the response to reducing or oxidizing gases, is therefore determined by the adsorption of these compounds and the subsequent trapping (or donation) of electrons by the adsorbed species. This modifies the space charge potential thus changing the sensor's conductivity. Similar working models also apply to thick- and thin-film semiconductor MOX gas sensors, including for instance $SnO_2 - RGTO^3$ gas sensors (Sberveglieri, 1992). The main reasons for the MOX sensor

[2] http://www.figaro.co.jp

[3] Rheotaxial Growth and Thermal Oxidation (RGTO) technique was developed in the early 90s at the Sensor Lab on purpose of growing metal oxide thin films. The films obtained with such technique show a structure characterized by polycrystalline agglomerates uniformly distributed and connected by necks. This highly porous structure leads to a large surface area well suited for gas absorption

drift has been attributed to different structural and morphological variations of the sensor as discussed in the following subsections.

2.1.1 Chemical diffusion of oxygen vacancies

Usually two chemical effects are considered as potential sources of drift in MOX sensors: a) the resistivity change induced by chemisorption of water on the sensor (Schierbaum et al., 1991), and b) the chemical diffusion of oxygen vacancies that was investigated by means of relaxation experiments on tin dioxide single crystals (Kamp et al., 2001).
Kamp et al. have proven that for TGS sensors the chemical diffusion of oxygen is fast enough to cause severe drift effects due to stoichiometry changes. This effect was in turn attributed to two main origins:

- the chemical diffusion of oxygen in SnO_2 consists of the simultaneous transport of oxygen vacancies and conduction electrons that induces conductivity changes in the bulk and/or space charge layer of SnO_2 grains. Changes in oxygen vacancy concentration over the sensor working time influences the space charge and the overall conductivity which reflects on the sensor baseline drift;
- the diffusion of oxygen vacancies can be induced by the space charge electric field alone (field induced migration). For the case of the TGS this process is induced by a signal change, via changes of the space charge potential. Since oxygen vacancies are usually majority carriers, the result of the redistribution of oxygen vacancies is a severe modification of the properties of the space charge region itself.

2.1.2 Physical changes of the MOX sensing elements

Scanning electron microscopy investigation of thin-film SnO_2 sensors integrated on CMOS micro-machined hotplates (Sharma et al., 2001) indicated that the sensing thin-film cracks after long operation cycles. Indeed, the thermal stress in the micro-hotplate induces cracks during a large number of heating and cooling cycles. Cracks in the thin-film can be considered a major physical cause of the sensor drift. Authors also suggest that Cu doping could be a manner to suppress (or strongly reduce) the film cracking.
Similar results are mentioned in other works, e.g., investigating the variation of electrical response to CO and CH_4 induced by the continuous thin-film SnO_2 sensor operation at 400degC for six months. A slowly varying decreasing of response to both baseline air and gases was observed (Fig. 1) and attributed to the coalescence of grains due to their intrinsic poor degree of crystallinity. In fact a significant improvement on the stability of the sensor platforms can be achieved using single crystalline materials such as novel quasi-one-dimensional MOX nanostructures (Comini & Sberveglieri, 2010).

2.1.3 Degradation of the electric contacts

Besides the preparation of stable grain structures, the proper choice of electric contact geometries (shape and thickness) is particularly important in the design of reliable conductometric MOX sensors.
Recently, thin-film $SnO_2 - RGTO$ sensors have been investigated at SENSOR lab by scanning electron microscopy including the study of morphology by secondary electron detection and back scattered electrons which allow for compositional information (unpublished results). Electrical tests, performed in parallel, predicted a drift time constant of about 85 days . The

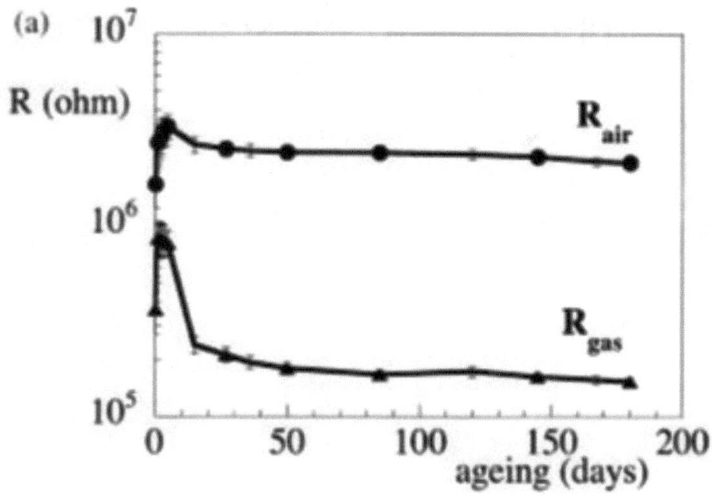

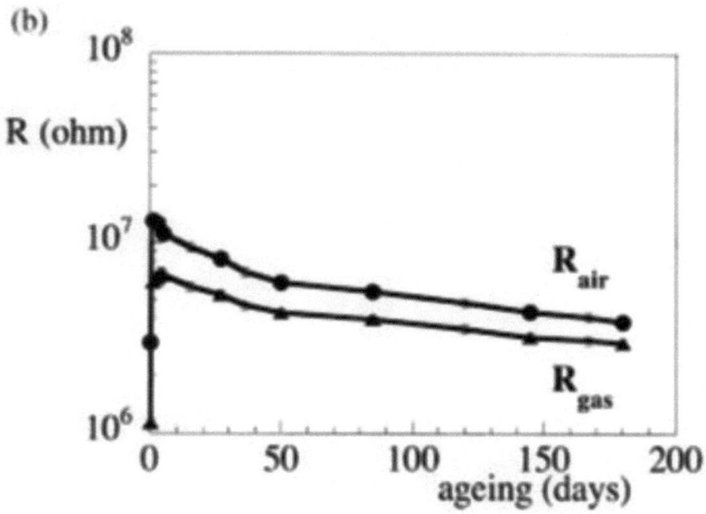

Fig. 1. R_{air} and R_{gas} for a SnO_2 (Au doped) thin film vs. the aging time; the sensor is operated at 400degC. (a): 500 ppm CO in 30% RH; (b): 5000 ppm CH_4 in 30% RH (reproduced from (Nelli et al., 2000))

correlation of these results show that the long term stability is mostly determined by the instability of platinum inter-digitized contacts (IDCs). Two concurrent effects were observed (Fig. 2):

- first, the erosion of the IDC, which appears thinner after aging, and the formation of Pt agglomerates. Pt agglomeration was observed only on the IDCs deposited over the $SnO_2 - RGTO$ layer, while the Pt structures deposited over the alumina substrate did not exhibit agglomeration. From this, we might argue that the larger RGTO roughness accelerates the degradation rate of the above deposited Pt layer;

- second, the formation of *Pt* agglomerates could be promoted by electro-migration phenomena. This second hypothesis has been confirmed observing that the formation of *Pt* agglomerates is more evident in the area in which the section of the electrodes is thinner, i.e., where the current density is bigger. Increasing the IDC thickness from 500nm to 900nm the drift time constant was increased to about 130 days.

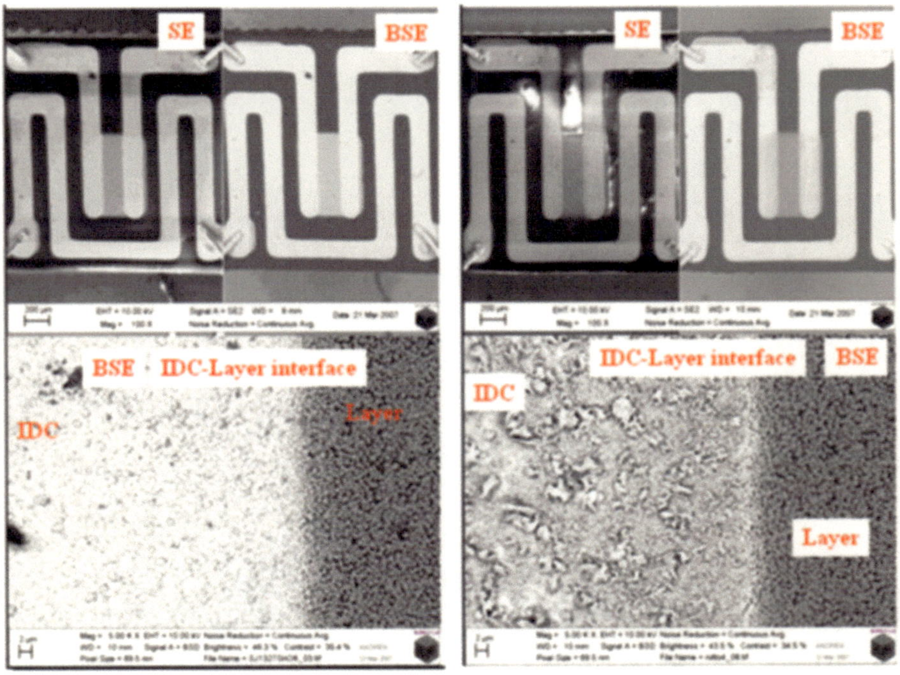

Fig. 2. Comparison between as prepared (left) and 120 days aged sensor (right) by SEM- EDX analysis. Aged sensor shows a larger transparency of the IDC (compare upper left and right images) and the formation of Pt agglomerates (compare bottom left with bottom right). Legend: SE=Secondary Electrons; BSE=Back Scattered Electrons (Courtesy of Dr. A. Ponzoni and Dr. M.Ferroni, University of Brescia and CNR-IDASC)

2.2 Overall considerations

In summary, even if we considered a well defined class of deeply investigated sensors such as MOX sensors, the drift phenomenology is still not totally understood since real polycrystalline samples show a variety of aspects and a high degree of complexity. Depending on the specific technology (e.g., TGS or thin-film) one aspect can be dominant over the others, but often concomitant causes are present.

Irreversible changes of the sensor response might also occur, one of the most common reasons being the sensor surface poisoning (Ruhland et al., 1998). This arises when the sensor is exposed to a gas (e.g., a corrosive acid) which strongly binds or interacts with the sensing material leading to a deep change in its physicochemical properties. Indeed, some authors have reported the occurrence of sensor faults during long measurement runs in monitoring odors on a landfill site (Romain & Nicolas, 2010). In other cases, the presence of strongly binding substances in the samples, such as sulphur compounds (Pratt & Williams, 1997) or some acids Schaller et al. (2000) led to irreversible poisoning. When this occurs, the sensor should be replaced and there is no possibility to compensate the effect. On the contrary, slowly varying drift phenomena, which are frequently reported in the literature, can be coped with proper "soft" methods that will be illustrated in the following sections.

3. Drift counteraction methods: a taxonomy and review

Attempts to mitigate the negative effect of gas sensor drift are not new. A great deal of work has been directed towards the development of drift correction methods and algorithms, which tackle the problem from different perspectives, depending on the situation. Nevertheless, the study of sensor drift still remains a challenging task for the chemical sensor community always looking for novel improved solutions. Fig. 3 provides a rough classification into four main categories of the solutions proposed in the literature, which are presented hereafter.

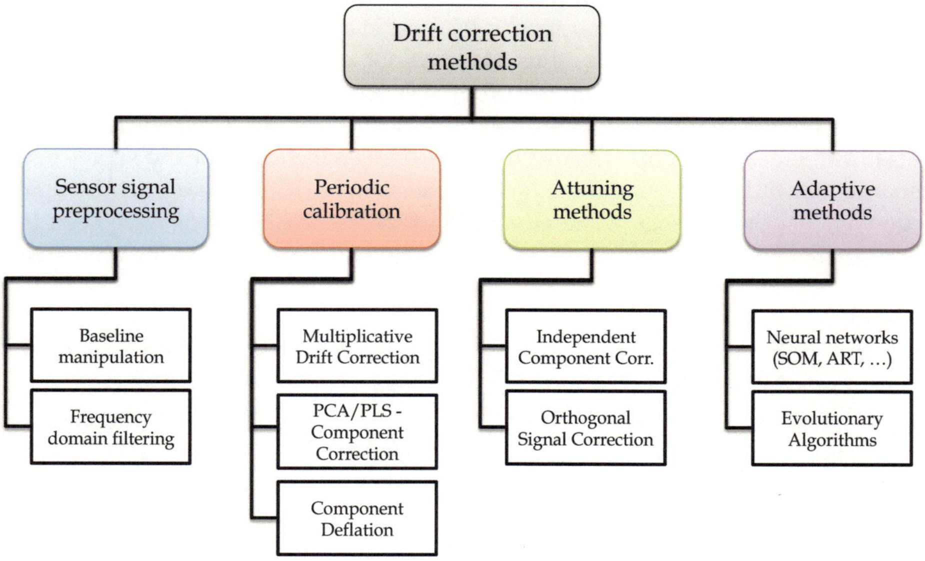

Fig. 3. Taxonomy of drift correction methodologies published in the scientific literature

3.1 Sensor signal preprocessing

3.1.1 Baseline manipulation

One of the simplest methods for drift compensation that has been proposed in the literature, which is also widely used as a pre-processing method, is the transformation of individual sensor signals (Gardner & Bartlett, 1999) based on the initial value of the transient response (the so called "baseline") thus the name of *baseline manipulation*.
Three basic transformations are common practice:

1. *Differential*: subtracts the baseline of each sensor and then can help compensating additive drift effects which are both present in the baseline and gas response:

$$\hat{s}(t) = y(t) - y(0) = [x(t) + \delta] - [x(0) + \delta] = x(t) - x(0) \quad (1)$$

where $\hat{s}$ is the transformed (corrected) response, y is the measured response, x is the ideal sensor response without drift and δ is the drift contribution which is assumed to be constant and uniform.

2. *Relative*: divides by the baseline and might correct for (constant and uniform) multiplicative drift effects:

$$\hat{s}(t) = \frac{y(t)}{y(0)} = \frac{x(t) + \delta x(t)}{x(0) + \delta x(0)} = \frac{x(t)(1+\delta)}{x(0)(1+\delta)} = \frac{x(t)}{x(0)} \tag{2}$$

3. *Fractional*: a combination of the previous two that works for multiplicative drift and has the advantage of providing dimensionless measurements and normalized sensor responses.

$$\hat{s}(t) = \frac{y(t) - y(0)}{y(0)} = \frac{x(t)(1+\delta) - x(0)(1+\delta)}{x(0)(1+\delta)} = \frac{x(t) - x(0)}{x(0)} \tag{3}$$

The first two transformations are too specific because in real applications the drift is generally not additive neither multiplicative, thus they are not able to correct the drift effect while they are (usually the second one) used to simply "normalize" the sensor response. The last manipulation obviously does not work if additive drift is present. Conversely it can amplify the noise in the measurements because the drift term, which is typically small, remains at the denominator thus degrading the quality of the sample. It therefore provides again poor correction against drift effects. More advanced preprocessing methods have been tried out based on sensor signal processing in the frequency domain more than in the time domain.

3.1.2 Filtering

Counteracting methods based on filtering strategies focus on the application of signal preprocessing techniques to filter out portions of the signal containing drift contaminations. Drift typically occurs in a different frequency domain with respect to interesting signals, being in general a slower process. Therefore, proper transformations of sensor signals from time to frequency domain and a careful removal of the lowest frequency components can filter the drift out.

Feature extraction techniques based on Discrete Wavelet Transform (DWT) can be a powerful tool to remove the drift contamination in the low-frequency behavior of the sensor responses (Hui et al., 2003; Llobet et al., 2002; Zuppa et al., 2003). A moving median filter and Fourier band-pass filters are some examples applied to removing either high-frequency fluctuations (such as noise, spikes) or low-frequency changes such as drift. In comparison to these filters, DWT technique provides a flexible analysis of the signal at different resolutions by applying iteratively high-pass and low pass filters. DWT technique allows to remove the selected low-frequency components easily and in such a way that the signal is not distorted.

The DWT provides a multiresolution signal decomposition of sensor response: it analyzes the signal at different frequency bands with different resolutions by successively projecting it down onto two basis of functions, which are obtained by applying shift and scaling operations to two prototype functions called the "scaling function" and the "wavelet function", respectively. The scaling function is associated with low-pass filter and the convolution between the signal and the scaling functions gives the low-frequency components of the signal. Conversely, the wavelet function is associated with high-pass filter and its

convolution with the signal gives the high-frequency components. These components are called approximation coefficients.
The multilevel wavelet decomposition and signal reconstruction is performed by the Mallat algorithm. It consists of iteratively applying high-pass and low-pass filters on the vector of approximation coefficients, obtaining a sequence of increasingly smoothed and halved versions of the original signal. Once a specific wavelet has been chosen (e.g. Daubechies orthonormal functions), the pair of low-pass and high-pass filters are defined.
The DWT decomposition level is fixed once analysis in the frequency domain has been carried out to single out the frequency domain (cut-off frequency) of the drifting trend. The approximation coefficients associated with the lowest frequencies which have drift contamination are then discarded and the wavelet reconstruction of the corrected signal is computed by using the remaining coefficients.

3.2 Periodic calibration

Different drift correction methods are based on the estimation of the drift effect on the system to be later removed. Drift effect estimation can be made, for instance, by measuring the change in the sensor responses to one (or more) reference gas, which is measured with some intervals along the experiment. This strategy can be applied in a univariate way (sensor-by-sensor) or in a multivariate way by removing the directions of dispersion of the reference data in the feature space.

3.2.1 Multiplicative drift correction

Univariate calibration is a straightforward method in which a reference value, i.e., the response to a reference gas, is used and all subsequent sensor readings are individually corrected to it. Fryder et al. (Fryder et al., 1995) and Haugen et al. (Haugen et al., 2000) proposed to model temporal variations of the system with a multiplicative drift correction (MDC) factor obtained by measuring the calibrant and then to apply the same correction to the actual samples.
In particular, Haugen et al. propose a re-calibration method performed in two steps: within a single measurement sequence to compensate for short terms trends and between measurement sequences to compensate for long term fluctuations. This strategy provides very good results and it is currently in use in commercial electronic noses. However, assumptions are too tight: MDC makes a supposition that the drift is multiplicative, which means that the perturbation is proportional to the signal level. Furthermore, it assumes that the relationship between the individual sensor response to the reference gas and the response to the test gas has to be strictly linear.

3.2.2 Multivariate component correction

One of the first attempts of performing robust drift correction by multivariate methods was proposed by Artursson et al. (Artursson et al., 2000) under the name of Component Correction (CC). Two correction methods, one based on Principal Component Analysis (PCA) and one based on Partial Least Square (PLS) are proposed in the paper.
Measurements performed with arrays of chemical sensors contain several redundant information since, in general, the different variables are collinear. PCA (Wold et al., 1987) is a common compression method used to efficiently represent this information. With PCA, the dominating variability in the measurement space can be captured by means of two matrices:

1. the loading matrix **P** that represents a new coordinate system onto which the measurement vector **X** must be projected, and
2. the score matrix **T** that represents the coordinate of the sample in the space represented by the columns of **P**.

Within the new space defined by **P**, dimensions (referred to as components) are ordered by decreasing variability in the input measurements. CC uses PCA in conjunction with the reference gas technique. If the sensor responses to a certain reference gas contain a significant amount of drift, the first component identified by the PCA analysis on these measurements, which is the one that describes the maximum variability, will likely define the direction of the drift. This is motivated by the fact that the sensors are always exposed to the same gas and thus they are expected to provide always the same response with the exception of some random noise. This direction is defined by a loading vector **p** obtained as the first column of the loading matrix **P**. Projecting (multiplying) the sample **X** on this loading vector gives a score vector **t** representing the drift:

$$\mathbf{t} = \mathbf{X} \cdot \mathbf{p} \quad (4)$$

Drift correction can be then implemented by subtracting from the original data the bilinear expression $\mathbf{t} \cdot \mathbf{p}^T$ which represent an approximation of the drift component in the original sample space (i.e., all other directions are preserved and the variance that distinguishes and separates classes of samples in the data space is preserved), thus obtaining a corrected sample :

$$\mathbf{X}_{corr} = \mathbf{X} - \mathbf{t} \cdot \mathbf{p}^T \quad (5)$$

Removing one component is usually enough whenever we are facing drift effects caused by aging of the sensors. However, if non-linear drift effects are observed (e.g., caused by both aging effects and chemical background) more than one component can be subtracted.
A similar correction strategy can be also obtained by using a regression model. Since the drift caused by aging effects has a preferred direction in the measurement space, it should be possible to describe this change as a function of time. The Partial Least Square (PLS) regression model (Wold et al., 1984) is able to infer the dependence between two set of variables (the sample matrix **X** and a matrix **Y** representing the time in our specific case) by using a set of orthogonal score vectors. Artursson et al. propose to compute a weight vector **w** and a loading vector **p** according to the PLS model on a set of measurements of a reference gas (as in the case of PCA). These vectors are first used to compute the drift component **t** according to the PLS regression model:

$$\mathbf{t} = \mathbf{X} \cdot \mathbf{w} \left(\mathbf{p}^T \mathbf{w}\right)^{-1} \quad (6)$$

Then, similarly to (5) the corrected sample is obtained as the residual after the first component has been subtracted from the original data:

$$\mathbf{X}_{corr} = \mathbf{X} - \mathbf{t} \cdot \mathbf{p}^T \quad (7)$$

Again several components can be removed by repeating equations (6) and (7).

3.2.3 Multivariate component deflation

Another attempt to perform multivariate drift correction has been proposed in Gutierrez-Osuna (2000). The overall idea is to introduce a set of variables defined by a vector $\mathbf{Y}$ whose variance can be attributed to drift or interferents. Examples of these variables can be the response to a wash/reference gas that is usually performed prior to each measurement, time stamps, temperature, pressure, humidity, etc. The basic idea is to measure the sensor/array response to an odor $\mathbf{X}$ and remove the variance in $\mathbf{X}$ that can be explained by the variables in $\mathbf{Y}$ (by means of regression/deflation).
The approach basically applies Canonical Correlation Analysis (CCA) or Partial Least Squares (PLS) to find two linear projections $\mathbf{Y}' = \mathbf{a} \cdot \mathbf{Y}$ and $\mathbf{X}' = \mathbf{b} \cdot \mathbf{X}$ that are maximally correlated. This can be formally expressed as:

$$\{\mathbf{a}, \mathbf{b}\} = argmax\left[corr\left(\mathbf{aY}, \mathbf{bX}\right)\right] \tag{8}$$

$\mathbf{Y}'$ and $\mathbf{X}'$ are in fact low-dimensional projections that summarize the linear dependencies between $\mathbf{Y}$ and $\mathbf{X}$. At this point Ordinary Least Squares (OLS) (Dillon & Goldstein, 1984) can be used to find a regression model $\mathbf{X}_{pred} = \mathbf{w} \cdot \mathbf{X}'$ able to minimize the difference between $\mathbf{X}$ and $\mathbf{X}_{pred}$. This is formally expressed as:

$$\mathbf{w} = argmin\left[\mathbf{X} - \mathbf{w} \cdot \mathbf{X}'\right] \tag{9}$$

The OLS prediction vector ($\mathbf{X}_{pred}$) in fact contains the variance of the odor vector $\mathbf{X}$ that can be explained by $\mathbf{X}'$ and indirectly by $\mathbf{Y}$ since $\mathbf{X}'$ and $\mathbf{Y}'$ are correlated. At this point it is enough to deflate $\mathbf{X}$ and use the residual $\mathbf{X}_{corr}$ as a drift-corrected vector:

$$\mathbf{X}_{corr} = \mathbf{X} - \mathbf{X}_{pred} \tag{10}$$

In summary, techniques based on recalibration of the system using a reference gas can give very good results, but care must be taken in the selection of such gas. It must be representative of all classes being measured, since they are supposed to drift in similar ways. It should also be stable along the time, available and easy to measure (Salit & Turk, 1998). Fitting all these constraints is in general complex and expensive, therefore methods overcoming this limitation have been explored.

3.3 Attuning methods

Attuning methods try to perform component correction without resorting to the use of calibration samples, but trying to deduce drift components directly from the training data. They can provide significant improvements in the classification rate over a fixed time period, and may also make possible to obtain real responses to be used in gas quantitative analysis.

3.3.1 Independent component correction

Di Natale et al. approached the problem of sensor drift with an attuning method considering also disturbances derived from the measurement environment (Di Natale et al., 2002a). In correction methods such as CC (Artursson et al., 2000) that are based on PCA, the computed principal components are mutually uncorrelated. However, this condition is not enough to guarantee that relevant signals are completely separated by disturbances (non-correlation does not necessarily implies statistical independence) and often a principal component carries information on both signal and disturbances. To tackle with this limitation Di Natale et al.

propose to exploit the Independent Component Analysis (ICA) (Cornon, 1994) as a technique to separate a data matrix into a series of components each independent from the others. In this case, independence means that the information carried by each component cannot be inferred from the others, i.e., the joint probability of independent quantities is obtained as the product of the probability of each item.
ICA is applied to EN data to preserve only those components correlated with the sample features relevant to the application. In fact, ICA is computed on the training set (in this case no reference gas is required). Contrarily to Gaussian based models such as PCA and PLS, in ICA it is not possible to determine the variance carried by each component, and it is therefore not possible to choose the component to eliminate based on this information. In fact it is not possible to establish an order among the different components.
The way the authors propose to select components to eliminate is a supervised method: the independent components that mostly correlate with the objective of the measurement are chosen, while those more correlated with the disturbances are discarded. This solution provides good results especially in removing drift effect due to external causes (e.g., temperature, pressure) that can be monitored during experiments and used as variables to select the components to discard. However, whenever the causes of the drift are not completely known, selecting the components to discard may result complex or in some cases not possible.

3.3.2 Orthogonal Signal Correction

Recently, Padilla et al. (Padilla et al., 2010) proposed a very interesting drift attuning method based on Orthogonal Signal Correction (OSC). OSC is a signal processing technique first introduced by Wold et al. in (Wold et al., 1998) for NIR spectra correction.
OSC analyzes a set of sensor-array data **X** that represent a set of independent variables, and a concentration vector, or class label vector **C** that represent a set of dependent variables. The main idea is to remove variance of **X** which is not correlated to the variables in **C**. This is achieved by constraining the deflation of non-relevant information of **X** so that only information orthogonal to **X** must be removed (Fig. 4). The condition of orthogonality therefore assures that the signal correction process removes as little information as possible form the original data. Several different variants of the OSC algorithm have been presented in the literature. Padilla et al. applied the Wise implementation of the OSC algorithm[4].
Even if OSC proved to be one of the most effective techniques for drift correction it does not completely solve the problem. One of the main drawbacks is the need for a set of training data containing a significant amount of drift making it possible to precisely identify the set of orthogonal components to be rejected. This may not be possible in industrial setups where training data are usually collected over a short period of time. Moreover, the introduction of new analytes to the recognition library represents a major problem since rejected components might be necessary to robustly identify these new classes.

3.4 Adaptive methods

Adaptive methods for drift correction try to adapt the PaRC model by taking into account pattern changes due to drift effects. The main benefit of the adaptation is an increasing time validity of the PaRC model which in turn reduces the request for calibration. Even if the first

[4] http://www.SIGE vector.com/MATLAB/OSC.HTML

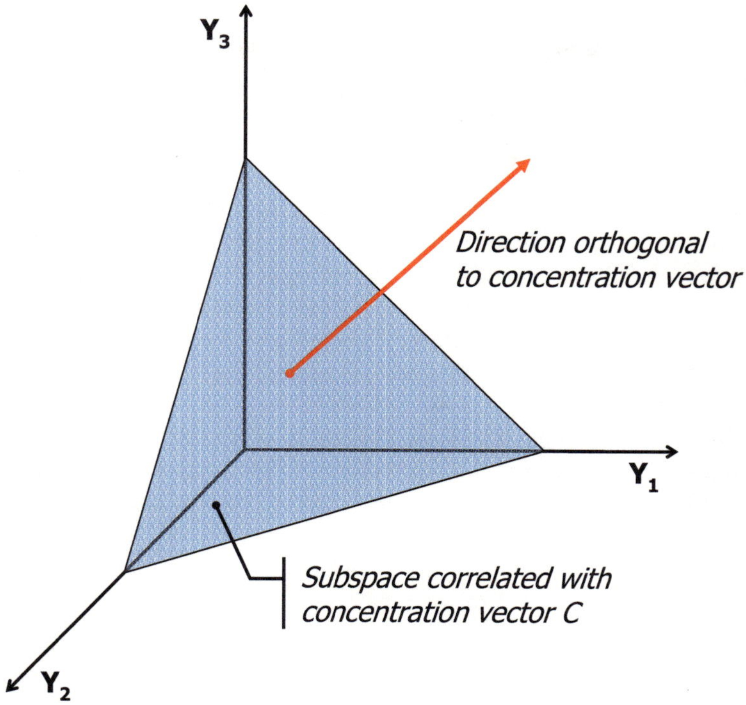

Fig. 4. Orthogonal signal correction overall idea

attempt of developing an adaptive drift correction method dates back to the late 90's (Marco et al., 1998; Vlachos et al., 1997), only few publications have been proposed in the literature, and this problem still remains a challenge for the sensor community.

3.4.1 Neural networks

The first adaptive approaches for drift correction have been developed resorting to artificial neural networks. Neural networks are an important tool for building PaRC systems with several interesting features. First, neural networks are data driven self-adaptive methods in that they can adjust themselves to the data without any explicit specification of functional or distributional form for the underlying model. Second, they are universal functional approximations in that neural networks can approximate any function with arbitrary accuracy (Cybenko, 1989; Devijver & Kittler, 1982). Since any classification procedure seeks a functional relationship between the group membership and the attributes of the object, accurate identification of this underlying function is doubtlessly important. Third, neural networks are nonlinear models, which make them flexible in modeling real world complex relationships such as those presented by gas chemical sensors. Finally, neural networks are able to estimate the posterior probabilities, which provide the basis for establishing classification rule and performing statistical analysis (Richard & Lippmann, 1991).
Several neural network based drift correction methods exploit Kohonen Self Organizing Maps (SOMs) (Kohonen, 1990). (Davide et al., 1994; Marco et al., 1998; Zuppa et al., 2004) use a single SOM common to all classes while (Distante et al., 2002) proposes to use a separate SOM for

each odor. A slightly different approach is proposed in (Vlachos et al., 1997) that exploits a different network architecture, namely the Adaptive Resonance Theory (ART) neural network (Carpenter et al., 1991) that allows for new classes to be created.

Regardless the type and the specific architecture of the neural network all proposed methods achieve drift correction by exploiting the inner way neural networks work. As common in pattern recognition, a preliminary training phase is used to train the neural network in the identification of samples from the sensor array. In this phase a set of samples is provided to the network which learns similarities according to different learning rules (Bishop, 1995). Each new sample slightly changes the way the nodes of the network behave. Both supervised learning in which training samples are already labeled into a set of classes and unsupervised learning in which training samples are not labeled and the set of classes is inferred by the network itself is possible. When training has converged, the net has learned the characteristics of the input patterns and can be used for classification. In this stage the learning capability of the network is usually disabled. The basic idea to allow for drift compensation is to maintain a certain learning rate also during the normal use of the network in order to learn changes of the input patterns due to drift effects. The learning rate must be kept to a low level in order to avoid over-fitting of the model.

Although neural networks represented the first attempt of implementing adaptive drift correction methods they have several drawbacks:

- Drift correction is possible only for slow phenomena. A discontinuity in response between consecutive exposures (regardless of the time interval between the exposures) would immediately invalidate the classification model and would prevent adaptation.
- Selecting the appropriate learning rate to keep during normal operation is complex and may strongly impact the correction capability. To the best of our knowledge no automatic methods have been proposed so far for efficiently tuning this parameter.
- The adaptive model is rather complex and typically requires a high number of training samples.
- Several classification algorithms have been presented in the literature other than neural networks (e.g., SVM, K-NN, Random Forests, etc.). The literature shows that, depending on the specific set of data, some methods may perform better than others. Having a very tight integration between the correction method and the classification system prevents from exploiting the best PaRC model for the specific problem.
- Finally, drift correction methods based on neural networks are mainly limited to gas classification applications. Whenever both classification and gas quantitative analysis are required, it would be difficult for current adaptive methods to be applied to obtain reliable gas concentration measurements (Hui et al., 2003).

3.4.2 Evolutionary algorithms

Recently, a new adaptive drift correction method based on the use of evolutionary algorithms has been presented by Di Carlo et al. (Di Carlo et al., 2010; 2011). The overall idea is to exploit the learning capabilities of evolutionary algorithms to compute a multiplicative correction factor **C** used to correct incoming samples. Under the hypothesis that, in the very short term, the variation imposed by the drift can be considered linear in time, the paper proposes to apply the correction exploiting a linear transformation:

$$\mathbf{X}_{corr} = \mathbf{X} + \mathbf{X} \cdot \mathbf{C} \tag{11}$$

Although this assumption is a limit for previous drift counteractions that do not allow for adaptation, it is a good approximation in this case since the correction factor is not a fixed quantity but it is continuously adapted to follow the variation imposed by the drift. The hypothesis of linearity is therefore assumed only within a restricted time window (or number of measurements) whose size can be adapted in order to respect this constraint.

The correction algorithm can be coupled with any selected PaRC model (e.g., SVM, K-NN, Random Forests, etc.) and elaborates groups of consecutive measurements, denoted as windows, according to the following steps:

1. The initial correction factor, immediately after training of the PaRC model, is set to the null matrix since no correction is required at this time;
2. For each window of samples the following operations must be performed:
   (a) Correct each sample of the window using the current correction factor according to equation (11);
   (b) Classify each corrected sample;
   (c) Use the corrected samples, and the classification results in an evolutionary process able to adapt the correction factor to the changes observed in the current window (see later for an explanation of how this process works);
   (d) Correct each sample again using the updated correction factor;
   (e) Classify the new corrected samples and provide the obtained results.

The adaptation of the correction factor exploits a Covariance Matrix Adaptation Evolutionary Strategy (CMA-ES) which is an optimization method first proposed by Hansen, Ostermeier, and Gawelczyk (Hansen et al., 1995) in mid 90s. It basically iteratively searches for a definite matrix which is able to minimize a given objective function. The approach is best suited for difficult non-linear, non-convex, and non-separable problems, of at least moderate dimensionality. In this specific case, the problem tackled by the CMA-ES is finding the best correction factor that makes possible to obtain similar distributions between corrected samples of a window compared to the samples used to train the PaRC model. Several metrics can be exploited to identify similarity.

In (Di Carlo et al., 2011), under the hypothesis of Gaussian distribution of samples around the related centroids, the CMA-ES is exploited to identify the best correction factor able to move each sample toward the centroid of the related class, thus compensating for drift effects that tend to move classes in the features space. The concept of proximity to the related centroid is introduced by testing the system using several types of objective functions that compute the distance of a sample from the related centroid with different metrics.

This method introduces a set of important improvements compared to previous adaptive methods based on neural networks:

- it can work in cooperation with any PaRC algorithm thus allowing for the selection of the best classification model depending on the specific application;
- it is more robust to discontinuity in the data;
- together with sample classifications it also provides corrected measures thus allowing also for gas quantitative analysis.

4. Comparisons and discussions

Only partial comparisons and limited considerations can be done based on literature reports. In fact, due to lack of shared sensor dataset and public software codes, only few comparison works that partially contrast different approaches and report the pros and cons of each solution have been published (Hui et al., 2003; Padilla et al., 2010; Romain & Nicolas, 2010; Sisk & Lewis, 2005). Every paper tends to show good results related to the newly proposed approach. However, in many cases, measurements are collected during few days, thus neglecting long term drift effects, or drift was simulated.

Signal preprocessing techniques seem attractive for their simplicity, but cannot be considered an effective drift correction approach for chemical sensors. Baseline manipulation works well only in special cases that fit the assumptions of the employed feature transformation. Filtering methods in the frequency domain appear more promising. (Zuppa et al., 2007) compared the DWT filter with the relative baseline manipulation technique. PCA shows that the DWT approach is superior in terms of cluster dispersion with clear improvements in samples discrimination. The DWT method has been often combined to adaptive drift correction strategies based on neural networks. Nevertheless, frequency decomposition remains mostly useful for diagnostics and analysis because strong dependence of results from user defined parameters (e.g. the cut-off frequency).

According to Sisk and Lewis (Sisk & Lewis, 2005) using a single calibrant or a set of calibrants is perhaps the only robust method to mitigate drift effects even in the presence of sensor drift over an extremely long period of time. Sisk and Lewis implemented a simple linear sensor-by-sensor calibration scheme which proved to be effective at restoring the classification performance of difficult binary separation tasks affected by drift. In many other cases this univariate approach failed to work. Thus, one relevant open point is the tradeoff between model simplicity and robustness.

Univariate calibration methods, like MDC, are first-order approximations that do not exploit sensor cross-correlations. Empirically they provided good performance. However, they have several intrinsic limitations. The most important limitation is that the MDC method assumes that the relative percentage (increase or decrease) of the sensor response due to the drift is the same for the reference samples and test samples. This can be true for sample chemical compositions and concentrations very close to the reference gas but not otherwise. Therefore the amount of drift may be different for different samples, i.e., the drift for the reference gas and some samples is not same. The consequence of this is that the sensor drift does not have to be the multiplicative factor estimated from the reference gas, which the MDC method assumes.

As pointed out by Artursson et al. (Artursson et al., 2000) multivariate techniques, for instance CC, better perform both in terms of drift correction performance and flexibility. The reader may notice that both the CC method and the MDC method are linear, but they differ in what is assumed to be linear. For the CC method, drift is assumed to mainly influence the sensors along a straight line, described by the drift in the reference gas. In the univariate MDC method it is assumed that the relative change in each sensor is the same for the reference gas and all test gases. There exist cases where the drift is linear for one of the methods but not for the other.

Both MDC and CC methods suffer from the same disadvantages when it comes to handling non-linearities under their respective restrictions. However, as long as the relationship between the reference gas and the test gas (for the MDC method) or the relationship between the sensors (CC method) remains linear, there is no problem for both methods in handling non-linearities.

Calibration is certainly the most time-intensive method for drift correction since it requires system retraining, additional measurements and labour. Hence, it must be used sparingly. Moreover, while this approach is quite simple to implement for physical sensors, where the quantity to be measured is exactly known, chemical sensors pose a series of challenging problems. Indeed, in chemical sensing, the choice of the calibrant strongly depends on the specific application especially when the sensing device is composed of a considerable number of cross-correlated sensors (Haugen et al., 2000; Hines et al., 1999). This leads to loss of generalization and lack of standardization which, on the contrary, is an important requirement for industrial systems.

Several attuning methods, which deduce drift components directly from the training data, have been proposed to perform drift correction without resorting to the use of calibration. Padilla et al. (Padilla et al., 2010) compared OSC and CC. According to this work, OSC outperforms PCA-CC for a limited period of time while on the long term the advantage was not clear. More complex OSC models (with higher number of components) resulted in a better correction of variance for shorter times, but they degraded faster than simpler OSC. It was also observed that both OSC and PCA-CC methods are relatively robust regarding small calibration set sizes and perform rather well with a reduced calibration set. OSC results showed a higher variance than PCA-CC, but this was attributed to the number of components chosen in the models. On the other hand, PCA-CC needed a smaller training set and a single chemical species. This advantage may turn into disadvantage if the reference class is not properly chosen.

Therefore, OSC seems to be a promising approach: it is time tested in chemometrics, uses multivariate and calibrant information, and it is simple to implement and interpret. However, although attuning methods represent a step forward in the definition of an efficient solution to the drift correction problem, the main limitation is that they do no contain provisions for updating the model and thus may ultimately be invalidated by time evolving drift effects. Conversely, they can dramatically fail when the drift direction changes. This might happen if one sensor gets stressed. In this case one must resolve to use adaptive methods.

Adaptive methods represent an important step forward in tackling the problem of sensor drift in artificial olfaction systems. Until now they have been not sufficiently investigated and contrasted with other methods in order to definitively assess their superior capabilities. In fact, current solutions still present some limitations that prevent their widespread application. In particular, they require equiprobable and frequent sampling of all classes to avoid that a single class drifts too much making it unrecognizable. Moreover they strongly rely on correct identification from the PaRC model to take track of how different classes change. Local errors in the classification may easily reduce the capability of adaptation, thus reducing the effectiveness of the correction system. Further research efforts are required in this direction.

5. Conclusions and future research directions

In this chapter, after a brief overview of drift phenomenology in gas chemical sensors, the main challenges faced when developing drift correction techniques have been presented and discussed. A deep review of state-of-the-art methodologies proposed in the scientific literature has been illustrated with a rational taxonomy of the various approaches.

Drift correction is perhaps one the most relevant challenges in the field of chemical sensors. Indeed, in spite of constant improvements in micro/nano fabrication techniques that allowed the production of sensing devices with superior stability, it is still impossible to fabricate chemical sensors without drift.

Much work has been done in the last fifteen years to develop adapted techniques and robust algorithms. In spite further research is still required to compare them on some benchmark data set. A complete comparison of performance should take into account several parameters, for instance: the type of sensors and likely the use of hybrid arrays; the presence/absence of a drift model for the given sensors; the short and long term drift behavior; the size of the data set (feature space dimension, number of measurements); the problem complexity (number of classes, degree of overlap, type of chemicals/odors).

The problem of data correction in presence of simultaneous sources of drift, other than sensor drift, should be also investigated since it is often the case in practical situations. To this, one idea could be combining semi-supervised methods able to learn the actual source of drift, which might clearly change with the measured samples, with adaptive drift correction strategies that can account for the continuous drift direction change in the feature space.

Finally, many algorithms of drift correction could be adapted to the problem of sensor failure and subsequent replacement, which has received little attention until now but represents a relevant problem in the long term. Indeed, only few works can be found in the literature aiming at detection and correction of sensor faults, like for example Tomic et al. (Tomic et al., 2004) that studied calibration transfer techniques like MDC and CC related to sensor replacement.

The current trend in the field of artificial olfaction is to enlarge both the size of data sets and the dimensionality of the sensor arrays, by building for instance huge micro-machined sensor arrays made of many thousands sensors. In the near future, this shall definitively originate challenging issues for data analysis requiring more powerful drift correction algorithms able to handle large volume of data as well as high-dimensional features with acceptable time and storage complexities. Very large sensor arrays will also pose problems connected to variables redundancy. Therefore, a further relevant issue is the selection of appropriate and meaningful features to combine with the drift correctors, which can greatly reduce the burden of subsequent designs of classification/regression systems.

6. References

Aliwell, S. R., Halsall, J. F., Pratt, K. F. E., O'Sullivan, J., Jones, R. L., Cox, R. A., Utembe, S. R., Hansford, G. M. & Williams, D. E. (2001). Ozone sensors based on wo3: a model for sensor drift and a measurement correction method, *Measurement Science & Technology* 12(6): 684–690.

Artursson, T., Eklov, T., Lundström, I., Mårtensson, P., Sjöström, M. & Holmberg, M. (2000). Drift correction for gas sensors using multivariate methods, *Journal of Chemometrics,Special Issue: Proceedings of the SSC6* 14(5-6): 711–723.

Bishop, C. M. (1995). *Neural Networks for Pattern Recognition*, Oxford Univ. Press.

Carpenter, G. A., Grossberg, S. & Reynolds, J. H. (1991). Artmap: Supervised real-time learning and classification of nonstationary data by a self-organizing neural network, *Neural Networks* 4(5): 565–588.

Chen, D. & Chan, P. (2008). An intelligent isfet sensory system with temperature and drift compensation for long-term monitoring, *Sensors Journal, IEEE* 8(12): 1948–1959.

Comini, E. & Sberveglieri, G. (2010). Metal oxide nanowires as chemical sensors, *Materials Today* 13(7-8): 36–44.

Cornon, P. (1994). Independent component analysis-a new concept?, *Signal Process* 36: 287.

Cybenko, G. (1989). Approximation by superpositions of a sigmoidal function, *Mathematics of Control, Signals, and Systems* 2(4): 303–314.

Davide, F., Di Natale, C. & D'Amico, A. (1994). Self-organizing multisensor systems for odour classification: Internal categorization, adaptation, and drift rejection, *Sensors and Actuators B* 18-19: 244–250.

Devijver, P. & Kittler, J. (1982). *Pattern recognition: A statistical approach*, Prentice/Hall International.

Di Carlo, S., Falasconi, M., Sánchez, E., Scionti, A., Squillero, G. & Tonda, A. (2010). Exploiting evolution for an adaptive drift-robust classifier in chemical sensing, *Applications of Evolutionary Computation* pp. 412–421.

Di Carlo, S., Falasconi, M., Sanchez, E., Scionti, A., Squillero, G. & Tonda, A. (2011). Increasing pattern recognition accuracy for chemical sensing by evolutionary based drift compensation, *Pattern Recognition Letters* 32(13): 1594–1603.

Di Natale, C., Martinelli, E. & D'Amico, A. (2002a). Counteraction of environmental disturbances of electronic nose data by independent component analys, *Sensors and Actuators B: Chemical* 82(2-3): 158–165.

Di Natale, C., Martinelli, E. & D'Amico, A. (2002b). Counteraction of environmental disturbances of electronic nose data by independent component analysis, *Sensors and Actuators B: Chemical* 82(2-3): 158–165.

Dillon, W. & Goldstein, M. (1984). *Multivariate analysis: Methods and applications*, New York Wiley.

Distante, C., P., S. & Persaud, K. (2002). Dynamic cluster recognition with multiple self-organising maps, *Pattern Analysis and Applications* 5: 306–315.

Fryder, M., Holmberg, M., Winquist, F. & Lündstrom, I. (1995). A calibration technique for an electronic nose, *Proc. of Eurosensors IX*, pp. 683–686.

Gardner, J. & Bartlett, P. (1999). *Electronic Noses. Principles and Applications*, Oxford University Press, New York, NY, USA.

Gardner, J. W. (1987). Pattern recognition in the warwick electronic nose, *Proc. of the 8th Int. Congress of European Chemoreception Res. Organisation.*

Gobbi, E., Falasconi, M., Concina, I., Mantero, G., Bianchi, F., Mattarozzi, M., Musci, M. & Sberveglieri, G. (2010). Electronic nose and alicyclobacillus spp. spoilage of fruit juices: An emerging diagnostic tool, *Food Control* 21(10): 1374–1382.

Goepel, W. & Schierbaum, K. (1995). Sno2 sensors: current status and future prospects, *Sensors and Actuators B: Chemical* 26(1-3): 1–12.

Gutierrez-Osuna, R. (2000). Drift reduction for metal-oxide sensor arrays using canonical correlation regression and partial least squares, *Proceedings of the 7th International Symposium On Olfaction and Electronic Nose*, Institute of Physics Publishing, pp. 147–152.

Hansen, N., Ostermeier, A. & Gawelczyk, A. (1995). On the adaptation of arbitrary normal mutation distributions in evolution strategies, *The Generating Set Adaptation. Proc. 6th Internat. Conf. Genetic Algorithms*, pp. 312–317.

Haugen, J.-E., Tomic, O. & Kvaal, K. (2000). A calibration method for handling the temporal drift of solid state gas-sensors, *Analytica Chimica Acta* 407(1-2): 23–39.

Haupt, K. & Mosbach, K. (2000). Molecularly imprinted polymers and their use in biomimetic sensors, *Chemical Reviews* 100(7): 2495–2504.

Hines, E., Llobet, E. & Gardner, J. (1999). Electronic noses: a review of signal processing techniques, *IEE Proceedings - Circuits, Devices and Systems* 146(6): 297–310.

Holmberg, M., Davide, F., Di Natale, C., D'Amico, A., Winquist, F. & Lundstr "om, I. (1997). Drift counteraction in odour recognition applications: lifelong calibration method, *Sensors and Actuators B: Chemical* 42(3): 185–194.

Hui, D., Jun-Hua, L. & Zhong-Ru, S. (2003). Drift reduction of gas sensor by wavelet and principal component analysis, *Sensors and Actuators B: Chemical* 96(1-2): 354–363.

International Union of Pure and Applied Chemistry (1991). Chemical sensors definitions and classification, *Pure&Appl. Chern.* 63(9).

Ionescu, R., Vancu, A. & Tomescu, A. (2000). Time-dependent humidity calibration for drift corrections in electronic noses equipped with sno2 gas sensors, *Sensors and Actuators B: Chemical* 69(3): 283–286.

Janata, J. (2009). *Principles of Chemical Sensors*, 2nd ed.2nd printing. edn, Springer Publishing Company, Incorporated.

Kamp, B., Merkle, R. & Maier, J. (2001). Chemical diffusion of oxygen in tin dioxide, *Sensors and Actuators B: Chemical* 77(1-2): 534–542.

Kohonen, T. (1990). The self-organizing map, *Proceedings of the IEEE* 78(9): 1464–1480.

Llobet, E., Brezmes, J., Ionescu, R., Vilanova, X., Al-Khalifa, S., Gardner, J. W., Barsan, N. & Correig, X. (2002). Wavelet transform and fuzzy artmap-based pattern recognition for fast gas identification using a micro-hotplate gas sensor, *Sensors and Actuators B: Chemical* 83(1-3): 238–244.

Marco, S., Ortega, A., Pardo, A. & Samitier, J. (1998). Gas identification with tin oxide sensor array and self-organizing maps: adaptive correction of sensor drifts, *IEEE Transactions on Instrumentation and Measurement* 47(1): 316–321.

Mielle, P. (1996). Managing dynamic thermal exchanges in commercial semiconduting gas sensors, *Sensors and Actuators B: Chemical* 34(1-3): 533–538.

Nelli, P., Faglia, G., Sberveglieri, G., Cereda, E., Gabetta, G., Dieguez, A., Romano-Rodriguez, A. & Morante, J. R. (2000). The aging effect on sno2-au thin film sensors: electrical and structural characterization, *Thin Solid Films* 371(1-2): 249–253.

Owens, W. B. & Wong, A. P. S. (2009). An improved calibration method for the drift of the conductivity sensor on autonomous ctd profiling floats by theta–s climatology, *Deep-Sea Research Part I-Oceanographic Research Papers* 56(3): 450–457.

Padilla, M., Perera, A., Montoliu, I., Chaudry, A., Persaud, K. & Marco, S. (2010). Drift compensation of gas sensor array data by orthogonal signal correction, *Chemometrics and Intelligent Laboratory Systems* 100(1): 28–35.

Pardo, M. & Sberveglieri, G. (2004). Electronic olfactory systems based on metal oxide semiconductor sensor arrays, *MRS Bulletin* 29(10): 703–708.

Pearce, T. C., Shiffman, S. S., Nagle, H. T. & Gardner, J. W. (2003). *Handbook of machine olfaction*, Weinheim: Wiley-VHC.

Persaud, K. & Dodd, G. (1982). Analysis of discrimination mechanisms in the mammalian olfactory system using a model nose, *Nature* 299: 352–355.

Polster, A., Fabian, M. & Villinger, H. (2009). Effective resolution and drift of paroscientific pressure sensors derived from long-term seafloor measurements, *Geochem. Geophys. Geosyst.* 10.

Pratt, K. & Williams, D. (1997). Self diagnostic gas sensitive resistors in sour gas applications, *Sensors and Actuators B: Chemical* 45(2): 147–153.

Richard, M. & Lippmann, R. (1991). Neural network classifiers estimate bayesian a posteriori probabilities, *Neural computation* 3(4): 461–483.

Röck, F., Barsan, N. & Weimar, U. (2008). Electronic nose: current status and future trends, *Chem Rev.* 108(2): 705–725.

Romain, A. & Nicolas, J. (2010). Long term stability of metal oxide-based gas sensors for e-nose environmental applications: An overview, *Sensors and Actuators B: Chemical* 146(2): 502–506.

Ruhland, B., Becker, T. & M
"uller, G. (1998). Gas-kinetic interactions of nitrous oxides with sno2 surfaces, *Sensors and Actuators B: Chemical* 50(1): 85–94.

Salit, M. & Turk, G. (1998). A drift correction procedure, *Analytical Chemistry* 70(15): 3184–3190.

Sberveglieri, G. (1992). Classical and novel techniques for the preparation of sno2 thin-film gas sensors, *Sensors and Actuators B: Chemical* 6(1-3): 239–247.

Schaller, E., Bosset, J. & Escher, F. (2000). Instability of conducting polymer sensors in an electronic nose system, *Analusis* 28(3): 217–227.

Schierbaum, K., Weimar, U., Goepel, W. & Kowalkowski, R. (1991). Conductance, work function and catalytic activity of sno2-based gas sensors, *Sensors and Actuators B: Chemical* 3(3): 205–214.

Sharma, R. K., Chan, P. C. H., Tang, Z., Yan, G., Hsing, I.-M. & Sin, J. K. O. (2001). Investigation of stability and reliability of tin oxide thin-film for integrated micro-machined gas sensor devices, *Sensors and Actuators B: Chemical* 81(1): 9–16.

Shurmer, H., Fard, A., Barker, J., Bartlett, P., Dodd, G. & Hayat, U. (1987). Development of an electronic nose, *Physics in Technology* 18(4): 170–176.

Sisk, B. C. & Lewis, N. S. (2005). Comparison of analytical methods and calibration methods for correction of detector response drift in arrays of carbon black-polymer composite vapor detector, *Sensors and Actuators B: Chemical* 104(2): 249–268.

Tomic, O., Eklov, T., Kvaal, K. & Haugena, J. (2004). Recalibration of a gas-sensor array system related to sensor replacement, *Analytica Chimica Acta* 512: 199–206.

Vezzoli, M., Ponzoni, A., Pardo, M., Falasconi, M., Faglia, G. & Sberveglieri, G. (2008). Exploratory data analysis for industrial safety application, *Sensors and Actuators B: Chemical* 131(1): 100–109.

Vlachos, D., Fragoulis, D. & Avaritsiotis, J. (1997). An adaptive neural network topology for degradation compensation of thin film tin oxide gas sensors, *Sensors and Actuators B: Chemical* 45(3): 223–228.

Wold, S., Albano, C., Dunn, W., Edlund, U., Esbensen, K., Geladi, P., Helberg, S., Johansson, E., Lindberg, W. & Sjostrom, M. (1984). *Dependencies between variables: multiple regression (MR), partial least square modeling with latent variables (PLS) and PARC levels 3 and 4*, Reidel, pp. 61–74.

Wold, S., Antti, H., Lindgren, F. & Öhman, J. (1998). Orthogonal signal correction of near-infrared spectra, *Chemometrics and Intelligent Laboratory Systems* 44: 175–185.

Wold, S., Esbensen, K. & Geladi, P. (1987). Principal component analysis, *Chemiometrics Intell. Lab. Syst.* 2: 37–52.

Zuppa, M., Distante, C., Persaud, K. C. & Siciliano, P. (2003). Recovery of drifting sensor responses by means of dwt analysis, *Sensors and Actuators B: Chemical* 120(2): 411–416.

Zuppa, M., Distante, C., Persaud, K. C. & Siciliano, P. (2007). Recovery of drifting sensor responses by means of dwt analysis, *Sensors and Actuators B: Chemical* 120(2): 411–416.

Zuppa, M., Distante, C., Siciliano, P. & Persaud, K. C. (2004). Drift counteraction with multiple self-organising maps for an electronic nose, *Sensors and Actuators B: Chemical* 98(2-3): 305–317.

15

Statistical Analysis of Chemical Sensor Data

Jeffrey C. Miecznikowski[1] and Kimberly F. Sellers[2]
[1]*SUNY University at Buffalo*
[2]*Georgetown University*
USA

1. Introduction

Chemical sensors measure and quantify substances via their associated chemical or physical response, thus providing data that can be analyzed to address a scientific question of interest (Eggins; 2002). Used in a variety of applications from monitoring to medicine, chemical sensors vary vastly by construction, style, format, size and dimension, and complexity. The common, underlying feature of these sensors lies in the associated data, which are abundant with technical and structural complexities, making statistical analysis a difficult task. These data further share a common need to be measured, analyzed, and interpreted properly so that the resulting inference is accurate.

There are many image analysis algorithms available and amenable to a myriad of chemical analysis problems, thus potentially applicable to chemical sensor data problems in particular. By applying these tools to chemical sensor data, we can optimize and evaluate a chemical sensor's ability to perform its intended tasks. This chapter is designed to give an overview of the modern statistical algorithms that are commonly used when designing and analyzing chemical sensor experiments. Without focusing the discussion around a specific chemical sensor platform, our goal is to provide a general framework that will be applicable, to some degree, for all chemical sensor data.

From the beginning to the end of an experiment, various statistical methods can be employed to improve or understand the current scientific analysis. We decompose a general experiment into several facets and provide the motivation for potential statistical methods that can be applied within each component. Section 2 describes the pre-processing techniques that are available for summarizing the low-level image or signal data so that the subsequent scientific questions can be properly addressed. In this section, we particularly focus on removing background noise in order to isolate the chemical sensor signal data, quantifying this data, and normalizing it so that the resulting data are scalable across conditions. Section 3 introduces the higher-level statistical approaches that are used to analyze the pre-processed data, and Section 4 describes the statistical computational tools available for use to perform the analyses. Finally, Section 5 demonstrates and motivates the significance of these methods via a chemical sensor case study, and Section 6 concludes the chapter with discussion and summary.

2. Pre-processing

For the analysis of chemical sensor data, many of the suggested means to resolve low-level analysis problems have been posed by computer scientists or engineers, with little statistical contribution or consideration, and they remain open problems because of significant drawbacks in the proposed approaches. Alternatively, problems have been addressed in direct association with chemical sensor data analysis without recognizing that the resulting data represent a special case from a larger context of image data whose structure has been considered by statisticians. Scientists, for example, are interested in better tools that allow for a completely automated approach to detect chemical changes. They, therefore, recognize the need for minimal inherent noise in order to gain in data reproducibility and trust the obtained summary information for subsequent statistical analysis. Statistical and/or data mining tools, and machine learning algorithms are all methods that scientists can use to remove the noise from the meaningful signal contained within an experiment.

Similar to other high throughput biological experiments, several initial steps are often necessary before analyzing the data for a scientific question. Natale et al. (2006) discuss several preprocessing steps including feature extraction, zero-centered scaling, autoscaling, and normalization. Jurs et al. (2000) outline these techniques for chemical sensor arrays, and further include discussion on background or baseline subtraction, and linearization. Meanwhile, a good introduction to the general notion of pre-processing is provided in Gentleman (2005). Although Gentleman (2005) introduce these methods motivated by a different data source, many of their techniques are general enough for chemical sensor data. These issues are all substantial problems that need to be addressed, because any subsequent chemical sensor data analyses are contingent particularly on appropriate and statistically sound low-level procedures.

2.1 Background correction

Background noise associated with chemical sensors can occur for any number of reasons, such as the nature of the chemical processes and the machines used to scan or quantify the sensor. Irrespective of the cause, the background effect must be removed in order for the data of interest to be accurate and informative.

A variety of background correction techniques exist for various data forms, depending on the dimensionality and structure of the data. A general approach for background correction is to simply subtract the blank sample response, i.e. the response which is obtained before any sample is placed within the sensor (see Jurs et al. (2000), or Sellers et al. (2007) for an analogous approach). Other general approaches subtract either the global minimum from the data, or perform some local Winsorization at a low-percentile value. Such approaches are generally accepted for sensor data that appear in spectral form (e.g. Coombes et al. (2005); Lin et al. (2005); Morris et al. (2005)). While analogous approaches can likewise be applied to two-dimensional data, other methods have also been suggested, which include filtering in the wavelet domain (Coombes et al.; 2003), and asymmetric least squares splines regression (Befekadu et al.; 2008).

2.2 Sensor detection and quantification

In the event of fluorescent- or imaging-based sensor data, there are numerous computer algorithms and summary statistics available to identify the location and size of these features in the raw image data. Peak detection and quantification, for example, are of interest as they relate to spectral data. Various methods to achieve peak detection have been proposed via simple to complex means. One can recognize these methodological developments over time as further study was devoted to this area. Coombes et al. (2003) first suggested that peaks be detected by noting the locations where a change in slope occurs, and later fine-tuned the approach by instead considering the maximum value within the kth nearest neighbors; see Coombes et al. (2005), and independent discussion by Fushiki et al. (2006). More advanced proposals are to either apply an undecimated discrete wavelet transform (Morris et al.; 2005), or a continuous wavelet transform (Du et al.; 2006). These methods better eliminate the risk of detecting false positive peaks (i.e. data believed to represent peaks from a true signal when the data are actually, say for example, representative of residual noise).

In a two-dimensional chemical sensor setting, there are several classes of algorithms that can be applied for spot detection and quantification. Determining the locations and boundaries for chemical sensors in two dimensions falls under the general research area of image segmentation. Within image segmentation there are four main approaches: threshold techniques, boundary-based techniques, region-based methods, and hybrid techniques that combine boundary and region criteria (Adams and Bischof; 1994). Threshold techniques are based on the theory that all pixels whose values lie within a certain range belong to one class. This method neglects spatial information within the image and, in general, does not work well with noisy or blurred images. Boundary-based methods are motivated by the postulate that pixel values change rapidly at the boundary between two regions. Such methods apply a gradient operator in order to determine rapid changes in intensity values. High values in a gradient image provide candidates for region boundaries which must then be modified to produce closed curves that delineate the spot boundaries. The conversion of edge pixel candidates to boundaries of the regions of interest is often a difficult task. The complement of the boundary-based approach is to work within the region of interest, e.g. the chemical sensor. Region-based methods work under the theory that neighboring pixels within the region have similar values. This leads to the class of algorithms known as "region growing", of which the "split and merge" techniques are popular. In this technique, the general procedure is to compare one pixel to its neighbor. If some criterion of homogeneity is satisfied, then that pixel is said to belong to the same class as one or more of its neighbors. As expected, the choice of the homogeneity criterion is critical for even moderate success and can be highly deceiving in the presence of noise.

Finally, the class of hybrid techniques that combine boundary and region criteria includes morphological watershed segmentation and variable-order surface fitting. The watershed method is generally applied to the gradient of the image. In this case, the gradient image can be viewed as a topography map with boundaries between the regions represented as "ridges". Segmentation is then equivalent to "flooding" the topography from local minima with region boundaries erected to keep water from different minima exclusive. Unlike the boundary-based methods above, the watershed is guaranteed to produce closed boundaries even if the transitions between regions are of variable strength or sharpness. Such hybrid techniques, like the watershed method, encounter difficulties with chemical sensor images in which regions are both noisy or have blurred or indistinct boundaries. A popular alternative is seeded region growing (SRG). This method is based on the similarity of pixels within regions but has an algorithm similar to the watershed method. SRG is controlled by choosing a small number of pixels or regions called "seeds". These seeds will control the location and formation of the regions in which the image will be segmented. The number of seeds determines what is a feature and what is irrelevant or noise-embedded. Once given the seeds, SRG divides the image into regions such that each connected region component intersects with exactly one of the seeds. The choice of the number of seeds is crucial to this algorithm's success. Fortunately, with many chemical sensor experiments, the number of chemical sensors, and thus the seed, is known beforehand; see Adams and Bischof (1994) for further details.

Feature quantification is also an important issue, as there are several options that aid in reducing data dimensionality and complexity. At the same time, one ideally wants to measure a feature in such a way that captures an optimal amount of sensor information. Pardo and

Sberveglieri (2007) compare the performance of five feature summaries in chemical sensor arrays: the relative change in resistance; the area under the curve over gas adsorption, and gas desorption; and the phase space integral over adsorption, and desorption. In their study, while they do not attain uniform results across the various datasets, they find (on average) that the phase integral over desorption performed best. Further, the integral and phase space integral over desorption performed better than the analogous computations associated with adsorption. These results are consistent with other applied fields where such feature quantification is performed by computing the associated area under the curve. Carmel et al. (2003) instead argue that focusing on such features (such as the difference between the peak and baseline, the area under the curve, the area under curve left of the peak, or the time from the beginning to the peak of a signal) does not fully capture certain sensor properties, thus limiting one's ability to perform analyses. Focusing on transient signals, the authors fit various parametric models to chemical sensors for electronic noses, namely exponential, Lorentzian, and double-sigmoid models. Their results show that the double-sigmoid models fit optimally, followed by the Lorentzian model, with the exponential model being the worst of the three but still with decent performance. The computational time needed to fit these models, however, showed that the Lorentzian and exponential models were estimated far more quickly than for the double-sigmoid model. This makes sense because the double-sigmoid model has nine parameters that require estimation, while the Lorentzian and exponential models only have two parameters. Given this tradeoff, the authors propose using the Lorentzian model to analyze such chemical sensor data.

2.3 Normalization

In a normalization step, the goal is to remove obscuring sources of variation to give accurate measurements of the desired signal. Normalization could proceed in a manner similar to that described in Sellers et al. (2007) to remove known possible sources of variation, where one can obtain associated response data based on the presence of these factors in the design.

Linearization can also be performed by considering the engineering-derived equations that drive the signal (see, e.g., Robins et al. (2005)). Some chemical sensors are ruled by a power law relationship between sensor signal and analyte concentration; this is often the case, for example, with metal-oxide semiconductor gas sensors (Natale et al.; 2006). Using least squares approaches, it is possible to estimate the parameters in a power law relationship. This is a popular approach when preprocessing chemical sensor data because many of the subsequent analyses (e.g. linear discriminant analysis, principal component analysis, principal component regression, partial least squares) assume a linear relationship between sensor response and sample class (Jurs et al.; 2000).

Relative scaling is a common practice in the normalization of chemical sensor arrays, however the approach by which it is performed may vary. Options include dividing the signal by either the maximum signal value, the Euclidean norm from the signal, or the maximum value from a reference signal. In any respect, relative scaling serves to normalize the data in order to be on the same scale.

2.3.1 Quantile normalization

Quantile normalization is a very popular normalization method, because of its generality; it does not require building (non)-linear models to describe the experimental system. Let each experimental unit (e.g. subject, patient, or sample) be measured via the proposed chemical sensor(s) which produce(s) a profile for this experimental unit, and assume that our chemical sensor is, in fact, a panel of many chemical sensors. The quantile normalization thus imposes the same empirical distribution of the chemical sensor intensity of each profile (e.g. the profile for each experimental unit will have the same quartiles, etc). The algorithm proposed in Bolstad et al. (2003) is designed so that all profiles are matched (aligned) with the empirical distribution of the averaged sample profiles.

2.4 Low-level analysis discussion

Any or all low-level analysis procedures can be performed to obtain summary information on the raw chemical sensor data. The order of operations for these algorithms, however, are inconsistent and generally unrecoverable. As a result, the resulting preprocessed chemical sensor data can vary, thus potentially causing severe repercussions in the high-level analysis (see Baggerly et al. (2004)). To this end, one should be mindful of the low-level analyses performed (along with their order of operations) and comfortable with their use in data preprocessing. Nonetheless, data preprocessing results in the $S \times I$ summary matrix, $\mathbf{X} = (x_{si})$, where x_{si} denotes the normalized measure of sensor s in sample i. This data matrix will be used for subsequent statistical analysis.

3. Data analysis

There are several approaches that can be pursued to analyze the preprocessed data, depending on the question of interest. This section introduces these high-level, downstream methods. Here, we assume that the resulting preprocessed data matrix has rows associated with the chemical sensors used for the analysis, while the columns refer to the samples or patients. Jurs et al. (2000) classifies several methods as either statistical (including linear discriminant analysis (LDA), and principal component analysis (PCA)), or using neural networks while cluster analysis tools are classified separately. Given the popularity of LDA and PCA, we focus on these statistical methods here; see Jurs et al. (2000) for added discussion regarding various alternatives.

Linear discriminant analysis (LDA) is a statistical method (credited to Fisher) for dimension reduction and potential classification in that it distinguishes between two or more groups. The discriminant functions are derived from means and covariance matrices, thus working to maximize the distance between groups (relative to the variance within respective groups). While LDA is a popular dimension reduction technique for its natural approach, it tends to overfit when the ratio of training samples to dimensionality is small; see Wang et al. (2004). Jurs et al. (2000) concur that one needs a "relatively large number of samples from each class in the training data" that is representative of the population.

Principal component analysis (PCA) is an alternative statistical approach for dimension reduction. Invented by Karl Pearson, PCA performs singular value decomposition on the data matrix, $\mathbf{X}$, where the resulting terms relate to the eigenvalue-eigenvector form of $\mathbf{X}'\mathbf{X}$ and $\mathbf{X}\mathbf{X}'$, respectively. PCA is a popular choice in chemical sensor analysis because the first two principal components often account for at least 80% of the chemical sensor data variance (Jurs et al.; 2000), and is more robust to overfitting than LDA (Wang et al.; 2004). This method, however, may not successfully classify groups. Low-variance sensors, or nonlinear or nonadditive sensors can make classification difficult (Jurs et al.; 2000; Wang et al.; 2004).

Recognizing the limitations of these statistical methods, Wang et al. (2004) propose a "hybrid" model, termed Principal Discriminant Analysis (PDA). The hybrid matrix,

$$H = (1-\epsilon)S_w^{-1}S_b + \epsilon S_T,$$

	Reject H_0	Fail to reject H_0
H_0 true	Type I error	Correct decision
H_0 false	Correct decision	Type II error

Table 1. Possible outcomes for a null hypothesis, H_0, and associated outcome (rejecting or failing to reject H_0).

can be interpreted as a weighted average of the within- and between-group matrices from the LDA solution, and the total data covariance associated with PCA. The optimal $\epsilon \in [0,1]$ is attained via cross-validation, where $\epsilon = 0$ attains the LDA eigenvalues, and $\epsilon = 1$ produces the PCA projection. As a result, PDA provides a compromise between the popular LDA and PCA.

Similar to Jurs et al. (2000), we explore supervised and unsupervised machine learning/ statistical techniques to understand large complex datasets. Specifically, we examine linear modeling techniques to determine significantly different chemical sensors between two or more populations (e.g. neural nets as in Hashem et al. (1995)[1]). We also explore classification (supervised) and clustering (unsupervised) techniques to explore the similarities and differences between samples or between sensors. Within classification methods, we explore methods to both build and validate (e.g. data splitting/ cross validation) the classification schemes. Ultimately, we use the concepts of sensitivity and specificity to choose among a class of classification schemes.

3.1 Multiple testing

While LDA, PCA, and PDA all work with the entire dataset, commonly researchers would like to identify a subset of chemical sensors that are associated with the outcome. In this sense, the researchers are performing a data reduction, where the goal is to choose a subset of chemical sensors that are related or associated with the outcome. In this setting, the researcher commonly uses hypothesis testing to choose the important subset of chemical sensors.

Hypothesis testing seeks to obtain statistically significant results regarding a question of interest. In this process, the null hypothesis (H_0) represents the status quo statement while the alternative hypothesis (usually denoted as H_1 or H_a) defines that which is to be potentially proven or determined. When performing a hypothesis test, one wants to make a correct decision. There are, however, four possible scenarios that can occur when performing such a test; see Table 1. Two scenarios represent correct decisions, while the other two are incorrect decisions or "errors": (1) when one rejects the null hypothesis when it is actually true, and (2) when one does not reject the null hypothesis when it is actually false. The probability associated with the first scenario is referred to as Type I error (denoted α), and the second scenario's probability is termed Type II error (denoted β). For completeness in this discussion, statistical power refers to the probability of rejecting the null hypothesis when (in fact) the null hypothesis is false. In other words, statistical power equals one minus the Type II error (i.e. $1 - \beta$). Even when performing one hypothesis test, one wants to minimize the error probabilities.

We assume that our chemical sensor is multivariate, in the sense that each experimental unit (i.e., subject, patient, sample, or animal) is measured with several chemical sensors. Thus each unit acquires a chemical sensor profile, that is a collection of signals acquired from the chemical sensor. The goal in hypothesis testing is to examine each chemical sensor in

[1] In the Appendix, we discuss neural networks as outlined in Hashem et al. (1995) as a form of regression models.

		Condition +	Condition -	
Test	+	TP	FP	*PPV*
Test	-	FN	TN	*NPV*
		Sensitivity	Specificity	

Table 2. Table displaying the summary measures to distinguish positives (cases) from negatives (controls). TP (FP) denotes the number of true (false) positives, while TN (FN) denotes the number of true (false) negatives. PPV and NPV denote positive and negative predictive value, respectively. See Section 3.2.2 for details.

	H_0 Retained	H_0 Rejected	Total
H_0 True	U	V	m_0
H_0 False	T	Q	$m - m_0$
	$m - R$	R	m

Table 3. A summary of results from analyzing multiple hypothesis tests, where each cell represents the number (counts) in each category with m total tests.

light of the multiple chemical sensor levels measured in each experimental unit. In this setting, researchers and statisticians usually design their tests such that rejecting H_0 will yield discoveries or chemical sensors of interest. For example, when testing case samples against control samples on a chemical sensor platform, we would like to configure our hypothesis tests such that we reject H_0 for chemical sensors that are distinct between the cases and controls; see Section 5.

Commonly these hypothesis tests are performed using (linear) regression models. In these regressions, we estimate parameters designed to measure the effects of a chemical sensor in relation to an outcome. Common outcomes might be survival times or group membership (case vs. control). Using estimates of these parameters for a given chemical sensor, we can determine its significance. The interested reader is referred to Rawlings et al. (1998) and Cohen (2003) for comprehensive discussion of linear models and associated hypothesis testing.

In light of this discussion for multiple sensors, we can define our Type I and Type II errors in terms of sensitivity and specificity. That is, sensitivity is defined as

$$\text{sensitivity} = \frac{\text{number of true positives}}{\text{number of true positives} + \text{number of false negatives}}, \quad (1)$$

while specificity is defined as

$$\text{specificity} = \frac{\text{number of true negatives}}{\text{number of true negatives} + \text{number of false positives}}. \quad (2)$$

This situation is also summarized in Table 2. Note that sensitivity and specificity are estimated values because, in any experiment, we do not know the number of true positives or true negatives. Our goal when performing multiple testing and classification is to maximize the sensitivity and specificity, thus limiting the number of errors committed.

Table 2 can be refined in light of performing multiple hypothesis tests. In Table 3, we generalize the hypothesis testing in light of performing m hypothesis tests. Note that in Table 3, U, V, T, Q denote random variables (unknowns), while we assume that m is a fixed unknown quantity of hypothesis tests.

In light of performing multiple hypothesis tests (one for each sensor), we need a method to control the Type I error across the multiple tests. A first attempt for controlling Type I error is to perform a Bonferroni correction where, given m hypothesis tests, the measure for statistical significance is now attained if the associated p-value is less than α/m. In other words, the significance level is now scaled by the number of hypothesis tests. While this approach successfully adjusts for multiple tests, the procedure is far too stringent! The following subsections detail alternative Type I error rates and the available methods to control those errors, where Table 3 provides the associated notation in terms of probabilities (Pr): the familywise error rate, FWER $= Pr(V \geq 1)$; the k-familywise error rate, k-FWER $= Pr(V \geq k)$; and the false discovery rate (FDR), which is $E(V/R)$ if $R > 0$, or otherwise 0 with $E()$ denoting the expected value function.

3.1.1 Familywise error rates

The k-FWER error rate is a generalized version of the familywise error rate (FWER). Control of FWER refers to controlling the probability of committing one or more false discoveries. If we let V denote the number of false positives from m hypothesis tests, then notationally (according to Lehmann and Romano (2005)), α control of FWER can be expressed as

$$Pr(V \geq 1) \leq \alpha, \tag{3}$$

or equivalently,

$$Pr(V = 0) \geq 1 - \alpha. \tag{4}$$

Note that α is usually chosen to be small, e.g., 0.05. Often Equation (3) is abbreviated as FWER $\leq \alpha$. In k-FWER, the equation becomes

$$Pr(V \geq k) \leq \alpha, \tag{5}$$

where $k \geq 1$ and α are usually determined prior to the analysis. Similar to FWER, control of k-FWER can be expressed as k-FWER $\leq \alpha$. Note that there is the potential for ambiguity in control of k-FWER since, occasionally (as in Gentleman (2005)), k-FWER may be expressed as $Pr(V > k) \leq \alpha$ for $k \geq 0$.

The adjusted Bonferroni method to control k-FWER is a generalized version of the Bonferroni correction designed to control FWER (Lehmann and Romano; 2005). The Bonferroni correction is designed to control the FWER at level α by doing each individual test at level α/m, where m is the number of tests. The adjustment given in Lehmann and Romano (2005) to control k-FWER at α is done by performing each test at level $k\alpha/m$. By performing each test at this level, the probability against k or more false positives is no larger than α; that is, k-FWER $\leq \alpha$. The proof is supplied in Lehmann and Romano (2005) and is a generalization of the proof for the original Bonferroni method designed to control FWER. For a description of other methods to control k-FWER and a power comparison of k-FWER methods, see Miecznikowski et al. (2011).

3.1.2 False discovery rate

Multiple statistical testing procedures began to be reexamined in the early 1990s with the advent of high-throughput genomic technologies. The Benjamini and Hochberg (BH) method was proposed to control the false discovery rate (FDR), or the expected rate of false test positives (see Benjamini and Hochberg (1995)). In the BH multiple testing procedure, the FDR is controlled by the following scheme:

1. Let $p_{(1)} < \cdots < p_{(m)}$ denote the m ordered p-values (smallest to largest).
2. Denote $\hat{t} = p_{(k)}$ for the largest k such that $p_{(k)} \leq \frac{k\alpha}{m}$.
3. Reject all null hypotheses, H_{0i}, for which $p_i \leq \hat{t}$.

Note that we define FDR such that

$$FDR \equiv E[V/R], \tag{6}$$

where E denotes the expected value function. Benjamini and Hochberg (1995) proves that, if the above procedure is applied, $FDR \leq \alpha$. Storey (2002) further show that, for p-value threshold t,

$$\text{FDR}(t) = \frac{(1-\pi)t}{(1-\pi)t + \pi F(t)}, \tag{7}$$

where π is the probability that an alternative hypothesis is true, and $F(t)$ is the distribution of p-values given the alternative. FDR performance has been evaluated for sensor detection given a variety of scenarios, for example, in the presence of correlation (Benjamini and Yekutieli; 2001; Shao and Tseng; 2007). Importantly, note that FDR analysis does not control what Genovese and Wasserman (2004) call "the realized FDR" (rFDR), i.e. the number of false rejections V divided by the number of rejections R (assuming at least one rejection) which, in fact, can be quite variable (as shown in Gold et al. (2009)).

3.2 Classification

This section provides an overview of classification models. Throughout this section, we assume that the outcome variable of interest is binary. This is commonly the situation with case/control experiments where, for example, the goal may be to predict the presence or absence of a disease.

The simplest and most direct approach to classification with a binary outcome variable is to estimate the regression function, $r(x) = E(Y|X = x)$, and use the classification rule,

$$\hat{h}(x) = \begin{cases} 1 & \text{if } \hat{r}(x) > 0.5 \\ 0 & \text{otherwise.} \end{cases} \tag{8}$$

Here, the simplest regression model is the linear regression model,

$$Y = r(x) + \epsilon = \beta_0 + \sum_{j=1}^{d} \beta_j X_j + \epsilon, \tag{9}$$

where the errors, ϵ, have mean 0. A simple example of this classifier is provided in Section 5. Other examples of classifiers include linear discriminant analysis, support vector machines, and ensemble classifiers using bootstrapping and bagging techniques; see Hastie et al. (2005) and Wasserman (2004) for a more complete treatment of classification models. Similar to hypothesis testing, we want an accurate classifier that commits relatively few errors; i.e. we would like a classifier with a high sensitivity and specificity (see Equations (1) and (2)). To estimate sensitivity and specificity for a given classification model, we commonly use cross validation methods, as described in the next section.

3.2.1 Cross validation

Cross validation can be classified under the general realm of sample splitting. Its objective is to obtain an estimate of the prediction qualities of the model when using that same data to build the prediction model. The simplest version of cross-validation involves randomly splitting the data into two pieces: the training set, and the validation set. The classifier is constructed from the training set, and the associated error is estimated using the validation set; the error is defined as

$$\text{Error} = \text{number of misclassifications/ number of predictions.} \tag{10}$$

Two extensions of this method are g-fold cross validation, and leave-one-out cross validation (LOOCV). Note that LOOCV is a special case of g-fold cross validation, where g is equal to the number of objects in the dataset. As described in Wasserman (2004) in a g-fold cross validation, we do the following:

1. Randomly divide the data into G groups of approximately equal size.
2. For $g = 1$ to G,
   (a) delete group g from the data.
   (b) fit or compute the classifier from the remaining data.
   (c) use the classifier to predict the data in group g, and let L denote that observed error rate.
3. Let the overall error rate be estimated from averaging over the error rates from the previous step.

See Section 3.2.2 for a discussion of other summary measures (e.g. sensitivity and specificity) that are commonly estimated with cross validation.

3.2.2 Summary measures in a population

Naturally, we want our classifiers to make accurate predictions. We have seen that specificity and sensitivity as estimated via cross validation are reasonable measures to summarize our classification models. In this section, however, we highlight some of the measures used to evaluate our classification models in a population. The measures introduced in this section are often crucial in deciding the utility of a chemical sensor.

When determining a classifier's effectiveness, analysts usually calculate the sensitivity and specificity using cross-validation. To understand the potential utility of the chemical sensor-derived classifier, however, it is important to calculate the positive predictive value (PPV) and negative predictive value (NPV). The PPV is the proportion of subjects with positive test results who are correctly diagnosed. It reflects the probability that a positive test reflects the truly positive underlying condition. The PPV depends heavily on the prevalence of the outcome of interest, which is usually unknown. Using Bayes Theorem (see Wasserman (2004)) and Table 2, we can derive the positive predictive value as

$$PPV = \frac{(\text{sensitivity})(\text{prevalence})}{(\text{sensitivity})(\text{prevalence}) + (1-\text{specificity})(1-\text{prevalence})}. \tag{11}$$

Note that we define the prevalence in terms of epidemiologic factors, i.e. prevalence. Prevalence (of disease) is the total number of (disease) cases in the population divided by the number of individuals in the population. Prevalence is (essentially) an estimate of how common the underlying condition is within a population over a certain period of time.

Defining a as the number of individuals in a given population with the disease at a given time, and b as the number of individuals in the same population at risk of developing the disease at this given time (not including those already with the disease), the prevalence is specified by

$$\text{prevalence} = \frac{a}{a+b}. \tag{12}$$

Similarly, we can define the NPV as the proportion of subjects with a negative test result who are correctly diagnosed. A high NPV means that, when the test yields a negative result, it is uncommon that the result should have been positive. Mathematically, the NPV is computed as

$$NPV = \frac{(\text{specificity})(\text{1-prevalence})}{(\text{specificity})(\text{1-prevalence}) + (1-\text{sensitivity})(\text{prevalence})}. \tag{13}$$

While sensitivity and specificity play a role in assessing a chemical sensor, we stress that NPV and PPV are often the measures used when deciding the clinical and medical utility of a potential chemical sensor panel. A thorough handling of the topic of estimation with regard to sensitivity, specificity, PPV, and NPV is provided in Pepe (2004), Cai et al. (2006), and Pepe et al. (2008).

4. Software development

`Bioconductor` (Gentleman et al.; 2004) and *R* (R Development Core Team; 2008) are two statistical computing tools that can be used for statistical programming and data analysis. Both are freeware tools that are downloadable from the internet. Researchers developing novel chemical sensors should consider writing a computational package for analyzing their chemical sensor data in *R*. This will enable the statistical methods and algorithms to be used by the scientific community. For various examples of *R* packages, see Gaile et al. (2010) and Gentleman (2005).

For example, Chandrasekhar et al. (2009) authored a software package specific to Xerogel chemical sensor images such as those shown in Figure 1 (A). These images and the Xerogel technology are further described in Chandrasekhar et al. (2009). The *Xerogel R* package consists of routines that import the tagged image file format (TIFF) image, read the image into a matrix, binarize the image matrix, identify the position and structure of the spots, and return statistics such as the mean, median, and total intensity for these spots. The summary statistics represent the results after preprocessing and, ultimately, provide information on how the intensity of light varies with varying amounts of the volatile organic compounds (VOCs). A summary set of images demonstrating the pre-processing of a representative Xerogel image is shown in Figure 1 (A)-(D).

5. Case study

In this case study, we examine a subset of the data from Schröder et al. (2010) which represents a chemical sensor dataset designed to classify pancreatic cancer patients from normal patients. This dataset is publicly available and can be downloaded from `ArrayExpress` (see Parkinson et al. (2009)) with ID accession number E-MEXP-3006; see `http://www.ebi.ac.uk/arrayexpress/`. This dataset is representative of a two dimensional chemical sensor dataset. The experiment employs protein antibody microarrays with two color channels on an array consisting of 1800 features (proteins).

The preprocessing for this data is consistent with the methods described in Section 2. The data in this study were preprocessed using the following scheme:

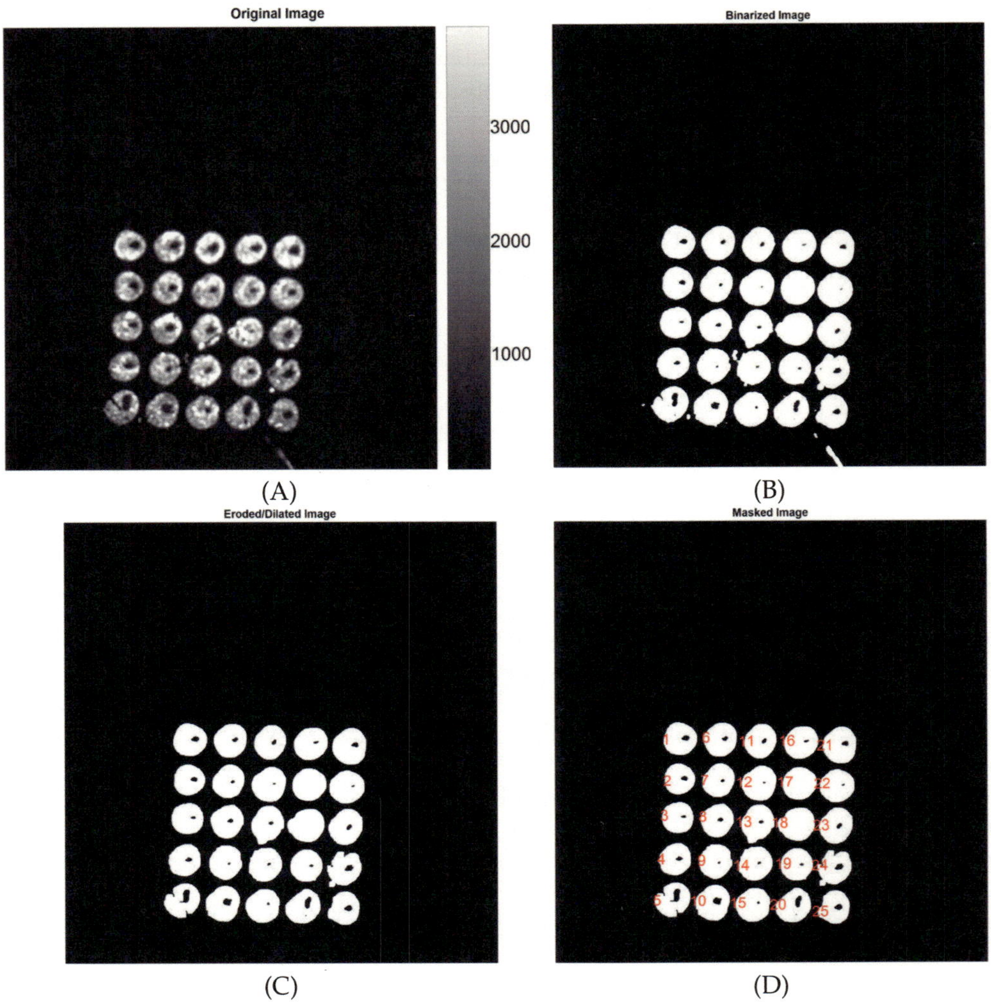

Fig. 1. **Xerogel Preprocessing:** Xerogel Images showing the preprocessing steps. (A) Original Image (B) Binarized Image (C) Eroded/Dilated Image (D) Masked Image. For more details on Xerogel chemical sensors see Chandrasekhar et al. (2009).

- Background Correction - as recommended in Ritchie et al. (2007), a convolution of normal and exponential distributions is fitted to the foreground intensities using the background intensities as a covariate, and the expected signal given the observed foreground becomes the corrected intensity.
- Normalization - `lowess` is applied as proposed in Yang et al. (2002) and Smyth and Speed (2003). Here, the signal in each array is adjusted to account for the intensity bias with a nonlinear curve fitting method, lowess.

All preprocessing steps were performed using the *limma* package in *R* (see Smyth (2004)). The experimental design for this experiment is fully described in Schröder et al. (2010). For our subset of data, we have a total of 12 subjects, specifically three patients in each of four groups (male controls, female controls, males with pancreatic cancer, and females with pancreatic cancer). The patients were named in terms of their disease status (healthy = h, cancer = c) and gender (male = m, female = f); e.g. hf_1 refers to the sample for the first healthy female, while cm_1 refers to the sample for the first pancreatic cancer male. Midstream urine samples were collected from each patient and pH was adjusted to 7. After sample preparation, the samples were dye-labeled and incubated to antibody microarrays containing 1,800 features. Figure 2 shows a representative fluorescence array from this study where a urine sample and reference consisting of a pool of samples from diseased and healthy subjects were labeled with different dyes.

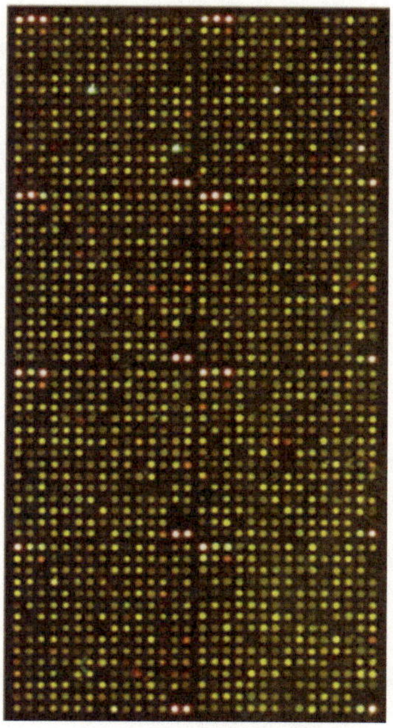

Fig. 2. Protein antibody image from our case study (figure taken from Schröder et al. (2010)).

After preprocessing, we arrive at a matrix containing 1678 rows (proteins) and 12 columns (samples), where each entry represents the logarithmic intensity (amount) of the protein present in the sample (relative to the reference channel). The heatmap representing this data is shown in Figure 3.

In performing an exploratory data analysis, we looked at clustering the samples as well as the distance between the samples. The clustering results in terms of a hierarchical clustering are shown in Figure 4. From this figure, we see that sample hf_2 may represent an outlier in this study; this is further confirmed in Figure 5. From this figure, we see that hf_2 is widely separated from the other samples in the study (white band). Note that the Euclidean distance metric was used to calculate the distance between each sample. For an overview of distance metrics, see Chapter 12 of Gentleman (2005). This potential outlier, hf_2, is also confirmed by studying the protein profile in Figure 3, where we see a pattern that is inconsistent with the other samples.

To further the analysis in this case study, we build linear model(s) to discover potential differentially expressed proteins between cancer patients and normal patients; see Section 3.1. We explored univariate protein models and multivariate protein models that adjusted

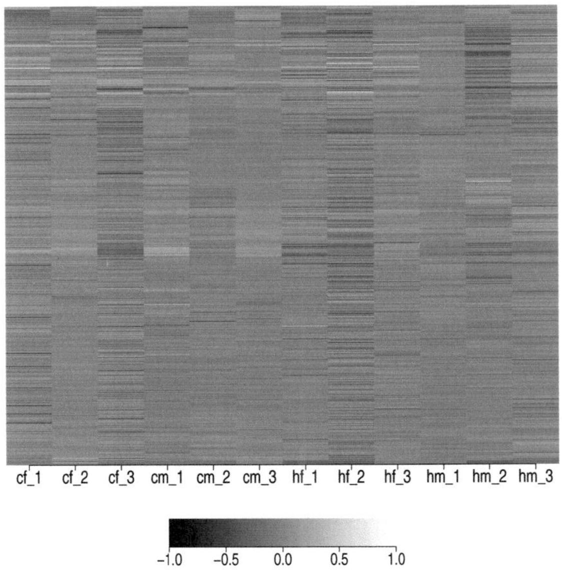

Fig. 3. Heatmap showing the proteins levels for each sample after preprocessing. Note sample hf_2 appears to be an outlier in this subset of data.

for gender. Building these models allows us to test each protein for association with disease status. In particular, we let

$$y_{ji} = \mu_j + \beta_{dz}x_{dz} + \epsilon_{ji} \tag{14}$$

denote the observed protein level for protein j in sample i ($i = 1, \ldots, 12$), with x_{dz} equaling 1 if sample j is a cancer sample and 0 otherwise and ϵ_{ji} is assumed to be normally distributed with mean 0 and variance σ_j^2. In this model, we are interested in estimating the parameters for each protein. For protein j, the pancreatic cancer samples have a mean of $\mu_j + \beta_{dz}$ while the healthy samples have a mean of μ_j. We are especially interested in proteins where β_{dz} is significantly different from 0, as this indicates proteins that are significantly different between diseased patients and healthy patients. Accordingly, for each protein, we wish to test the hypothesis,

$$H_0 : \beta_{dz} = 0. \tag{15}$$

Using an empirical Bayes test described in Smyth (2004), we obtain a test statistic and p-value for each protein corresponding to the test in Equation (15). Using a Šidàk control method described in Miecznikowski et al. (2011), we control k-FWER such that the probability of committing no more than five false positives is no larger than 0.05. Under this scheme, we discover three significant proteins; see Table 4.

In Equation (16), we introduce a more complex model. We include the explanatory variable x_{gen}, which is 1 if sample j is a female and 0 otherwise. By incorporating a variable for the patient's gender, this model is more complex than the model described in Equation (14). This model is described as

$$y_{ji} = \mu_j + \beta_{gen}x_{gen} + \beta_{dz}x_{dz} + \epsilon_{ji}. \tag{16}$$

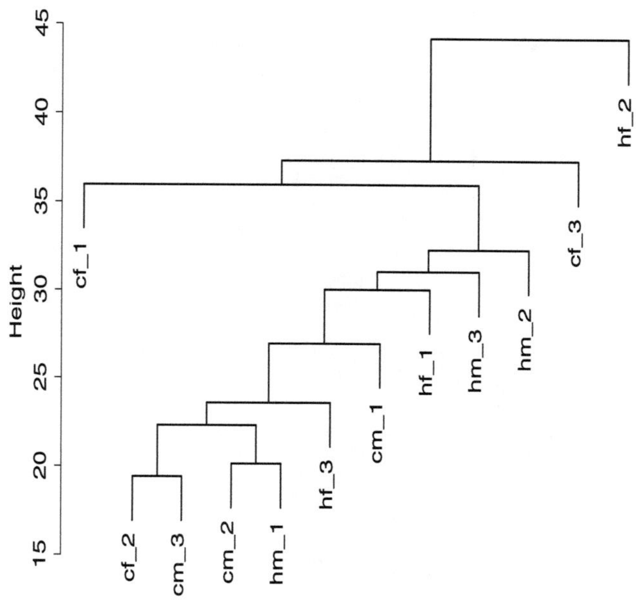

Fig. 4. Hierarchical clustering showing the similarity of the samples. From this figure it appears that sample hf_2 is an outlier.

	k-FWER Control via Šidàk Method		
	k	α	# Sig Proteins
Univariate Model	5	0.05	2
	10	0.05	3
Multivariate Model	5	0.05	1
	10	0.05	4

Table 4. Table displaying the significant proteins from an analysis with univariate and multivariate models as described in Equations (14) and (16), respectively.

After estimating the parameters in Equation (16), we are also interested in testing the null hypothesis in Equation (15). With this model, however, the parameter β_{dz} represents proteins that are significantly different in disease states (case/control) after adjusting for potential gender biases. For protein j, the female pancreatic cancer patients have estimates, $\mu_j + \beta_{gen} + \beta_{dz}$; while the female healthy patients have estimates, $\mu_j + \beta_{gen}$. Similarly, the male pancreatic cancer patients have estimates $\mu_j + \beta_{dz}$, while the male healthy patients have estimates, μ_j.

Using the model described in Equation (16) for each protein, we obtain the test statistic and p-value for β_{dz} from an empirical Bayes test (Smyth; 2004). As with the model in Equation (14), we use a Šidàk method to control k-FWER such that the probability of committing 10 or more false positives to be no larger than 0.05. Under this scheme, we obtain four significant proteins. Figure 6 displays the heatmap of the significant proteins under this setting and Table 4 displays the number of significant proteins under different configurations for controlling k-FWER.

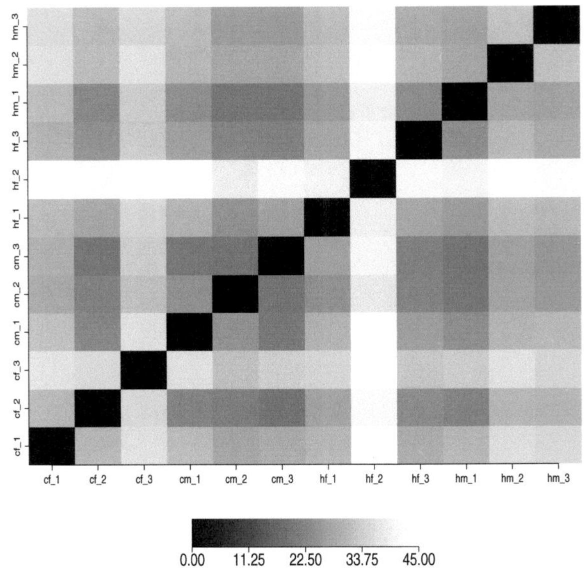

Fig. 5. The distance matrix heatmap showing the measured Euclidean distance between the samples using the protein profile for each sample. From this figure, the profile for sample hf_2 is the furthest in terms of Euclidean distance from the other samples.

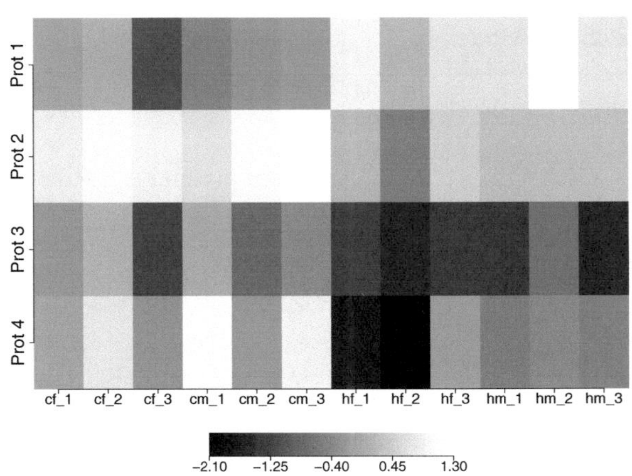

Fig. 6. Heatmap of the significant proteins as determined in case study using Equation (16) and a k-FWER error controlling scheme.

After determining the significant proteins in this study, it is reasonable to examine models for prediction. We explored using a logistic regression model as described in Section 3.2. We choose a classification model using the most significant protein ($Prot1$) as determined from fitting the model in Equation (14). The fitted logistic regression equation is

$$\hat{r}(Prot1) = \frac{exp(-10.303 * Prot1)}{1 + exp(-10.303 * Prot1)}, \tag{17}$$

where $Prot1$ is the intensity of the most significant protein after fitting Equation (14); the profile for this protein is shown in Row 1 in Figure 6. Our logistic classifier is then specified by

$$\hat{h}(x) = \begin{cases} \text{Disease} & \text{if } \hat{r}(x) > 0.5 \\ \text{Healthy} & \text{if } \hat{r}(x) \leq 0.5. \end{cases} \tag{18}$$

Using a leave-one-out cross-validation method as described in Section 3.2.1, we obtain a sensitivity estimate of 83.3% (5/6) and a specificity estimate of 100% (6/6); the misclassified sample is cf_2. As seen in Row 1 of Figure 6, cf_2 has a value for $Prot1$ indicating a pattern more aligned with the healthy samples. Similar to Schröder et al. (2010), our conclusion from this analysis is that a urine proteomic profile as measured on antibody arrays shows promise in diagnosing pancreatic cancer. Due to the limited sample size (12 samples) of our case study, however, we caution the reader not to rely heavily on these estimates of sensitivity and specificity.

6. Summary

Chemical sensor data appear in a variety of contexts, from breathalyzers to carbon monoxide or smoke detectors. Given their pervasive existence in various aspects of life, it is essential that these sensors work and provide proper analysis to accurately assess chemical questions of interest. This can only happen with proper statistical insight and tools to provide accurate assessments that can lead to appropriate decision-making. Hirschfeld et al. (1984) noted the importance of chemometrics with sensor arrays, particularly the use of pattern recognition strategies and learning algorithms. As a result, calibration, quantification, and reproducibility are all attainable. In this chapter, we highlight some of the state-of-the-art statistical methods that are used to calibrate, quantify, and ultimately, benchmark modern chemical sensors. We stress that further advancements in the use and utility of chemical sensors will require input from statisticians to determine the accuracy/reproducibility of the sensors as well as their ability to make inferences on a population.

7. Appendix

In this section, we provide a brief overview of neural networks as they are commonly used for prediction with chemical sensor data (see Hashem et al. (1995)). As discussed in Wasserman (2004), neural networks often take the form,

$$Y = \beta_0 + \sum_{j=1}^{p} \beta_j \sigma \left(\alpha_0 + \alpha^T X \right) \tag{19}$$

where σ is a smooth function. When compared to models such as in Equations (14) and (16), the neural networks model is obviously more complex. As such, these models are often difficult to fit to datasets and often require large datasets and heavy computational power. Besides Wasserman (2004), other references for neural networks can be found in Bhadeshia (1999); MacKay (2003).

8. References

Adams, R. and Bischof, L. (1994). Seeded region growing, *Pattern Analysis and Machine Intelligence, IEEE Transactions on* 16(6): 641–647.

Baggerly, K., Morris, J. and Coombes, K. (2004). Reproducibility of seldi-tof protein patterns in serum: comparing datasets from different experiments, *Bioinformatics* 20(5): 777–785.

Befekadu, G., Tadesse, M., Hathout, Y. and Ressom, H. (2008). Multi-class alignment of lc-ms data using probabilistic-based mixture regression models, *Engineering in Medicine and Biology Society, 2008. EMBS 2008. 30th Annual International Conference of the IEEE*, IEEE, pp. 4094–4097.

Benjamini, Y. and Hochberg, Y. (1995). Controlling the false discovery rate: a practical and powerful approach to multiple testing, *Journal of the Royal Statistical Society. Series B (Methodological)* pp. 289–300.

Benjamini, Y. and Yekutieli, D. (2001). The control of the false discovery rate in multiple testing under dependency, *The Annals of Statistics* 29(4): 1165–1188.

Bhadeshia, H. (1999). Neural networks in materials science, *ISIJ international* 39(10): 966–979.

Bolstad, B., Irizarry, R., Åstrand, M. and Speed, T. (2003). A comparison of normalization methods for high density oligonucleotide array data based on variance and bias, *Bioinformatics* 19(2): 185.

Cai, T., Pepe, M., Zheng, Y., Lumley, T. and Jenny, N. (2006). The sensitivity and specificity of markers for event times, *Biostatistics* 7(2): 182.

Carmel, L., Levy, S., Lancet, D. and Harel, D. (2003). A feature extraction method for chemical sensors in electronic noses, *Sensors and Actuators B: Chemical* 93(1-3): 67 – 76. Proceedings of the Ninth International Meeting on Chemical Sensors.
URL: *http://www.sciencedirect.com/science/article/pii/S0925400503002478*

Chandrasekhar, R., Miecznikowski, J., Gaile, D., Govindaraju, V., Bright, F. and Sellers, K. (2009). Xerogel package, *Chemometrics and Intelligent Laboratory Systems* 96(1): 70–74.

Cohen, J. (2003). *Applied multiple regression/correlation analysis for the behavioral sciences*, Vol. 1, Lawrence Erlbaum.

Coombes, K., Fritsche Jr, H., Clarke, C., Chen, J., Baggerly, K., Morris, J., Xiao, L., Hung, M. and Kuerer, H. (2003). Quality control and peak finding for proteomics data collected from nipple aspirate fluid by surface-enhanced laser desorption and ionization, *Clinical Chemistry* 49(10): 1615.

Coombes, K., Tsavachidis, S., Morris, J., Baggerly, K., Hung, M. and Kuerer, H. (2005). Improved peak detection and quantification of mass spectrometry data acquired from surface-enhanced laser desorption and ionization by denoising spectra with the undecimated discrete wavelet transform, *Proteomics* 5: 4107–4117.

Du, P., Kibbe, W. and Lin, S. (2006). Improved peak detection in mass spectrum by incorporating continuous wavelet transform-based pattern matching, *Bioinformatics* 22(17): 2059.

Eggins, B. (2002). *Chemical sensors and biosensors*, Wiley.

Fushiki, T., Fujisawa, H. and Eguchi, S. (2006). Identification of biomarkers from mass spectrometry data using a "common" peak approach, *BMC Bioinformatics* 7: 358.

Gaile, D., Shepherd, L., Bruno, A., Liu, S., Morrison, C., Sucheston, L. and Miecznikowski, J. (2010). iGenomicViewer: R package for visualisation of high dimension genomic data, *International Journal of Bioinformatics Research and Applications* 6(6): 584–593.

Genovese, C. and Wasserman, L. (2004). A stochastic process approach to false discovery control, *The Annals of Statistics* 32(3): 1035–1061.

Gentleman, R. (2005). *Bioinformatics and computational biology solutions using R and Bioconductor*, Springer Verlag.

Gentleman, R. C., Carey, V. J., Bates, D. M., Bolstad, B., Dettling, M., Dudoit, S., Ellis, B., Gautier, L., Ge, Y., Gentry, J., Hornik, K., Hothorn, T., Huber, W., Iacus, S., Irizarry, R., Leisch, F., Li, C., Maechler, M., Rossini, A. J., Sawitzki, G., Smith, C., Smyth, G., Tierney, L., Yang, J. Y. H. and Zhang, J. (2004). Bioconductor: Open software development for computational biology and bioinformatics, *Genome Biology* 5: R80. URL: *http://genomebiology.com/2004/5/10/R80*

Gold, D., Miecznikowski, J. and Liu, S. (2009). Error control variability in pathway-based microarray analysis, *Bioinformatics* 25(17): 2216–2221.

Hashem, S., Keller, P., Kouzes, R. and Kangas, L. (1995). Neural-network-based data analysis for chemical sensor arrays, *Proceedings of SPIE*, Vol. 2492, p. 33.

Hastie, T., Tibshirani, R., Friedman, J. and Franklin, J. (2005). The elements of statistical learning: data mining, inference and prediction, *The Mathematical Intelligencer* 27(2): 83–85.

Hirschfeld, T., Callis, J. B. and Kowalski, B. R. (1984). Chemical sensing in process analysis, *Science* 226(4672): 312–318.
URL: *http://www.sciencemag.org/content/226/4672/312.abstract*

Jurs, P. C., Bakken, G. A. and McClelland, H. E. (2000). Computational methods for the analysis of chemical sensor array data from volatile analytes, *Chemical Reviews* 100(7): 2649–2678.
URL: *http://pubs.acs.org/doi/abs/10.1021/cr9800964*

Lehmann, E. and Romano, J. (2005). Generalizations of the familywise error rate, *The Annals of Statistics* 33(3): 1138–1154.

Lin, S., Haney, R., Campa, M., Fitzgerald, M. and Patz, E. (2005). Characterising phase variations in maldi-tof data and correcting them by peak alignment, *Cancer Informatics* 1(1): 32–40.

MacKay, D. (2003). *Information theory, inference, and learning algorithms*, Cambridge Univ Pr.

Miecznikowski, J., Gold, D., Shepherd, L. and Liu, S. (2011). Deriving and comparing the distribution for the number of false positives in single step methods to control k-fwer, *Statistics & Probability Letters* (in press).

Morris, J., Coombes, K., Koomen, J., Baggerly, K. and Kobayashi, R. (2005). Feature extraction and quantification for mass spectrometry in biomedical applications using the mean spectrum, *Bioinformatics* 21(9): 1764.

Natale, C., Martinelli, E., Pennazza, G., Orsini, A. and Santonico, M. (2006). Data Analysis for Chemical Sensor Arrays, *Advances in Sensing with Security Applications* pp. 147–169.

Pardo, M. and Sberveglieri, G. (2007). Comparing the performance of different features in sensor arrays, *Sensors and Actuators B: Chemical* 123(1): 437 – 443.
URL: *http://www.sciencedirect.com/science/article/pii/S0925400506006411*

Parkinson, H., Kapushesky, M., Kolesnikov, N., Rustici, G., Shojatalab, M., Abeygunawardena, N., Berube, H., Dylag, M., Emam, I., Farne, A. et al. (2009). Arrayexpress update-from an archive of functional genomics experiments to the atlas of gene expression, *Nucleic acids research* 37(suppl 1): D868.

Pepe, M. (2004). *The statistical evaluation of medical tests for classification and prediction*, Oxford University Press, USA.

Pepe, M., Feng, Z., Huang, Y., Longton, G., Prentice, R., Thompson, I. and Zheng, Y. (2008). Integrating the predictiveness of a marker with its performance as a classifier, *American journal of epidemiology* 167(3): 362.

R Development Core Team (2008). *R: A Language and Environment for Statistical Computing*, R Foundation for Statistical Computing, Vienna, Austria. ISBN 3-900051-07-0. URL: *http://www.R-project.org*

Rawlings, J., Pantula, S., Dickey, D. and MyiLibrary (1998). *Applied regression analysis: a research tool*, Springer New York, NY, US.

Ritchie, M., Silver, J., Oshlack, A., Holmes, M., Diyagama, D., Holloway, A. and Smyth, G. (2007). A comparison of background correction methods for two-colour microarrays, *Bioinformatics* 23(20): 2700.

Robins, P., Rapley, V. and Thomas, P. (2005). A probabilistic chemical sensor model for data fusion, *Information Fusion, 2005 8th International Conference on*, Vol. 2, IEEE, pp. 7–pp.

Schröder, C., Jacob, A., Tonack, S., Radon, T., Sill, M., Zucknick, M., Rüffer, S., Costello, E., Neoptolemos, J., Crnogorac-Jurcevic, T. et al. (2010). Dual-color proteomic profiling of complex samples with a microarray of 810 cancer-related antibodies, *Molecular & Cellular Proteomics* 9(6): 1271.

Sellers, K., Miecznikowski, J., Viswanathan, S., Minden, J. and Eddy, W. (2007). Lights, Camera, Action! Systematic variation in 2-D difference gel electrophoresis images, *Electrophoresis* 28(18): 3324–3332.

Shao, Y. and Tseng, C. (2007). Sample size calculation with dependence adjustment for fdr-control in microarray studies, *Statistics in medicine* 26(23): 4219–4237.

Smyth, G. (2004). Linear models and empirical Bayes methods for assessing differential expression in microarray experiments, *Statistical applications in genetics and molecular biology* 3(1): 3.

Smyth, G. and Speed, T. (2003). Normalization of cdna microarray data, *Methods* 31(4): 265–273.

Storey, J. (2002). A direct approach to false discovery rates, *Journal of the Royal Statistical Society. Series B, Statistical Methodology* pp. 479–498.

Wang, M., Perera, A. and Gutierrez-Osuna, R. (2004). Principal discriminants analysis for small-sample-size problems: application to chemical sensing, *Sensors, 2004. Proceedings of IEEE*, IEEE, pp. 591–594.

Wasserman, L. (2004). *All of statistics: a concise course in statistical inference*, Springer Verlag.

Yang, Y., Dudoit, S., Luu, P., Lin, D., Peng, V., Ngai, J. and Speed, T. (2002). Normalization for cdna microarray data: a robust composite method addressing single and multiple slide systematic variation, *Nucleic acids research* 30(4): e15.